Teubner Skripten zur
Mathematischen Stochastik

Konrad Behnen and Georg Neuhaus
Rank Tests with Estimated Scores
and Their Application

# Teubner Skripten zur Mathematischen Stochastik

Herausgegeben von
Prof. Dr. rer. nat. Jürgen Lehn, Technische Hochschule Darmstadt
Prof. Dr. rer. nat. Norbert Schmitz, Universität Münster
Prof. Dr. phil. nat. Wolfgang Weil, Universität Karlsruhe

Die Texte dieser Reihe wenden sich an fortgeschrittene Studenten, junge Wissenschaftler und Dozenten der Mathematischen Stochastik. Sie dienen einerseits der Orientierung über neue Teilgebiete und ermöglichen die rasche Einarbeitung in neuartige Methoden und Denkweisen; insbesondere werden Überblicke über Gebiete gegeben, für die umfassende Lehrbücher noch ausstehen. Andererseits werden auch klassische Themen unter speziellen Gesichtspunkten behandelt. Ihr Charakter als Skripten, die nicht auf Vollständigkeit bedacht sein müssen, erlaubt es, bei der Stoffauswahl und Darstellung die Lebendigkeit und Originalität von Vorlesungen und Seminaren beizubehalten und so weitergehende Studien anzuregen und zu erleichtern.

# Rank Tests with Estimated Scores and Their Application

By Prof. Dr. rer. nat. Konrad Behnen
and Prof. Dr. rer. nat. Georg Neuhaus
Universität Hamburg

B. G. Teubner Stuttgart 1989

Prof. Dr. rer. nat. Konrad Behnen

Geboren 1941 in Werpeloh. Von 1962 bis 1967 Studium der Mathematik und Physik an der Universität Münster. 1967 Diplom in Mathematik, 1969 Promotion an der Universität Münster. 1970/71 Gastaufenthalt an der Universität Berkeley. 1974 Habilitation an der Universität Freiburg. Professuren für Mathematische Stochastik an den Universitäten Karlsruhe (1974/75), Bremen (1975/78) und Hamburg (seit 1978).

Prof. Dr. rer. nat. Georg Neuhaus

Geboren 1943 in Banfe/Wittgenstein. Von 1962 bis 1967 Studium der Mathematik und Physik in Münster. 1967 Diplom, 1969 Promotion in Münster. 1970 Medizinische Abteilung der Farbwerke Hoechst. Von 1970 bis 1974 Assistent in Münster und Freiburg. 1974 Habilitation im Fach Mathematik. Von 1974 bis 1977 Professor in Gießen. Seit 1977 Professor in Hamburg.

CIP-Titelaufnahme der Deutschen Bibliothek

**Behnen, Konrad:**
Rank tests with estimated scores and their application / by
Konrad Behnen and Georg Neuhaus. – Stuttgart : Teubner, 1989
  (Teubner-Skripten zur mathematischen Stochastik)

  ISBN 978-3-519-02728-7    ISBN 978-3-322-94762-8 (eBook)
  DOI 10.1007/978-3-322-94762-8

NE: Neuhaus, Georg:



Herstellung: Druckhaus Beltz, Hemsbach/Bergstraße
Umschlaggestaltung: M. Koch, Reutlingen

# Preface

The general aim of this book is to present a new class of nonlinear rank tests for a variety of important testing problems. The need for such new procedures stems from the fact that the classical *linear* rank tests are sensitive only for small classes of alternatives, while our *nonlinear* rank tests are designed to be sensitive for broad classes of alternatives. The development of the new procedures is strongly influenced by the opinion that in many real world situations the classical shift assumption is too idealized. By introducing general nonparametric models we get rid of the shift assumption. For the two-sample situation a detailed motivation is given in Chapter 1.

Our theoretical results and many Monte Carlo simulations have convinced us that the proposed procedures are of real practical importance and should be used in statistical applications. Therefore in Chapter 2 we present a simple algorithmic description of the new rank tests—without stressing any mathematical theory—and a step by step evaluation of numerical examples, whereas in Part II of the book (Chapters 2–6) we give a rigorous asymptotic theory of all proposals of Chapter 2.

These chapters discuss different models and motivate the special proposals. Most of the concepts and of the theoretical results are based on our own work. The material has been published in journals only partially. As far as we know there are no comparable results published by other authors. Naturally this material is presented at a graduate level of mathematics. It may be discussed in graduate courses on asymptotic statistics.

An important feature of our exposition is the consequent and elegant treatment of ties, i.e. we allow for discontinuous distributions throughout the book, especially our tests are valid even for categorical data. While constructing the new procedures we've developed a complete theory under the assumption of arbitrary distribution functions. This leads to a simple and satisfactory treatment of tied observations, also for classical rank procedures.

Readers who aren't interested in the theoretical background may immediately start with Chapter 2. In this chapter (Sections 2.1–2.7) we explain and discuss the different testing procedures together with their applications. The chapter is readable without previous theoretical knowledge. For the continuous models tables of critical values are provided in Section 7.10 (Appendix), whereas for the general models (tied observations) computing programs for the test statistics and the conditional p-values are available, cf. p. 417.

Werner Schütt did uncounted Monte Carlo simulations in order to compare the power properties of various procedures and to check the proposed bandwidth. Günter Heimann read most parts of the manuscript. To both of them we are grateful and also to the DFG (Deutsche Forschungsgemeinschaft), which granted the necessary financial support. Finally, thanks are due to Mrs. R. Witt, who typed most parts of the final manuscript.

Hamburg, January 1989               K. Behnen       G. Neuhaus

# Contents

# Some General Symbols

| | |
|---|---|
| $1(X \in B)$ | indicator of the event $X \in B$ |
| $\mathcal{L}(X)$ | law of the random variable $X$ |
| $\mathcal{L}(X) \sim f$ | $\mathcal{L}(X)$ is distributed according to $f$ |
| $[x]$ | integer part of $x$ |
| $\Phi$ | standard normal distribution function |
| $u_\alpha = \Phi^{-1}(1 - \alpha)$ | upper $\alpha$-quantile of $\Phi$ |
| $\varphi(u, f_0)$ | shift score function, cf. formula (1.1.10) |
| $c_{Ni}$ | two-sample coefficients, cf. formula (1.1.16) |
| $\varphi(u, f_0, D)$ | generalized shift score function, cf. formula (1.2.4) |
| $\lambda$ | Lebesgue measure on the interval $(0, 1)$ |
| $R_i$ | rank of $X_i$, cf. formula (1.1.1) |
| $\xrightarrow{\mathcal{L}}$ | convergence in distribution |
| $a := b$ | $a$ equals $b$ by definition |
| $\#$ | number of elements in a set |
| $\chi_\varrho^2$ | chi-square distribution with $\varrho$ degrees of freedom |
| $\otimes$ | Kronecker product of matrices |
| $L_2(0, 1)$, $L_2^0(0, 1)$ | square integrable functions on $(0, 1)$, cf. p. 105 |
| $< \cdot, \cdot >$, $\| \cdot \|$ | inner product and norm on $L_2(0, 1)$ |
| $C[0, 1]$ | continuous real valued functions on $[0, 1]$ |
| $\mathcal{B}(C[0, 1])$ | Borel $\sigma$-algebra on $C[0, 1]$ |
| $\| \cdot \|_\infty$ | sup-norm, cf. p. 121 |
| $W_0$ | Brownian bridge, cf. p. 121 |
| $< x, y >_\Gamma$, $|x|_\Gamma$ | inner product $x^T \Gamma y$ and corresp. norm, cf. (3.2.54) |
| $\mathbb{R}_J^r$, $\mathbb{R}_J^{r+}$ | cf. formula (3.2.58) |
| $\varphi_1(u, f_0)$ | scale score function, cf. formula (4.1.6) |
| $F_-$ | distr. funct. of $-X$, if $F$ is the distr. funct. of $X$ |
| $R_i^+$ | rank of $|X_i|$, cf. formula (5.1.3) |
| $\text{sign}(x)$ | the sign of $x$, where $\text{sign}(0) = 0$, cf. formula (5.1.4) |
| $o_{P_N}(1)$ | remainder term tending to zero in $P_N$-probability |

# Part I

# Motivation and Applications

# Chapter 1

# Introduction and Motivation

Since the monograph of Hájek and Šidák (1967) it's indisputable that linear
rank tests form an attractive alternative to the classical tests based on normality assumptions, cf. Hollander and Wolfe (1973), Lehmann (1975), Conover
(1980), and others. But meanwhile it has been demonstrated that the asymptotic optimality of linear rank tests for special types of alternatives is connected
with rather low power for other types of alternatives. Therefore there have
been some attempts to increase the power for certain classes of alternatives by
adaptation, e.g. Randles and Hogg (1973) and others.

The first aim of the present chapter will be to show that the gain of power of
those tests is completely due to the assumption of strict shift models. Additionally, their power may be much lower than the power of a standard linear
rank test, e.g. the Wilcoxon rank test, if there is a small deviation from the
assumption of shift models.

The second aim will be the identification of the underlying optimal score function in a general nonparametric model which contains the shift model as a
special submodel. Starting from the linear rank test with the appropriate
nonparametric score function we shall construct new nonlinear rank tests for
various testing problems and investigate the asymptotic power behaviour, cf.
Chapter 3 to Chapter 6. The application of the new rank tests and a Monte
Carlo power comparison are discussed in Chapter 2, cf. Figure 2.1.c to Figure 2.1.h.

Because the problem of comparing a new treatment or drug with a standard
is one of the prominent problems in applied as well as in theoretical statistics

and because optimal tests under various special assumptions are well-known, we'll take this problem in order to explain our ideas and to demonstrate the motivation of the new testing procedures. As soon as the motivation and the construction principles are clear in this case, there is no difficulty to discuss other testing problems, too.

The following review of linear rank tests under the shift model will use the results of Hájek and Šidák (1967) without specifying the exact reference. It may be helpful if the reader is somewhat familiar with the theory of linear rank statistics.

A simple algorithmic description of the proposed new rank tests and a step by step evaluation of numerical examples—without stressing any mathematical theory—is given in Chapter 2. Therefore readers who aren't interested in the theoretical background may skip Chapter 1 and immediately start with Chapter 2.

## 1.1 The shift model: A review

In the usual shift model for comparing two treatments the $n$ observations of the standard treatment are represented by $n$ independent and identically distributed (i.i.d.) real random variables $Y_1, ..., Y_n$ with absolutely continuous density $f_0$ , and the $m$ observations of the new treatment are represented by $m$ real i.i.d. random variables $X_1, ..., X_m$ with the shifted density $f_\vartheta(x) = f_0(x - \vartheta)$ , $x \in \mathbb{R}$ . Additionally $(X_1, ..., X_m)$ and $(Y_1, ..., Y_n)$ are assumed to be stochastically independent.

If it's desirable to observe large values, then $\vartheta > 0$ means that the new treatment is better than the standard treatment, whereas $\vartheta \leq 0$ means that the standard treatment is (weakly) better than the new treatment. Therefore we'll consider the one-sided problem of testing the *hypothesis* $\mathcal{H} : \vartheta \leq 0$ versus the *alternative* $\mathcal{A} : \vartheta > 0$ .

Putting $X_{m+j} = Y_j$, $j = 1, ..., n$ , the *rank* $R_i$ of $X_i$ in the pooled sample $(X_1, ..., X_N)$ , $N := m + n$ , is defined by ( $\forall\, 1 \leq i \leq N$ )

$$R_i = \sum_{j=1}^{N} 1(X_j \leq X_i), \tag{1.1.1}$$

where $1(C)$ denotes the *indicator* of condition $C$ , i.e. $1(C) = 1$ if $C = $ True and $1(C) = 0$ if $C = $ False .

Under the *hypothesis of randomness* $\mathcal{H}_0 : \vartheta = 0$ the vector of ranks $R = (R_1, ..., R_N)$ is uniformly distributed on the $N!$ permutations of $(1, ..., N)$ , independently of the underlying continuous distribution function $F_0$ . Using a *linear rank statistic* of the form

$$S_N = \sqrt{\frac{mn}{N}} \left( \frac{1}{m} \sum_{i=1}^{m} a_N(R_i) - \frac{1}{n} \sum_{i=m+1}^{N} a_N(R_i) \right) \tag{1.1.2}$$

with increasing scores $a_N(1) \leq a_N(2) \leq \cdots \leq a_N(N)$ it's well-known that any corresponding linear rank test of the form

$$\psi_N = 1\big( S_N > c_N \big) + \gamma_N\, 1\big( S_N = c_N \big) \tag{1.1.3}$$

has a monotone power function, i.e.

$$\vartheta_1 < \vartheta_2 \quad \Longrightarrow \quad E_{\vartheta_1}(\psi_N) \leq E_{\vartheta_2}(\psi_N). \tag{1.1.4}$$

Thus, given the level $\alpha \in (0, 1)$ and fixing the critical value $c_N$ according to $P_0\{S_N > c_N\} + \gamma_N P_0\{S_N = c_N\} = \alpha$ the linear rank test $\psi_N$ is an *unbiased* level $\alpha$ test for testing the hypothesis $\mathcal{H} : \vartheta \leq 0$ versus the alternative

$\mathcal{A} : \vartheta > 0$. Since $\mathcal{L}_0(S_N)$ and $c_N$ are the same for any underlying density $f_0$, we might consider $f_0$ as a nuisance parameter and use the same $\psi_N$ for all $f_0$. But unfortunately the test $\psi_N$ will be optimal only if there is a correspondence between $f_0$ and the scores $a_N(1), ..., a_N(N)$ of $\psi_N$, whereas the power of $\psi_N$ may decrease substantially if $f_0$ doesn't correspond to $a_N(1), ..., a_N(N)$.

In order to make this statement precise we'll consider local asymptotic alternatives of the following form:

Assume $m = m_N$ and $n = n_N$ such that $m \to \infty$ and $n \to \infty$ as $N \to \infty$. For any $\varrho \geq 0$ define local asymptotic shift alternatives $\big(Q_{N\varrho}(f_0),\ N \geq 1\big)$ according to

$$\mathcal{L}(X_1) = \cdots = \mathcal{L}(X_m) \ \sim\ f_0\big(x - \varrho\,\sqrt{\tfrac{N}{mn}}\,\big),$$

$$\mathcal{L}(X_{m+1}) = \cdots = \mathcal{L}(X_N) \sim f_0(x), \tag{1.1.5}$$

$$Q_{N\varrho}(f_0) = \mathcal{L}(X_1, ..., X_N).$$

Using the local asymptotic parameter $\varrho$ the null hypothesis of randomness corresponds to $\varrho = 0$, whereas $\varrho > 0$ means that the new treatment is better than the standard treatment. However, the asymptotic model assumes that the amount of shift will depend on the sample sizes $m$ and $n$ in a way which makes the decision between $\varrho = 0$ and $\varrho = \varrho_0 > 0$ comparably hard for any sample size. Therefore the asymptotic power under such sequences of alternatives may be used in order to compare the qualities of different tests.

If we assume that the step function $a_N\big(1 + [Nu]\big)$, $0 < u < 1$, of the scores $a_N(1), ..., a_N(N)$, where $[x]$ denotes the integer part of $x$, can be approximated in quadratic mean by some fixed square-integrable *score function* $h : (0, 1) \to \mathbb{R}$,

$$\int_0^1 \Big(a_N\big(1 + [Nu]\big) - h(u)\Big)^2 du \ \overset{N\to\infty}{\longrightarrow}\ 0, \tag{1.1.6}$$

with the properties $\int_0^1 h(u)\, du = 0$ and $\|h\| := \big(\int_0^1 h^2(u)\, du\big)^{1/2} > 0$, then the corresponding linear rank test

$$\psi_{N\alpha} \ =\ 1\big(\, S_N \geq u_\alpha \|h\|\, \big), \tag{1.1.7}$$

where $0 < \alpha < 1$ is a given level and $u_\alpha = \Phi^{-1}(1 - \alpha)$ is the upper $\alpha$-quantile of the standard normal distribution, is a test of asymptotic level $\alpha$ for testing $\varrho = 0$ versus $\varrho > 0$, and the asymptotic power of $\psi_{N\alpha}$ under $Q_{N\varrho}(f_0)$ is given by

$$\lim_{N\to\infty} E\big[\, \psi_{N\alpha} \mid Q_{N\varrho}(f_0)\, \big] \ =\ 1 - \Phi\big(u_\alpha - c(h, f_0)\varrho\big), \tag{1.1.8}$$

where

$$c(h, f_0) = \frac{1}{\|h\|} \int_0^1 h(u)\, \varphi(u, f_0)\, du \qquad (1.1.9)$$

and

$$\varphi(u, f_0) = -\frac{f_0'\big(F_0^{-1}(u)\big)}{f_0\big(F_0^{-1}(u)\big)}\,, \qquad 0 < u < 1. \qquad (1.1.10)$$

In order to make all assumptions explicit we have to pose the additional assumption

$$\|\varphi(\cdot, f_0)\|^2 = \int_0^1 \varphi^2(u, f_0)\, du = \int_{-\infty}^{+\infty} \left(\frac{f_0'}{f_0}\right)^2 f_0\, dx < \infty, \qquad (1.1.11)$$

i.e. the density $f_0$ has to have *finite Fisher-information.*

For any given $\varrho > 0$ the asymptotic power (1.1.8) achieves its maximum $1 - \Phi(u_\alpha - \|\varphi(\cdot, f_0)\|\varrho)$ if the score function $h$ has the property

$$h(u) = c\, \varphi(u, f_0), \qquad 0 < u < 1, \qquad (1.1.12)$$

for some constant $c > 0$. Therefore $\varphi(\cdot, f_0)$ is called the *optimal score function* for testing $Q_{N0}(f_0)$ versus the shift alternatives $Q_{N\varrho}(f_0)$, $\varrho > 0$. The corresponding asymptotically optimal linear rank statistic according to (1.1.2) and (1.1.6) is denoted by $S_N(f_0)$, i.e. the linear rank test

$$\psi_{N\alpha}(f_0) = 1\big( S_N(f_0) \geq u_\alpha\|\varphi(\cdot, f_0)\| \big) \qquad (1.1.13)$$

is asymptotically optimal for testing the null hypothesis of randomness $\mathcal{H}_0$ versus the local asymptotic shift alternatives $\big(Q_{N\varrho}(f_0), N \geq 1\big)$, $\varrho > 0$, in the family of all asymptotic level $\alpha$ tests.

### 1.1.1 Example

The *normal shift model* is defined by formula (1.1.5) where $F_0$ is the standard normal distribution function,

$$F_0(x) = \Phi(x) = \int_{-\infty}^x f_0(t)\, dt, \qquad x \in \mathbb{R},$$

$$(1.1.14)$$

$$\text{where} \quad f_0(t) = \frac{1}{\sqrt{2\pi}}\, \exp(-\frac{1}{2}t^2).$$

According to Table 1.1 the optimal score function $\varphi(\cdot, f_0)$ is $\Phi^{-1}$ and the corresponding Fisher-information $\|\varphi(\cdot, f_0)\|^2$ is 1. The standardized optimal score function $\varphi(\cdot, f_0)/\|\varphi(\cdot, f_0)\|$ is plotted in Figure 1.1. The function $\varphi(\cdot, f_0)$ corresponding to the normal density is very smooth. Therefore the simple

**Figure 1.1**

*The standardized optimal score functions* $\varphi(\cdot, f_0)/\|\varphi(\cdot, f_0)\|$ *for the normal shift model* ( $\star\ \star\ \star$ ) , *the logistic shift model* ( $\cdot\ \cdot\ \cdot$ ) , *and the Cauchy shift model* ( $\circ\ \circ\ \circ$ ) .

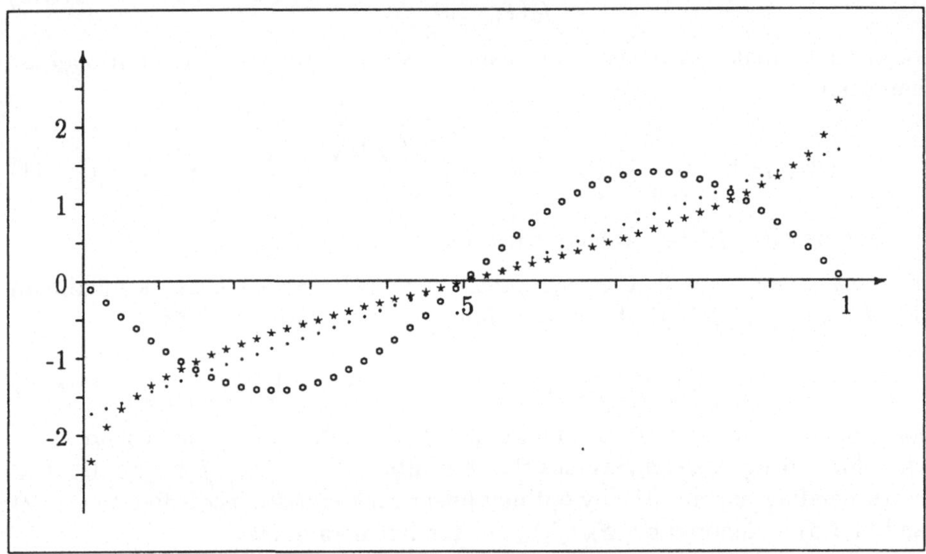

scores $a_N(i) = \varphi\big(i/(N+1), f_0\big)$ can be proved to fulfil condition (1.1.6) and the optimal linear rank statistic (1.1.2) is the *van der Waerden rank statistic*

$$S_N(f_0) = \sum_{i=1}^{N} c_{Ni} \ \Phi^{-1}\big(\frac{R_i}{N+1}\big), \qquad (1.1.15)$$

where the two-sample coefficients $c_{Ni}$ are defined by

$$c_{Ni} = \sqrt{\frac{mn}{N}} \begin{cases} 1/m \, , & \text{if} \quad 1 \leq i \leq m, \\ -1/n \, , & \text{if} \quad m+1 \leq i \leq N. \end{cases} \qquad (1.1.16)$$

### 1.1.2 Example

The *logistic shift model* is defined by formula (1.1.5) where $F_0$ is the logistic distribution function,

$$F_0(x) = \frac{\exp(x)}{1 + \exp(x)}, \qquad x \in \mathbb{R}. \qquad (1.1.17)$$

According to Table 1.1 the optimal score function $\varphi(u, f_0)$ is $2u - 1$ and the corresponding Fisher-information $\|\varphi(\cdot, f_0)\|^2$ is $1/3$. The standardized optimal score function is plotted in Figure 1.1. The function $\varphi(\cdot, f_0)$ corresponding to the logistic density is very smooth. Therefore the simple scores $a_N(i) = \varphi(i/(N+1), f_0)$ can be proved to fulfil condition (1.1.6) and the optimal linear rank statistic (1.1.2) is the *Wilcoxon rank statistic*

$$S_N(f_0) = \frac{2}{N+1} \sum_{i=1}^{N} c_{Ni}\, R_i\,, \tag{1.1.18}$$

where the two sample coefficients $c_{Ni}$ are defined in (1.1.16).

### 1.1.3 Example

The *Cauchy shift model* is defined by formula (1.1.5) where $F_0$ is the Cauchy distribution function,

$$F_0(x) = \int_{-\infty}^{x} \frac{1}{\pi(1+t^2)}\, dt = \frac{1}{2} + \frac{1}{\pi} \arctan(x), \quad x \in \mathbb{R}. \tag{1.1.19}$$

The optimal score function $\varphi(u, f_0)$ and the corresponding Fisher-information $\|\varphi(\cdot, f_0)\|^2$ are given in Table 1.1. The standardized optimal score function is plotted in Figure 1.1. The function $\varphi(\cdot, f_0)$ corresponding to the Cauchy density is very smooth. Therefore the simple scores $a_N(i) = \varphi(i/(N+1), f_0)$ can be proved to fulfil condition (1.1.6) and the optimal linear rank statistic (1.1.2) is

$$S_N(f_0) = 2 \sum_{i=1}^{N} c_{Ni}\, \frac{T_i}{1+T_i^2}\,, \tag{1.1.20}$$

$$\text{where} \quad T_i = \tan\left( \pi\, \left(\frac{R_i}{N+1} - \frac{1}{2}\right) \right),$$

and where the two sample coefficients $c_{Ni}$ are defined in (1.1.16).

The above optimality result proves every linear rank test $\{\psi_{N\alpha}(f_0)\}$ to be asymptotically optimal. Unfortunately this means optimality only for the *corresponding* shift alternative $Q_{N\varrho}(f_0)$, $\varrho > 0$, while the asymptotic power for *other* shift alternatives $Q_{N\varrho}(f_1)$, $\varrho > 0$, say, depends on the value of $c(\varphi(\cdot, f_0), f_1)$, cf. (1.1.8) and (1.1.9). The resulting (asymptotic) power may be rather small if the shape of the assumed $f_0$ differs from the shape of the (unknown) underlying $f_1$.

In order to apply a linear rank test we are forced (at least implicitly) to choose some suitable $f_0$ in advance. Often the logistic density $f_0$ is chosen since on

## Table 1.1

|  | Normal | Logistic | Cauchy |
|---|---|---|---|
| $f(x)$ | $\dfrac{1}{\sqrt{2\pi}}\exp(\dfrac{-x^2}{2})$ $= \varphi(x)$ | $\dfrac{\exp(x)}{(1+\exp(x))^2}$ | $\dfrac{1}{\pi(1+x^2)}$ |
| $F(x)$ | $\Phi(x)$ | $\dfrac{\exp(x)}{1+\exp(x)}$ | $\dfrac{1}{2}+\dfrac{1}{\pi}\arctan(x)$ |
| $F^{-1}(u)$ | $\Phi^{-1}(u)$ | $\ln(\dfrac{u}{1-u})$ | $\tan[\pi(u-\tfrac{1}{2})]$ |
| $f\circ F^{-1}(u)$ | $\varphi\circ\Phi^{-1}(u)$ | $u(1-u)$ | $\dfrac{1}{\pi(1+\tan^2[\pi(u-\tfrac{1}{2})])}$ |
| $\varphi(u,f)$ | $\Phi^{-1}(u)$ | $2u-1$ | $\dfrac{2\tan[\pi(u-\tfrac{1}{2})]}{1+\tan^2[\pi(u-\tfrac{1}{2})]}$ |
| $\\|\varphi(\cdot,f)\\|^2$ | $1$ | $1/3$ | $1/2$ |
| $\varphi_1(u,f)$ | $(\Phi^{-1}(u))^2-1$ | $(2u-1)\ln(\dfrac{u}{1-u})-1$ | $\dfrac{\tan^2[\pi(u-\tfrac{1}{2})]-1}{\tan^2[\pi(u-\tfrac{1}{2})]+1}$ |
| $\\|\varphi_1(\cdot,f)\\|^2$ | $2$ | $1.4300$ | $1/2$ |

one hand the corresponding optimal test is the very simple Wilcoxon test, and on the other hand the power of the Wilcoxon test is relatively high also under normal shift alternatives, because of the large correlation

$$\int_0^1 \frac{\varphi(u, f_0)}{\|\varphi(\cdot, f_0)\|} \frac{\varphi(u, f_1)}{\|\varphi(\cdot, f_1)\|} \, du \; = \; \sqrt{\frac{3}{\pi}} \; = \; 0.977 \; , \qquad (1.1.21)$$

if $f_0$ is the logistic density and $f_1$ is the normal density.

If we *really* could trust the *shift* models, then we should construct a data based selection rule in order to estimate the underlying $f_0$ or the corresponding score function $\varphi(\cdot, f_0)$ .

In principle such construction may be found in Section VII.1.6 of Hájek and Šidák (1967): Only depending on the order statistic $X_N^{(1)}, ..., X_N^{(N)}$ of the pooled sample they construct a sequence of estimators $\tilde{\varphi}_N$ for $\varphi(\cdot, f_0)$ with unknown $f_0$ . But Hájek and Šidák don't recommend their estimators for practical application, since they expect very slow convergence. In our view *not even a very good estimator* of $\varphi(\cdot, f_0)$ will produce satisfactory results in practical application, since the assumption of *strict* shift usually is not realistic in real problems. It seems much more plausible that e.g. the extreme parts of a population react in quite another way to a treatment than the central part of that population. Therefore a good testing procedure shouldn't be too sensitive to such deviations from exact shift.

More simple procedures are proposed by Randles and Hogg (1973): Starting from a set of three score functions $\varphi(\cdot, f_i)$ , $i = 1, 2, 3$ , which correspond to light tails, medium tails, and heavy tails, they select one of the three score functions on the basis of the observed order statistic of the pooled sample and apply the corresponding linear rank test.

Both approaches fully exploit the assumption of *shift alternatives* , i.e. at least some features of the corresponding *shift score function* $\varphi(\cdot, f_0)$ are estimated on the basis of the order statistic of the pooled sample.

**Our starting point is just here:** In Section 1.2 we use the notion of *generalized shift models* in order to demonstrate that even very plausible deviations from the shift model will produce optimal score functions which are completely different from $\varphi(\cdot, f_0)$ , i.e., the shift score function $\varphi(\cdot, f_0)$ has no statistical interpretation if the shift model is violated. Since in reality we cannot trust in exact shift models, the consequence will be that the estimation of $\varphi(\cdot, f_0)$ may lead to a complete power break down.

However, e.g. the estimator $\tilde{\varphi}_N$ of Hájek and Šidák invariantly estimates $\varphi(\cdot, f_0)$ , even in generalized shift models where this function by no means is the optimal score function. Therefore the Hájek and Šidák procedure will fail in practical application, since in fact the "wrong" score function is estimated.

Similar objections apply to the procedures which are comparable to the proposals of Randles and Hogg (1973), since the selection of the score function is based on features of the ordered pooled sample, e.g. on tail classification of the pooled sample. However, only in the strict shift model there is a connection between the type of the tails and the shape of the optimal score function. In the case of generalized shift models the order statistic of the pooled sample contains almost no information about the shape of the optimal score function, cf. Section 1.2. Therefore suitable estimators of the scores shouldn't rely on the pooled order statistic but on the *ranks*, cf. Chapter 3.

In Section 1.3 we'll identify the underlying optimal score function in a general nonparametric model, i.e. we identify the score function which should be estimated instead of $\varphi(\cdot, f_0)$ . Naturally, if the exact shift model corresponding to $f_0$ is assumed, the score function of Section 1.3 coincides with $\varphi(\cdot, f_0)$ . The technical details of the estimation procedures are given in Chapter 3. The corresponding tests with estimated scores and their applications are presented in Chapter 2 for quite a number of different models, since the final results are very simple.

## 1.2  Generalized shift models

In this section we discuss the concept of generalized shift alternatives and derive the corresponding asymptotically optimal score function for testing a fixed underlying distribution versus generalized shift alternatives. Our aim is an explicit discussion of the problems arising, when the ideal shift model is not present, rather than the development of a theory of rank tests for such generalized shift models.

Let $F_0$ be the underlying (cumulative) distribution function of the standard treatment (second sample), and assume $F_0$ to have an absolutely continuous density $f_0$ with finite Fisher-information $\|\varphi(\cdot, f_0)\|^2 < \infty$.

Under the exact shift model of Section 1.1 the underlying (cumulative) distribution function $F_\vartheta$ of the new treatment (first sample) coincides with the $\vartheta$–shifted distribution of the standard treatment, i.e.

$$F_\vartheta(x) = F_0(x - \vartheta), \qquad x \in \mathrm{IR}. \tag{1.2.1}$$

Thus, under the exact shift model, any part of the population is forced to react by the same amount of shift if the new treatment is applied. But it seems much more realistic to allow different parts of the population to show different rates of reaction to the new treatment. Especially it's plausible that the extreme parts of the population will react in quite another way than the central part of that population.

A very simple *generalized shift model* of this kind is achieved if the distribution function of the first sample (new treatment) is given by

$$F_{\vartheta D}(x) = F_0\big(x - \vartheta D(x)\big), \qquad x \in \mathrm{IR}, \tag{1.2.2}$$

where the *shift function* $D : \mathrm{IR} \to \mathrm{IR}$ is assumed to be bounded with $D \geq 0$, $D \not\equiv 0$, and bounded derivative $d : \mathrm{IR} \to \mathrm{IR}$. Obviously the side condition $\vartheta d(x) \leq 1 \ \forall \ x \in \mathrm{IR}$ has to be assumed in order to make $F_{\vartheta D}$ a proper distribution function with corresponding density

$$f_{\vartheta D}(x) = \big(1 - \vartheta d(x)\big) f_0\big(x - \vartheta D(x)\big), \qquad x \in \mathrm{IR}. \tag{1.2.3}$$

Again $\vartheta > 0$ means that the new treatment is better than the standard treatment, and the shift function $D$ describes the different reaction of different parts of the population.The exact shift model (1.2.1) corresponds to $D \equiv 1$ and $d \equiv 0$.

The optimal score function $\varphi(\cdot, f_0)$ for testing $\varrho = 0$ versus $\varrho > 0$ under the exact shift model (1.1.5) has a natural analogue $\varphi(\cdot, f_0, D)$ in case of the generalized shift model (1.2.2), namely,

$$\varphi(u, f_0, D) = \frac{\partial \ln f_{\vartheta D}\left(F_0^{-1}(u)\right)}{\partial \vartheta}\bigg|_{\vartheta=0} \tag{1.2.4}$$

$$= \varphi(u, f_0)\, D\left(F_0^{-1}(u)\right) - d\left(F_0^{-1}(u)\right), \qquad 0 < u < 1.$$

In order to make this statement precise we have to consider local asymptotic alternatives of the following form:

Assume $m = m_N$ and $n = n_N$ such that $m \to \infty$ and $n \to \infty$ as $N \to \infty$. For any $\varrho \geq 0$ define local asymptotic alternatives ( $Q_{N\varrho}(F_0, D)$, $N \geq 1$ ) according to

$$\mathcal{L}(X_1) = \cdots = \mathcal{L}(X_m) \sim F_0\left(x - \varrho\sqrt{\frac{N}{mn}}\, D(x)\right),$$

$$\mathcal{L}(X_{m+1}) = \cdots = \mathcal{L}(X_N) \sim F_0(x),$$

$$Q_{N\varrho}(F_0, D) = \mathcal{L}(X_1, ..., X_N). \tag{1.2.5}$$

Then $\varrho = 0$ corresponds to the null hypothesis of randomness and $\varrho > 0$ means that the new treatment is better than the standard treatment since we have assumed $D \geq 0$, $D \neq 0$. Again the asymptotic model (1.2.5) assumes that the amount of generalized shift will depend on the sample sizes $m$, $n$, and $N = m + n$ in a way which makes the decision between $\varrho = 0$ and $\varrho = \varrho_0 > 0$ comparably hard for any sample size $N$. Therefore the asymptotic power under generalized shift alternatives $Q_{N\varrho}(F_0, D)$ may be used in order to compare different tests.

If the scores $a_N(1), ..., a_N(N)$ of the linear rank statistic $S_N$ given in (1.1.2) fulfil condition (1.1.6) then the linear rank test $\psi_{N\alpha} = 1(\,S_N \geq u_\alpha\|h\|\,)$ defined in (1.1.7) is a test of asymptotic level $\alpha$ for testing $\varrho = 0$ versus $\varrho > 0$, and the asymptotic power of $\psi_{N\alpha}$ under the generalized shift alternative $Q_{N\varrho}(F_0, D)$ is given by

$$\lim_{N \to \infty} E[\,\psi_{N\alpha} \mid Q_{N\varrho}(F_0, D)\,] = 1 - \Phi\left(u_\alpha - \varrho\, c(h, f_0, D)\right), \tag{1.2.6}$$

where

$$c(h, f_0, D) = \frac{1}{\|h\|} \int_0^1 h(u)\, \varphi(u, f_0, D)\, du. \tag{1.2.7}$$

For any given $\varrho > 0$ the asymptotic power (1.2.6) achieves its maximum $1 - \Phi(u_\alpha - \varrho\,\|\varphi(\cdot, f_0, D)\|)$ if the score function $h$ has the property

$$h(u) = c\, \varphi(u, f_0, D), \qquad 0 < u < 1, \tag{1.2.8}$$

for some constant $c > 0$. Thus $\varphi(\cdot, f_0, D)$ is the asymptotically optimal score function for testing the hypothesis of randomness versus the generalized shift alternatives $Q_{N\varrho}(F_0, D)$, $\varrho > 0$.

In the present case of generalized shift the shape of the optimal score function $\varphi(\cdot, f_0, D)$ is influenced by *two independent quantities,* namely the density $f_0$ *and* the shift function $D$. The following lemma demonstrates how drastically $\varphi(\cdot, f_0, D)$ may vary if the usual shift $D \equiv 1$ is replaced by some general shift function $D$.

### 1.2.1 Lemma

*Let $F$ and $G$ be distribution functions on $\mathbb{R}$ with respective densities $f$ and $g$. Assume $f$ and $g$ to be strictly positive and absolutely continuous (on all finite intervals) with finite Fisher information. Define the shift function $D$ by*

$$D(x) = \frac{\int_0^{F(x)} \varphi(y, g)\, dy}{\int_0^{F(x)} \varphi(y, f)\, dy} = \frac{g \circ G^{-1}(F(x))}{f \circ F^{-1}(F(x))}. \tag{1.2.9}$$

*Then the following equation holds true*

$$\varphi(\cdot, f, D) = \varphi(\cdot, g). \tag{1.2.10}$$

The *proof* is immediate from the following two formulae,

$$\int_0^t \varphi(y, f)\, dy = -f \circ F^{-1}(t), \quad 0 < t < 1, \tag{1.2.11}$$

$$\int_0^t \varphi(y, f, D)\, dy = \left(-f \circ F^{-1}(t)\right)\left(D \circ F^{-1}(t)\right), \qquad 0 < t < 1, \tag{1.2.12}$$

which may be proved by differentiating both sides.

Using the results listed in Table 1.1 we may compute the following examples.

### 1.2.2 Example

a) Let $f$ be the normal density $\varphi$ and let $g$ be the logistic density. Let $D_{NL}$ denote the shift function (1.2.9) in this case. Then we get

$$D_{NL}(x) = \frac{\Phi(x)\left(1 - \Phi(x)\right)}{\varphi(x)}, \qquad x \in \mathbb{R}. \tag{1.2.13}$$

**Figure 1.2**

*The shift functions $D_{NL}$ $(\ast\,\ast\,\ast)$, $D_{NC}$ $(\cdot\,\cdot\,\cdot)$, $D_{LC}$ $(\circ\,\circ\,\circ)$.*

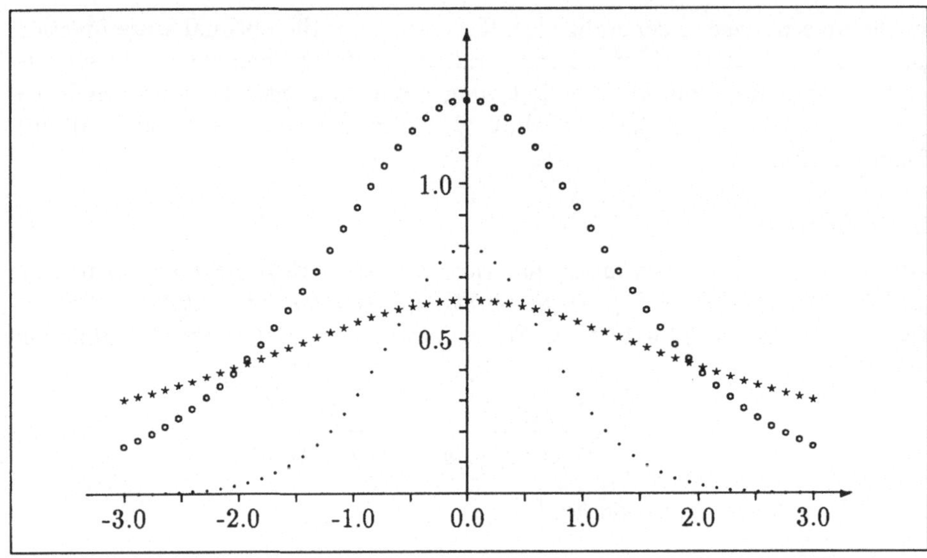


b) Let $f$ be the normal density $\varphi$ and let $g$ be the Cauchy density. Let $D_{NC}$ denote the shift function (1.2.9) in this case. Then we get

$$D_{NC}(x) = \frac{1}{\pi\,\varphi(x)\left(1 + \tan^2\left(\pi\left(\Phi(x) - 1/2\right)\right)\right)}, \qquad x \in \mathbb{R}. \qquad (1.2.14)$$

c) Let $f$ be the logistic density and let $g$ be the Cauchy density. Let $D_{LC}$ denote the shift function (1.2.9) in this case. Then we get

$$D_{LC}(x) \;=\; \frac{1}{\pi\,F(x)\left(1 - F(x)\right)\left(1 + \tan^2\left(\pi\left(F(x) - 1/2\right)\right)\right)},$$
$$(1.2.15)$$
$$F(x) \;=\; \frac{\exp(x)}{1 + \exp(x)}, \qquad x \in \mathbb{R}.$$

We realize that e.g. the optimal score function $\varphi(\,\cdot\,,\text{ normal density}, D_{NL})$ for *generalized shift alternatives with normal density and shift function $D_{NL}$* is exactly the same as the optimal score function $\varphi(\,\cdot\,,\text{ logistic density})$ for *usual logistic shift alternatives,*

$$\varphi(\,\cdot\,,\text{ normal density}, D_{NL}) \;=\; \varphi(\,\cdot\,,\text{ logistic density}). \qquad (1.2.16)$$

Thus, instead of the van der Waerden test, which is optimal for usual normal shift alternatives, the Wilcoxon test is optimal for the generalized normal shift alternatives with shift function $D_{NL}$ . Similarly we have

$$\varphi(\,\cdot\,,\text{ normal density}, D_{NC}) = \varphi(\,\cdot\,,\text{ Cauchy density}), \qquad (1.2.17)$$

and

$$\varphi(\,\cdot\,,\text{ logistic density}, D_{LC}) = \varphi(\,\cdot\,,\text{ Cauchy density}). \qquad (1.2.18)$$

Additionally, Example 1.2.2 shows that generalized shift alternatives corresponding to some given distribution function $F_0$ and to some shift function $D$ which concentrates the shift to the middle part of the distribution $F_0$ have the same effect as usual shift alternatives with respect to a distribution function $F_1$ with "heavier" tails than $F_0$ . A similar "conversion to lighter tails" by a suitable shift function $D$ may lead to difficulties, since—at least in the examples given above—the derivative of $D$ would be unbounded. Nevertheless it's obvious that the shape of the optimal score function has no sound connection with "heavy" or "light" tails as long as generalized shift alternatives are considered. We will have reached the goal of the present section, if this phenomenon is clear.

## 1.3   General alternatives

Let $X_1, ..., X_m, X_{m+1}, ..., X_{m+n}$ be independent real random variables and assume that each $X_i$ of the first sample $X_1, ..., X_m$ has the continuous distribution function $F$ whereas each $X_j$ of the second sample $X_{m+1}, ..., X_{m+n}$ has the continuous distribution function $G$ . Since the sample sizes $m$ , $n$ , and $N = m + n$ are given quantities the parameter space of this two-sample model is

$$\{(F, G) : F, G \in \mathcal{F}_1^c\} = \mathcal{F}_1^c \times \mathcal{F}_1^c, \tag{1.3.1}$$

where

$$\mathcal{F}_1^c := \{H : H \text{ any continuous distribution function on } \mathbb{R}\} \tag{1.3.2}$$

is the set of all continuous distribution functions on the real line $\mathbb{R}$ . The classical shift model as well as the generalized shift models of Section 1.2 are obvious submodels of (1.3.1). In the classical case we have

$$G = F_0 \qquad \text{and} \qquad F = F_\vartheta, \quad \vartheta \in \mathbb{R}, \tag{1.3.3}$$

where $F_\vartheta(x) = F_0(x - \vartheta)$ $\forall x \in \mathbb{R}$ . Similarly in the generalized case we have

$$G = F_0 \qquad \text{and} \qquad F = F_{\vartheta D}, \quad \vartheta \in \mathbb{R}, \tag{1.3.4}$$

where $F_{\vartheta D}(x) = F_0(x - \vartheta D(x))$ $\forall x \in \mathbb{R}$ with given shift function $D$ as defined in formula (1.2.2).

In both parametric cases the underlying parameter space of the model is the real line $\mathbb{R}$ , and we want to test the null hypothesis $\mathcal{H}_0 : \vartheta = 0$ versus the alternative $\mathcal{A} : \vartheta > 0$ (new treatment better than standard treatment).

A natural generalization of the parametric alternative $\mathcal{A} : \vartheta > 0$ in case of the *nonparametric model* (1.3.1) is given by

$$\mathcal{A}_2^0 = \{(F, G) \in \mathcal{F}_1^c \times \mathcal{F}_1^c : F \leq G, F \neq G\}, \tag{1.3.5}$$

which means ( $\forall i = 1, ..., m$, $\forall j = m + 1, ..., N$ )

$$\Pr\{X_i > y\} = 1 - F(y) \geq 1 - G(y) = \Pr\{X_j > y\} \quad \forall y \in \mathbb{R},$$
$$\Pr\{X_i > y\} = 1 - F(y) > 1 - G(y) = \Pr\{X_j > y\} \quad \exists y \in \mathbb{R}, \tag{1.3.6}$$

i.e. the distribution of the first sample is *stochastically larger* than the distribution of the second sample.

Similarly, the parametric null hypothesis $\mathcal{H}_0 : \vartheta = 0$ is generalized to the *nonparametric hypothesis of randomness*

$$\mathcal{H}_0^r = \{(F, G) \in \mathcal{F}_1^c \times \mathcal{F}_1^c : F = G\}. \tag{1.3.7}$$

Since rank statistics are distribution-free under the null hypothesis $\mathcal{H}_0^r$ the linear rank tests of the previous sections are asymptotically optimal for testing $\mathcal{H}_0^r$ versus special shift-subclasses of $\mathcal{A}_2^0$ . But unfortunately the (asymptotic) power may decrease substantially, if there are deviations from the special shift assumptions, cf. Section 1.2.

In order to construct tests for $\mathcal{H}_0^r$ versus $\mathcal{A}_2^0$ with high power even in cases where the shift assumption is substantially violated, we'll proceed in the following nonparametric way:

Starting with an arbitrary alternative point $(F, G) \in \mathcal{A}_2^0$ we'll construct a corresponding "least favorable" null hypothesis point $(H, H) \in \mathcal{H}_0^r$ and identify the (asymptotically) optimal score function $b$ for testing the simple hypothesis $(H, H)$ versus the simple alternative $(F, G)$ . In a first step this will lead to a linear rank test with score function $b$ which is asymptotically optimal for testing the null hypothesis of randomness $\mathcal{H}_0^r$ versus the simple alternative $(F, G)$ , cf. Theorem 3.0.1. In a second step the optimal score function $b$ will be substituted by suitable estimators in order to construct tests with high power for large classes of alternatives $(F, G) \in \mathcal{A}_2^0$ , cf. Sections 3.1. and 3.2.

In the sequel the sample sizes $m$ , $n$ and thus $N = m + n$ are fixed and $(F, G) \in \mathcal{F}_1^c \times \mathcal{F}_1^c$ is any given pair of continuous distribution functions. Corresponding to $F$ and $G$ we define the continuous distribution function $H$ as the mixture of $F$ and $G$ with respect to the fractions of the sample sizes,

$$H = \frac{m}{N} F + \frac{n}{N} G. \tag{1.3.8}$$

Here and in the sequel we'll identify the distribution functions with the corresponding probability measures. Apparently the mixture $H$ dominates the measures $F$ and $G$ and there are Radon-Nikodym derivatives $dF/dH$ and $dG/dH$ with the property

$$\frac{m}{N} \frac{dF}{dH} + \frac{n}{N} \frac{dG}{dH} = 1. \tag{1.3.9}$$

Considering the problem of testing the null hypothesis $\mathcal{H}_0^r$ versus the alternative $\mathcal{A}_2^0$ and using (1.3.9) we'll demonstrate that the mixture $H$ of $F$ and $G$ may be viewed as a nuisance parameter, whereas the distribution of the transformed random variables $H(X_i)$ may be viewed as the parameter under test.

As a first step we'll compute the distribution function of the random variable $H(X_i)$ , which takes values only in the unit interval $[0, 1]$ :

For each $i = 1, ..., m$ and each $0 < t < 1$ we get

$$\Pr\{ H(X_i) \le t \} = F\{ x : H(x) \le t \} = F\{ x : H(x) < t \}$$

$$= F\{ x : x < H^{-1}(t) \} = F \circ H^{-1}(t), \tag{1.3.10}$$

and similarly for each $j = m+1, ..., N$ and each $0 < t < 1$

$$\Pr\{\, H(X_j) \leq t \,\} \;=\; G \circ H^{-1}(t) \,. \tag{1.3.11}$$

The distribution functions $F \circ H^{-1}$ and $G \circ H^{-1}$ on $(0,1)$ are dominated by the Lebesgue measure $\lambda$ on the interval $(0,1)$, which is proved by the following chain of equalities ( $\forall\, 0 < t < 1$ )

$$\int_{(0,t]} \frac{dF}{dH} \circ H^{-1} \; d\lambda \;=\; \int_{(0,t)} \frac{dF}{dH} \circ H^{-1} \; d\lambda$$

$$=\; \int \left(1_{(0,t)} \circ H\right) \left(\frac{dF}{dH} \circ H^{-1} \circ H\right) dH \;=\; \int 1_{(-\infty, H^{-1}(t))} \frac{dF}{dH} \; dH$$

$$=\; F \circ H^{-1}(t) \,. \tag{1.3.12}$$

Obviously formula (1.3.12) proves that the distribution functions of $H(X_1)$ and $H(X_N)$ are absolutely continuous with respective $\lambda$–densities $f$ and $g$ according to

$$f := \frac{d(F \circ H^{-1})}{d\lambda} = \frac{dF}{dH} \circ H^{-1}, \qquad g := \frac{d(G \circ H^{-1})}{d\lambda} = \frac{dG}{dH} \circ H^{-1} \,. \tag{1.3.13}$$

Because of formula (1.3.9) we get in addition

$$\frac{m}{N} f \;+\; \frac{n}{N} g \;=\; 1. \tag{1.3.14}$$

These densities $f$ and $g$ are used in order to define a *nonparametric score function* $b : (0,1) \to \mathbb{R}$ pertaining to $(F, G)$, namely,

$$b \;:=\; \sqrt{\frac{mn}{N}} \, (f - g) \;=\; \sqrt{\frac{mn}{N}} \left(\frac{dF}{dH} - \frac{dG}{dH}\right) \circ H^{-1} \,. \tag{1.3.15}$$

The factor $\sqrt{mn/N}$ in the definition of $b$ has been included in order to have a suitable scaling for the local asymptotic theory of Section 3.1. Clearly the inequalities $f \geq 0$, $g \geq 0$ and formula (1.3.14) imply

$$\int_0^1 b \; d\lambda = 0 \qquad \text{and} \qquad -\sqrt{\frac{m}{n}} \leq \frac{b}{\sqrt{N}} \leq \sqrt{\frac{n}{m}} \,. \tag{1.3.16}$$

Defining the absolutely continuous function $B : [0,1] \to \mathbb{R}$ as the integral of the nonparametric score function $b$, i.e.

$$B(t) \;=\; \int_0^t b \; d\lambda \qquad \forall\, t \in [0,1], \tag{1.3.17}$$

we may represent $F$ and $G$ in terms of $B$ and $H$ , namely,

$$F = H + \frac{1}{m}\sqrt{\frac{mn}{N}}\,B \circ H, \qquad G = H - \frac{1}{n}\sqrt{\frac{mn}{N}}\,B \circ H. \qquad (1.3.18)$$

For the proof of (1.3.18) notice that definition (1.3.15) implies

$$\begin{aligned}
B \circ H(x) &= \sqrt{\frac{mn}{N}}\int_0^{H(x)}\left(\frac{dF}{dH} - \frac{dG}{dH}\right)\circ H^{-1}\,d\lambda \\[2mm]
&= \sqrt{\frac{mn}{N}}\int_{-\infty}^{x}\left(\frac{dF}{dH} - \frac{dG}{dH}\right)dH \\[2mm]
&= \sqrt{\frac{mn}{N}}\,(F - G)(x). \qquad (1.3.19)
\end{aligned}$$

Therefore the definition (1.3.8) of $H$ yields (1.3.18).

The continuity of $H$ and (1.3.19) imply the following additional representation of $B$ in terms of $F$ and $G$ ,

$$B = \sqrt{\frac{mn}{N}}\,(F - G)\circ H^{-1}. \qquad (1.3.20)$$

Now let's start with any continuous distribution function $H$ and any absolutely continuous function $B : [0,1] \to \mathbb{R}$ such that $B(0) = B(1) = 0$ and $-\sqrt{m/n} \leq B'/\sqrt{N} \leq \sqrt{n/m}$ . Define the functions $F^*$ and $G^*$ on the unit interval $[0,1]$ according to

$$F^*(t) = t + \frac{1}{m}\sqrt{\frac{mn}{N}}\,B(t), \qquad G^*(t) = t - \frac{1}{n}\sqrt{\frac{mn}{N}}\,B(t). \qquad (1.3.21)$$

Obviously the assumptions on $B$ imply that $F^*$ and $G^*$ are absolutely continuous distribution functions on [0,1]. Therefore

$$F^*\circ H = H + \frac{1}{m}\sqrt{\frac{mn}{N}}\,B\circ H \quad\text{and}\quad G^*\circ H = H - \frac{1}{n}\sqrt{\frac{mn}{N}}\,B\circ H \qquad (1.3.22)$$

are continuous distribution functions on $\mathbb{R}$ with the properties

$$\frac{m}{N}\,F^* \circ H + \frac{n}{N}\,G^* \circ H = H,$$

$$\sqrt{\frac{mn}{N}}\left(F^* \circ H - G^* \circ H\right)\circ H^{-1} = B. \qquad (1.3.23)$$

Combining the results we've proved the following reparametrization of the two-sample model (1.3.1),

$$\mathcal{F}_1^c \times \mathcal{F}_1^c = \big\{ (H + c_{N1} B \circ H, \, H + c_{NN} B \circ H) : \; B \in \mathcal{B}_N^0, \; H \in \mathcal{F}_1^c \big\}, \quad (1.3.24)$$

where

$$\mathcal{B}_N^0 := \left\{ B : \begin{array}{c} B \text{ is an absolutely continuous function on } [0,1], \\ B(0) = B(1) = 0, \quad \text{and} \\ -\sqrt{m/n} \le B'/\sqrt{N} \le \sqrt{n/m} \quad [\lambda - a.e.] \end{array} \right\} \quad (1.3.25)$$

and where the coefficients $c_{Ni}$ have been defined in formula (1.1.16), i.e.

$$c_{N1} = \frac{1}{m}\sqrt{\frac{mn}{N}} \quad \text{and} \quad c_{NN} = -\frac{1}{n}\sqrt{\frac{mn}{N}}.$$

Additionally the equalities (1.3.19) and (1.3.20) imply the following set of equivalences,

$$
\begin{aligned}
F \le G &\iff B \le 0 \,, \\
F \ge G &\iff B \ge 0 \,, \\
F = G &\iff B = 0 \,, \\
F \ne G &\iff B \ne 0 \,.
\end{aligned}
\quad (1.3.26)
$$

Therefore the $(B, H)$–parametrization (1.3.24) of the general model induces the following $(B, H)$–parametrizations of the nonparametric hypothesis of randomness (1.3.7),

$$\mathcal{H}_0^r = \big\{ (H + c_{N1} B \circ H, \, H + c_{NN} B \circ H) : \; B \in \mathcal{B}_N^0, \; H \in \mathcal{F}_1^c, \; B = 0 \big\}, \quad (1.3.27)$$

and of the nonparametric one-sided alternative (1.3.5),

$$
\begin{aligned}
\mathcal{A}_2^0 = \big\{ (H + c_{N1} B \circ H, \, &H + c_{NN} B \circ H) : \\
&B \in \mathcal{B}_N^0, \; H \in \mathcal{F}_1^c, \; B \le 0, \; B \ne 0 \big\}.
\end{aligned}
\quad (1.3.28)
$$

In the $(B, H)$–parametrization the $H \in \mathcal{F}_1^c$ obviously is a *nuisance parameter*, whereas $B \in \mathcal{B}_N^0$ is the *parameter under test*, i.e. only the parameter $B$ contains the information whether the corresponding pair $(F, G)$ belongs to the null hypothesis $\mathcal{H}_0^r$ or to the alternative $\mathcal{A}_2^0$. From (1.3.24) to (1.3.28) it may be expected that the shape of $B$ will determine (at least approximately) the form of the respective optimal test.

In fact, if the function $B$ is fixed, Theorem 3.0.1 will prove the linear rank test

$$\psi_{N\alpha}(b) = 1\big( S_N(b) \ge u_\alpha \|b\| \big) \quad (1.3.29)$$

to be asymptotically optimal for testing the null hypothesis $\mathcal{H}_0^r$ versus local asymptotic alternatives of the form

$$( H + \varrho\, c_{N1}B \circ H, \; H + \varrho\, c_{NN}B \circ H ), \quad 0 < \varrho \le 1, \; H \in \mathcal{F}_1^c, \qquad (1.3.30)$$

where $b$ is the derivative (1.3.15) of $B$ and where $S_N(b)$ is the two-sample linear rank statistic with the score function $b$, i.e.

$$S_N(b) = \sqrt{\frac{mn}{N}} \left( \frac{1}{m} \sum_{i=1}^{m} b_N(R_i) - \frac{1}{n} \sum_{i=m+1}^{N} b_N(R_i) \right),$$

$$= \sum_{i=1}^{N} c_{Ni}\, b_N(R_i), \qquad (1.3.31)$$

where

$$b_N(i) := N \int_{(i-1)/N}^{i/N} b(x)\, dx = N \left( B(\frac{i}{N}) - B(\frac{i-1}{N}) \right). \qquad (1.3.32)$$

For each alternative point (1.3.30) the corresponding $B$–function $B_{H,\varrho}$ has the same shape as the given $B$, namely $B_{H,\varrho} = \varrho\, B$. Because of the equality

$$(H + \varrho\, c_{N1}B \circ H, \; H + \varrho\, c_{NN}B \circ H)$$
$$= (1 - \varrho)(H, H) + \varrho(H + c_{N1}B \circ H, \; H + c_{NN}B \circ H) \qquad (1.3.33)$$

each of the parameter sets

$$\left\{ (H + \varrho\, c_{N1}B \circ H, \; H + \varrho\, c_{NN}B \circ H) : \; 0 \le \varrho \le 1 \right\} \qquad (1.3.34)$$

may be viewed as a line starting at the null hypothesis point $(H, H)$ and ending at the alternative point $(H + c_{N1}B \circ H, \; H + c_{NN}B \circ H)$. Therefore the function $B$ is called the *direction* of the alternatives (1.3.30), and $\psi_{N\alpha}(b)$ is asymptotically optimal for the corresponding direction $B$.

If we consider $\psi_{N\alpha}(b)$ under local asymptotic alternatives of some other direction $B_1$, then the second part of Theorem 3.0.1 proves

$$\lim_{N \to \infty} E\big[\, \psi_{N\alpha}(b) \mid (H + \varrho\, c_{N1}B_1 \circ H, \; H + \varrho\, c_{NN}B_1 \circ H) \,\big]$$

$$= 1 - \Phi\big( u_\alpha - \varrho\, c(b, b_1)\, \|b_1\| \big), \qquad (1.3.35)$$

where $\|b_1\| = \left( \int_0^1 b_1^2(x)\, dx \right)^{1/2}$ and

$$c(b, b_1) = \frac{1}{\|b\|\, \|b_1\|} \int_0^1 b(x)\, b_1(x)\, dx. \qquad (1.3.36)$$

Thus, in principle the situation here is similar to the first two sections. For any given direction $B \in \mathcal{B}_N^0$, $B \neq 0$, the corresponding linear rank test $\psi_{N\alpha}(b)$ is asymptotically optimal for the alternatives (1.3.30) of direction $B$. But under alternatives (1.3.30) of some other direction $B_1 \in \mathcal{B}_N^0$, $B_1 \neq 0$, the maximum power will be achieved only if $B = c_0 B_1$ for some constant $c_0 > 0$, whereas the (asymptotic) power of $\psi_{N\alpha}(b)$ decreases according to the decrease of the correlation $c(b, b_1)$, cf. (1.3.36) and (1.3.35).

In practice we run into the same problem as in the shift model of Section 1.1: Since $b = B'$ is unknown we need some rule in order to select a suitable score function. As in Section 1.1 the selection should be based on the estimation of some features of the optimal score function $b$. But from Section 1.2 we know that the nonparametric score function $b$ is not related to the shift score function $\varphi(\cdot, f_0)$ if the shift model doesn't apply. Therefore, in general, estimators of $\varphi(\cdot, f_0)$ won't be suitable estimators of $b$.

Even if the basic problem is analogous to the the shift model, there is in fact a big difference: While under the assumption of shift alternatives (1.2.1) the optimal score function $\varphi(\cdot, f_0)$ is a *nuisance parameter,* in the general nonparametric model (1.3.24) the optimal score function $b$, i.e. the direction $B$, is the *parameter under test.* Therefore the problem of estimating the optimal score function $b$ is heavily correlated with the problem of testing $\mathcal{H}_0^r$ versus $\mathcal{A}_2^0$. This difference has far-reaching consequences: In the local asymptotic shift model (1.1.5) there is a *consistent* estimator of the shift score function $\varphi(\cdot, f_0)$ producing an *adaptive* test, i.e. the resulting test has the same asymptotic power as if $f_0$ were known. In the nonparametric local asymptotic model (1.3.30) *there is no consistent estimator* of the nonparametric score function $b$ since the existence of a consistent estimator would contradict the fact ( cf. formula (7.1.22) of Appendix 7.1 ) that the distribution of $(X_1, ..., X_N)$ under $( H + \varrho c_{N1} B \circ H, \ H + \varrho c_{NN} B \circ H )$ is *contiguous* to the distribution of $(X_1, ..., X_N)$ under $(H, H)$. In order to prove this contradiction assume $\hat{b}_N$ to be a consistent estimator of the underlying $b$. Under the null hypothesis $(H, H)$ we have $b = 0$. Therefore $\hat{b}_N$ must be a consistent estimator of the zero function under any alternative contiguous to the given null hypothesis, too, which contradicts the consistency of $\hat{b}_N$.

As a consequence we have to be content with estimators $\hat{b}_N$ of $b$ which have weaker convergence properties under $( H + \varrho c_{N1} B \circ H, \ H + \varrho c_{NN} B \circ H )$ than consistency. In Section 3.1 we will construct estimators $\hat{b}_N$ such that, for any $0 < t < 1$, the sequence of real random variables $\hat{b}_N(t)$ will be asymptotically normal with the mean approximately equal to $b(t)$.

The final construction principle of the proposed tests is very simple: If the direction $B$ were known, the linear rank statistic $S_N(b)$ would be (asymptotically) optimal for testing $\mathcal{H}_0^r$ versus alternatives of the direction $B$. Since, in reality, the underlying direction $B$ is unknown, we'll construct suitable estimators $\hat{b}_N$ of $b = B'$ and substitute the unknown $b$ by its estimator $\hat{b}_N$ , i.e. we propose the test statistic $S_N(\hat{b}_N)$ for testing $\mathcal{H}_0^r$ versus $\mathcal{A}_2^0$ .

The explicit construction of the estimators $\hat{b}_N$ and the discussion of the properties of the resulting statistics $S_N(\hat{b}_N)$ is given in Chapter 3. For suitable $\hat{b}_N$ the statistics $S_N(\hat{b}_N)$ are quite simple and the corresponding tests can be used without explicit evaluation of $\hat{b}_N$ , cf. Chapter 2.

Usually this construction principle is applied in order to substitute nuisance parameters. The following classical parametric example will demonstrate that the same principle may be used in order to substitute parameters under test.

### 1.3.1 Example

Let $X_1, ..., X_N$ be i.i.d. real random variables with $X_i \sim \mathcal{N}(\vartheta, \sigma^2)$ , $\vartheta \in \mathbb{R}$ , $\sigma^2 > 0$ , and consider the null hypothesis $H_0 : \vartheta = 0, \sigma^2 > 0$ versus the two-sided alternative $H_1 : \vartheta \neq 0, \sigma^2 > 0$ . As an analogue to the local asymptotic alternatives (1.3.30) we consider local asymptotic alternatives defined by $(\varrho/\sqrt{N}, \sigma^2)$ , $\varrho \in \mathbb{R}$ , $\sigma^2 > 0$ . Then we may rewrite the testing problem in terms of the local parameter $\varrho$ as

$$H_0 : \varrho = 0, \ \sigma^2 > 0 \qquad \text{versus} \qquad H_1 : \varrho \neq 0, \ \sigma^2 > 0. \qquad (1.3.37)$$

Apparently $\varrho$ plays the role of the direction $B \in \mathcal{B}_N^0$ while $\sigma^2 > 0$ is a nuisance parameter resembling the nuisance parameter $H \in \mathcal{F}_1^c$ .

For any $\varrho_0 \neq 0$ and $\sigma^2 > 0$ we consider the problem of testing the simple hypothesis $\varrho = 0$ versus the simple alternative $\varrho = \varrho_0$ . Clearly the optimal Neyman–Pearson test rejects the null hypothesis, if the statistic

$$S_N(\varrho_0, \sigma^2) \ = \ \frac{\sqrt{N} \ \bar{X}_N \ \varrho_0}{\sigma^2}. \qquad (1.3.38)$$

is sufficiently large.

Without loss of asymptotic power the nuisance parameter $\sigma^2 > 0$ may be substituted by the sample variance $s_N^2$ . Similarly, it will be shown in Section 3.1 that the nuisance parameter $H \in \mathcal{F}_1^c$ of the model (1.3.24) can be eliminated by using rank statistics. As far as the remaining local parameter $\varrho_0$ is concerned the situation is quite similar to the situation of the parameter $B \in \mathcal{B}_N^0$ . Firstly, $\varrho_0$ as well as $B$ are the parameters under test. Secondly, also $\varrho_0$ can't be estimated consistently by some sequence of estimators $\hat{\varrho}_N$ under local asymptotic alternatives $\vartheta_N = \varrho_0/\sqrt{N}$ , which again follows from a contiguity argument. Therefore we must be content with estimators $\hat{\varrho}_N$ converging *in distribution* under $(\varrho_0/\sqrt{N}, \sigma^2)$ . Obviously the optimal unbiased estimator of $\varrho_0$ is $\hat{\varrho}_N = \sqrt{N}\,\bar{X}_N$ and

$$\mathcal{L}\left[\sqrt{N}\,\bar{X}_N \mid \left(\tfrac{\varrho_0}{\sqrt{N}}, \sigma^2\right)\right] \xrightarrow{\mathcal{L}} \mathcal{N}(\varrho_0, \sigma^2). \qquad (1.3.39)$$

A substitution of the unknown $\varrho_0$ and $\sigma^2$ in formula (1.3.38) by the estimators $\hat{\varrho}_N$ and $s_N^2$ yields the test statistic

$$S_N(\hat{\varrho}_N, s_N^2) \;=\; \frac{N\,\bar{X}_N^2}{s_N^2}\,, \qquad (1.3.40)$$

i.e. the square of *Student's t–statistic* which has well-known optimality properties for testing $H_0$ versus $H_1$ .

The main difference between the general two-sample situation and this classical example is the fact that the parameter $B$ varies in an infinite dimensional function space, while $\vartheta$ is a one-dimensional parameter.

We'll conclude the introduction by briefly discussing the connection between the generalized shift score function $\varphi(\cdot, f_0, D)$ defined in (1.2.4) and the nonparametric score function $b$ defined in (1.3.15), if we assume the generalized shift model (1.2.5) with underlying $F_0$ , shift function $D$ , and local parameter $\varrho > 0$ , i.e. we assume $G = F_0$ with absolutely continuous density $f_0$ and

$$F(x) \;=\; F_N(x) \;=\; F_0\!\left(x - \varrho\sqrt{\frac{N}{mn}}\,D(x)\right) \qquad (1.3.41)$$

with density

$$f_N(x) \;=\; \left(1 - \varrho\sqrt{\frac{N}{mn}}\,d(x)\right) f_0\!\left(x - \varrho\sqrt{\frac{N}{mn}}\,D(x)\right). \qquad (1.3.42)$$

The corresponding mixture $H = H_N = (m/N)F_N + (n/N)F_0$ obviously has the density $h_N = (m/N)f_N + (n/N)f_0$ . Therefore the nonparametric score function (1.3.15) has the representation

$$b_N \;=\; \sqrt{\frac{mn}{N}}\left(\frac{dF_N}{dH_N} - \frac{dF_0}{dH_N}\right)\circ H_N^{-1}$$

$$\;=\; \sqrt{\frac{mn}{N}}\,\frac{(f_N - f_0)\circ H_N^{-1}}{\left(\frac{m}{N}\,f_N + \frac{n}{N}\,f_0\right)\circ H_N^{-1}} \qquad (1.3.43)$$

and suitable regularity conditions for $f_0$ and $D$ imply the convergence

$$\int_0^1 \Big(b_N(x) - \varrho\,\varphi(x, f_0, D)\Big)^2\,dx \xrightarrow{N\to\infty} 0. \qquad (1.3.44)$$

Thus, in case of generalized shift with any smooth underlying shift function $D$ , the nonparametric score function $b_N$ automatically has the approximate shape of $\varphi(\cdot, f_0, D)$ , i.e. any suitable estimator of $b_N$ is a suitable estimator of $\varrho\,\varphi(\cdot, f_0, D)$ if the generalized shift $(F_0, \varrho\,\sqrt{N/(mn)}\,D)$ applies. Especially, any suitable estimator of the nonparametric score function $b_N$ also estimates the shift score function $\varrho\,\varphi(\cdot, f_0)$ if the exact shift model $(F_0, \varrho\,\sqrt{N/(mn)})$ applies.

Figure 1.3.a to Figure 1.3.c show the graphs of some examples of $b_N$ together with the graphs of the respective limiting functions $\varphi(\cdot, f_0, D)$ corresponding to (1.3.41) and (1.3.43) for $\varrho = 1$ . The examples correspond to lower shift, central shift, upper shift, and exact shift as defined in formula (3.1.111) of Section 3.1 for the normal distribution (Figure 1.3.a), the logistic distribution (Figure 1.3.b), and the Cauchy distribution (Figure 1.3.c).

If, however, the shift function $D$ were really known (e.g. if usual shift alternatives were present), then it would be better (at least theoretically) not to estimate the nonparametric score function $b_N$ but to construct consistent estimators of $\varphi(\cdot, f_0, D)$ , cf. Section 1.1 and Section 1.2. But such estimators, e.g. the estimators of Hájek and Šidák (1967) and of Randles and Hogg (1973), will lead to nonsense if the special shift model isn't true. In order to see this let's assume that the underlying parameter $(F, G)$ is given in the $(B, H)$ –version,

$$(F, G) \,=\, (H + c_{N1} B \circ H,\ H + c_{NN} B \circ H), \qquad (1.3.45)$$

and that the distribution function $H$ has an absolutely continuous density $h$ with finite Fisher-information. Since the usual estimators of the shift score function $\varphi(\cdot, f_0)$ are based on the ordered pooled sample, they will—in the present case—estimate properties of the function $\varphi(\cdot, h)$ , whereas $b$ is the function that should be estimated. Since $H$ is a nuisance parameter in the general situation, there is no connection between $\varphi(\cdot, h)$ and $b$ .

**Figure 1.3.a**

*The score function $b_N$ for $m = 10$, $n = 20$ and its approximation $\varphi(\cdot, f, D)$ for the **normal** distribution function $F$ in case of lower shift $[D = 1 - F]$, upper shift $[D = F]$, central shift $[D = 4F(1-F)]$, and exact shift $[D = 1]$.*

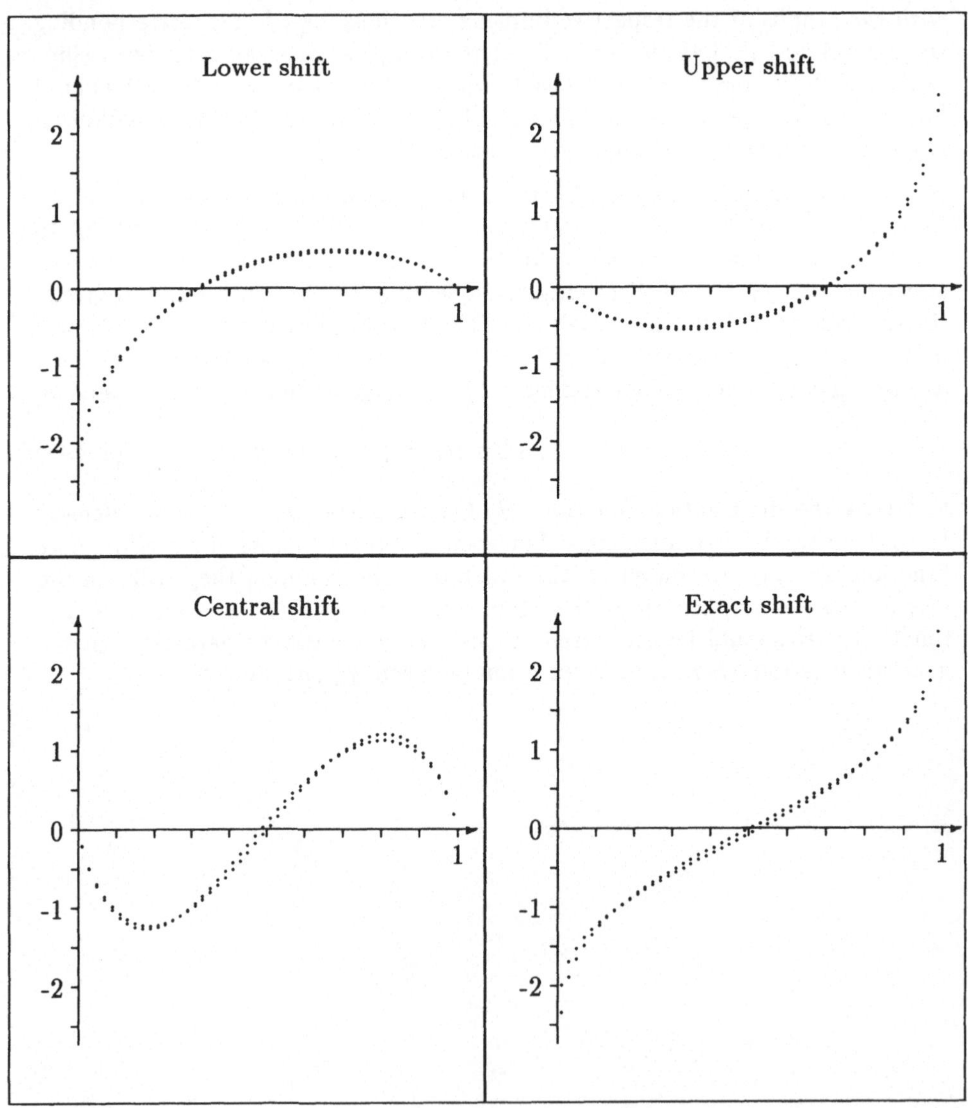

**Figure 1.3.b**

*The score function $b_N$ for $m = 10$, $n = 20$ and its approximation $\varphi(\cdot, f, D)$ for the* **logistic** *distribution function $F$ in case of lower shift $[D = 1 - F]$, upper shift $[D = F]$, central shift $[D = 4F(1-F)]$, and exact shift $[D = 1]$.*

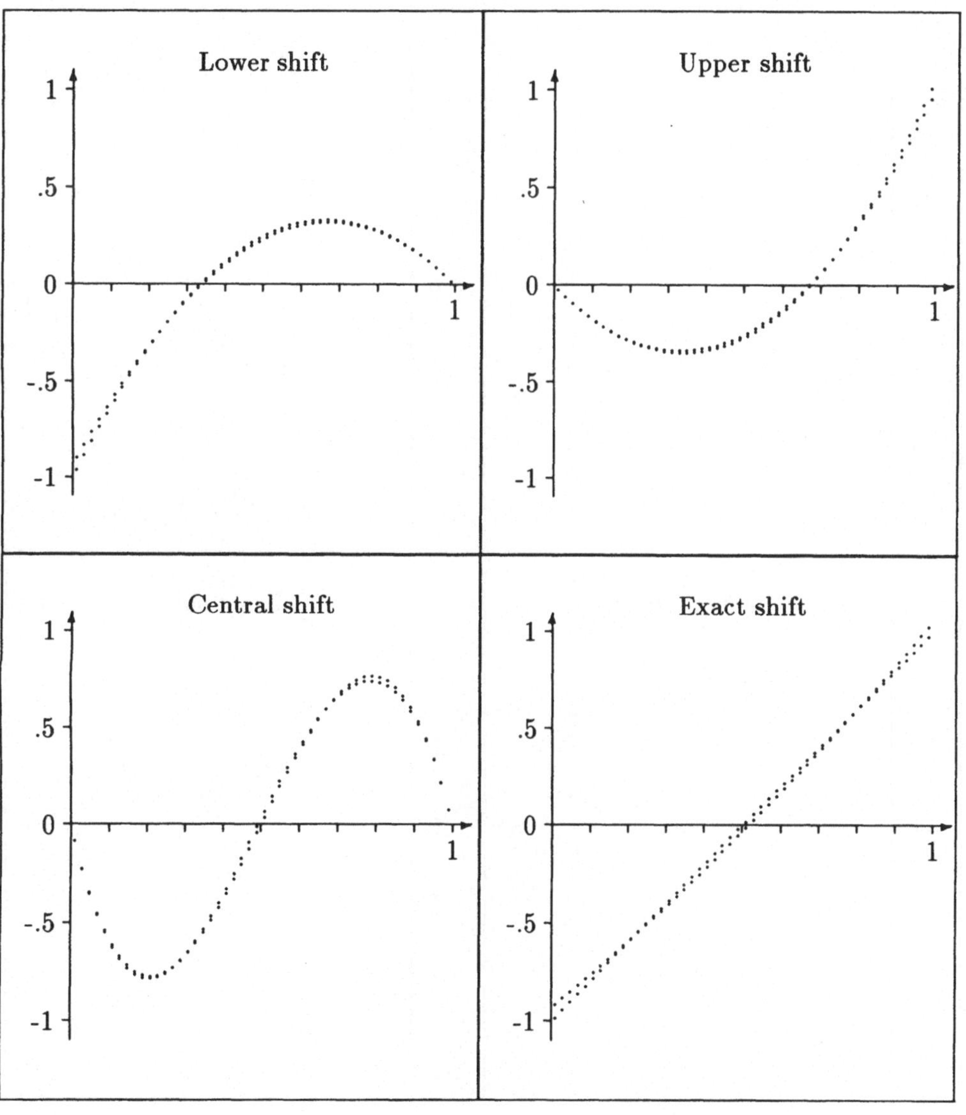

**Figure 1.3.c**

*The score function $b_N$ for $m = 10$, $n = 20$ and its approximation $\varphi(\cdot, f, D)$ for the **Cauchy** distribution function $F$ in case of lower shift $[D = 1 - F]$, upper shift $[D = F]$, central shift $[D = 4F(1-F)]$, and exact shift $[D = 1]$.*

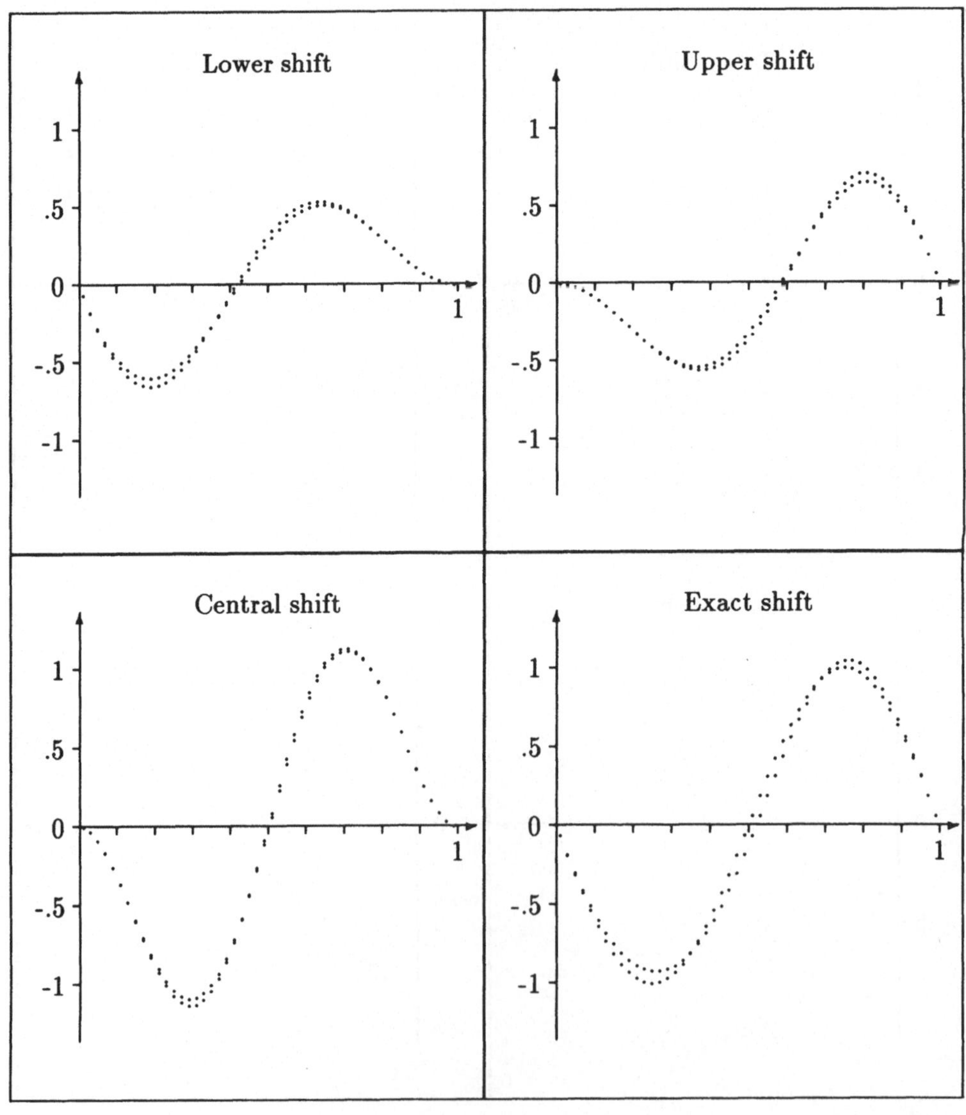

# Chapter 2

# Applications

In the introduction and motivation of Chapter 1 we've discussed the merits and the shortcomings of linear rank tests by means of the two-sample problem. Similar arguments apply to other testing problems, too. Tests based on linear rank statistics $S_N(b)$ have turned out to be optimal only for alternatives determined by the score function $b$ . Since the alternative and thus the optimal $b$ are unknown in reality, we suggest the following *heuristic adaptation procedure:*

Instead of using a linear rank statistic $S_N(b)$ with fixed score function $b$ , we'll estimate the optimal score function by some estimator $\hat{b}_N$ based on ranks, and use the *nonlinear* rank statistic $S_N(\hat{b}_N)$ as the new test statistic.

The actual construction of the estimators $\hat{b}_N$ and the evaluation of the finite and asymptotic properties of the resulting nonlinear rank statistics $S_N(\hat{b}_N)$ is given in Chapter 3 for the two-sample model, and in Chapters 4 to 6 for a variety of other models.

In this chapter we'll present the proposed tests without stressing the underlying mathematics and asymptotics. Therefore Chapter 2 may serve as a 'cook-book' for direct application. Especially we'll illustrate the applicability and the scope of the new rank procedures by suitable examples.

Using the asymptotic results of Chapters 3 to 6 as a guide we'll present some Monte Carlo results which compare the new procedures with classical competitors in a variety of situations.

# 2.1  Two samples differing in location

The problem is the comparison of two treatments on the basis of two independent samples. We'll discuss testing procedures for testing the null hypothesis of randomness (both treatments behave equally) versus the alternative that the *first* treatment produces (stochastically) larger values than the *second* treatment (one-sided model) or versus the alternative that the two treatments behave differently (omnibus model).

Notice, if you want to prove that the *second* treatment produces (stochastically) larger values than the *first* treatment you have to interchange the role of the two samples.

**Data:**

- The random sample $X_1, ..., X_m$ of size $m$ from the first treatment.

- The random sample $X_{m+1}, ..., X_{m+n}$ of size $n$ from the second treatment.

There is no restriction with respect to tied observations, i.e. the underlying model may allow for discontinuities or even may be purely discrete. In case of tied observations the given tables of critical values are not valid, but there are simulation programs for evaluating the conditional $p$-values of the proposed tests.

All proposed tests will be *rank tests*, i.e. they are based exclusively on the vector of ranks $(R_1, ..., R_N)$ of the pooled sample $(X_1, ..., X_N)$, where $N = m + n$ is the total sample size and where the rank $R_i$ of $X_i$ in the pooled sample is defined as

$$R_i = \# \left\{ j \in \{1, ..., N\} \mid X_j \leq X_i \right\} \qquad i = 1, ..., N. \tag{1}$$

Notice, the rank of $X_i$ counts all observations of the pooled sample which are less than or equal to $X_i$, also in case of tied observations.

**Model Assumptions:**

- Both samples are random samples from their respective populations.

- Both samples are mutually independent.

- The respective underlying (cumulative) distribution functions $F$ and $G$ of the random variables of the first sample and of the second sample are arbitrary distribution functions on the real line, i.e. they may have discontinuities.

- The measurement scale is the real line or at least ordinal in order to have well-defined ranks.

## Testing Problems:

- *One-sided problem:* Testing the null hypothesis of randomness,

$$\mathcal{H}_0^r : F = G,$$

versus the one-sided alternative that the first treatment produces stochastically larger values than the second treatment,

$$\mathcal{A}_2^0 : F \leq G, F \neq G.$$

- *Omnibus problem:* Testing the null hypothesis of randomness,

$$\mathcal{H}_0^r : F = G,$$

versus the omnibus alternative that the two treatments behave differently,

$$\mathcal{A}_2 : F \neq G.$$

## Two-Sample Rank Process:

Essential for the data inspection and for the definition of the test statistics is the following rank process $W_N = (W_N(t), 0 \leq t \leq 1)$ :

Let $d$ denote the number of different values among the ranks $R_1, ..., R_N$ , which is the same as the number of different values among the original observations $X_1, ..., X_N$ . In the continuous model (no ties) we have $d = N$ . In the general model (ties are possible) $d$ is a random quantity.

Let $T_1 < \cdots < T_d$ denote the ordered values of the different values in $R_1, ..., R_N$ . If no ties are present we have $d = N$ and $T_i = i$ for $i = 1, ..., N$ . If all observations are tied, we have $d = 1$ and $T_1 = N$ . In any case we have $1 \leq d \leq N$ and $T_d = N$ .

Using $d$ , $T = (T_1, ..., T_d)$ , and the additional definition $T_0 := 0$ we define the rank process $W_N$ according to, cf. Proposition 3.3.12,

$$W_N\left(\frac{T_j}{N}\right) = n \sum_{i=1}^{m} 1(R_i \leq T_j) - m \sum_{i=m+1}^{N} 1(R_i \leq T_j), \quad 0 \leq j \leq d, \quad (2)$$

and according to linear interpolation between the points $(T_j/N, W_N(T_j/N))$ , $j = 0, 1, ..., d$ , in the $(x, y)$ –plane. Especially we have $W_N(0) = W_N(1) = 0$ .

For easy computation of $W_N$ we use the following representation, cf. (3.3.102) to (3.3.104):

Let $\tau_j = T_j - T_{j-1}$, $j = 1, ..., d$, denote the *lengths of the ties* in the ordered pooled sample $X^{(1)} \leq \cdots \leq X^{(N)}$. For any $j = 1, ..., d$ let $\tau_{1j}$ and $\tau_{2j}$ denote the *length of the $j$-th tie corresponding to the respective first and second sample*, i.e.

$$\tau_{1j} = \#\left\{ i \in \{1, ..., m\} \mid X_i = X^{(T_j)} \right\} = \#\left\{ i \in \{1, ..., m\} \mid R_i = T_j \right\},$$

$$\tau_{2j} = \#\left\{ i \in \{m+1, ..., m+n\} \mid X_i = X^{(T_j)} \right\}$$

$$= \#\left\{ i \in \{m+1, ..., m+n\} \mid R_i = T_j \right\}. \tag{3}$$

Obviously this implies $\tau_{1j} + \tau_{2j} = \tau_j$, $j = 1, ..., d$. Using $\tau_{1j}$, $\tau_{2j}$, and $\tau_j$ we get

$$W_N\left(\frac{T_j}{N}\right) - W_N\left(\frac{T_{j-1}}{N}\right) = n\,\tau_{1j} - m\,\tau_{2j}, \qquad j = 1, ..., d. \tag{4}$$

Since $W_N(\cdot)$ is linear on the interval $[T_{j-1}/N, T_j/N]$ for each $j = 1, ..., d$, and since $W_N(0) = 0$ holds true, the process $W_N$ is completely determined by formula (4). Especially we have for each $j = 1, ..., d$

$$\Delta W_N\left(\frac{i}{N}\right) := W_N\left(\frac{i}{N}\right) - W_N\left(\frac{i-1}{N}\right)$$

$$= \frac{n\,\tau_{1j} - m\,\tau_{2j}}{\tau_j} \qquad \text{if } T_{j-1} < i \leq T_j. \tag{5}$$

Obviously the tie-lengths vectors $(\tau_{11}, ..., \tau_{1d})$ and $(\tau_{21}, ..., \tau_{2d})$ of the first and second sample with respect to the pooled sample contain the same information as the ordered ranks of the first sample and the ordered ranks of the second sample. For the actual computation of the rank process $W_N$ formula (5) seems to be most convenient.

If *no ties are present,* we get $d = N$, $\tau_i = 1$ for $i = 1, ..., N$, and

$$\tau_{1i} = \begin{cases} 1, & \text{if the value of } X^{(i)} \text{ belongs to the first sample,} \\ 0, & \text{if the value of } X^{(i)} \text{ belongs to the second sample,} \end{cases}$$

$$\tau_{2i} = \begin{cases} 0, & \text{if the value of } X^{(i)} \text{ belongs to the first sample,} \\ 1, & \text{if the value of } X^{(i)} \text{ belongs to the second sample.} \end{cases}$$

**The Omnibus Test:**

The test statistic $S$ for testing the null hypothesis of randomness $\mathcal{H}_0^r$ versus the omnibus alternative $\mathcal{A}_2$ is, cf. formula (3.1.25) in the continuous case or formulas (3.3.88) and (3.3.89) in the general case,

$$S = \frac{1}{mnN} \sum_{i=1}^{N}\sum_{j=1}^{N} \Delta W_N(\frac{i}{N})\, \Delta W_N(\frac{j}{N})\, k(i,j), \qquad (6)$$

where $\Delta W_N(i/N)$, $1 \le i \le N$, is defined in formula (5) and where

$$k(i,j) = K_a\Big(\frac{i-1/2}{N}, \frac{j-1/2}{N}\Big), \qquad 1 \le i,j \le N, \qquad (7)$$

and

$$K_a(s,t) = \frac{1}{a}\Big(K(\frac{t+s}{a}) + K(\frac{t-s}{a}) + K(\frac{t+s-2}{a})\Big), \quad s,t \in [0,1], \qquad (8)$$

with fixed bandwidth $a = 0.40$ and Parzen-2 kernel

$$K(t) = \begin{cases} 4/3 - 8\,|t|^2 + 8\,|t|^3, & \text{if} \quad |t| < 1/2, \\ 8\,(1-|t|)^3/3, & \text{if} \quad 1/2 \le |t| < 1, \\ 0, & \text{if} \quad |t| \ge 1. \end{cases} \qquad (9)$$

The omnibus test will reject the null hypothesis $\mathcal{H}_0^r : F = G$ at level $\alpha$ in favor of the omnibus alternative $\mathcal{A}_2 : F \ne G$ if the value of the test statistic $S$ is larger than the upper $\alpha$–quantile $k_\alpha(\tau)$ of the conditional distribution of $S$ under $\mathcal{H}_0^r$ given the tie-lengths vector $\tau = (\tau_1, ..., \tau_d)$ of the ordered pooled sample, i.e. the omnibus test may be written in the form

$$\psi = 1\Big(S > k_\alpha(\tau)\Big), \qquad (10)$$

where $k_\alpha(\tau)$ is defined by

$$P_{\mathcal{H}_0^r}\{S > k_\alpha(\tau) \mid \tau\} \le \alpha < P_{\mathcal{H}_0^r}\{S \ge k_\alpha(\tau) \mid \tau\}. \qquad (11)$$

The upper $\alpha$–quantiles $k_\alpha$ for the continuous model are given in Table 2.1.A of the Appendix.

**Simulation of the conditional $p$–values:**

In principle there is no difficulty to evaluate the conditional critical values $k_\alpha(\tau)$ or the conditional $p$–values:

The conditional distribution of the rank vector $(R_1, ..., R_N)$ given the vector $\tau = (\tau_1, ..., \tau_d)$ is the same for any null hypothesis point $(F, G) = (F, F) \in \mathcal{H}_0^r$, cf. Lemma 3.3.10. Moreover, each rank $R_i$ may be written as a simple function of the randomized ranks $R^* = (R_1^*, ..., R_N^*)$ and of the vector $\tau$, cf. formula (3.3.99),

$$R_i = T_j, \quad \text{if} \quad T_{j-1} < R_i^* \leq T_j \quad \text{for some} \quad 1 \leq j \leq d, \tag{12}$$

and $R^*$ and $\tau$ are stochastically independent under the hypothesis of randomness $\mathcal{H}_0^r$. Finally, under $\mathcal{H}_0^r$ the vector $R^*$ is uniformly distributed on the permutations of $(1, ..., N)$. Therefore a generator of random permutations may be used in order to simulate the conditional distribution of $(R_1, ..., R_N)$ given $\tau$ via (12).

The exact evaluation of the critical values via (12) is limited to the unconditional case (no ties present) and to moderate sizes of $N$.

**The One-sided Test:**

The test statistic $S^0$ for testing the null hypothesis of randomness $\mathcal{H}_0^r$ versus the one-sided alternative $\mathcal{A}_2^0$ (first treatment better than second treatment) is, cf. formula (3.1.30) in the continuous case or formula (3.3.97) in the general case,

$$S^0 = \frac{1}{mnN} \sum_{i=1}^{N} \sum_{j=1}^{N} \Delta W_N^0\left(\frac{i}{N}\right) \Delta W_N\left(\frac{j}{N}\right) k(i, j), \tag{13}$$

where

$$\Delta W_N^0\left(\frac{i}{N}\right) = W_N^0\left(\frac{i}{N}\right) - W_N^0\left(\frac{i-1}{N}\right), \quad 1 \leq i \leq N, \tag{14}$$

and

$$W_N^0(t) = \min\left(0, W_N(t)\right) \quad 0 \leq t \leq 1. \tag{15}$$

The weights $k(i, j)$, $\Delta W_N(j/N)$, and the process $W_N$ are defined as in the omnibus case, cf. (2), (5), and (7) to (9).

The one-sided test will **reject** the null hypothesis $\mathcal{H}_0^r : F = G$ at level $\alpha$ in favor of the one-sided alternative $\mathcal{A}_2^0 : F \leq G, F \neq G$, if the value of the test statistic $S^0$ is larger than the upper $\alpha$–quantile $k_\alpha^0(\tau)$ of the conditional distribution of $S^0$ under $\mathcal{H}_0^r$ given the tie-lengths vector $\tau = (\tau_1, ..., \tau_d)$ of the ordered pooled sample, i.e. the one-sided test may be written in the form

$$\psi^0 = 1\left(S^0 > k_\alpha^0(\tau)\right) \tag{16}$$

where $k_\alpha^0(\tau)$ is defined by

$$P_{\mathcal{H}_0^r}\left\{ S^0 > k_\alpha^0(\tau) \mid \tau \right\} \leq \alpha < P_{\mathcal{H}_0^r}\left\{ S^0 \geq k_\alpha^0(\tau) \mid \tau \right\}. \tag{17}$$

The upper $\alpha$–quantiles $k_\alpha^0$ for the continuous model are given in Table 2.1.B of the Appendix.

The evaluation or simulation of the conditional critical values $k_\alpha^0(\tau_1, ..., \tau_d)$ is completely similar to the omnibus case since the statistic $S^0$ is a function of the two-sample rank process $W_N$, too.

Notice, the only difference between the omnibus statistic $S$ and the one-sided statistic $S^0$ is the substitution of the weights $\Delta W_N(i/N)$ by the weights $\Delta W_N^0(i/N)$. This substitution is reasonable, since the process $W_N/\sqrt{mnN}$ is an estimator of the underlying parameter $B \in \mathcal{B}_N^0$, and since the one-sided model is characterized by $B \leq 0$, cf. (1.3.27) and (1.3.28), i.e. the rank process $W_N^0 = W_N 1(W_N < 0)$ is a 'projection' of the original rank process $W_N$ onto the one-sided model $B \leq 0$.

**Numerical Example 2.1.1:**

- The sample sizes are $m = 7$ and $n = 9$, i.e. the total sample size is $N = 16$.

- The observations of the first sample are
  $(x_1, ..., x_7) = (9.0, \ 6.0, \ 3.0, \ 6.0, \ 10.0, \ 2.0, \ 7.0)$.
  The ordered first sample is 2.0, 3.0, 6.0, 6.0, 7.0, 9.0, 10.0 .

- The observations of the second sample are
  $(x_8, ..., x_{16}) = (8.0, \ 3.0, \ 3.0, \ 2.0, \ 9.0, \ 5.0, \ 7.0, \ 5.0, \ 9.0)$.
  The ordered second sample is 2.0, 3.0, 3.0, 5.0, 5.0, 7.0, 8.0, 9.0, 9.0 .

- The ordered pooled sample is
  2.0, 2.0, 3.0, 3.0, 3.0, 5.0, 5.0, 6.0, 6.0, 7.0, 7.0, 8.0, 9.0, 9.0, 9.0, 10.0.

- The vector of ranks is
  $(r_1, ..., r_{16}) = (15; \ 9, \ 5, \ 9, \ 16, \ 2, \ 11; \ 12, \ 5, \ 5, \ 2, \ 15, \ 7, \ 11, \ 7, \ 15)$.

- The number of different values among the observations (ranks) is $d = 8$.

- The ordered different values among the ranks are
  $T = (T_1, ..., T_d) = (2, \ 5, \ 7, \ 9, \ 11, \ 12, \ 15, \ 16)$.

- The lengths of the ties in the ordered pooled sample are
  $\tau = (\tau_1, ..., \tau_d) = (2, \ 3, \ 2, \ 2, \ 2, \ 1, \ 3, \ 1)$.

- The lengths of the ties corresponding to the first and second sample are
  $\tau_1 = (\tau_{11}, ..., \tau_{1d}) = (1, \ 1, \ 0, \ 2, \ 1, \ 0, \ 1, \ 1)$,
  $\tau_2 = (\tau_{21}, ..., \tau_{2d}) = (1, \ 2, \ 2, \ 0, \ 1, \ 1, \ 2, \ 0)$.

**Figure 2.1.a**

*Numerical Example 2.1.1: The graph of $\left( W_N(t),\, 0 \le t \le 1 \right)$ .*

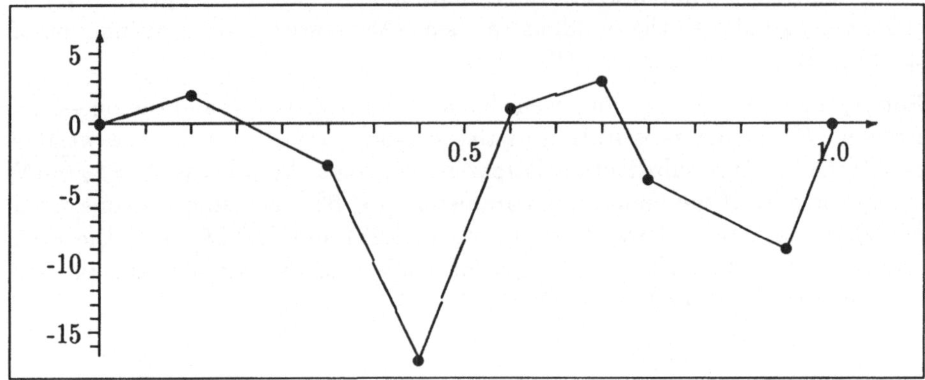

- The defining values $W_N(T_j/N) - W_N(T_{j-1}/N) = n\,\tau_{1j} - m\,\tau_{2j}$ , $j = 1,...,d$ , of the rank process $W_N$ are
  2, $-5$, $-14$, 18, 2, $-7$, $-5$, 9 .

- According to formula (5) the values of $\Delta W_N(i/N)$ , $i = 1,...,N$ , are
  1, 1, $-5/3$, $-5/3$, $-5/3$, $-7$, $-7$, 9, 9, 1, 1, $-7$, $-5/3$, $-5/3$, $-5/3$, 9 .

- Therefore the values of $W_N(i/N)$ , $i = 0,1,...,N$ , are
  0, 1, 2, 1/3, $-4/3$, $-3$, $-10$, $-17$, $-8$, 1, 2, 3, $-4$, $-17/3$, $-22/3$, $-9$, 0 .

- The graph of $W_N$ is given in Figure 2.1.a.

- We use (6) to (9) with the bandwidth $a = 0.40$ in order to evaluate the value

$$s = S(r_1,...,r_{16}) = 1.2515$$

of the omnibus statistic $S$ .

- Since ties are present, there is no table with the corresponding exact conditional critical value $k_\alpha(2,3,2,2,2,1,3,1)$ but we utilize formula (12) in order to evaluate the actual (conditional) $p$–values by Monte Carlo simulation. With a Monte Carlo sample size of 10,000 we get

$$P_{\mathcal{H}_0^r}\left\{ S > 1.2515 \mid \tau = (2,3,2,2,2,1,3,1) \right\} = 0.828 ,$$

$$P_{\mathcal{H}_0^r}\left\{ S \ge 1.2515 \mid \tau = (2,3,2,2,2,1,3,1) \right\} = 0.830 .$$

- This means that the omnibus test (10) will reject the null hypothesis of randomness $\mathcal{H}_0^r : F = G$ in favor of the omnibus alternative $\mathcal{A}_2 : F \ne G$

**Figure 2.1.b**

*Numerical Example 2.1.1: The graph of* $(W_N^0(t), 0 \leq t \leq 1)$ .

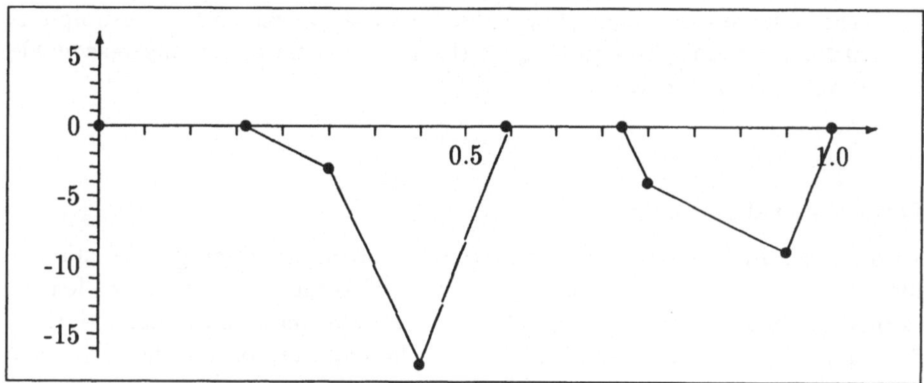

exactly for those levels $\alpha$ which are larger than or equal to the observed $p$-value. Since the observed $p$-value of the example is $0.828$ , the data will **not** reject the null hypothesis for any reasonable level $\alpha$ , e.g. for $\alpha = 0.10$ .

- The values of $W_N^0(i/N)$ , $i = 0, 1, ..., N$ , are

$$0, 0, 0, 0, -4/3, -3, -10, -17, -8, 0, 0, 0, -4, -17/3, -22/3, -9, 0 .$$

- The graph of $W_N^0$ is given in Figure 2.1.b.

- The values of $\Delta W_N^0(i/N)$ , $i = 1, ..., N$ , are

$$0, 0, 0, -4/3, -5/3, -7, -7, 9, 8, 0, 0, -4, -5/3, -5/3, -5/3, 9 .$$

- Using (13), (14), (5), and (7) to (9) with the bandwidth $a = 0.40$ we evaluate the value $s^0$ of the one-sided statistic $S^0$ as

$$s^0 = S^0(r_1, ..., r_{16}) = 1.1173 .$$

- Since ties are present, there is no table with the corresponding exact conditional critical value $k_\alpha^0(2,3,2,2,2,1,3,1)$ but we utilize formula (12) in order to evaluate the actual (conditional) $p$-values by Monte Carlo simulation. With a Monte Carlo sample size of 10,000 we get

$$P_{\mathcal{H}_0^\tau}\{ S^0 > 1.1173 \mid \tau = (2,3,2,2,2,1,3,1) \} = 0.458 ,$$

$$P_{\mathcal{H}_0^\tau}\{ S^0 \geq 1.1173 \mid \tau = (2,3,2,2,2,1,3,1) \} = 0.461 .$$

- This means that the one-sided test will reject the null hypothesis of randomness $\mathcal{H}_0^r : F = G$ in favor of the one-sided alternative $\mathcal{A}_2^0 : F \leq G, F \neq G$ exactly for those levels $\alpha$ which are larger than or equal to the observed $p$-value. Since the observed $p$-value of the example is 0.458, the data will **not** reject the null hypothesis for any reasonable level $\alpha$, e.g. for $\alpha = 0.10$.

## Some general remarks:

As discussed in Chapter 3 the estimator $\hat{b}_N$ of the underlying optimal score function $b_N$ is constructed along the lines of kernel estimators for density estimation, but in contrast to usual density estimation here the bandwidth $a$ does not depend on the sample size $N$. The main reason for this difference is that the resulting nonlinear rank statistic $S_N(\hat{b}_N)$ has better asymptotic power properties, if the bandwidth stays away from zero when the sample size $N$ tends to infinity, cf Section 3.1.D. The asymptotic power behavior for local alternatives, cf. Chapter 3, and extensive Monte Carlo simulations have led us to propose the rather big bandwidth $a = 0.40$.

According to Chapter 3 the two-sample rank process $W_N/\sqrt{mnN}$ is a good estimator of the underlying parameter $B \in \mathcal{B}_N^0$. Therefore Figure 2.1.a and Figure 2.1.b indicate that the given data will not reject the null hypothesis $\mathcal{H}_0^r : F = G\,(B = 0)$ in favor of the omnibus alternative $\mathcal{A}_2 : F \neq G\,(B \neq 0)$ or in favor of the one-sided alternative $\mathcal{A}_2^0 : F \leq G, F \neq G\,(B \leq 0, B \neq 0)$.

In any case you should have a look at the graph of the two-sample rank process $W_N$ for the given data. This graph contains all the information which is relevant for testing the null hypothesis of randomness $\mathcal{H}_0^r$ versus the one-sided alternative $\mathcal{A}_2^0$ (first treatment better than second treatment), cf. Section 1.3 and Chapter 3. In case of the alternative $\mathcal{A}_2^0 : B \leq 0, B \neq 0$ the rank process $(W_N(t)/\sqrt{mnN}, 0 \leq t \leq 1)$ should be negative for most values of $t \in [0,1]$.

Even if the $p$-value of the test statistic $S^0$ leads to the rejection of $\mathcal{H}_0^r$ in favor of $\mathcal{A}_2^0$, the corresponding graph of the observed two-sample rank process may indicate that the rejection should not be in favor of $\mathcal{A}_2^0 : B \leq 0, B \neq 0$ but only in favor of $\mathcal{A}_2 : B \neq 0$.

If it's very important that the rejection of $\mathcal{H}_0^r$ really means the acceptance of $\mathcal{A}_2^0 : B \leq 0, B \neq 0$, i.e. if you really want to be sure that the first treatment is better than the second treatment, then you should use a very conservative test. The highest security would be achieved by using a test at level $\alpha$ for testing the very large null hypothesis $\mathcal{H}_2^0 : B \notin \mathcal{A}_2^0$ versus (its complement) $\mathcal{A}_2^0$. It's shown in Subsection 3.2.3 of Section 3.2 that **Galton's test** may be viewed (at least approximately) as a test for testing $\mathcal{H}_2^0$ versus $\mathcal{A}_2^0$. Of course we have to pay for the higher security with a substantial loss of power. For details we refer to Section 3.2 and to Behnen and Neuhaus (1983).

**The Projection Rank Test:**

Here we propose a second nonlinear rank statistic $S^\pi$ for testing the null hypothesis of randomness $\mathcal{H}_0^r$ versus the **one-sided alternative** $\mathcal{A}_2^0$ (first treatment better than the second treatment).

The basic idea is to replace the single score function of a classical linear rank statistic by a finite number of sensibly chosen score functions $b_1, ..., b_r$. As discussed in subsection C) of Section 3.2 we propose $r = 3$ and

$$
\begin{aligned}
b_1(u) &= (1 - u)(3u - 1) \\[2mm]
b_2(u) &= 8\,u\,(1 - u)(2u - 1) \\[2mm]
b_3(u) &= u\,(3u - 2) = -b_1(1 - u), \qquad 0 \le u \le 1.
\end{aligned}
\tag{18}
$$

The motivation is that the score functions $b_1$, $b_2$ and $b_3$ will take care of potential lower shift, central shift, upper shift, and exact shift, too, in a situation where the Wilcoxon test is asymptotically optimal for the exact shift only, c.f. part C) of Section 3.2.

Defining $B_j(t) = \int_0^t b_j(u)\,du$, $0 \le t \le 1$, $j = 1, 2, 3$, we get

$$
\begin{aligned}
B_1(t) &= B_3(1 - t) = -t^3 + 2t^2 - t, \\[2mm]
B_2(t) &= -4t^4 + 8t^3 - 4t^2, \qquad 0 \le t \le 1.
\end{aligned}
\tag{19}
$$

With $0 = T_0 < T_1 < \cdots < T_d$, $\tau_j = T_j - T_{j-1}$, and $\tau_{1j}$, $\tau_{2j}$, $j = 1, ..., d$, as in (2) and (3) we define the vector $\vec{S} = (S_1, S_2, S_3)^T$ of linear rank statistics $S_\varrho$, $\varrho = 1, 2, 3$, by

$$
\begin{aligned}
S_\varrho &= \sqrt{\frac{mn}{N}} \left( \frac{1}{m} \sum_{i=1}^{m} b_{\varrho N}^\tau(R_i) - \frac{1}{n} \sum_{i=m+1}^{N} b_{\varrho N}^\tau(R_i) \right) \\[3mm]
&= \sqrt{\frac{N}{mn}} \sum_{j=1}^{d} \left( B_\varrho(\tfrac{T_j}{N}) - B_\varrho(\tfrac{T_{j-1}}{N}) \right) \frac{(n\,\tau_{1j} - m\,\tau_{2j})}{\tau_j},
\end{aligned}
\tag{20}
$$

where the $b_{\varrho N}^\tau(i)$ are the averaged scores corresponding to $b_\varrho$, cf. (3.3.43) and (3.3.44). In the next step we'll define the $3 \times 3$ –matrix

$$
\Gamma = \left( \gamma_{\varrho\sigma} \right)_{\varrho,\sigma = 1,2,3}
\tag{21}
$$

according to ( $\forall\, \varrho, \sigma = 1, 2, 3$ )

$$\gamma_{\varrho\sigma} \;=\; N \sum_{j=1}^{d} \Big( B_\varrho(\tfrac{T_j}{N}) - B_\varrho(\tfrac{T_{j-1}}{N}) \Big) \Big( B_\sigma(\tfrac{T_j}{N}) - B_\sigma(\tfrac{T_{j-1}}{N}) \Big) \, \tfrac{1}{\tau_j} \,. \qquad (22)$$

For each subset $J$ of $\{1, 2, 3\}$ such that $J \neq \emptyset$ we put

$$\vec{S}_J \;=\; (S_\varrho : \varrho \in J)^T \qquad \text{and} \qquad \Gamma_J \;=\; \Big( \gamma_{\varrho,\sigma} \Big)_{\varrho,\sigma \in J} \,,$$

where the elements of $J$ are arranged in increasing order. Then the proposed *projection rank statistic* $S^\pi$ is defined by

$$S^\pi \;=\; \max\Big\{ \vec{S}_J^T (\Gamma_J)^{-1} \vec{S}_J : (\Gamma_J)^{-1} \vec{S}_J \geq 0,\, \emptyset \neq J \subset \{1, 2, 3\} \Big\} \qquad (23)$$

with $S^\pi := 0$ if $(\Gamma_J)^{-1} \vec{S}_J \not\geq 0 \;\; \forall\, J \subset \{1, 2, 3\}$, $J \neq \emptyset$. Of course the definition (23) requires a nonsingular $\Gamma$, otherwise $S^\pi$ is not defined.

The projection rank test will reject the null hypothesis $\mathcal{H}_0^\tau : F = G$ at level $\alpha$ in favor of the one-sided alternative $\mathcal{A}_2^0 : F \leq G, F \neq G$ if the value of the test statistic $S^\pi$ is larger than the upper $\alpha$–quantile $k_\alpha^\pi(\tau_1, ..., \tau_d)$ of the conditional distribution of $S^\pi$ under $\mathcal{H}_0^\tau$ given the tie-lengths vector $\tau = (\tau_1, ..., \tau_d)$ of the ordered pooled sample, i.e. the projection rank test may be written in the form

$$\psi^\pi \;=\; 1\Big( S^\pi > k_\alpha^\pi(\tau) \Big) \qquad (24)$$

where $k_\alpha^\pi(\tau)$ is defined by

$$P_{\mathcal{H}_0^\tau}\Big\{ S^\pi > k_\alpha^\pi(\tau) \mid \tau \Big\} \leq \alpha < P_{\mathcal{H}_0^\tau}\Big\{ S^\pi \geq k_\alpha^\pi(\tau) \mid \tau \Big\} \,. \qquad (25)$$

The upper $\alpha$–quantiles $k_\alpha^\pi$ for the continuous model are given in Table 2.1.C of the Appendix.

In the Numerical Example 2.1.1 we evaluate the value $s^\pi$ of the projection rank statistic $S^\pi$ as

$$s^\pi \;=\; S^\pi(r_1, ..., r_{16}) \;=\; 0.2826 \,.$$

Since ties are present there is no table with the exact conditional critical value $k_\alpha^\pi(2, 3, 2, 2, 2, 1, 3, 1)$, but similar to the evaluation of the conditional $p$–values of the statistic (13) we may use a Monte Carlo simulation program in order to get the actual (conditional) $p$–values of $S^\pi$. With a Monte Carlo sample size of 10,000 we get

$$P_{\mathcal{H}_0^\tau}\Big\{ S^\pi > 0.2826 \mid \tau = (2, 3, 2, 2, 2, 1, 3, 1) \Big\} \;=\; 0.533 \,,$$

$$P_{\mathcal{H}_0^\tau}\Big\{ S^\pi \geq 0.2826 \mid \tau = (2, 3, 2, 2, 2, 1, 3, 1) \Big\} \;=\; 0.536 \,.$$

This means that the projection test will reject the null hypothesis of randomness $\mathcal{H}_0^r : F = G$ in favor of the one-sided alternative $\mathcal{A}_2^0 : F \leq G,\ F \neq G$ exactly for those levels $\alpha$ which are larger than or equal to the observed $p$-value. Since the observed $p$-value of the example is $0.53$, the data will not reject the null hypothesis for any reasonable level $\alpha$, e.g. for $\alpha = 0.10$.

Another approximate asymptotic computation of the $p$-values is provided by the following formula:

For $t \geq 0$ we get from (3.3.123) and (3.2.78)

$$P_{\mathcal{H}_0^r}\{ S^\pi > t \mid \tau \} \approx w_1 P\{\chi_1^2 > t\} + w_2 P\{\chi_2^2 > t\} + w_3 P\{\chi_3^2 > t\}, \qquad (26)$$

where $\chi_\varrho^2$, $\varrho = 1,2,3$, denote random variables with respective $\chi_\varrho^2$-distributions and where $w_\varrho$, $\varrho = 1,2,3$, are defined by

$$w_1 = \frac{1}{4\pi}\left( \arccos(\varrho_{12}^*) + \arccos(\varrho_{13}^*) + \arccos(\varrho_{23}^*) \right),$$

$$w_2 = \frac{1}{4\pi}\left( \arccos(\varrho_{12}) + \arccos(\varrho_{13}) + \arccos(\varrho_{23}) \right), \qquad (27)$$

$$w_3 = \frac{1}{2} - w_1.$$

Here $\left( \varrho_{ij} \right)_{ij=1,2,3}$ is the correlation matrix corresponding to the covariance matrix $\Gamma$ defined in (22) and $\left( \varrho_{ij}^* \right)_{ij=1,2,3}$ is the correlation matrix corresponding to $\Gamma^{-1}$, c.f. (3.2.93) and (3.2.96).

In the above numerical example we get

$$w_1 = 0.4720, \qquad w_2 = 0.2286, \qquad w_3 = 0.0280.$$

Therefore the approximate observed (conditional) $p$-value computed from (26) with $t = s^\pi = 0.2826$ is

$$P_{\mathcal{H}_0^r}\left\{ S^\pi > 0.2826 \mid \tau = (2,3,2,2,2,1,3,1) \right\} \approx 0.506.$$

**Monte Carlo Results:**

In Figure 2.1.c to Figure 2.1.e we've displayed the results of a Monte Carlo power comparison between the one-sided Wilcoxon test and the new one-sided test (16) under generalized shift alternatives of the form (3.1.110) with shift functions according to (3.1.111) and (3.1.112). The graphs of the corresponding optimal score functions are displayed in Figure 1.3.a to Figure 1.3.c.

Only in the exact logistic shift model, where the Wilcoxon test is asymptotically optimal, and in the exact normal shift model, where the Wilcoxon test is nearly asymptotically optimal, the Wilcoxon power is higher than the power of the test (16). For all other cases the power of the new test is well above the Wilcoxon power. (Note, there is a substantial difference in the scaling of the 12 graphs.) It's quite easy to construct other (nonparametric) types of alternatives where the difference is even bigger.

Similarly, in Figure 2.1.f to Figure 2.1.h we've displayed the results of a Monte Carlo power comparison between the new one-sided test (16) and the new projection rank test (24) under the same generalized shift alternatives as before. Obviously all graphs are rather similar, with some tendency in favor of the projection rank test. But notice, the projection rank test has been defined on the basis of the generalized logistic shift functions (3.2.89), which are approximations of the generalized normal shift functions and of the generalized Cauchy shift functions, too, cf. Figure 1.3.a to Figure 1.3.c and Figure 3.2.b. Therefore the comparison is biased in favor of the projection rank test (24). Since the test (16) doesn't use any prior information about the underlying alternatives used in the Monte Carlo simulation, we may expect the test (16) to be superior to the test (24) in other situations.

## Figure 2.1.c

Empirical power at level $\alpha = 0.01, 0.02, ..., 0.10$ under generalized **normal** shift alternatives $F(x - D(x))$ for the Wilcoxon rank test ( ∘ ∘ ∘ ), and the rank test (16) with Parzen-2 kernel and bandwidth $a = 0.40$ (• • •) .

The sample sizes are $m = 30$ , $n = 40$ , the Monte Carlo sample size is $10\,000$ .

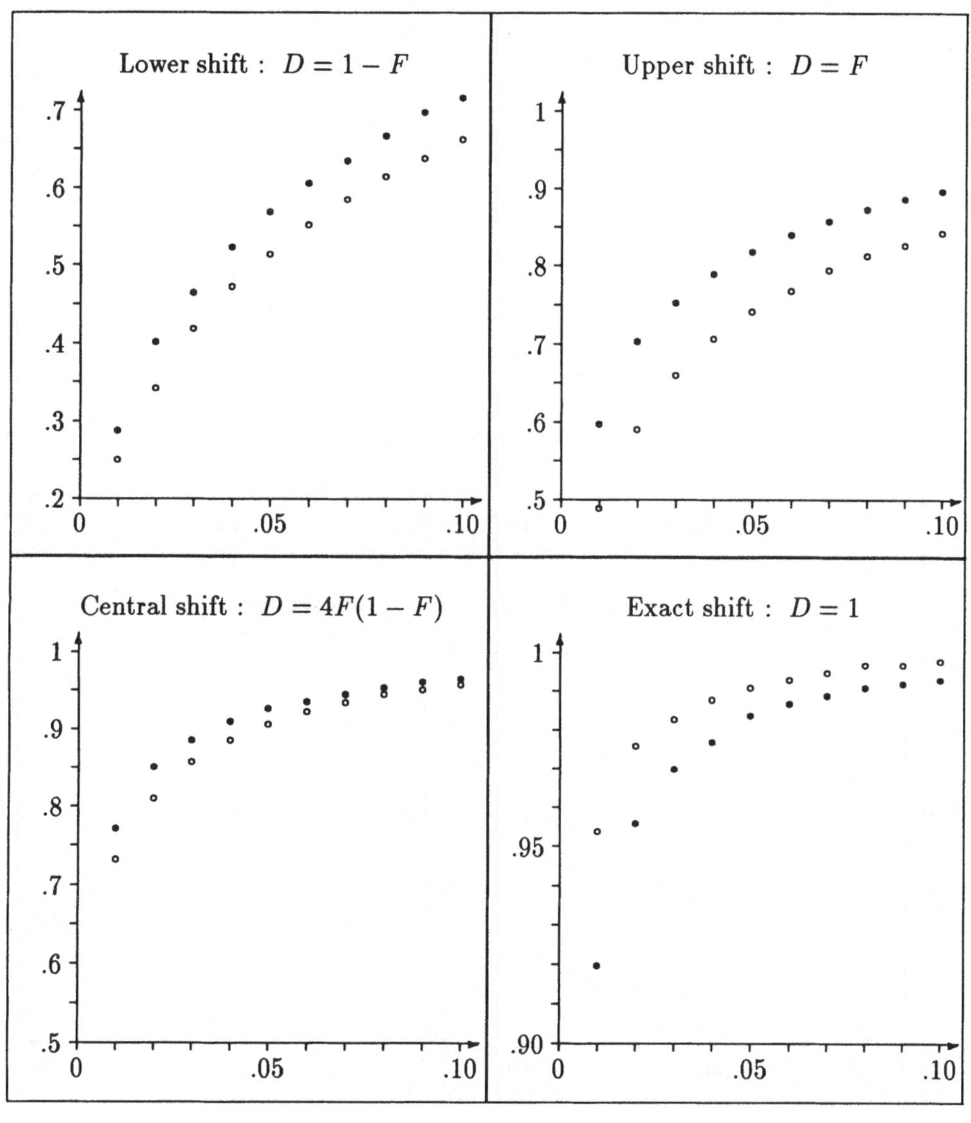

### Figure 2.1.d

Empirical power at level  $\alpha = 0.01, 0.02, ..., 0.10$  under generalized **logistic** shift alternatives  $F\big(x - D(x)\big)$  for the Wilcoxon rank test (  $\circ\circ\circ$  ), and the rank test (16) with Parzen-2 kernel and bandwidth  $a = 0.40$  ($\bullet\bullet\bullet$) .

The sample sizes are  $m = 30$ ,  $n = 40$ , the Monte Carlo sample size is  $10\,000$ .

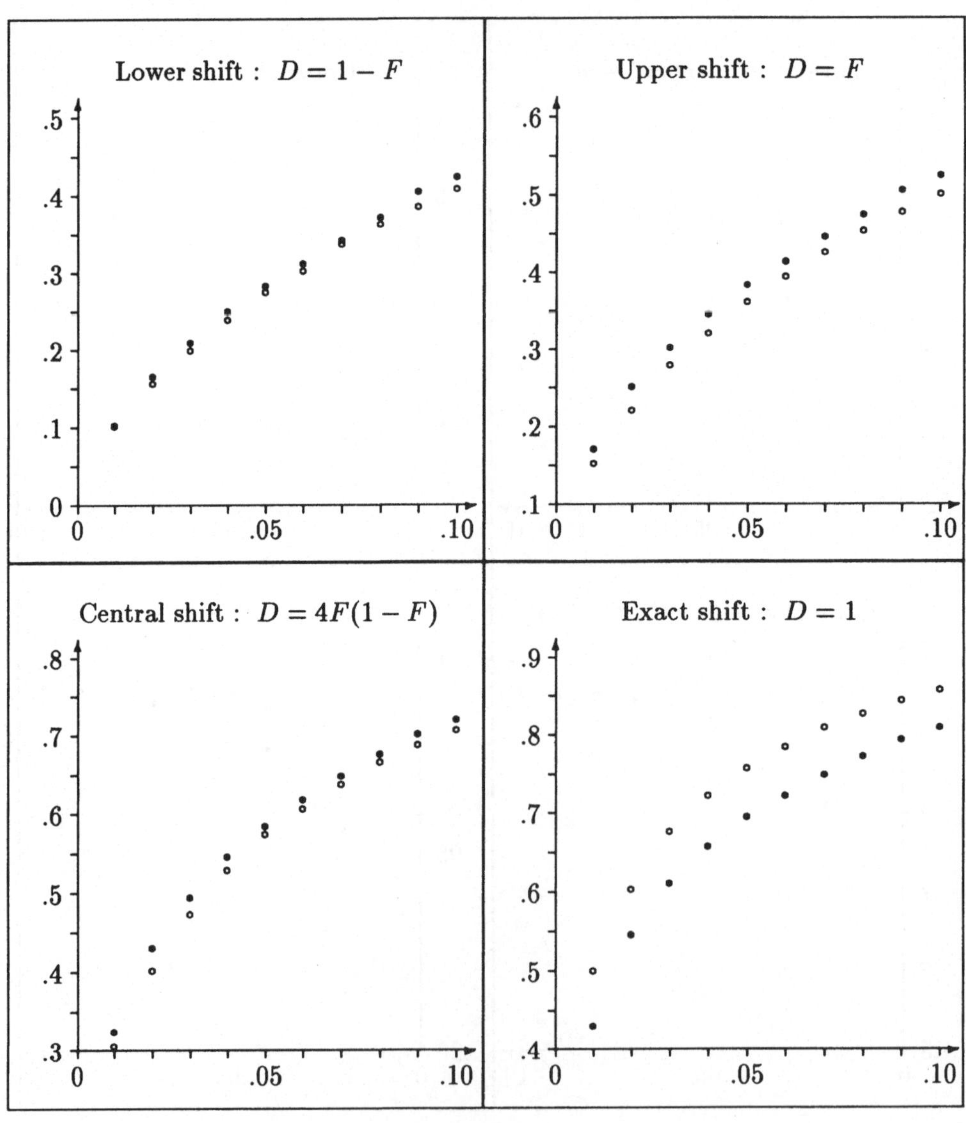

## Figure 2.1.e

Empirical power at level $\alpha = 0.01, 0.02, ..., 0.10$ under generalized **Cauchy** shift alternatives $F(x - D(x))$ for the Wilcoxon rank test ( ∘ ∘ ∘ ), and the rank test (16) with Parzen-2 kernel and bandwidth $a = 0.40$ (• • •) .

The sample sizes are $m = 30$ , $n = 40$ , the Monte Carlo sample size is $10\,000$ .

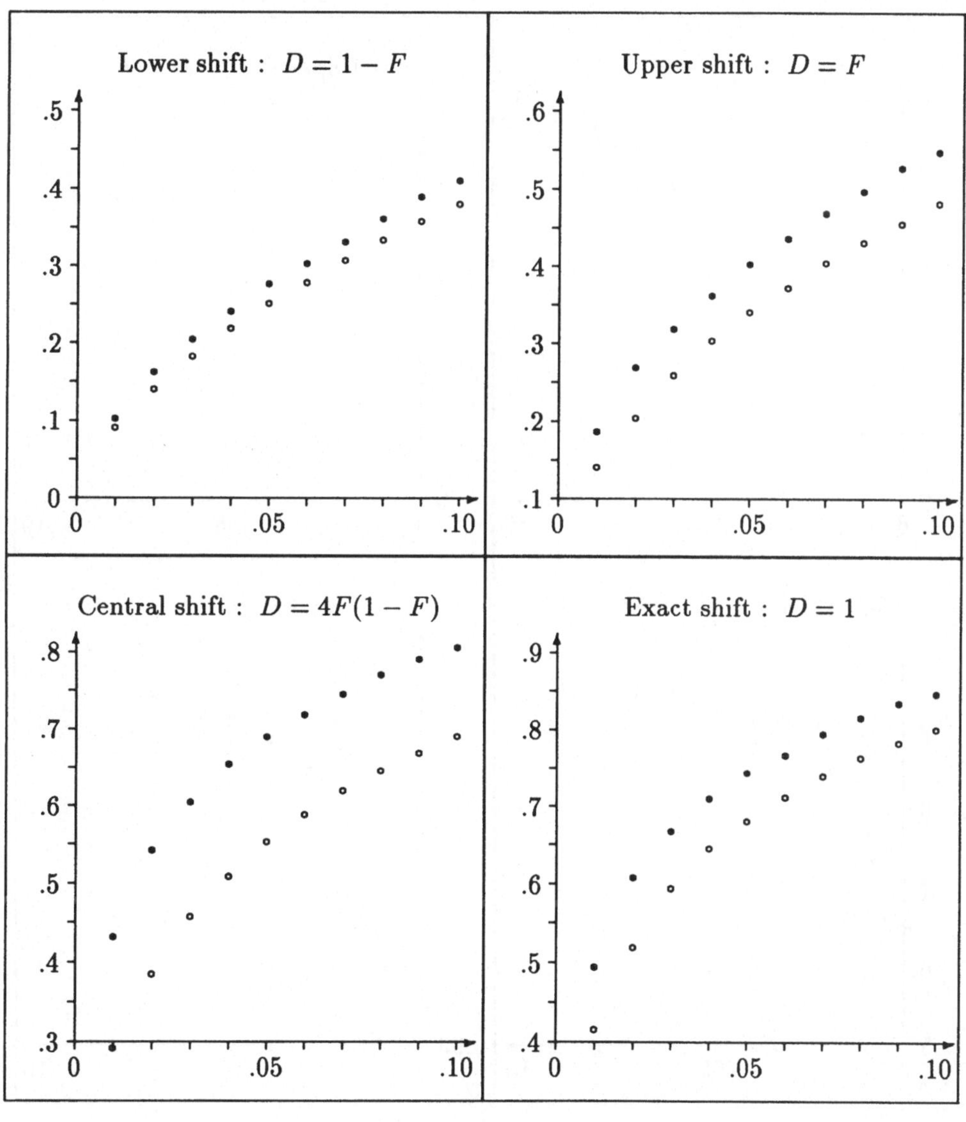

### Figure 2.1.f

Empirical power at level $\alpha = 0.01, 0.02, ..., 0.10$ under generalized **normal** shift alternatives $F\bigl(x - D(x)\bigr)$ for the projection rank test (24) ( $\circ\circ\circ$ ), and the rank test (16) with Parzen-2 kernel and bandwidth $a = 0.40$ ($\bullet\bullet\bullet$).

The sample sizes are $m = 30$, $n = 40$, the Monte Carlo sample size is $10\,000$.

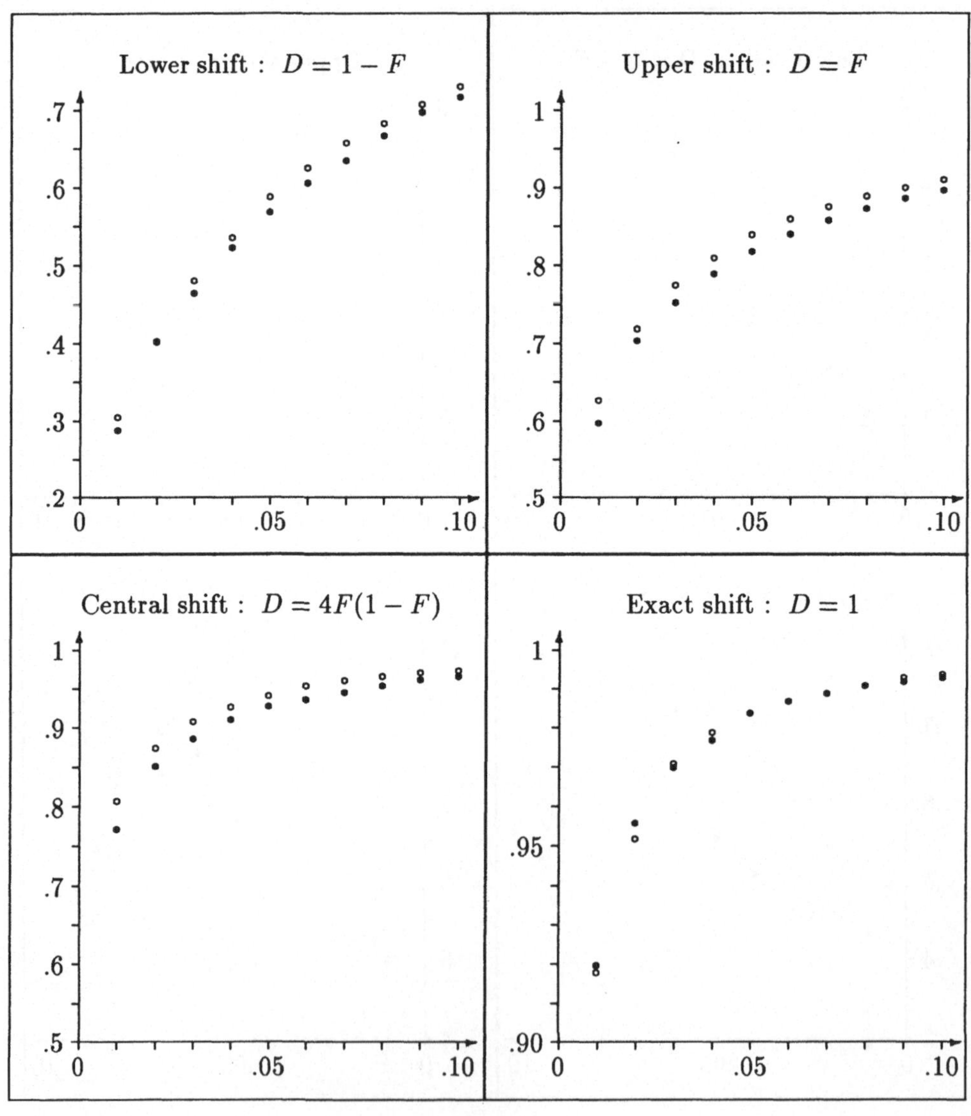

## Figure 2.1.g

Empirical power at level  $\alpha = 0.01, 0.02, ..., 0.10$  under generalized **logistic** shift alternatives  $F(x - D(x))$  for the projection rank test (24) ( ∘ ∘ ∘ ), and the rank test (16) with Parzen-2 kernel and bandwidth  $a = 0.40$  (● ● ●) .

The sample sizes are  $m = 30$ ,  $n = 40$ , the Monte Carlo sample size is  $10\,000$ .

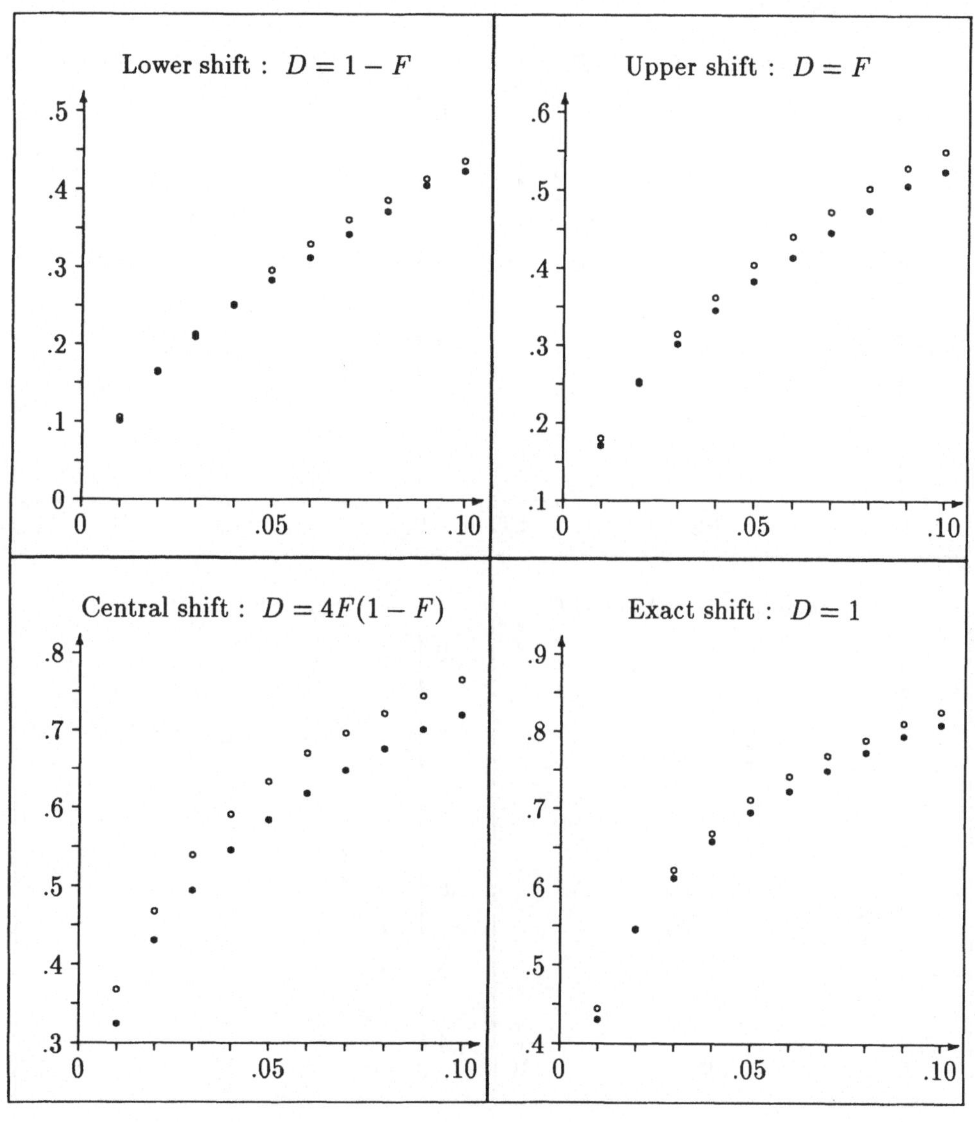

### Figure 2.1.h

Empirical power at level $\alpha = 0.01, 0.02, ..., 0.10$ under generalized **Cauchy** shift alternatives $F\big(x - D(x)\big)$ for the projection rank test (24) ( $\circ\,\circ\,\circ$ ), and the rank test (16) with Parzen-2 kernel and bandwidth $a = 0.40$ ($\bullet\,\bullet\,\bullet$) .

The sample sizes are $m = 30$ , $n = 40$ , the Monte Carlo sample size is $10\,000$ .

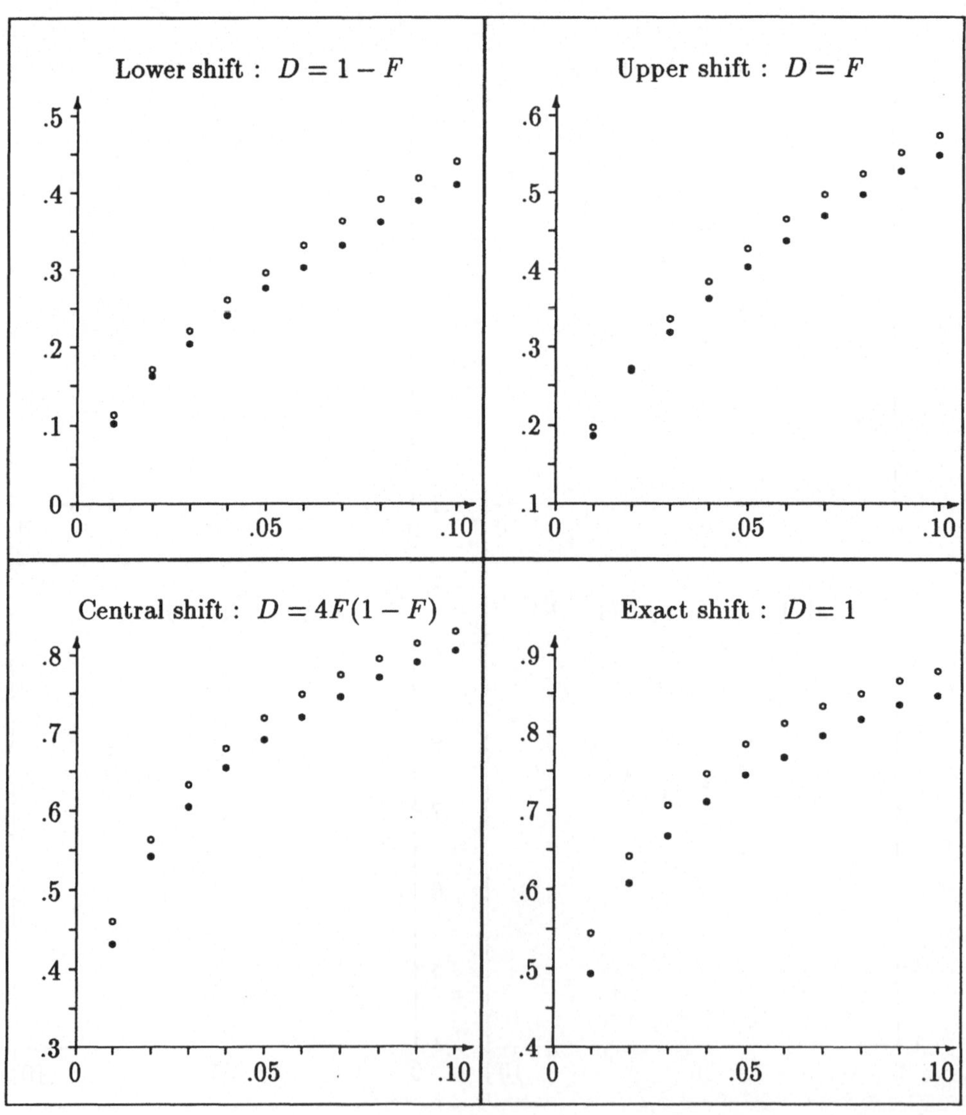

## 2.2  Two samples differing in scale

The framework and notation of the present section is the same as in the preceding section and will not be repeated here. The only difference is that now we will consider the so-called *dispersion about some prefixed number* $\mu$ , $0 < \mu < 1$ , instead of the stochastic ordering of Section 2.1. The dispersion problem is a nonparametric generalization of the classical (parametric) scale problem, see Section 4.1 for details.

The most interesting case is the *dispersion about the median*, i.e. the case $\mu = 1/2$ .

**Data and Model Assumptions:**

Exactly the same as in Section 2.1.

**Testing Problem (dispersion about $\mu$) :**

Testing the null hypothesis of randomness

$$\mathcal{H}_0^r : F = G$$

versus the alternative that $F$ is more dispersed about $\mu$ than $G$ for the given $0 < \mu < 1$ , i.e:

$$\mathcal{A}_2^\mu \; : \quad \begin{cases} F \neq G & \text{such that} \\ F(x) \geq G(x), & \text{if } \frac{m}{N} F(x) + \frac{n}{N} G(x) < \mu , \\ F(x) \leq G(x), & \text{if } \frac{m}{N} F(x) + \frac{n}{N} G(x) \geq \mu . \end{cases} \qquad (1)$$

**The Dispersion Test :**

The test statistic $S^\mu$ for testing the null hypothesis of randomness $\mathcal{H}_0^r$ versus the dispersion alternative $\mathcal{A}_2^\mu$ is constructed in the same way as the one-sided test statistic $S^0$ defined in formula (13) of Section 2.1, with the only difference that the rank process $W_N^0 = \min(0, W_N)$ of Section 2.1 has to be replaced by the $\mu$ –dispersion rank process $W_N^\mu$ which is defined by

$$W_N^\mu(t) \;=\; \begin{cases} W_N^+(t), & \text{if } 0 \leq t \leq \mu_1, \\ \frac{\mu_2 - t}{\mu_2 - \mu_1} W_N^+(\mu_1) + \frac{t - \mu_1}{\mu_2 - \mu_1} W_N^0(\mu_2), & \text{if } \mu_1 \leq t \leq \mu_2, \\ W_N^0(t), & \text{if } \mu_2 \leq t \leq 1, \end{cases} \qquad (2)$$

with $W_N^+ = \max(0, W_N)$ and

$$\begin{aligned} \mu_1 &= \max\{ T_i/N : T_i/N \leq \mu, \, i = 0, ..., d \}, \\ \mu_2 &= \min\{ T_i/N : T_i/N > \mu, \, i = 0, ..., d \}. \end{aligned} \qquad (3)$$

For the (unrestricted) two-sample rank process $W_N$ see (2) or (5) of Section 2.1.

Using the weights $\Delta W_N(j/N)$ and $k(i,j)$ as defined in formulae (5) and (7) of Section 2.1 the dispersion rank statistic $S^\mu$ is

$$S^\mu \;=\; \frac{1}{mnN} \sum_{i=1}^{N}\sum_{j=1}^{N} \Delta W_N^\mu(\frac{i}{N})\,\Delta W_N(\frac{j}{N})\,k(i,j) \tag{4}$$

with

$$\Delta W_N^\mu(\frac{i}{N}) \;=\; W_N^\mu(\frac{i}{N}) - W_N^\mu(\frac{i-1}{N}), \qquad 1 \le i \le N, \tag{5}$$

c.f. formula (13) of Section 2.1.

The dispersion rank test will reject the nullhypothesis $\mathcal{H}_0^\tau : F = G$ at level $\alpha$ in favor of the dispersion alternative $\mathcal{A}_2^\mu$, if the value of the statistic $S^\mu$ is larger than the upper $\alpha$–quantile $k_\alpha^\mu(\tau)$ of the conditional distribution of $S^\mu$ under $\mathcal{H}_0^\tau$ given the tie-lengths vector $\tau = (\tau_1,...,\tau_d)$ of the ordered pooled sample, i.e. the dispersion rank test may be written in the form

$$\psi^\mu \;=\; 1\Big( S^\mu > k_\alpha^\mu(\tau) \Big), \tag{6}$$

where $k_\alpha^\mu(\tau)$ is defined by

$$P_{\mathcal{H}_0^\tau}\big\{ S^\mu > k_\alpha^\mu(\tau) \mid \tau \big\} \;\le\; \alpha \;<\; P_{\mathcal{H}_0^\tau}\big\{ S^\mu \ge k_\alpha^\mu(\tau) \mid \tau \big\}. \tag{7}$$

The upper $\alpha$–quantiles $k_\alpha^\mu$ for the continuous model and $\mu = 1/2$ (median) are given in Table 2.2.A of the Appendix.

The evaluation or simulation of the conditional critical values $k_\alpha^\mu(\tau)$ is completely similar to the omnibus case of Section 2.1.

Notice, for the case $\mu = 0$ the statistic $S^\mu$ coincides with the one-sided statistic $S^0$ defined in (13) of Section 2.1.

**The Projection Rank Test For Dispersion About The Median:**

Here we propose a second nonlinear rank statistic $S^\pi$ for testing the null hypothesis of randomness $\mathcal{H}_0^\tau$ versus the dispersion alternative $\mathcal{A}_2^\mu$ with $\mu = 1/2$ (the first treatment is more dispersed about the median than the second treatment).

The form of the statistic $S^\pi$, its computation, and the evaluation or simulation of the conditional critical values $k_\alpha^\pi(\tau)$ are exactly the same as in formulae (20) to (27) of Section 2.1, the only difference being that the score functions (18) of Section 2.1 have to be replaced by the following score functions, cf.

(4.1.42) to (4.1.45) of Section 4.1,

$$b_1(u) = (1 - u)\left(3(1 - u) - 2\right)\ln(\frac{1-u}{u}) - (1 - u),$$

$$b_2(u) = 4\,u\,(1 - u)\left(2\,(2u - 1)\ln(\frac{u}{1-u}) - 1\right), \tag{8}$$

$$b_3(u) = b_1(1 - u), \qquad 0 < u < 1,$$

with corresponding integrals $B_j(t) = \int_0^t b_j(u)\,du$, $0 \le t \le 1$, $j = 1,2,3$, according to

$$B_1(t) = t\,(1 - t)^2\,\ln(\frac{1-t}{t}),$$

$$B_2(t) = 4\,t^2\,(1 - t)^2\,\ln(\frac{1-t}{t}), \tag{9}$$

$$B_3(t) = -\,B_1(1 - t) = t^2\,(1 - t)\,\ln(\frac{1-t}{t}).$$

The upper $\alpha$–quantiles $k_\alpha^\pi$ for the continuous model are given in Table 2.2.B of the Appendix.

**Numerical Example 2.2.1:**

- The sample sizes are $m = 7$ and $n = 9$, i.e. the total sample size is $N = 16$.

- The observations of the first sample are
  $(x_1, ..., x_7) = (3.0,\ 9.0,\ 3.0,\ 6.0,\ 10.0,\ 2.0,\ 8.0)$.
  The ordered first sample is 2.0, 3.0, 3.0, 6.0, 8.0, 9.0, 10.0 .

- The observations of the second sample are
  $(x_8, ..., x_{16}) = (7.0,\ 3.0,\ 6.0,\ 2.0,\ 9.0,\ 5.0,\ 7.0,\ 5.0,\ 9.0)$.
  The ordered second sample is 2.0, 3.0, 5.0, 5.0, 6.0, 7.0, 7.0, 9.0, 9.0 .

- The ordered pooled sample is
  2.0, 2.0, 3.0, 3.0, 3.0, 5.0, 5.0, 6.0, 6.0, 7.0, 7.0, 8.0, 9.0, 9.0, 9.0, 10.0.

- The vector of ranks is
  $(r_1, ..., r_{16}) = (5,\ 15,\ 5,\ 9,\ 16,\ 2,\ 12;\ \ 11,\ 5,\ 9,\ 2,\ 15,\ 7,\ 11,\ 7,\ 15)$.

- The number of different values among the observations (ranks) is $d = 8$.

- The ordered different values among the ranks are
  $T = (T_1, ..., T_d) = (2,\ 5,\ 7,\ 9,\ 11,\ 12,\ 15,\ 16)$.

- The lengths of the ties in the ordered pooled sample are
  $\tau = (\tau_1, ..., \tau_d) = (\,2,\ 3,\ 2,\ 2,\ 2,\ 1,\ 3,\ 1\,)$ .

- The lengths of the ties corresponding to the first and second sample are
  $\tau_1 = (\tau_{11}, ..., \tau_{1d}) = (\,1,\ 2,\ 0,\ 1,\ 0,\ 1,\ 1,\ 1\,)$ ,
  $\tau_2 = (\tau_{21}, ..., \tau_{2d}) = (\,1,\ 1,\ 2,\ 1,\ 2,\ 0,\ 2,\ 0\,)$ .

- The defining values $W_N(T_j/N) - W_N(T_{j-1}/N) = n\,\tau_{1j} - m\,\tau_{2j}$ , $j = 1, ..., d$ , of the rank process $W_N$ are $2, 11, -14, 2, -14, 9, -5, 9$ .

- According to (5) of Sect.2.1 the values of $\Delta W_N(i/N)$ , $i = 1, ..., N$ , are
  $1, 1, 11/3, 11/3, 11/3, -7, -7, 1, 1, -7, -7, 9, -5/3, -5/3, -5/3, 9$ .

- Therefore the values of $W_N(i/N)$ , $i = 0, 1, ..., N$ , are
  $0, 1, 2, 17/3, 28/3, 13, 6, -1, 0, 1, -6, -13, -4, -17/3, -22/3, -9, 0$ .

- The graph of $W_N$ is given in Figure 2.2.a.

- In case of dispersion about the median, i.e. $\mu = 1/2$ , we get $\mu_1 = 7/16$ and $\mu_2 = 9/16$ .

- Therefore the values of $W_N^{1/2}(i/N)$ , $i = 0, 1, ..., N$ , are
  $0, 1, 2, 17/3, 28/3, 13, 6, 0, 0, 0, -6, -13, -4, -17/3, -22/3, -9, 0$ .

- The graph of $W_N^{1/2}$ is given in Figure 2.2.b.

- The values of $\Delta W_N^{1/2}(i/N)$ , $i = 1, ..., N$ , are
  $1, 1, 11/3, 11/3, 11/3, -7, -6, 0, 0, -6, -7, 9, -5/3, -5/3, -5/3, 9$ .

- Using (4) and (5) with $\mu = 1/2$ and the bandwidth $a = 0.40$ we evaluate the value $s^\mu$ of the dispersion statistic $S^\mu$ as

$$s^\mu = S^\mu(r_1, ..., r_{16}) = 1.1329 .$$

- Since ties are present there is no table with the corresponding exact conditional critical value $k_\alpha^\mu(2, 3, 2, 2, 2, 1, 3, 1)$ but we utilize formula (12) of Sect.2.1 in order to evaluate the actual (conditional) $p$ –values by Monte Carlo simulation. With a Monte Carlo sample size of 10,000 we get

$$P_{\mathcal{H}_0^\tau}\big\{\, S^\mu > 1.1329 \mid \tau = (2,3,2,2,2,1,3,1)\,\big\} = 0.433 ,$$

$$P_{\mathcal{H}_0^\tau}\big\{\, S^\mu \geq 1.1329 \mid \tau = (2,3,2,2,2,1,3,1)\,\big\} = 0.436 .$$

- This means that the dispersion test will reject the null hypothesis of randomness $\mathcal{H}_0^\tau : F = G$ in favor of the dispersion alternative $\mathcal{A}_2^{1/2}$ exactly for those levels $\alpha$ which are larger than or equal to the observed $p$ –value. Since the observed $p$ –value of the example is $0.433$ the data will **not** reject the null hypothesis for any reasonable level $\alpha$ , e.g. for $\alpha = 0.10$ .

**Figure 2.2**

*Numerical Example 2.2.1: The graph of* $\left( W_N(t),\, 0 \le t \le 1 \right)$.

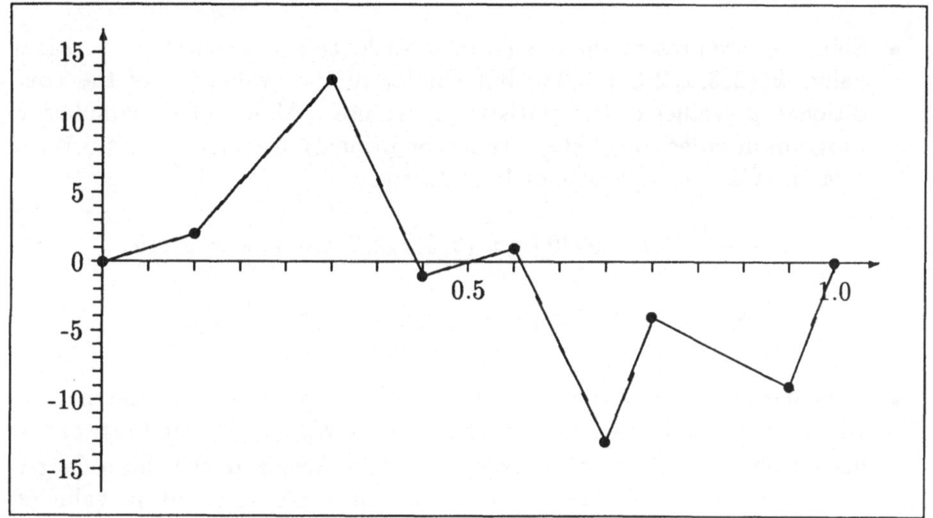

*Numerical Example 2.2.1: The graph of* $\left( W_N^\mu(t),\, 0 \le t \le 1 \right)$ *for* $\mu = 1/2$.

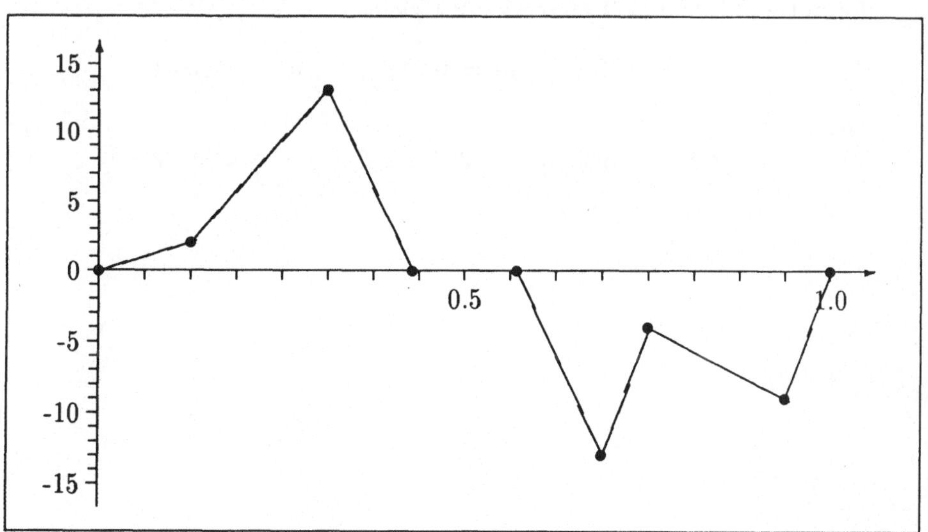

- Using (8), (9), and the corresponding formulae (20) to (23) of Section 2.1 we evaluate the value $s^\pi$ of the projection rank statistic $S^\pi$ for dispersion about the median as

$$s^\pi = S^\pi(r_1, ..., r_{16}) = 1.9039.$$

- Since ties are present there is no table with the exact conditional critical value $k_\alpha^\pi(2,3,2,2,2,1,3,1)$, but similar to the evaluation of the conditional $p$–values of the statistic (4) we use a Monte Carlo simulation program in order to get the actual (conditional) $p$–values of $S^\pi$. With a Monte Carlo sample size of 10,000 we get

$$P_{\mathcal{H}_0^r}\Big\{ S^\pi > 1.9039 \mid \tau = (2,3,2,2,2,1,3,1) \Big\} = 0.216,$$

$$P_{\mathcal{H}_0^r}\Big\{ S^\pi \geq 1.9039 \mid \tau = (2,3,2,2,2,1,3,1) \Big\} = 0.219.$$

- This means that the projection rank test for dispersion about the median will reject the null hypothesis of randomness $\mathcal{H}_0^r : F = G$ in favor of the dispersion alternative $\mathcal{A}_2^{1/2}$ exactly for those levels $\alpha$ which are larger than or equal to the observed $p$–value. Since the observed $p$–value of the example is $0.216$ the data will **not** reject the null hypothesis for any reasonable level $\alpha$, e.g. for $\alpha = 0.10$.

- The approximate asymptotic computation of the $p$-value according to formulae (26) and (27) of Section 2.1 yields

$$w_1 = 0.4649, \qquad w_2 = 0.2428, \qquad w_3 = 0.0351,$$

and

$$P_{\mathcal{H}_0^r}\Big\{ S^\pi > 1.9039 \mid \tau = (2,3,2,2,2,1,3,1) \Big\} \approx 0.1925.$$

## 2.3 Several samples on the real line

In this section the comparison of two treatments is generalized to the comparison of $k \geq 2$ treatments. We'll discuss testing procedures for testing the null hypothesis of randomness (all the $k$ treatments behave equally) versus the one-sided alternative of trend (the first treatment produces stochastically larger values than the second treatment, the second treatment produces stochastically larger values than the third treatment etc.) and also testing procedures for testing the null hypothesis of randomness versus the omnibus alternative (at least two of the treatments behave differently).

Notice, in accordance with the one-sided two-sample model of Section 2.1 the trend alternative implies that the *first treatment* produces *stochastically larger* values than the other treatments (decreasing trend).

**Data:**

- The random sample $X_{11}, ..., X_{1n_1}$ of size $n_1$ from treatment 1 .

- The random sample $X_{21}, ..., X_{2n_2}$ of size $n_2$ from treatment 2 .

- $\cdots$

- The random sample $X_{k1}, ..., X_{kn_k}$ of size $n_k$ from treatment $k$ .

There is no restriction with respect to tied observations, i.e. the underlying model may allow for discontinuities or even may be purely discrete. In case of tied observations there are no tables of the (conditional) critical values, but there are simulation programs for evaluating the conditional $p$ –values of the proposed tests.

All proposed tests will be *rank tests*, i.e. they are based exclusively on the vector of ranks
$$(R_{11}, ..., R_{1n_1}, R_{21}, ..., R_{2n_2}, ..., R_{k1}, ..., R_{kn_k}) = (R_1, ..., R_N) = R$$
of the pooled sample
$$(X_{11}, ..., X_{1n_1}, X_{21}, ..., X_{2n_2}, ..., X_{k1}, ..., X_{kn_k}) = (X_1, ..., X_N) = X ,$$
where $N = n_1 + n_2 + \cdots + n_k$ is the total sample size and where the rank $R_{qi}$ of $X_{qi}$ in the pooled sample $X$ is defined as

$$R_{qi} = \#\left\{ j \in \{1, ..., N\} \mid X_j \leq X_{qi} \right\}, \qquad i = 1, ..., n_q, \quad q = 1, ..., k. \qquad (1)$$

Notice, the rank of $X_{qi}$ counts all observations of the pooled sample which are less than or equal to $X_{qi}$ , also in case of tied observations.

**Model Assumptions:**

- Each of the $k$ samples is a random sample from the respective population.

- The $k$ samples are mutually independent.

- The respective underlying (cumulative) distribution functions $F_q$ of the random variables $X_{qi}$ are arbitrary distribution functions on the real line, i.e. the $F_q$ may have discontinuities.

- The measurement scale is the real line or at least ordinal in order to have well-defined ranks.

**Testing Problems:**

- *Trend problem:* Testing the null hypothesis of randomness,

$$\mathcal{H}_0^r : \ F_1 = F_2 = \cdots = F_k \,,$$

versus the one-sided alternative of decreasing trend

$$\mathcal{A}_k^0 : \ F_1 \leq F_2 \leq \cdots \leq F_k, \ F_1 \neq F_k \,,$$

i.e. the first sample is stochastically larger than the second sample, the second sample is stochastically larger than the third sample etc..

- *Omnibus problem:* Testing the null hypothesis of randomness,

$$\mathcal{H}_0^r : \ F_1 = F_2 = \cdots = F_k \,,$$

versus the omnibus alternative that at least two treatments behave differently,

$$\mathcal{A}_k : \ F_q \neq F_r \quad \text{for some } \ q, r \in \{1, ..., k\} \,.$$

Essential for the data inspection and for the definition of the test statistics are the following rank processes $W_{qN} = \big(W_{qN}(t), 0 \leq t \leq 1\big)\,, \ q = 1, ..., k$ :

Let $d$ denote the number of different values among the ranks $R_1, ..., R_N$, which is the same as the number of different values among the original (pooled) observations $X_1, ..., X_N$ . In the continuous model (no ties) we have $d = N$ . In the general model (ties are possible) $d$ is a random quantity.

Let $T_1 < \cdots < T_d$ denote the ordered values of the different values in $R_1, ..., R_N$ . If no ties are present, we have $d = N$ and $T_i = i$ for $i = 1, ..., N$ . If all observations are tied, we have $d = 1$ and $T_1 = N$ . In any case we have $1 \leq d \leq N$ and $T_d = N$ .

Using $d$, $T = (T_1, ..., T_d)$, and the additional definition $T_0 := 0$ we define for each $q = 1, ..., k$ the rank process $W_{qN}$ according to

$$W_{qN}\left(\frac{T_j}{N}\right) = \frac{1}{n_q} \sum_{i=1}^{n_q} 1(R_{qi} \leq T_j) - \frac{T_j}{N}, \quad 0 \leq j \leq d, \tag{2}$$

and according to linear interpolation between the points $\left(T_j/N, W_{qN}(T_j/N)\right)$, $j = 0, 1, ..., d$, in the $(x, y)$–plane. Especially we have for each $q = 1, ..., d$ the property $W_{qN}(0) = W_{qN}(1) = 0$.

Notice, in the two-sample case of Section 2.1 we have $n_1 = m$, $n_2 = n$, $W_{1N} = W_N/(mN)$, and $W_{2N} = -W_N/(nN)$. For general $k \geq 2$ we have the property

$$n_1 W_{1N} + n_2 W_{2N} + \cdots + n_k W_{kN} = 0. \tag{3}$$

For easy computation of $W_{1N}, ..., W_{qN}$ we use the following representation, cf. (4.2.136) to (4.2.138):

Let $\tau_j = T_j - T_{j-1}$, $j = 1, ..., d$, denote the *lengths of the ties* in the ordered pooled sample $X^{(1)} \leq \cdots \leq X^{(N)}$. For any $q = 1, ..., k$ and any $j = 1, ..., d$ let $\tau_{qj}$ denote the *length of the $j$–th tie corresponding to the respective $q$–th sample*, i.e.

$$\tau_{qj} = \#\left\{ i \in \{1, ..., n_q\} \mid X_{qi} = X^{(T_j)} \right\}. \tag{4}$$

Obviously this implies $\tau_{1j} + \cdots + \tau_{kj} = \tau_j$, $j = 1, ..., d$. Using $\tau_{qj}$ and $\tau_j$ we get

$$W_{qN}\left(\frac{T_j}{N}\right) - W_{qN}\left(\frac{T_{j-1}}{N}\right) = \frac{\tau_{qj}}{n_q} - \frac{\tau_j}{N}, \qquad j = 1, ..., d. \tag{5}$$

Since $W_{qN}(\cdot)$ is linear on the interval $[T_{j-1}/N, T_j/N]$ for each $j = 1, ..., d$, and since $W_{qN}(0) = 0$ holds true, the process $W_{qN}$ is completely determined by formula (5). Especially we have for each $q = 1, ..., k$ and each $j = 1, ..., d$

$$\Delta W_{qN}\left(\frac{i}{N}\right) := W_{qN}\left(\frac{i}{N}\right) - W_{qN}\left(\frac{i-1}{N}\right)$$

$$= \frac{1}{N} \frac{N \tau_{qj} - n_q \tau_j}{n_q \tau_j} \qquad \text{if } T_{j-1} < i \leq T_j. \tag{6}$$

Obviously the tie-lengths vectors $\vec{\tau}_q := (\tau_{q1}, ..., \tau_{qd})$, $q = 1, ..., k$, of the $k$ samples with respect to the pooled sample contain the same information as the ordered (within each sample) ranks of the $k$ samples. For the actual computation of the rank processes $W_{qN}$ formula (6) seems to be most convenient.

If *no ties are present*, we get $d = N$ and $\tau_i = 1$ $\forall i = 1, ..., N$. This implies

$$\tau_{qi} = \begin{cases} 1, & \text{if the value of } X^{(i)} \text{ belongs to the q–th sample,} \\ 0, & \text{if the value of } X^{(i)} \text{ belongs to some other sample,} \end{cases}$$

$$i = 1, ..., N, \quad q = 1, ..., k \,.$$

### The Omnibus Test:

The test statistic $S$ for testing the null hypothesis of randomness $\mathcal{H}_0^r$ versus the omnibus alternative $\mathcal{A}_k$ is, cf. formula (4.2.48),

$$S \;=\; \sum_{i=1}^{N}\sum_{j=1}^{N} k(i,j) \sum_{q=1}^{k} n_q \,\Delta W_{qN}\!\left(\frac{i}{N}\right) \Delta W_{qN}\!\left(\frac{j}{N}\right), \qquad (7)$$

where

$$\Delta W_{qN}\!\left(\frac{i}{N}\right) \;=\; W_{qN}\!\left(\frac{i}{N}\right) - W_{qN}\!\left(\frac{i-1}{N}\right), \qquad 1 \le i \le N, \qquad (8)$$

$$k(i,j) \;=\; K_a\!\left(\frac{i-1/2}{N}, \frac{j-1/2}{N}\right), \qquad 1 \le i,j \le N, \qquad (9)$$

and

$$K_a(s,t) \;=\; \frac{1}{a}\left( K(\frac{t+s}{a}) + K(\frac{t-s}{a}) + K(\frac{t+s-2}{a}) \right), \quad s,t \in [0,1], \quad (10)$$

with fixed bandwidth $a = 0.40$ and Parzen-2 kernel

$$K(t) = \begin{cases} 4/3 - 8\,|t|^2 + 8\,|t|^3, & \text{if} \quad |t| < 1/2\,, \\[4pt] 8\,(1 - |t|)^3/3\,, & \text{if} \quad 1/2 \le |t| < 1\,, \\[4pt] 0\,, & \text{if} \quad |t| \ge 1\,. \end{cases} \qquad (11)$$

The omnibus test will **reject the null hypothesis** $\mathcal{H}_0^r$ ($F_1 = \cdots = F_k$) at level $\alpha$ in favor of the omnibus alternative $\mathcal{A}_k^0$ ($F_q \ne F_r$ for some $q, r \in \{1, ..., k\}$) **if** the value of the test statistic $S$ is larger than the upper $\alpha$–quantile $k_\alpha(\tau)$ of the conditional distribution of $S$ under $\mathcal{H}_0^r$ given the tie-lengths vector $\tau = (\tau_1, ..., \tau_d)$ of the ordered pooled sample, i.e. the omnibus test may be written in the form

$$\psi \;=\; 1\!\left( S > k_\alpha(\tau) \right), \qquad (12)$$

where $k_\alpha(\tau)$ is defined by

$$P_{\mathcal{H}_0^r}\{ S > k_\alpha(\tau) \mid \tau \} \;\le\; \alpha \;<\; P_{\mathcal{H}_0^r}\{ S \ge k_\alpha(\tau) \mid \tau \}\,. \qquad (13)$$

The upper $\alpha$–quantiles $k_\alpha$ for the continuous model and $k = 3$ are given in Table 2.3.A of the Appendix.

**Simulation of the conditional $p$-values:**

In principle there is no difficulty to evaluate the conditional critical values $k_\alpha(\tau)$ or the conditional $p$-values:

The conditional distribution of the rank vector $(R_1, ..., R_N)$ given the vector $\tau = (\tau_1, ..., \tau_d)$ is the same for any null hypothesis point $(F_1, ..., F_k) = (F, ..., F) \in \mathcal{H}_0^r$, cf. Lemma 3.3.10. Moreover, each rank $R_i$ may be written as a simple function of the randomized ranks $R^* = (R_1^*, ..., R_N^*)$ and of the vector $\tau$, cf. formula (3.3.99),

$$R_i = T_j, \qquad \text{if} \quad T_{j-1} < R_i^* \leq T_j \quad \text{for some } 1 \leq j \leq d, \qquad (14)$$

and $R^*$ and $\tau$ are stochastically independent under the hypothesis of randomness $\mathcal{H}_0^r$. Finally, under $\mathcal{H}_0^r$ the vector $R^*$ is uniformly distributed on the permutations of $(1, ..., N)$. Therefore a generator of random permutations may be used in order to simulate the conditional distribution of $(R_1, ..., R_N)$ given $\tau$ via (14).

The exact evaluation of the critical values via (14) is limited to the unconditional case (no ties present) and to moderate sizes of $N = n_1 + \cdots + n_k$.

In the two-sample case $(n_1 = m, n_2 = n)$ of Section 2.1 we have $W_{1N} = W_N/(mN)$ and $W_{2N} = -W_N/(nN)$. Therefore the statistic (7) coincides with the two-sample omnibus statistic (6) of Section 2.1 in this case.

**The Trend Test:**

The test statistic $S^0$ for testing the null hypothesis of randomness $\mathcal{H}_0^r$ versus the one-sided alternative $\mathcal{A}_k^0$ of decreasing trend is, cf. formula (4.2.49),

$$S^0 = \sum_{i=1}^{N}\sum_{j=1}^{N} k(i,j) \sum_{q=1}^{k} n_q \, \Delta W_{qN}^0\left(\frac{i}{N}\right) \Delta W_{qN}\left(\frac{j}{N}\right), \qquad (15)$$

where

$$\Delta W_{qN}^0\left(\frac{i}{N}\right) = W_{qN}^0\left(\frac{i}{N}\right) - W_{qN}^0\left(\frac{i-1}{N}\right), \qquad 1 \leq i \leq N, \qquad (16)$$

and

$$W_{qN}^0(t) = \max_{1 \leq r \leq q} W_{rN}(t), \qquad 0 \leq t \leq 1. \qquad (17)$$

The weights $k(i,j)$, $\Delta W_{qN}(j/N)$, and the processes $W_{qN}$ are defined as in the omnibus case, cf. (2) and (8) to (11).

The trend test will **reject** the null hypothesis $\mathcal{H}_0^r$ $(F_1 = \cdots = F_k)$ at level $\alpha$ in favor of the trend alternative $\mathcal{A}_k^0$ $(F_1 \leq F_2 \leq \cdots \leq F_k, F_1 \neq F_k)$ if the value of the test statistic $S^0$ is larger than the upper $\alpha$-quantile $k_\alpha^0(\tau)$ of the conditional distribution of $S^0$ under $\mathcal{H}_0^\tau$ given the tie-lengths vector $\tau = (\tau_1, ..., \tau_d)$ of the ordered pooled sample, i.e. the trend test may be written in the form

$$\psi^0 \;=\; 1\Big( S^0 > k_\alpha^0(\tau) \Big) \tag{18}$$

where $k_\alpha^0(\tau)$ is defined by

$$P_{\mathcal{H}_0^\tau}\big\{ S^0 > k_\alpha^0(\tau) \mid \tau \big\} \;\leq\; \alpha \;<\; P_{\mathcal{H}_0^\tau}\big\{ S^0 \geq k_\alpha^0(\tau) \mid \tau \big\}. \tag{19}$$

The upper $\alpha$-quantiles $k_\alpha^0$ for the continuous model and $k = 3$ are given in Table 2.3.B of the Appendix.

The evaluation or simulation of the conditional critical values $k_\alpha^0(\tau)$ is completely similar to the omnibus case since the statistic $S^0$ is a functional of the rank processes $W_{qN}$ , $q = 1, ..., k$ , too.


**Numerical Example 2.3.1:**

- In order to have a simple example we choose $k = 3$ with sample sizes $n_1 = 10$ , $n_2 = 10$ , and $n_3 = 10$ , i.e. the total sample size is $N = n_1 + n_2 + n_3 = 30$ .

- The ordered observations of the first sample $(x_{11}, ..., x_{1n_1})$ are
  2.0, 2.0, 3.0, 3.0, 5.0, 6.0, 8.0, 8.0, 10.0, 11.0 .

- The ordered observations of the second sample $(x_{21}, ..., x_{2n_2})$ are
  0.0, 1.0, 2.0, 2.0, 3.0, 4.0, 7.0, 8.0, 9.0, 9.0 .

- The ordered observations of the third sample $(x_{31}, ..., x_{3n_3})$ are
  0.0, 0.0, 1.0, 2.0, 4.0, 4.0, 4.0, 7.0, 7.0, 8.0 .

- The number of different values among the pooled observations is $d = 12$ , and the different values of the ordered pooled observations are
  0.0, 1.0, 2.0, 3.0, 4.0, 5.0, 6.0, 7.0, 8.0, 9.0, 10.0, 11.0 .

- The ordered different values of the ranks are
  $T = (T_1, ..., T_d) = (3, 5, 10, 13, 17, 18, 19, 22, 26, 28, 29, 30)$.

- The ordered ranks of the three samples are
  $(10, 10, 13, 13, 18, 19, 26, 26, 29, 30)$ for the first sample,
  $(3, 5, 10, 10, 13, 17, 22, 26, 28, 28)$ for the second sample,
  $(3, 3, 5, 10, 17, 17, 17, 22, 22, 26)$ for the third sample.

- The lengths of the ties in the ordered pooled sample are
  $\tau = (\tau_1, ..., \tau_d) = (3, 2, 5, 3, 4, 1, 1, 3, 4, 2, 1, 1)$ .

- The lengths of the ties corresponding to the different samples are
  $(\tau_{11}, ..., \tau_{1d}) = (\, 0,\ 0,\ 2,\ 2,\ 0,\ 1,\ 1,\ 0,\ 2,\ 0,\ 1,\ 1\,)$ for the first sample,
  $(\tau_{21}, ..., \tau_{2d}) = (\, 1,\ 1,\ 2,\ 1,\ 1,\ 0,\ 0,\ 1,\ 1,\ 2,\ 0,\ 0\,)$ for the second sample,
  $(\tau_{31}, ..., \tau_{3d}) = (\, 2,\ 1,\ 1,\ 0,\ 3,\ 0,\ 0,\ 2,\ 1,\ 0,\ 0,\ 0\,)$ for the third sample.

- The defining values $W_{qN}(T_j/N) - W_{qN}(T_{j-1}/N) = \tau_{qj}/n_q - \tau_j/N$,
  $j = 1, ..., d$, of the rank processes $W_{qN}$ are
  $(\, -3,\ -2,\ 1,\ 3,\ -4,\ 2,\ 2,\ -3,\ 2,\ -2,\ 2,\ 2\,)/30$     for $q = 1$,
  $(\, 0,\ 1,\ 1,\ 0,\ -1,\ -1,\ -1,\ 0,\ -1,\ 4,\ -1,\ -1\,)/30$     for $q = 2$,
  $(\, 3,\ 1,\ -2,\ -3,\ 5,\ -1,\ -1,\ 3,\ -1,\ -2,\ -1,\ -1\,)/30$     for $q = 3$.

- The values of $\Delta W_{qN}(i/N)$, $i = 1, ..., N$, $q = 1, 2, 3$, are evaluated according to formula (6).

- The graphs of $W_{qN}$, $q = 1, 2, 3$, are given in Figure 2.3.a.


**Numerical Evaluation of the Omnibus Test:**

In the above Numerical Example 2.3.1 the evaluation of the omnibus statistic
(7) to (11) with the bandwidth $a = 0.40$ yields

$$s = S(r_1, ..., r_{30}) = 4.4531.$$

Since ties are present, there is no table with the corresponding exact conditional
critical value $k_\alpha(3, 2, 5, 3, 4, 1, 1, 3, 4, 2, 1, 1)$ but we may use a Monte Carlo
simulation program in order to get the actual (conditional) $p$-values:

A uniform random generator is used in order to simulate the vector of randomized ranks $R^* = (R_1^*, ..., R_N^*)$ under $\mathcal{H}_0^\tau$. Given the observed lengths of
ties $\tau = (3, 2, 5, 3, 4, 1, 1, 3, 4, 2, 1, 1)$ and the simulated vector $R^*$ the representation (14) is used in order to evaluate the (conditional) vector of ranks
$R = (R_1, ..., R_N)$ and the corresponding $S(R)$. With a Monte Carlo sample
size of 10,000 we get the following conditional $p$-values of the rank statistic
$S$,

$$P_{\mathcal{H}_0^\tau}\big\{ S > 4.4531 \mid \tau = (3, 2, 5, 3, 4, 1, 1, 3, 4, 2, 1, 1) \big\} = 0.613,$$

$$P_{\mathcal{H}_0^\tau}\big\{ S \geq 4.4531 \mid \tau = (3, 2, 5, 3, 4, 1, 1, 3, 4, 2, 1, 1) \big\} = 0.613.$$

This means that the omnibus test (18) will reject the null hypothesis of randomness $\mathcal{H}_0^\tau$ in favor of the omnibus alternative $\mathcal{A}_k$ exactly for those levels
$\alpha$ which are larger than or equal to the observed $p$-value. Since the observed
$p$-value of the example is $0.61$, the data will **not** reject the null hypothesis
for any reasonable level $\alpha$, e.g. for $\alpha = 0.10$.

## Figure 2.3

*Numerical Example 2.3.1: The graphs of $\left(W_{qN}(t),\ 0 \leq t \leq 1\right)$, $q = 1, 2, 3$.*

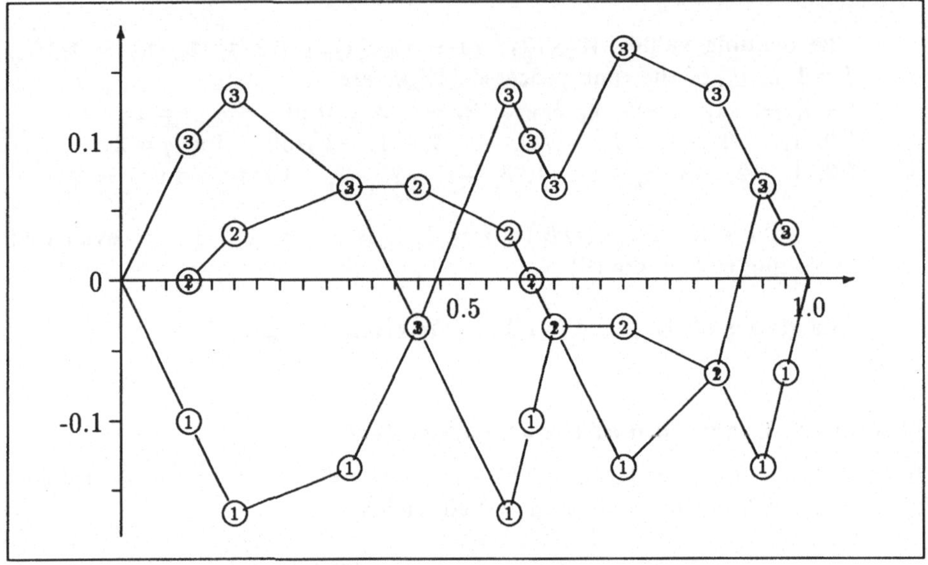

*Numerical Example 2.3.1: The graphs of $\left(W_{qN}^{0}(t),\ 0 \leq t \leq 1\right)$, $q = 1, 2, 3$.*

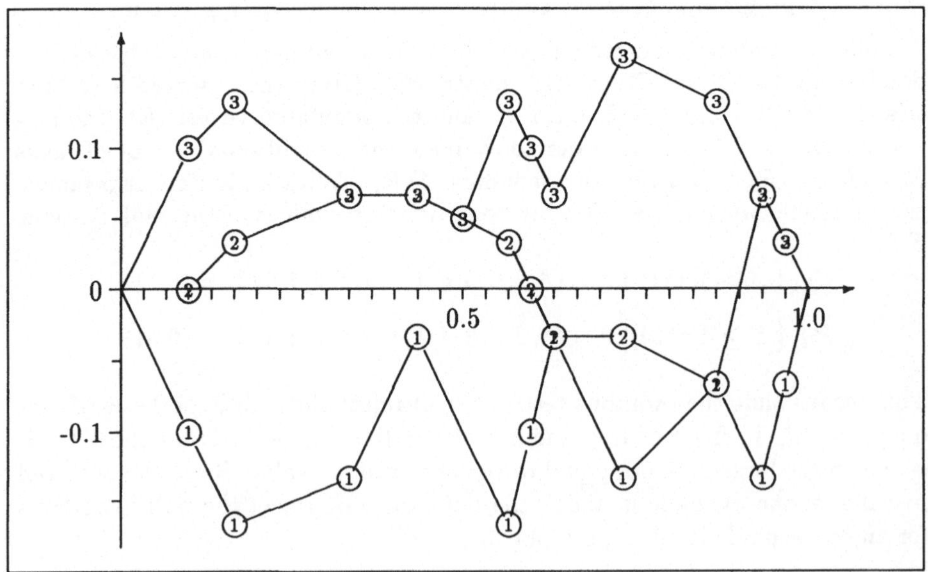

**Numerical Evaluation of the Trend Test:**

In the above Numerical Example 2.3.1 the evaluation of the trend statistic (15) to (17) with the bandwidth $a = 0.40$ yields

$$s^0 = S^0(r_1, ..., r_{30}) = 4.2062.$$

The graphs of the observed processes $W_{qN}^0$, $q = 1, 2, 3$, are given in Figure 2.3.b.

Since ties are present, there is no table with the corresponding exact conditional critical value $k_\alpha^0(3, 2, 5, 3, 4, 1, 1, 3, 4, 2, 1, 1)$, but we may use a Monte Carlo simulation program in order to get the actual (conditional) $p$-values, cf. the numerical evaluation of the omnibus test. With a Monte Carlo sample size of 10,000 we get

$$P_{\mathcal{H}_0^r}\left\{ S^0 > 4.2062 \mid \tau = (3, 2, 5, 3, 4, 1, 1, 3, 4, 2, 1, 1) \right\} = 0.165,$$

$$P_{\mathcal{H}_0^r}\left\{ S^0 \geq 4.2062 \mid \tau = (3, 2, 5, 3, 4, 1, 1, 3, 4, 2, 1, 1) \right\} = 0.165.$$

This means that the trend test will reject the null hypothesis of randomness $\mathcal{H}_0^r$ in favor of the trend alternative $\mathcal{A}_k^0$ exactly for those levels $\alpha$ which are larger than or equal to the observed $p$-value. Since the observed $p$-value of the example is $0.165$, the data will **not** reject the null hypothesis for any reasonable level $\alpha$, e.g. for $\alpha = 0.10$, but at least there is some indication of trend, cf. Figure 2.3.a and Figure 2.3.b.

**Some Remarks:**

Using the representation $F_q = H_N + N^{-1/2}B_{qN} \circ H_N$, $q = 1, ..., k$, of the underlying distribution functions $(F_1, ..., F_k)$ which is defined in (4.2.114) to (4.2.117), the rank process $W_{qN}$ is the natural estimator of the function $B_{qN}/\sqrt{N}$. Therefore the $k$ processes $W_{1N}, ..., W_{kN}$ will give some indication whether the null hypothesis $\mathcal{H}_0^r$ $(B_{1N} = B_{2N} = \cdots = B_{kN})$ may be true, or whether the data will support the one-sided alternative of trend $\mathcal{A}_k^0$ $(B_{1N} \leq B_{2N} \leq \cdots \leq B_{kN}, B_{1N} \neq B_{kN})$ or some other type of alternative.

Notice, the only difference between the omnibus statistic $S$ and the trend statistic $S^0$ is the substitution of the weights $\Delta W_{qN}(i/N)$ by the weights $\Delta W_{qN}^0(i/N)$. This substitution is reasonable, since the process $W_{qN}$ is an estimator of the underlying parameter $B_{qN}/\sqrt{N}$, and since the one-sided trend model is characterized by $B_{1N} \leq B_{2N} \leq \cdots \leq B_{kN}$, cf. (4.2.16) and (4.2.120), i.e. the rank processes $W_{qN}^0 = \max\{W_{rN} : 1 \leq r \leq q\}$, $q = 1, ...k$, may be viewed as a 'projection' of the original rank processes $(W_{1N}, ..., W_{kN})$ onto the one-sided trend model $B_{1N} \leq \cdots \leq B_{kN}$. In order to fulfil the additional side condition $n_1 B_{1N} + \cdots + n_k B_{kN} = 0$, cf. (4.2.11), we may center the processes $W_{qN}^0$, $q = 1, ..., k$, by $(n_1 W_{1N}^0 + \cdots + n_k W_{kN}^0)/N$ . If we do so, the corresponding trend statistic (15) doesn't change because of property (3). Therefore we'll use the simpler 'projection' (17).

**The Projection Rank Test of Trend:**

Here we propose a second nonlinear rank statistic $S^\tau$ for testing the null hypothesis of randomness $\mathcal{H}_0^r$ versus the one-sided alternative of decreasing trend $\mathcal{A}_k^0$ ( the first treatment is better than the second treatment, the second treatment is better than the third treatment, etc. ).

The basic idea is to replace the single score function of a classical linear rank statistic for testing decreasing trend by a finite number of sensibly chosen score functions $b_1, ..., b_r$ .

As discussed in Section 4.2 we propose $r = 3$ and choose the score functions $b_1$ , $b_2$ , $b_3$ as defined in formula (18) of Section 2.1 with corresponding integrals $B_\varrho(t) = \int_0^t b_\varrho(u)\, du$ , $\varrho = 1, 2, 3$ , given in Section 2.1 (19). This choice will take care of potential lower shift, central shift, upper shift, and also exact shift, especially in the generalized logistic shift model, where the Wilcoxon test is asymptotically optimal only for the exact shift alternatives, c.f. part C) of Section 3.2.

In a second step we design the expected relative amount of trend in the $k$ samples by the *regression constants* $\varrho_i$ , $1 \le i \le k$ , with the property

$$\varrho_1 \ge \varrho_2 \ge \cdots \ge \varrho_k, \qquad \varrho_1 \ne \varrho_k, \tag{20}$$

cf. formula (4.2.81). If no prior information is available we propose

$$\varrho_1 = k, \quad \varrho_2 = k - 1, \cdots, \varrho_k = 1, \tag{21}$$

which represents the expectation of equal amount of trend for all $k$ samples. With $0 = T_0 < T_1 < ... < T_d$ , $\tau_j = T_j - T_{j-1}$ , and $\tau_{qj}$ , $q = 1, ..., k$ , $j = 1, ..., d$ , as in formula (4) we define a vector $\vec{S} = (S_1, S_2, S_3)^T$ of linear rank statistics $S_\varrho$ , $\varrho = 1, 2, 3$ , according to

$$S_\varrho = \frac{\sqrt{N}}{\sigma} \sum_{j=1}^{d} \left( B_\varrho(\frac{T_j}{N}) - B_\varrho(\frac{T_{j-1}}{N}) \right) \sum_{i=1}^{k} n_i\, \varrho_i \left( W_{iN}(\frac{T_j}{N}) - W_{iN}(\frac{T_{j-1}}{N}) \right) \frac{1}{\tau_j}$$

$$= \frac{\sqrt{N}}{\sigma} \sum_{j=1}^{d} \left( B_\varrho(\frac{T_j}{N}) - B_\varrho(\frac{T_{j-1}}{N}) \right) \frac{1}{\tau_j} \sum_{i=1}^{k} n_i\, \varrho_i \left( \frac{\tau_{ij}}{n_i} - \frac{\tau_j}{N} \right) \tag{22}$$

with, cf. formulae (4.2.107) and (6),

$$\sigma = \Big( \sum_{i=1}^{k} \frac{n_i}{N}\, (\varrho_i - \bar{\varrho})^2 \Big)^{1/2}, \qquad \bar{\varrho} = \sum_{i=1}^{k} \frac{n_i}{N}\, \varrho_i . \tag{23}$$

Now the projection rank statistic $S^\pi$ for testing $\mathcal{H}_0^r$ versus the trend alternative $\mathcal{A}_k^0$ is defined as in Section 2.1 (23), but with the above definition (22) of $\vec{S}$ instead of Section 2.1 (20). Therefore the actual computation of the upper $\alpha$-quantile $k_\alpha^\pi(\tau_1, ..., \tau_d)$ of the conditional distribution of $S^\pi$ under $\mathcal{H}_0^r$ given the tie-lengths vector $\tau = (\tau_1, ..., \tau_d)$ of the ordered pooled sample, the form of the corresponding projection rank test of trend $\psi^\pi = 1\bigl(S^\pi > k_\alpha^\pi(\tau)\bigr)$, and the approximation

$$P_{\mathcal{H}_0^r}\bigl\{ S^\pi > t \,|\, \tau \bigr\} \approx w_1 P\{\chi_1^2 > t\} + w_2 P\{\chi_2^2 > t\} + w_3 P\{\chi_3^2 > t\} \qquad (24)$$

are literally the same as in Section 2.1 (21) to Section 2.1 (27). Especially the matrices $\Gamma$ and $\Gamma_J$ are defined by Section 2.1 (22), and the weigths $w_1$, $w_2$, $w_3$ are given in Section 2.1 (27). We omit an explicit repetition.

The upper $\alpha$-quantiles $k_\alpha^\pi$ for the continuous model and $k = 3$ are given in Table 2.3.C of the Appendix.

In the two-sample case $(k = 2)$ with regression constants (21) the respective statistics $S_\varrho$ defined in formula (22) and in Section 2.1 (20) coincide.

**Numerical Evaluation of the Projection Rank Test of Trend:**

In the above Numerical Example 2.3.1 the evaluation of the projection statistic $S^\pi$ yields

$$s^\pi = S^\pi(r_1, ..., r_{30}) = 2.0865\,.$$

Since ties are present there is no table with the corresponding exact conditional critical value $k_\alpha^\pi(3, 2, 5, 3, 4, 1, 1, 3, 4, 2, 1, 1)$ but we may use a Monte Carlo simulation program in order to get the actual (conditional) $p$-values, cf. the numerical evaluation of the omnibus test. With a Monte Carlo sample size of 10,000 we get

$$P_{\mathcal{H}_0^r}\bigl\{ S^\pi > 2.0865 \,|\, \tau = (3, 2, 5, 3, 4, 1, 1, 3, 4, 2, 1, 1) \bigr\} = 0.177\,,$$

$$P_{\mathcal{H}_0^r}\bigl\{ S^\pi \geq 2.0865 \,|\, \tau = (3, 2, 5, 3, 4, 1, 1, 3, 4, 2, 1, 1) \bigr\} = 0.177\,.$$

This means that the projection test will reject the null hypothesis of randomness $\mathcal{H}_0^r$ in favor of the trend alternative $\mathcal{A}_k^0$ exactly for those levels $\alpha$ which are larger than or equal to the observed $p$-value. Since the observed $p$-value of the example is $0.177$ the data will **not** reject the null hypothesis for any reasonable level $\alpha$, e.g. for $\alpha = 0.10$.

The approximate asymptotic computation of the $p$-value according to formula (24) yields

$$w_1 = 0.4694\,, \qquad w_2 = 0.2352\,, \qquad w_3 = 0.0306\,,$$

and

$$P_{\mathcal{H}_0^r}\bigl\{ S^\pi > 2.0865 \,|\, \tau = (3, 2, 5, 3, 4, 1, 1, 3, 4, 2, 1, 1) \bigr\} \approx 0.1696\,.$$

## 2.4   Several samples on the circle

In this section the comparison of $k$ treatments on the real line is carried over to samples on the circle, which will be represented by the interval $[0, 2\pi)$. The framework and the notations are the same as in the previous Section 2.3 with the following exceptions:

- All observations $X_{ij}$ take their values in $[0, 2\pi)$.

- Only the *omnibus problem* is treated, namely the problem of testing the null hypothesis of randomness,

$$\mathcal{H}_0^r : F_1 = F_2 = ... = F_k,$$

  versus the omnibus alternative that at least two treatments behave differently,

$$\mathcal{A}_k : F_q \neq F_r \quad \text{for some} \quad q, r \in \{1, ..., k\}.$$

- The kernel $K_a(s, t)$ from Section 2.3 (10) is replaced by the *periodic kernel* (4.3.15),

$$K_a(s,t) = \frac{1}{a}\left( K\big(\frac{t-s-1}{a}\big) + K\big(\frac{t-s}{a}\big) + K\big(\frac{t-s+1}{a}\big) \right), \quad (1)$$

  $s, t \in [0, 1]$, with fixed bandwidth $a = 0.40$ and Parzen-2 kernel $K$ given in Section 2.3 (11).

Then the test statistic $S$ for testing $\mathcal{H}_0^r$ versus $\mathcal{A}_k$ is defined as in Section 2.3 (7),

$$S = \sum_{i=1}^{N}\sum_{j=1}^{N} k(i,j) \sum_{q=1}^{k} n_q \, \Delta W_{qN}\big(\frac{i}{N}\big) \, \Delta W_{qN}\big(\frac{j}{N}\big), \quad (2)$$

with the weights

$$k(i,j) = K_a\Big(\frac{i-1/2}{N}, \frac{j-1/2}{N}\Big), \quad 1 \le i, j \le N, \quad (3)$$

evaluated from the periodic kernel (1).

Using the periodic kernel (1) instead of the kernel (10) from Section 2.3 makes the test statistic $S$ invariant under rotations of the circle, i.e. its value doesn't depend on the choice of the zero-position.

The upper $\alpha$-quantiles $k_\alpha$ for the continuous model and $k = 2$ are given in Table 2.4 of the Appendix.

**Numerical Example 2.4.1:**

For reasons of comparison let's consider the data of Example 2.1.1 as data on the circle. Since the zero-position and the scaling don't matter we may use the evaluation of the rank process $W_N$ given in Example 2.1.1. Notice, in the two-sample case $(n_1 = m, \ n_2 = n)$ we have $W_{1N} = W_N/(mN)$ and $W_{2N} = -W_N/(nN)$ . Therefore the evaluation of the rank statistic $S$ defined in (2) and (3) with the bandwidth $a = 0.40$ yields

$$s = S(r_1, ..., r_{16}) = 1.1490 .$$

(The corresponding value of Example 2.1.1 is $1.2515$ .)

Since ties are present, there is no table with the exact conditional critical value $k_\alpha(2, 3, 2, 2, 2, 1, 3, 1)$ but we may use a Monte Carlo simulation program in order to get the actual (conditional) $p$ –values, cf. the numerical evaluation of the omnibus test in Section 2.3. With a Monte Carlo sample size of 10,000 we get

$$P_{\mathcal{H}_0^r}\big\{ S > 1.1490 \mid \tau = (2, 3, 2, 2, 2, 1, 3, 1) \big\} = 0.734 ,$$

$$P_{\mathcal{H}_0^r}\big\{ S \geq 1.1490 \mid \tau = (2, 3, 2, 2, 2, 1, 3, 1) \big\} = 0.736 .$$

(The corresponding $p$ –values of Example 2.1.1 are $0.828$ and $0.830$ .)

This means that the test will reject the null hypothesis of randomness $\mathcal{H}_0^r$ in favor of the omnibus alternative $\mathcal{A}_2 : F_1 \neq F_2$ exactly for those levels $\alpha$ which are larger than or equal to the observed $p$ –value. Since the observed $p$ –value of the example is $0.734$ , the data will **not** reject the null hypothesis for any reasonable level $\alpha$ , e.g. for $\alpha = 0.10$ .

## 2.5   Two samples under type II censoring

In this section the two-sample model of Section 2.1 is extended to the case of censored data (type II censoring). The framework and the notations are the same as in Section 2.1. In case of type II censoring, however, we assume the potential observations $X_1, ..., X_N$ to be obtained sequentially in increasing order. Thus, if $X^{(1)} \leq X^{(2)} \leq ... \leq X^{(N)}$ denote the ordered values of $X_1, ..., X_N$, the first observed value is $X^{(1)}$, the second observed value is $X^{(2)}$, $\cdots$, but only those $X^{(i)}$ with $X^{(i)} \leq X^{(r)}$ will be observed in reality, where $r \in \{1, ..., N-1\}$ is the prefixed number of actual observations. The random variables $X_i$ with $X_i > X^{(r)}$ are *unobservable* and will be called *censored.*

Therefore, in addition to the assumptions of Section 2.1, we assume:

- For the given $r \in \{1, ..., N-1\}$ and for any $i = 1, ..., N$ the random variable $X_i$ will be observed only if $X_i \leq X^{(r)}$ holds true.

If all $X_i$ are different, there are exactly $r$ observable random variables. If ties are present, the number of observable random variables may be greater than $r$.

In any case we can use all the test statistics of Section 2.1 in the same way as in Section 2.1, if we consider the unobservable random variables as *tied to some value greater than* $X^{(r)}$, e.g. we (formally) replace all unobservable random variables by $X^{(r)} + 1$.

The upper $\alpha$-quantiles $k_\alpha$, $k_\alpha^0$, and $k_\alpha^\pi$ for the continuous model are given in Table 2.5.A to Table 2.5.C of the Appendix for different fractions of censored observations.

**Numerical Example 2.5.1:**

Let's use the observed values of Example 2.1.1, but under the additional assumption of type II censoring with $r = 10$. Therefore we have to consider all observations $X_i$ with $X_i > X^{(10)}$ as censored.

- The sample sizes ( observable + unobservable ) are $m = 7$ and $n = 9$, i.e. the total sample size is $N = 16$.

- Because of $x^{(10)} = 7.0$ the observations of the first sample are $(x_1, ..., x_7) = ( *,\ 6.0,\ 3.0,\ 6.0,\ *,\ 2.0,\ 7.0 )$, where $*$ means a censored observation. The ordered observed values of the first sample are $2.0,\ 3.0,\ 6.0,\ 6.0,\ 7.0$.

- Similarly the observations of the second sample are
  $(x_8, ..., x_{16}) = (\ast,\ 3.0,\ 3.0,\ 2.0,\ \ast,\ 5.0,\ 7.0,\ 5.0,\ \ast)$ and the ordered observable values of the second sample are $2.0,\ 3.0,\ 3.0,\ 5.0,\ 5.0,\ 7.0$ .

- The ordered pooled sample is
  $2.0,\ 2.0,\ 3.0,\ 3.0,\ 3.0,\ 5.0,\ 5.0,\ 6.0,\ 6.0,\ 7.0,\ 7.0,\ \ast,\ \ast,\ \ast,\ \ast,\ \ast$ .

- The vector of ranks is
  $(r_1, ..., r_{16}) = (\,16,\ 9,\ 5,\ 9,\ 16,\ 2,\ 11;\ 16,\ 5,\ 5,\ 2,\ 16,\ 7,\ 11,\ 7,\ 16\,)$ ,
  where the rank $N = 16$ has been assigned to all censored data $\ast$ .

- The number of different values among the ranks is $d = 6$ .

- The ordered different values among the ranks are
  $T = (T_1, ..., T_d) = (\,2,\ 5,\ 7,\ 9,\ 11,\ 16\,)$ .

- The lengths of the ties in the ordered pooled sample are
  $\tau = (\tau_1, ..., \tau_d) = (\,2,\ 3,\ 2,\ 2,\ 2,\ 5\,)$ .

- The lengths of the ties corresponding to the first and second sample are
  $\tau_1 = (\tau_{11}, ..., \tau_{1d}) = (\,1,\ 1,\ 0,\ 2,\ 1,\ 2\,)$ ,
  $\tau_2 = (\tau_{21}, ..., \tau_{2d}) = (\,1,\ 2,\ 2,\ 0,\ 1,\ 3\,)$ .

- The defining values $W_N(T_j/N) - W_N(T_{j-1}/N) = n\,\tau_{1j} - m\,\tau_{2j}$ ,
  $j = 1, ..., d$ , of the rank process $W_N$ are $2,\ -5,\ -14,\ 18,\ 2,\ -3$ .

- According to formula (5) of Section 2.1 the values of $\Delta W_N(i/N)$ ,
  $i = 1, ..., N$ , are $\quad 1,\ 1,\ -5/3,\ -5/3,\ -5/3,\ -7,\ -7,\ 9,\ 9,\ 1,\ 1,\ -3/5,$
  $-3/5,\ -3/5,\ -3/5,\ -3/5$ .

- Therefore the values $W_N(i/N)$ , $i = 0, 1, ..., N$ , are
  $0,\ 1,\ 2,\ 1/3,\ -4/3,\ -3,\ -10,\ -17,\ -8,\ 1,\ 2,\ 3,\ 12/5,\ 9/5,\ 6/5,\ 3/5,\ 0$ .

- The graph of $W_N$ is given in Figure 2.5.a.

- The evaluation of the omnibus statistic $S$ with the bandwidth $a = 0.40$ yields, cf. (6) to (9) of Section 2.1,

$$ s = S(r_1, ..., r_{16}) = 0.9549 . $$

- Since ties are present there is no table with the conditional critical value $k_\alpha(2, 3, 2, 2, 2, 5)$ but we utilize Section 2.1(12) in order to evaluate the actual (conditional) $p$-value by Monte Carlo simulation. With a Monte Carlo sample size of 10,000 we get

$$ P_{\mathcal{H}_0^\tau}\{\, S > 0.9549 \mid \tau = (2,3,2,2,2,5) \,\} = 0.804 , $$

$$ P_{\mathcal{H}_0^\tau}\{\, S \geq 0.9549 \mid \tau = (2,3,2,2,2,5) \,\} = 0.815 . $$

**Figure 2.5.a**

*Numerical Example 2.5.1: The graph of* $(W_N(t), 0 \le t \le 1)$ .

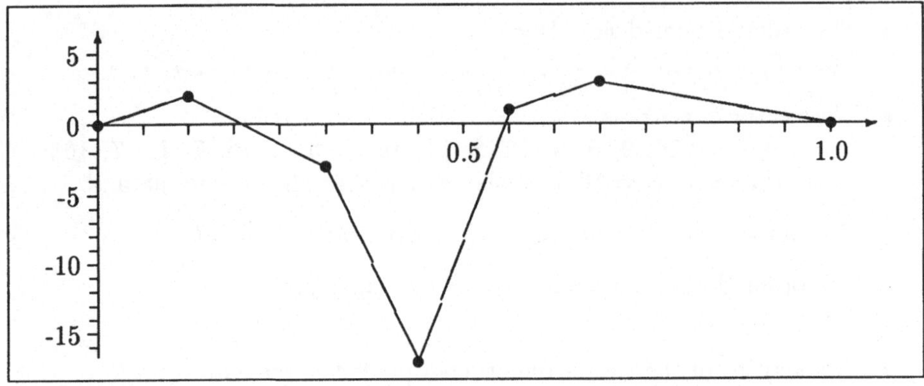

- This means that the omnibus test will reject the null hypothesis of randomness $\mathcal{H}_0^r : F = G$ in favor of the omnibus alternative $\mathcal{A}_2 : F \ne G$ exactly for those levels $\alpha$ which are larger than or equal to the observed $p$ –value. Since the observed $p$ –value of the example is $0.804$ the data will **not** reject the null hypothesis for any reasonable level $\alpha$ , e.g. for $\alpha = 0.10$ .

- The values of $W_N^0(i/N)$ , $i = 0, 1, ..., N$ , are
  $0, 0, 0, 0, -4/3, -3, -10, -17, -8, 0, 0, 0, 0, 0, 0, 0, 0$ .

- The graph of $W_N^0$ is given in Figure 2.5.b.

- The values of $\Delta W_N^0(i/N)$ , $i = 1, ..., N$ , are
  $0, 0, 0, -4/3, -5/3, -7, -7, 9, 8, 0, 0, 0, 0, 0, 0, 0$ .

- The evaluation of the one-sided rank statistic $S^0$ with the bandwidth $a = 0.40$ yields, cf. (13) and (14) of Section 2.1,

$$s^0 = S^0(r_1, ..., r_{16}) = 0.8075 .$$

- Since ties are present there is no table with the conditional critical value $k_\alpha^0(2, 3, 2, 2, 2, 5)$ but we utilize Section 2.1(12) in order to evaluate the actual (conditional) $p$ –value by Monte Carlo simulation. With a Monte Carlo sample size of 10,000 we get

$$P_{\mathcal{H}_0^r}\{ S^0 > 0.8075 \mid \tau = (2, 3, 2, 2, 2, 5) \} = 0.507 ,$$

$$P_{\mathcal{H}_0^r}\{ S^0 \ge 0.8075 \mid \tau = (2, 3, 2, 2, 2, 5) \} = 0.518 .$$

**Figure 2.5.b**

*Numerical Example 2.5.1: The graph of* $\left(W_N^0(t),\, 0 \le t \le 1\right)$.

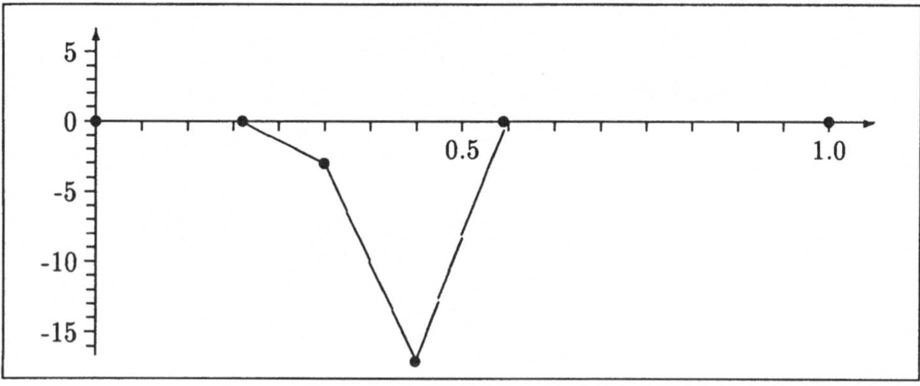

- This means that the one-sided rank test will reject the null hypothesis of randomness $\mathcal{H}_0^r : F = G$ in favor of the one-sided alternative $\mathcal{A}_2^0 :$ $F \le G, F \ne G$ exactly for those levels $\alpha$ which are larger than or equal to the observed $p$-value. Since the observed $p$-value of the example is 0.507 the data will **not** reject the null hypothesis for any reasonable level $\alpha$, e.g. for $\alpha = 0.10$.

- The evaluation of the projection rank statistic $S^\pi$ yields

$$s^\pi = S^\pi(r_1, ..., r_{16}) = 0.2548.$$

- Since ties are present there is no table with the conditional critical value $k_\alpha^\pi(2,3,2,2,2,5)$, but similar to the evaluation of the conditional $p$-value of the omnibus statistic we use a Monte Carlo simulation program in order to get the actual (conditional) $p$-value of $S^\pi$. With a Monte Carlo sample size of 10,000 we get

$$P_{\mathcal{H}_0^r}\left\{ S^\pi > 0.2548 \mid \tau = (2,3,2,2,2,5) \right\} = 0.521,$$

$$P_{\mathcal{H}_0^r}\left\{ S^\pi \ge 0.2548 \mid \tau = (2,3,2,2,2,5) \right\} = 0.543.$$

- This means that the projection rank test will reject the null hypothesis of randomness $\mathcal{H}_0^r : F = G$ in favor of the one-sided alternative $\mathcal{A}_2^0 :$ $F \le G, F \ne G$ exactly for those levels $\alpha$ which are larger than or equal to the observed $p$-value. Since the observed $p$-value of the example is 0.521 the data will not reject the null hypothesis for any reasonable level $\alpha$, e.g. for $\alpha = 0.10$.

- The approximate asymptotic computation of the $p$-value according to Section 2.1(26) yields

$$w_1 = 0.4817, \qquad w_2 = 0.2044, \qquad w_3 = 0.0183,$$

and

$$P_{\mathcal{H}_0^r}\left\{ S^\pi > 0.2548 \mid \tau = (2,3,2,2,2,5) \right\} \approx 0.4933.$$

## 2.6  The hypothesis of symmetry

As in 2.1 the basic problem is the comparison of two treatments, but in contrast to 2.1 rather than to have two independent samples we have paired observations $(Y_i, Z_i), i = 1, ..., n$, where the $Y$ - components are the measurements under the standard treatment whereas the $Z$ -components are the measurements under the new treatment. If the measurements are taken at matched pairs, the components $Y_i$ and $Z_i$ of $(Y_i, Z_i)$ are usually not stochastically independent. Then the equality of the two treatments may be modelled by the assumption

$$\mathcal{L}(Y_i, Z_i) = \mathcal{L}(Z_i, Y_i), \qquad 1 \leq i \leq n \tag{1}$$

implying the *symmetry* about zero of the differences $X_i = Z_i - Y_i$ . If the new treatment is better (produces larger measurements) than the standard treatment, the distribution of $X_i$ should be stochastically larger than the distribution of $-X_i$ . Thus, the comparison is based on i.i.d. real random variables $X_1, ..., X_n$ .

**Data:**

- The random sample $X_1, ..., X_n$ of size $n$ .

There is no restriction with respect to tied observations, i.e., the underlying model may allow for discontinuities or even may be purely discrete. In case of tied observations the tables of critical values are not valid, but there are simulation programs for evaluating the $p$ -values of the proposed tests.

All proposed tests well be *rank tests* in the following sense. They are based on the vector of ranks $R^+ = (R_1^+, ..., R_n^+)$ of the absolute values of $|X_1|, ..., |X_n|$ , i.e.

$$R_i^+ = \# \left\{ j \in \{1, ..., n\} \mid |X_j| \leq |X_i| \right\}, \qquad 1 \leq i \leq n, \tag{2}$$

and on the sign-vector $\operatorname{sign}(X) = (\operatorname{sign}(X_1), ..., \operatorname{sign}(X_n))$ with

$$\operatorname{sign}(X_i) = \begin{cases} -1, & \text{if} \quad X_i < 0, \\ 0, & \text{if} \quad X_i = 0, \qquad 1 \leq i \leq n. \\ 1, & \text{if} \quad X_i > 0. \end{cases} \tag{3}$$

Notice, the rank of $|X_i|$ counts all observations of the sample $(|X_1|, ..., |X_n|)$ of absolute values which are less than or equal to $|X_i|$ , also in case of tied observations.

**Model Assumptions:**

- The underlying (cumulative) distribution function $F$ of the random variables is an arbitrary distribution function on the real line, i.e., $F$ may have discontinuities.

- The measurement scale is the real line or at least ordinal with the possibility to compare with zero.

**Testing Problems:**

- *One-sided problem:* Testing the null hypothesis of symmetry

$$\mathcal{H}_0^s : \ F = F_-$$

where $F_-$ denotes the distribution function of $-X_i$ , versus the one-sided alternative of positive unsymmetry

$$\mathcal{A}_1^0 : \ F \leq F_- \, , F \neq F_-$$

corresponding to the case that the first treatment (the $Z$ 's) produces stochastically larger values than the second treatment (the $Y$ 's).

- *Omnibus problem:* Testing the null-hypothesis of symmetry

$$\mathcal{H}_0^s : \ F = F_-$$

versus the omnibus alternative that the two treatments (the $Z$ 's and $Y$ 's) behave differently,

$$\mathcal{A}_1 : \ F \neq F_- \, .$$

**The One-Sample Rank Process:**

Essential for the data inspection and for the definition of the test statistics is the following rank process $W_n = (W_n(t) : 0 \leq t \leq 1)$ :

Let $d$ denote the number of different *non-zero* values among the absolute observations $|X_1|, ..., |X_n|$ . In the continuous model (no ties, no zeros) we have $d = n$ . In the general model (ties and zeros are possible) $d$ is a random variable.

Let $T_0$ denote the number of zeros among $X_1, ..., X_n$ and let $T_1 < ... < T_d$ denote the different values of those ranks $R_1^+, ..., R_n^+$ which correspond to non-zero observations. If no ties and no zeros are present, we have $T_i = i$ for

$i = 1, ..., n$ . If all observations are zero, we have $d = 0$ and $T_0 = n$ . In any case we have $0 \leq d \leq n$ and $T_d = n$ .

Using $d$ and $T = (T_0, ..., T_d)$ we define the rank process $W_n$ according to

$$W_n(\frac{T_j}{n}) = -\sum_{i=1}^{n} 1(R_i^+ > T_j)\,\text{sign}(X_i), \qquad j = 0, 1, ..., d,$$

$$W_n(0) = W_n(\frac{T_0}{n}),$$

$$(4)$$

and according to linear interpolation between the points $(0, W_n(0))$ and $(T_j, W_n(T_j/n))$ , $j = 0, 1, ..., d$ , in the $(x, y)$ –plane, cf. Section 5.4.

Especially we have the equalities $W_n(0) = W_n(T_0/n) = -\sum_{j=1}^{n} \text{sign}(X_j)$ and $W_n(1) = 0$ .

For actual computation of $W_n$ we use an easier representation, c.f. (5.4.92) to (5.4.95):

Put $\tau^+ = (\tau_1^+, ..., \tau_d^+)$ , $\tau^- = (\tau_1^-, ..., \tau_d^-)$ and $\tau = \tau^+ + \tau^-$ with $\tau_j^+$ resp. $\tau_j^-$ the number of positive resp. negative observations with rank of their absolute value equal to $T_j$ , $j = 1, ..., d$ , i.e.

$$\tau_j^+ = \#\{\, i : X_i > 0,\ R_i^+ = T_j \,\},$$

$$\tau_j^- = \#\{\, i : X_i < 0,\ R_i^+ = T_j \,\},$$

$$(5)$$

and define

$$\Delta W_n(\frac{i}{n}) = 0, \qquad\qquad \text{if } 1 \leq i \leq T_0,$$

$$\Delta W_n(\frac{i}{n}) = W_n(\frac{i}{n}) - W_n(\frac{i-1}{n})$$

$$= \frac{\tau_j^+ - \tau_j^-}{\tau_j}, \qquad \text{if } T_{j-1} < i \leq T_j,\ j = 1, ..., d.$$

$$(6)$$

Then, because of $W_n(1) = 0$ or

$$W_n(0) = -\sum_{i=1}^{n} \text{sign}(X_j) = \sum_{i=1}^{d} (\tau_j^- - \tau_j^+), \qquad (7)$$

the process $(W_n(t), 0 \leq t \leq 1)$ is completely determined by (6) and by linear interpolation.

**The Omnibus Test:**

The test statistic $S$ for testing the null hypothesis of symmetry $\mathcal{H}_0^s$ versus the omnibus alternative $\mathcal{A}_1$ is, c.f. formula (5.4.83),

$$S = \frac{1}{n} \sum_{i=1}^{n} \sum_{j=1}^{n} \Delta W_n(\tfrac{i}{n})\, \Delta W_n(\tfrac{j}{n})\, k(i,j)\,, \tag{8}$$

where

$$k(i,j) = K_a\left(\frac{i-1/2}{n}, \frac{j-1/2}{n}\right), \qquad 1 \le i,j \le n, \tag{9}$$

and

$$K_a(s,t) = \frac{1}{a}\left( K(\frac{t+s}{a}) + K(\frac{t-s}{a}) + K(\frac{t+s-2}{a}) \right), \quad s,t \in [0,1], \tag{10}$$

with fixed bandwidth $a = 0.40$ and Parzen-2 kernel

$$K(t) = \begin{cases} 4/3 - 8\,|t|^2 + 8\,|t|^3, & \text{if} \quad |t| < 1/2\,, \\[2mm] 8\,(1-|t|)^3/3\,, & \text{if} \quad 1/2 \le |t| < 1\,, \\[2mm] 0\,, & \text{if} \quad |t| \ge 1\,. \end{cases} \tag{11}$$

The omnibus test will reject the null hypothesis of symmetry $\mathcal{H}_0^s$ at level $\alpha$ in favor of the omnibus alternative $\mathcal{A}_1$, if the value of the test statistic $S$ is larger than the upper $\alpha$–quantile $k_\alpha(\tau)$ of the conditional distribution of $S$ under $\mathcal{H}_0^s$ given the tie-lengths vector $\tau = (\tau_1, ..., \tau_d)$, i.e. the omnibus test may be written in the form

$$\psi = 1\Big( S > k_\alpha(\tau) \Big)\,, \tag{12}$$

where $k_\alpha(\tau)$ is defined by

$$P_{\mathcal{H}_0^s}\big\{ S > k_\alpha(\tau) \mid \tau \big\} \le \alpha < P_{\mathcal{H}_0^s}\big\{ S \ge k_\alpha(\tau) \mid \tau \big\}\,. \tag{13}$$

The upper $\alpha$–quantiles $k_\alpha$ for the continuous model are given in Table 2.6.A of the Appendix.


**Simulation of the conditional $p$–values:**

In principle there is no difficulty to evaluate the conditional critical values $k_\alpha(\tau)$ or the conditional $p$–values:

The conditional distribution of the random vectors $R^+ = (R_1^+, ..., R_n^+)$ and $\mathrm{sign}(X) = (\mathrm{sign}(X_1), ..., \mathrm{sign}(X_n))$ given the tie-lengths vector $\tau = (\tau_1, ..., \tau_d)$ is the same for any null hypothesis point $F = F_- \in \mathcal{H}_0^s$, c.f. Proposition 5.4.1.

Since the case $T_0 = n$ is equivalent to the degenerate case $X_1 = ... = X_n = 0$, we'll restrict the further discussion to the case $T_0 < n$.

Given the observed $\tau = (\tau_1, ... \tau_d)$ we define $T_j = T_0 + \tau_1 + ... + \tau_j$ and

$$\tau_j^+ = \# \left\{ i \in \{1, ..., n\} : T_{j-1} < i \leq T_j, \ \Delta_i = 1 \right\},$$

$$\tau_j^- = \tau_j - \tau_j^+, \qquad j = 1, ..., d, \tag{14}$$

where $\Delta_1, ..., \Delta_n$ are independent random variables with $P\{\Delta_i = 1\} = P\{\Delta_i = 0\} = 1/2$, $1 \leq i \leq n$. Then we combine formulae (6) and (8) for computing a statistic $S$, which is distributed according to the above conditional distribution.

For exact evaluation we enumerate all values of $(\Delta_1, ..., \Delta_n) \in \{0, 1\}^n$, while for Monte Carlo simulation we use a random number generator for procducing the values of $(\Delta_1, ..., \Delta_n)$.

**The One-sided test:**

The test statistic $S^0$ for testing the null hypothesis of symmetry $\mathcal{H}_0^s$ versus the one-sided alternative $\mathcal{A}_1^0 : F \leq F_-, F \neq F_-$ is, c.f. formula (5.4.85),

$$S^0 = \frac{1}{n} \sum_{i=1}^{n} \sum_{j=1}^{n} \Delta W_n^0(\tfrac{i}{n}) \, \Delta W_n(\tfrac{j}{n}) \, k(i,j), \tag{15}$$

where

$$\Delta W_n^0(\tfrac{i}{n}) = W_n^0(\tfrac{i}{n}) - W_n^0(\tfrac{i-1}{n}), \qquad 1 \leq i \leq n, \tag{16}$$

and

$$W_n^0(t) = \min\left(0, W_n(t)\right) \qquad 0 \leq t \leq 1. \tag{17}$$

The weights $k(i,j)$, $\Delta W_n(\tfrac{i}{n})$, and the process $W_n$ are defined as in the omnibus case, c.f. (6) and (9).

The one-sided test will reject the null hypothesis of symmetry $\mathcal{H}_0^s$ at level $\alpha$ in favor of the one-sided alternative $\mathcal{A}_1^0$, if the value of the test statistic $S^0$ is larger than the upper $\alpha$–quantile $k_\alpha^0(\tau)$ of the conditional distribution of $S^0$ under $\mathcal{H}_0^s$ given the tie-lengths vector $\tau = (\tau_1, ..., \tau_d)$, i.e. the one-sided test may be written in the form

$$\psi^0 = 1\left(S^0 > k_\alpha^0(\tau)\right), \tag{18}$$

where $k_\alpha^0(\tau)$ is defined by

$$P_{\mathcal{H}_0^s}\left\{ S^0 > k_\alpha^0(\tau) \mid \tau \right\} \leq \alpha < P_{\mathcal{H}_0^s}\left\{ S^0 \geq k_\alpha^0(\tau) \mid \tau \right\}. \tag{19}$$

The upper $\alpha$–quantiles $k_\alpha^0$ for the continuous model are given in Table 2.6.B of the Appendix.

The evaluation or simulation of the conditional critical values $k_\alpha^0(\tau)$ is completely similar to the omnibus case, since the statistic $S^0$ is a function of the one-sample rank process $W_n$ , too.

Notice, the only difference between the omnibus statistic $S$ and the one-sided statistic $S^0$ is the substitution of the weights $\Delta W_n(i/n)$ by the weights $\Delta W_n^0(i/n)$ . This substitution is reasonable since the process $W_n/n$ is an estimator of the underlying parameter $\tilde{B}$ , $\tilde{B}(t) = 2\,B(1/2+t/2)$ , $0 \le t \le 1$ , c.f. formula (5.2.6), with $B \in \mathcal{B}_n^s$ , c.f. formula (5.1.20), and since the one sided model is characterized by $B \le 0$ , c.f. formula (5.1.25), i.e. the rank process $W_n^0 = W_n 1(W_n < 0)$ is a 'projection' of the original rank process $W_n$ onto the one-sided model $B \le 0$ .

**Numerical Example 2.6.1:**

- The sample size is $n = 16$ .

- The observations $x_1, ..., x_{16}$ are
  -6.9, 7.4, 2.6, 4.1, -2.6, -1.3, 0.0, 9.1, -8.9, -8.1, 7.1, 7.4, 8.1, -8.9, 4.1, 0.0

- The observations ordered to increasing values of their absolute values are
  0.0, 0.0, -1.3, -2.6, 2.6, 4.1, 4.1, -6.9, 7.1, 7.4, 7.4, -8.1, 8.1, -8.9, -8.9, 9.1

- The number of zeros is $T_0 = 2$ .

- The number of different non-zero values among the absolute observations is $d = 9$ .

- The values of $T_1, ..., T_d$ are  3 , 5 , 7 , 8 , 9 , 11 , 13 , 15 , 16 .

- The vectors of ties corresponding to strictly positive resp. strictly negative resp. non-zero observations among the ordered absolute values are

$$\tau^+ = (\tau_1^+, ..., \tau_d^+) = (\,0,\ 1,\ 2,\ 0,\ 1,\ 2,\ 1,\ 0,\ 1\,),$$

$$\tau^- = (\tau_1^-, ..., \tau_d^-) = (\,1,\ 1,\ 0,\ 1,\ 0,\ 0,\ 1,\ 2,\ 0\,),$$

$$\tau = (\tau_1, ..., \tau_d) = (\,1,\ 2,\ 2,\ 1,\ 1,\ 2,\ 2,\ 2,\ 1\,).$$

- The vector of differences $\Delta W_n(i/n) = W_n(i/n) - W_n((i-1)/n)$ , $i = 1, ..., 16$ , is
  $(\,0,\ 0,\ -1,\ 0,\ 0,\ 1,\ 1,\ -1,\ 1,\ 1,\ 1,\ 0,\ 0,\ -1,\ -1,\ 1\,)$ .

- The initial value of the rank process is $W_n(0) = -2$ .

**Figure 2.6.a**

*Numerical Example 2.6.1: The graph of* $(W_n(t), 0 \le t \le 1)$.

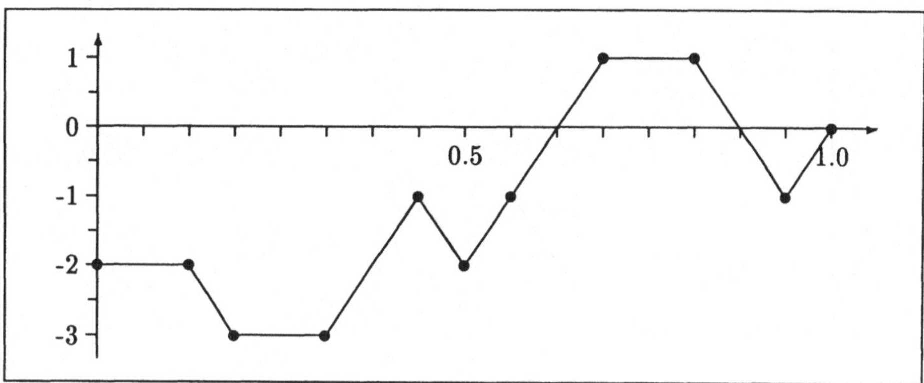

- The values of $W_n(0)$ and $W_n(\frac{i}{n}) = W_n(0) + \Delta W_n(\frac{1}{n}) + \cdots + \Delta W_n(\frac{i}{n})$ for $i = 1, ..., n$ are
$(-2, \ -2, \ -2, \ -3, \ -3, \ -3, \ -2, \ -1, \ -2, \ -1, \ 0, \ 1, \ 1, \ 1, \ 0, \ -1, \ 0)$.

- The graph $W_n$ is given in Figure 2.6.a.

- Using (8) to (11) with bandwidth $a = 0.40$ the value $s$ of the omnibus statistic $S$ is evaluated as

$$s \ = \ S\big(r_1^+, ..., r_{16}^+, \, \text{sign}(x_1), ..., \text{sign}(x_{16})\big) \ = \ 1.9638\,.$$

- Since ties and zeros are present, there is no table of the exact (conditional) critical value $k_\alpha(1, 2, 2, 1, 1, 2, 2, 2, 1)$ but we utilize formula (14) in order to evaluate the actual (conditional) $p$-values by Monte Carlo simulation. With a Monte Carlo sample size of 10,000 we get

$$P_{\mathcal{H}_0^s}\big\{ S > 1.9638 \mid \tau = (1, 2, 2, 1, 1, 2, 2, 2, 1) \big\} \ = \ 0.679\,,$$

$$P_{\mathcal{H}_0^s}\big\{ S \ge 1.9638 \mid \tau = (1, 2, 2, 1, 1, 2, 2, 2, 1) \big\} \ = \ 0.679\,.$$

- This means that the omnibus test (12) will reject the null hypothesis of symmetry $\mathcal{H}_0^s$ in favor of the omnibus alternative $\mathcal{A}_1$ exactly for those levels $\alpha$ which are larger than or equal to the observed $p$-value $0.68$. Thus, the present data will **not** reject the null hypothesis for any reasonable level $\alpha$, e.g. for $\alpha = 0.10$.

- The values of $(W_n^0(i/n), \, i = 0, 1, ..., n)$ are
$(-2, \ -2, \ -2, \ -3, \ -3, \ -3, \ -2, \ -1, \ -2, \ -1, \ 0, \ 0, \ 0, \ 0, \ 0, \ -1, \ 0)$.

**Figure 2.6.b**

*Numerical Example 2.6.1: The graph of* $\left(W_n^0(t),\, 0 \le t \le 1\right)$.

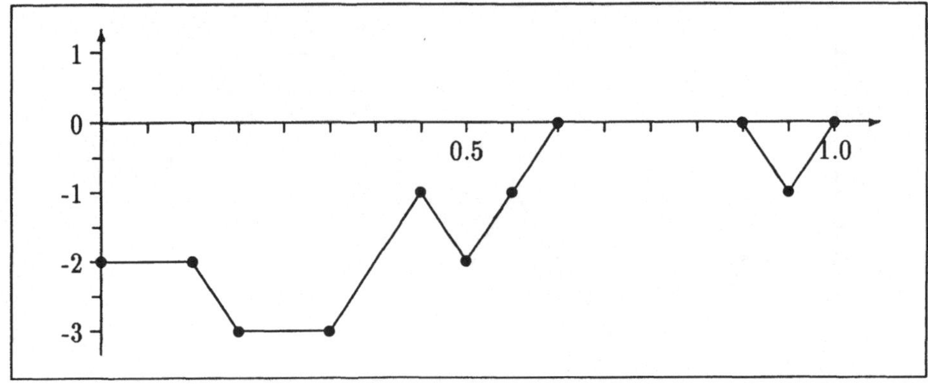


- The graph of $W_n^0$ is given in Figure 2.6.b.

- The values of $(\Delta W_n^0(i/n),\, i = 1, ..., n)$ are
  $(0,\ 0,\ -1,\ 0,\ 0,\ 1,\ 1,\ -1,\ 1,\ 1,\ 0,\ 0,\ 0,\ 0,\ -1,\ 1)$.

- Using (15) to (17) with the bandwidth $a = 0.40$ we evaluate the value
  $s^0$ of the one-sided statistic $S^0$ as

$$s^0 \;=\; S^0\left(r_1^+, ..., r_{16}^+,\, \mathrm{sign}(x_1), ..., \mathrm{sign}(x_{16})\right) \;=\; 1.4228\,.$$

- Since ties and zeros are present, there is no table with the corresponding
  exact conditional critical value $k_\alpha^0(1,2,2,1,1,2,2,2,1)$ but we utilize
  formula (14) in order to evaluate the actual (conditional) $p$–values by
  Monte Carlo simulation. With a Monte Carlo sample size of 10,000 we
  get

$$P_{\mathcal{H}_0^s}\left\{ S^0 > 1.4228 \,|\, \tau = (1,2,2,1,1,2,2,2,1) \right\} \;=\; 0.383\,,$$

$$P_{\mathcal{H}_0^s}\left\{ S^0 \ge 1.4228 \,|\, \tau = (1,2,2,1,1,2,2,2,1) \right\} \;=\; 0.383\,.$$

- This means that the one-sided test will reject the null hypothesis of sym-
  metry $\mathcal{H}_0^s$ in favor of the one-sided alternative $\mathcal{A}_1^0$ (first treatment better than second treatment) exactly for those levels $\alpha$ which are larger
  than or equal to the observed $p$–value 0.38 . Thus, the present data
  will **not** reject the null hypothesis for any reasonable level $\alpha$ , e.g. for
  $\alpha = 0.10$ .

**The Projection Rank Test:**

Here we propose a second nonlinear rank statistic $S^\pi$ for testing the null hypothesis of symmetry $\mathcal{H}_0^s$ versus the one-sided alternative of positive unsymmetry $\mathcal{A}_1^0$ corresponding to the case that the first treatment (the $Z$ 's ) produce stochastically larger values than the second treatment (the $Y$ 's ).

The basic idea is to replace the single score function $\tilde{b}$ of a classical linear rank statistic by a finite number of sensibly chosen score functions $\tilde{b}_1, ..., \tilde{b}_r$ . As discussed in Section 5.3, we propose $r = 2$ and

$$\tilde{b}_1(u) = u, \qquad \tilde{b}_2(u) = 2u(1-u^2), \qquad 0 \le u \le 1. \tag{20}$$

The motivation is that the score functions $\tilde{b}_1$ and $\tilde{b}_2$ will take care of central shift and exact shift in a situation where the Wilcoxon test is asymptotically optimal for the exact shift only, c.f. Section 5.3.

Defining $\tilde{B}_j(t) = \int_0^t \tilde{b}_j(u)\, du$ , $0 \le t \le 1$ , $j = 1, 2$ , we get

$$\tilde{B}_1(t) = t^2/2, \qquad \tilde{B}_2(t) = t^2 - t^4/2, \qquad 0 \le t \le 1. \tag{21}$$

With $T_0 < T_1 < ... < T_d = n$ , $\tau_j = T_j - T_{j-1}$ , and $\tau_j^+$ , $\tau_j^-$ , $j = 1, ..., d$ , as in (5) we define the vector $\vec{S} = (S_1, S_2)^T$ of averaged scores linear rank statistics $S_1$ , $S_2$ by

$$S_\varrho = \frac{1}{\sqrt{n}} \sum_{i=1}^{n} b_{\varrho n}^\tau(R_i^+)\, \mathrm{sign}(X_i)$$

$$= \sqrt{n} \sum_{j=1}^{d} \left( \tilde{B}_\varrho(\frac{T_j}{n}) - \tilde{B}_\varrho(\frac{T_{j-1}}{n}) \right) \frac{\tau_j^+ - \tau_j^-}{\tau_j}. \tag{22}$$

In the next step we'll define the $2 \times 2$ –matrix

$$\Gamma = \left( \gamma_{\varrho\sigma} \right)_{\varrho,\sigma=1,2} \tag{23}$$

according to ( $\forall\, \varrho, \sigma = 1, 2$ )

$$\gamma_{\varrho\sigma} = n \sum_{j=1}^{d} \left( \tilde{B}_\varrho(\frac{T_j}{n}) - \tilde{B}_\varrho(\frac{T_{j-1}}{n}) \right) \left( \tilde{B}_\sigma(\frac{T_j}{n}) - \tilde{B}_\sigma(\frac{T_{j-1}}{n}) \right) \frac{1}{\tau_j}. \tag{24}$$

Then the proposed projection rank statistic $S^\pi$ is defined by

$$S^\pi = \max\left\{ \vec{S}^T \Gamma^{-1} \vec{S}\, 1(\Gamma^{-1}\vec{S} \ge 0),\ \frac{S_1^2}{\gamma_{11}} 1(S_1 \ge 0),\ \frac{S_2^2}{\gamma_{22}} 1(S_2 \ge 0) \right\}. \tag{25}$$

Of course the definition (25) requires a nonsingular $\Gamma$ , otherwise $S^\pi$ is not defined.

The projection rank test will reject the null hypothesis $\mathcal{H}_0^s : F = F_-$ at level $\alpha$ in favor of the one-sided alternative $\mathcal{A}_1^0 : F \leq F_-, F \neq F_-$ , if the value of the test statistic $S^\pi$ is larger than the upper $\alpha$ –quantile $k_\alpha^\pi(\tau)$ of the conditional distribution of $S^\pi$ under $\mathcal{H}^s$ given the tie-lengths vector $\tau = (\tau_1, ..., \tau_d)$ , i.e. the projection rank test may be written in the form

$$\psi^\pi = 1\left( S^\pi > k_\alpha^\pi(\tau) \right), \tag{26}$$

where $k_\alpha^\pi(\tau)$ is defined by

$$P_{\mathcal{H}_0^s}\left\{ S^\pi > k_\alpha^\pi(\tau) \mid \tau \right\} \leq \alpha < P_{\mathcal{H}_0^s}\left\{ S^\pi \geq k_\alpha^\pi(\tau) \mid \tau \right\}. \tag{27}$$

The upper $\alpha$ –quantiles $k_\alpha^\pi$ for the continuous model are given in Table 2.6.C of the Appendix.

In the Numerical Example 2.6.1 we evaluate the value $S^\pi$ of the projection rank statistic $S^\pi$ as

$$s^\pi = S^\pi\left( r_1^+, ..., r_{16}^+, \operatorname{sign}(X_1), ..., \operatorname{sign}(X_{16}) \right) = 0.7013.$$

Since ties are present, there is no table with the exact conditional critical value $k_\alpha^\pi(1, 2, 2, 1, 1, 2, 2, 2, 1)$ , but similar to the evaluation of the conditional $p$ –values of the omnibus statistic (8), c.f. representation (14) , we may use a Monte Carlo simulation programm in order to get the actual (conditional) $p$ –values of $S^\pi$ . With a Monte Carlo sample size of 10.000 we get

$$P_{\mathcal{H}_0^s}\left\{ S^\pi > 0.7013 \mid \tau = (1, 2, 2, 1, 1, 2, 2, 2, 1) \right\} = 0.277,$$

$$P_{\mathcal{H}_0^s}\left\{ S^\tau \geq 0.7013 \mid \tau = (1, 2, 2, 1, 1, 2, 2, 2, 1) \right\} = 0.278.$$

This means that the projection rank test (26) will reject the null hypothesis of symmetry $\mathcal{H}_0^s : F = F_-$ in favor of the one-sided alternative $\mathcal{A}_1^0 : F \leq F_-, F \neq F_-$ exactly for those levels $\alpha$ which are larger than or equal the observed $p$ –value. Since the observed $p$ –value of the example is $0.28$ , the data will not reject the null hypothesis for any reasonable level $\alpha$ , e.g. for $\alpha = 0.10$ .

An approximate asymptotic computation of the $p$ –value is provided by the following formula:

For $t \geq 0$ we get from Theorem 3.2.7, (3.2.95), and (5.4.112) the asymptotic approximation

$$P_{\mathcal{H}_0^s}\left\{ S^\pi > t \mid \tau \right\} \approx \frac{1}{2} P\left\{ \chi_1^2 > t \right\} + w P\left\{ \chi_2^2 > t \right\}, \tag{28}$$

where $\chi^2_\varrho$, $\varrho = 1, 2$, denote random variables with respective $\chi^2_\varrho$ distributions, and where

$$w = \frac{1}{2} - \frac{1}{2\pi} \arccos\left(\frac{-\gamma_{12}}{\sqrt{\gamma_{11}\,\gamma_{22}}}\right).\tag{29}$$

In the above numerical example we get $w = 0.0906$. Therefore the approximate observed (conditional) $p$-value computed from (28) and (29) with $t = s^\pi = 0.7013$ is

$$P_{\mathcal{H}_0}\left\{S^\pi > 0.7013 \mid \tau = (1, 2, 2, 1, 1, 2, 2, 2, 1)\right\} \approx 0.265.$$

## 2.7   The hypothesis of independence

The problem is to test whether the components of a two-dimensional random variable are independent. We'll discuss testing procedures for testing the null hypothesis of independence versus the alternative of positive quadrant dependence or versus the omnibus alternative of dependence.

**Data:**

- The random sample $X_1 = (Y_1, Z_1), ..., X_n = (Y_n, Z_n)$ of size $n$ .

There is no restriction with respect to tied observations, i.e. the underlying model may allow for discontinuities or even may be purely discrete. In case of tied observations the given tables of critical values are not valid, but there are simulation programs for evaluating the conditional $p$ –values of the proposed tests.

All proposed tests will be *rank tests*, i.e. they are based exclusively on the two rank vectors $\vec{R}_1 = (R_{11}, ..., R_{1n})$ and $\vec{R}_2 = (R_{21}, ..., R_{2n})$ of the $Y$ 's and $Z$ 's, i.e.

$$R_{1i} = \#\left\{ j \in \{1, ..., n\} \mid Y_j \le Y_i \right\}, \qquad i = 1, ..., n,$$

$$R_{2i} = \#\left\{ j \in \{1, ..., n\} \mid Z_j \le Z_i \right\}, \qquad i = 1, ..., n. \tag{1}$$

Notice, the rank of $Y_i$ [ $Z_i$ ] counts all $Y$ –observations [ $Z$ –observations] which are less than or equal to $Y_i$ [ $Z_i$ ], also in case of tied observations.

**Model Assumptions:**

- The underlying bivariate (cumulative) distribution function $F(y, z)$ of the random variables $X_i = (Y_i, Z_i)$ is dominated by the product distribution function $G(y)H(z)$ , where $G(y) = F(y, \infty)$ is the distribution function of the $Y$ 's and $H(z) = F(\infty, z)$ is the distribution function of the $Z$ 's. Since no other restriction is assumed, the underlying distribution functions may have discontinuities.

- The measurement scale for the $Y$ 's and the $Z$ 's is the real line or at least ordinal in order to have well-defined ranks.

**Testing Problems:**

- *One-sided problem:* Testing the null hypothesis of independence,

$$\mathcal{H}_0^i : F = G \times H , \quad \text{i.e.} \quad F(y, z) = G(y)H(z) \quad \forall \, (y, z) \in \mathbb{R}^2,$$

versus the one-sided alternative of *positive quadrant dependence*,

$$\mathcal{A}^0 : \ F \geq G \times H, \ F \neq G \times H \,.$$

- *Omnibus problem:* Testing the null hypothesis of independence,

$$\mathcal{H}_0^i : \ F = G \times H \,,$$

versus the omnibus alternative of *dependence*,

$$\mathcal{A} : \ F \neq G \times H \,.$$

**Independence Rank Process:**

Essential for the data inspection and for the definition of the test statistics is the following rank process $W_n = \big(W_n(s,t), \ 0 \leq s,t \leq 1\big)$ :

Let $d_1$ denote the number of different values among the ranks $R_{11}, ..., R_{1n}$ of the $Y$ 's, which is the same as the number of different values among the $Y-$ observations. Similarly, let $d_2$ denote the number of different values among the ranks $R_{21}, ..., R_{2n}$ of the $Z$ 's, which is the same as the number of different values among the $Z-$observations. In the continuous model (no ties) we have $d_1 = d_2 = n$ . In the general model (ties are possible) $d_1$ and $d_2$ are random variables.

Let $T_{11} < \cdots < T_{1d_1}$ denote the ordered different values in $R_{11}, ..., R_{1n}$. Similarly, let $T_{21} < \cdots < T_{2d_2}$ denote the ordered different values in $R_{21}, ..., R_{2n}$. If no ties are present, we have $d_1 = d_2 = n$ and $T_{1i} = T_{2i} = i$ for $i = 1, ..., n$ . If all $Y-$observations [ $Z-$observations ] are tied we have $d_1 = 1$ and $T_{11} = n$ [ $d_2 = 1$ and $T_{21} = n$ ]. In any case we have $1 \leq d_1 \leq n$ , $1 \leq d_2 \leq n$ , and $T_{1d_1} = T_{2d_2} = n$ .

Using $d_k$ , $T_{k1}, ..., T_{kd_k}$ , and the additional definition $T_{k0} := 0$ , $k = 1, 2$ , we define the independence rank process $W_n$ at the points $(T_{1q}/n, T_{2r}/n)$ , $q = 0, 1, ..., d_1$ , $r = 0, 1, ..., d_2$ , according to, cf. (6.4.73) to (6.4.76),

$$W_n\left(\frac{T_{1q}}{n}, \frac{T_{2r}}{n}\right) = n\,\#\left\{ i \in \{1, ..., n\} \mid R_{1i} \leq T_{1q}, \ R_{2i} \leq T_{2r} \right\} - T_{1q}\,T_{2r} \,. \quad (2)$$

The definition of $W_n(s,t)$ for arbitrary $(s,t) \in [0,1]^2$ is completed by bi-linear interpolation in $s-$ and in $t-$direction, i.e. the value of $W_n(s,t)$ for $T_{1(q-1)}/n \leq s \leq T_{1q}/n$ and $T_{2(r-1)}/n \leq t \leq T_{2r}/n$ is

$$W_n(s,t) = \left(\frac{T_{1q} - ns}{\tau_{1q}}\right)\left(\frac{T_{2r} - nt}{\tau_{2r}}\right) W_n\left(\frac{T_{1(q-1)}}{n}, \frac{T_{2(r-1)}}{n}\right)$$

$$+ \left(\frac{ns - T_{1(q-1)}}{\tau_{1q}}\right)\left(\frac{T_{2r} - nt}{\tau_{2r}}\right) W_n\left(\frac{T_{1q}}{n}, \frac{T_{2(r-1)}}{n}\right)$$

$$+ \left(\frac{T_{1q} - ns}{\tau_{1q}}\right) \left(\frac{nt - T_{2(r-1)}}{\tau_{2r}}\right) W_n\left(\frac{T_{1(q-1)}}{n}, \frac{T_{2r}}{n}\right)$$

$$+ \left(\frac{ns - T_{1(q-1)}}{\tau_{1q}}\right) \left(\frac{nt - T_{2(r-1)}}{\tau_{2r}}\right) W_n\left(\frac{T_{1q}}{n}, \frac{T_{2r}}{n}\right), \qquad (3)$$

where

$$\tau_{1q} = T_{1q} - T_{1(q-1)}, \qquad q = 1, ..., d_1,$$

$$\tau_{2r} = T_{2r} - T_{2(r-1)}, \qquad r = 1, ..., d_2, \qquad (4)$$

are the tie-lengths of the ordered $Y$ –observations and of the ordered $Z$ – observations, respectively.

Notice, for any $0 \leq s,t \leq 1$ we have

$$W_n(s,0) = W_n(0,t) = W_n(s,1) = W_n(1,t) = 0. \qquad (5)$$

For easy computation of $W_n$ we may use the representation

$$W_n\left(\frac{k}{n}, \frac{l}{n}\right) = \sum_{i=1}^{k} \sum_{j=1}^{l} w_n(i,j), \qquad k,l = 1, ..., n, \qquad (6)$$

with

$$w_n(i,j) = n\, \frac{s(q,r)}{\tau_{1q}\, \tau_{2r}} - 1\,, \quad \text{if } T_{1(q-1)} < i \leq T_{1q},\ T_{2(r-1)} < j \leq T_{2r}, \qquad (7)$$

and

$$s(q,r) = \#\left\{ i \in \{1, ..., n\} \mid (R_{1i}, R_{2i}) = (T_{1q}, T_{2r}) \right\}. \qquad (8)$$

The tie-lengths vector $\tau = (\tau_{11}, ..., \tau_{1d_1};\ \tau_{21}, ..., \tau_{2d_2})$ and the $d_1 \times d_2$ –matrix $\left( s(q,r),\ q = 1, ..., d_1,\ r = 1, ..., d_2 \right)$ obviously contain the same information (up to arbitrary permutations of $X_1, ..., X_n$) as the two rank vectors $\vec{R}_1 = (R_{11}, ..., R_{1n})$ and $\vec{R}_2 = (R_{21}, ..., R_{2n})$. For the actual computation of the rank process $W_n$ formula (6) seems to be most convenient.

If *no ties are present*, we have $s(q,r) = 1$ iff $(R_{1i}, R_{2i}) = (q,r)$ for some $i \in \{1, ..., n\}$, and $s(q,r) = 0$ iff there is no such $i$.


**The Omnibus Test:**

The test statistic $S$ for testing the null hypothesis of independence $\mathcal{H}_0^i$ versus the omnibus alternative $\mathcal{A}$ is, cf. formula (6.2.11),

$$S = n^{-3} \sum_{i=1}^{n} \sum_{j=1}^{n} \sum_{q=1}^{n} \sum_{r=1}^{n} w_n(i,j)\, w_n(q,r)\, k(i,q)\, k(j,r), \qquad (9)$$

where

$$k(i,j) \; = \; K_a\Big(\frac{i - 1/2}{n}, \frac{j - 1/2}{n}\Big), \qquad 1 \le i,j \le n, \tag{10}$$

and

$$K_a(s,t) \; = \; \frac{1}{a}\Big( K(\frac{t+s}{a}) + K(\frac{t-s}{a}) + K(\frac{t+s-2}{a})\Big), \quad s,t \in [0,1], \tag{11}$$

with fixed bandwidth  $a = 0.40$  and Parzen-2 kernel

$$K(t) \; = \; \begin{cases} 4/3 - 8\,|t|^2 + 8\,|t|^3, & \text{if} \quad |t| < 1/2\,, \\[2mm] 8\,(1 - |t|)^3/3\,, & \text{if} \quad 1/2 \le |t| < 1\,, \\[2mm] 0\,, & \text{if} \quad |t| \ge 1\,. \end{cases} \tag{12}$$

The omnibus test will **reject the null hypothesis** $\mathcal{H}_0^i$ ( $F = G \times H$ ) at level $\alpha$ in favor of the omnibus alternative $\mathcal{A}$ ( $F \ne G \times H$ ) **if** the value of the test statistic $S$ is larger than the upper $\alpha$ –quantile $k_\alpha(\tau)$ of the conditional distribution of $S$ under $\mathcal{H}_0^i$ given the tie-lengths vector $\tau = (\tau_{11}, ..., \tau_{1d_1}; \tau_{21}, ..., \tau_{2d_2})$ corresponding to the ordered $Y$ – and $Z$ –sample, i.e. the omnibus test may be written in the form

$$\psi \; = \; 1\Big( S > k_\alpha(\tau) \Big), \tag{13}$$

where $k_\alpha(\tau)$ is defined by

$$P_{\mathcal{H}_0^i}\{ S > k_\alpha(\tau) \,|\, \tau \} \; \le \; \alpha \; < \; P_{\mathcal{H}_0^i}\{ S \ge k_\alpha(\tau) \,|\, \tau \}\,. \tag{14}$$

The upper $\alpha$ –quantiles $k_\alpha$ for the continuous model are given in Table 2.7.A of the Appendix.


**Simulation of the conditional $p$ –values:**

In principle there is no difficulty to evaluate the conditional critical values $k_\alpha(\tau)$ or the conditional $p$ –values:

The conditional distribution of the rank vector $\vec{R}_1 = (R_{11}, ..., R_{1n})$ given the tie-lengths vector $\vec{\tau}_1 := (\tau_{11}, ..., \tau_{1d_1})$ is the same for any underlying distribution function $F$. The same holds true for the conditional distribution of the rank vector $\vec{R}_2 = (R_{21}, ..., R_{2n})$ given the tie-lengths vector $\vec{\tau}_2 := (\tau_{21}, ..., \tau_{2d_2})$. Additionally, for arbitrary $F = G \times H \in \mathcal{H}_0^i$ the rank vectors $\vec{R}_1$ and $\vec{R}_2$ are conditionally independent for given $\tau = (\vec{\tau}_1; \vec{\tau}_2)$, i.e.

$$\mathcal{L}_{\mathcal{H}_0^i}\big[\, \vec{R}_1, \vec{R}_2 \,|\, \tau \,\big] \; = \; \mathcal{L}_{\mathcal{H}_0^i}\big[\, \vec{R}_1 \,|\, \vec{\tau}_1 \,\big] \otimes \mathcal{L}_{\mathcal{H}_0^i}\big[\, \vec{R}_2 \,|\, \vec{\tau}_2 \,\big]\,.$$

Given the observed $\tau = (\vec{\tau}_1;\ \vec{\tau}_2)$ the actual computation of $\mathcal{L}[\,S\mid\tau\,]$ under $\mathcal{H}_0^i$ works as follows:

Since the underlying rank process $W_n$ and the statistic $S$ are invariant under any permutation of $(R_{11}, R_{21}), ..., (R_{1n}, R_{2n})$, cf. definition (2), we may rearrange the pairs $(R_{1i}, R_{2i})$, $1 \leq i \leq n$, according to increasing values of the first components. Let's denote the resulting pairs by $(\tilde{R}_{11}, \tilde{R}_{21}), ..., (\tilde{R}_{1n}, \tilde{R}_{2n})$. Then formula (6.4.82) implies (for given $\tau$ )

$$\begin{aligned}
\tilde{R}_{1i} &= T_{1j}, &&\text{if}\quad T_{1(j-1)} < i \leq T_{1j} \quad\text{for some}\ \ 1 \leq j \leq d_1, \\[2mm]
\tilde{R}_{2i} &= T_{2j}, &&\text{if}\quad T_{2(j-1)} < Q_i \leq T_{2j} \quad\text{for some}\ \ 1 \leq j \leq d_2,
\end{aligned} \tag{15}$$

where $(Q_1, ..., Q_n)$ is a random permutation of $(1, ..., n)$.

For the exact evaluation of the conditional critical values we enumerate all permutations $(Q_1, ..., Q_n)$ of $(1, ..., n)$, for the Monte Carlo simulation of the conditional critical values we use a random number generator for producing random permutations $(Q_1, ..., Q_n)$ of $(1, ..., n)$. Obviously the exact evaluation is limited to rather small sizes of $n$.

**The One-sided Test:**

According to formulae (6.2.21) and (6.4.80) of Chapter 6 the test statistic for testing the null hypothesis of independence $\mathcal{H}_0^i$ versus the one-sided alternative $\mathcal{A}^0$ of positive quadrant dependence is

$$n^{-3} \sum_{i=1}^{n}\sum_{j=1}^{n}\sum_{q=1}^{n}\sum_{r=1}^{n} \Delta W_n^0(i,j)\, w_n(q,r)\, k(i,q)\, k(j,r), \tag{$*$}$$

where $w_n(\cdot,\cdot)$ and $k(\cdot,\cdot)$ are defined as in (9) and where

$$\Delta W_n^0(i,j) = W_n^0\Big(\frac{i}{n}, \frac{j}{n}\Big) - W_n^0\Big(\frac{i}{n}, \frac{j-1}{n}\Big) - W_n^0\Big(\frac{i-1}{n}, \frac{j}{n}\Big) + W_n^0\Big(\frac{i-1}{n}, \frac{j-1}{n}\Big)$$

and

$$W_n^0(s,t) = \max\big(0,\ W_n(s,t)\big), \qquad \forall\ s,t \in [0,1]. \tag{$**$}$$

In case of tied observations the exact evaluation or Monte Carlo simulation of the conditional critical values of the rank statistic $(*)$ is very time consuming, if the above definition of $W_n^0$ is used.

Therefore we define a modified test statistic $S^0$ for testing the null hypothesis of independence $\mathcal{H}_0^i$ versus the one-sided alternative $\mathcal{A}^0$ of positive quadrant dependence by

$$S^0 = n^{-3} \sum_{i=1}^{n}\sum_{j=1}^{n}\sum_{q=1}^{n}\sum_{r=1}^{n} w_n^0(i,j)\, w_n(q,r)\, k(i,q)\, k(j,r), \tag{16}$$

where $w_n(\cdot,\cdot)$ and $k(\cdot,\cdot)$ are defined as in (9), where

$$w_n^0(i,j) \;=\; W_n^0\Big(\frac{i}{n},\frac{j}{n}\Big) - W_n^0\Big(\frac{i}{n},\frac{j-1}{n}\Big)$$

$$- W_n^0\Big(\frac{i-1}{n},\frac{j}{n}\Big) + W_n^0\Big(\frac{i-1}{n},\frac{j-1}{n}\Big),$$

$$(17)$$

and where the modified 'projection' $W_n^0$ of the rank process $W_n$ onto the side conditions $B \geq 0$ and $B(\cdot,0) = B(0,\cdot) = B(\cdot,1) = B(1,\cdot) = 0$ is defined in the following way:

For $q = 0,1,...,d_1$ and $r = 0,1,...,d_2$ we define

$$W_n^0\big(T_{1q}/n, T_{2r}/n\big) \;=\; \max\big(0,\, W_n(T_{1q}/n, T_{2r}/n)\big). \qquad (18a)$$

For arbitrary $(s,t) \in [0,1]^2$ the definition of $W_n^0(s,t)$ is completed by bilinear interpolation in $s-$ and in $t-$direction, i.e. the value of $W_n^0(s,t)$ for $T_{1(q-1)}/n \leq s \leq T_{1q}/n$ and $T_{2(r-1)}/n \leq t \leq T_{2r}/n$ is

$$W_n^0(s,t) = \Big(\frac{T_{1q}-ns}{T_{1q}}\Big)\Big(\frac{T_{2r}-nt}{T_{2r}}\Big)\, W_n^0\Big(\frac{T_{1(q-1)}}{n},\frac{T_{2(r-1)}}{n}\Big)$$

$$+ \Big(\frac{ns-T_{1(q-1)}}{T_{1q}}\Big)\Big(\frac{T_{2r}-nt}{T_{2r}}\Big)\, W_n^0\Big(\frac{T_{1q}}{n},\frac{T_{2(r-1)}}{n}\Big)$$

$$+ \Big(\frac{T_{1q}-ns}{T_{1q}}\Big)\Big(\frac{nt-T_{2(r-1)}}{T_{2r}}\Big)\, W_n^0\Big(\frac{T_{1(q-1)}}{n},\frac{T_{2r}}{n}\Big)$$

$$+ \Big(\frac{ns-T_{1(q-1)}}{T_{1q}}\Big)\Big(\frac{nt-T_{2(r-1)}}{T_{2r}}\Big)\, W_n^0\Big(\frac{T_{1q}}{n},\frac{T_{2r}}{n}\Big). \qquad (18b)$$

Notice, the original statistic $(*)$ and the modified statistic $S^0$ coincide, if no ties are present. The difference between the original 'projection' $(**)$ and the modified 'projection' $W_n^0$ is shown in Example 2.7.1, cf. Figure 2.7.b and Figure 2.7.c.

The one-sided test will reject the null hypothesis $\mathcal{H}_0^i : F = G \times H$ at level $\alpha$ in favor of the one-sided alternative $\mathcal{A}^0 : F \geq G \times H, F \neq G \times H$, if the value of the test statistic $S^0$ is larger than the upper $\alpha-$quantile $k_\alpha^0(\tau)$ of the conditional distribution of $S^0$ under $\mathcal{H}_0^i$ given the tie-lengths vector $\tau = (\vec{\tau}_1;\, \vec{\tau}_2)$ of the $Y$'s and the $Z$'s, i.e. the one-sided test may be written in the form

$$\psi^0 \;=\; 1\Big(S^0 > k_\alpha^0(\tau)\Big) \qquad (19)$$

where $k_\alpha^0(\tau)$ is defined by

$$P_{\mathcal{H}_0^i}\big\{ S^0 > k_\alpha^0(\tau) \mid \tau \big\} \;\leq\; \alpha \;<\; P_{\mathcal{H}_0^i}\big\{ S^0 \geq k_\alpha^0(\tau) \mid \tau \big\}. \qquad (20)$$

The upper $\alpha$–quantiles $k_\alpha^0$ for the continuous model are given in Table 2.7.B of the Appendix.

The evaluation or simulation of the conditional critical values $k_\alpha^0(\tau)$ is completely similar to the omnibus case, since the statistic $S^0$ is a function of the independence rank process $W_n$ , too.

Notice, the only difference between the omnibus statistic $S$ and the one-sided statistic $S^0$ is the substitution of the weights $w_n(i,j)$ by the weights $w_n^0(i,j)$ . This substitution is reasonable, since the process $W_n/n^2$ is an estimator of the underlying parameter $B \in \mathcal{B}_n^i$ , cf. (6.1.14), and since the one-sided model is characterized by $B \geq 0$ , cf. (6.1.18), i.e. the rank process $W_n^0$ defined in (18) is a 'projection' onto the one-sided model $B \geq 0$ .

**Numerical Example 2.7.1:**

- The sample size is $n = 12$ and the observations $(y_1, z_1), ..., (y_n, z_n)$ are

$$(3.0, 3.0), \ (1.0, 7.0), \ (4.0, 6.0), \ (5.0, 8.0), \ (3.0, 5.0), \ (4.0, 8.0),$$
$$(2.0, 3.0), \ (1.0, 5.0), \ (3.0, 4.0), \ (1.0, 5.0), \ (2.0, 4.0), \ (3.0, 7.0).$$

- The rank vectors $\vec{R}_1 = (R_{11}, ..., R_{1n})$ and $\vec{R}_2 = (R_{21}, ..., R_{2n})$ are

$$\vec{R}_1 = (\, 9, \ 3, \ 11, \ 12, \ 9, \ 11, \ 5, \ 3, \ 9, \ 3, \ 5, \ 9 \,),$$
$$\vec{R}_2 = (\, 2, \ 10, \ 8, \ 12, \ 7, \ 12, \ 2, \ 7, \ 4, \ 7, \ 4, \ 10 \,).$$

- The number of different values in $\vec{R}_1$ and $\vec{R}_2$ are $d_1 = 5$ and $d_2 = 6$ , respectively.

- The ordered different values in $\vec{R}_1$ resp. $\vec{R}_2$ are

$$(T_{11}, ..., T_{1d_1}) = (\, 3, \ 5, \ 9, \ 11, \ 12 \,),$$
$$(T_{21}, ..., T_{2d_2}) = (\, 2, \ 4, \ 7, \ 8, \ 10, \ 12 \,).$$

- The lengths of the ties in the ordered $Y$ – observations respectively in the ordered $Z$ – observations are

$$\vec{\tau}_1 = (\tau_{11}, ..., \tau_{1d_1}) = (\, 3, \ 2, \ 4, \ 2, \ 1 \,),$$
$$\vec{\tau}_2 = (\tau_{21}, ..., \tau_{2d_2}) = (\, 2, \ 2, \ 3, \ 1, \ 2, \ 2 \,).$$

- The counts $s(q,r)$ , $q = 1, ..., d_1$ , $r = 1, ..., d_2$ , defined in formula (8) are

| $r =$ | 1 | 2 | 3 | 4 | 5 | 6 |
|---|---|---|---|---|---|---|
| $q = 1:$ | 0 | 0 | 2 | 0 | 1 | 0 |
| $q = 2:$ | 1 | 1 | 0 | 0 | 0 | 0 |
| $q = 3:$ | 1 | 1 | 1 | 0 | 1 | 0 |
| $q = 4:$ | 0 | 0 | 0 | 1 | 0 | 1 |
| $q = 5:$ | 0 | 0 | 0 | 0 | 0 | 1 |

- Therefore the defining weights $w_n(T_{1q}, T_{2r})$ , $q = 1, ..., d_1$ , $r = 1, ..., d_2$ , of formula (7) are

| $r =$ | 1 | 2 | 3 | 4 | 5 | 6 |
|---|---|---|---|---|---|---|
| $q = 1:$ | $-1.0$ | $-1.0$ | $5/3$ | $-1.0$ | $1.0$ | $-1.0$ |
| $q = 2:$ | $2.0$ | $2.0$ | $-1.0$ | $-1.0$ | $-1.0$ | $-1.0$ |
| $q = 3:$ | $0.5$ | $0.5$ | $0.0$ | $-1.0$ | $0.5$ | $-1.0$ |
| $q = 4:$ | $-1.0$ | $-1.0$ | $-1.0$ | $5.0$ | $-1.0$ | $-1.0$ |
| $q = 5:$ | $-1.0$ | $-1.0$ | $-1.0$ | $-1.0$ | $-1.0$ | $5.0$ |

- The defining values $W_n(T_{1q}/n, T_{2r}/n)$ , $q = 1, ..., d_1$ , $r = 1, ..., d_2$ , of the rank process $W_n$ are, cf. (2) or (6),

| $r =$ | 1 | 2 | 3 | 4 | 5 | 6 |
|---|---|---|---|---|---|---|
| $q = 1:$ | $-6$ | $-12$ | 3 | 0 | 6 | 0 |
| $q = 2:$ | 2 | 4 | 13 | 8 | 10 | 0 |
| $q = 3:$ | 6 | 12 | 21 | 12 | 18 | 0 |
| $q = 4:$ | 2 | 4 | 7 | 8 | 10 | 0 |
| $q = 5:$ | 0 | 0 | 0 | 0 | 0 | 0 |

- Thus the defining values $W_n^0(T_{1q}/n, T_{2r}/n)$ , $q = 1, ..., d_1$ , $r = 1, ..., d_2$ , of the rank process $W_n^0$ are, cf. (18a),

| $r =$ | 1 | 2 | 3 | 4 | 5 | 6 |
|---|---|---|---|---|---|---|
| $q = 1:$ | 0 | 0 | 3 | 0 | 6 | 0 |
| $q = 2:$ | 2 | 4 | 13 | 8 | 10 | 0 |
| $q = 3:$ | 6 | 12 | 21 | 12 | 18 | 0 |
| $q = 4:$ | 2 | 4 | 7 | 8 | 10 | 0 |
| $q = 5:$ | 0 | 0 | 0 | 0 | 0 | 0 |

**Figure 2.7.a**

*Numerical Example 2.7.1: The values of* $W_n(i/n, j/n)$, $\quad i, j = 0, 1, ..., n$.

| $j$ | | | | | | | | | | | | | |
|---|---|---|---|---|---|---|---|---|---|---|---|---|---|
| 12 | 0 | 0 | 0 | 0 | 0 | 0 | 0 | 0 | 0 | 0 | 0 | 0 | 0 |
| 11 | 0 | 1 | 2 | 3 | 4 | 5 | 6 | 7 | 8 | 9 | 7 | 5 | 0 |
| 10 | 0 | 2 | 4 | 6 | 8 | 10 | 12 | 14 | 16 | 18 | 14 | 10 | 0 |
| 9 | 0 | 1 | 2 | 3 | 6 | 9 | $\frac{21}{2}$ | 12 | $\frac{27}{2}$ | 15 | 12 | 9 | 0 |
| 8 | 0 | 0 | 0 | 0 | 4 | 8 | 9 | 10 | 11 | 12 | 10 | 8 | 0 |
| 7 | 0 | 1 | 2 | 3 | 8 | 13 | 15 | 17 | 19 | 21 | 14 | 7 | 0 |
| 6 | 0 | $-\frac{2}{3}$ | $-\frac{4}{3}$ | $-2$ | 4 | 10 | 12 | 14 | 16 | 18 | 12 | 6 | 0 |
| 5 | 0 | $-\frac{7}{3}$ | $-\frac{14}{3}$ | $-7$ | 0 | 7 | 9 | 11 | 13 | 15 | 10 | 5 | 0 |
| 4 | 0 | $-4$ | $-8$ | $-12$ | $-4$ | 4 | 6 | 8 | 10 | 12 | 8 | 4 | 0 |
| 3 | 0 | $-3$ | $-6$ | $-9$ | $-3$ | 3 | $\frac{9}{2}$ | 6 | $\frac{15}{2}$ | 9 | 6 | 3 | 0 |
| 2 | 0 | $-2$ | $-4$ | $-6$ | $-2$ | 2 | 3 | 4 | 5 | 6 | 4 | 2 | 0 |
| 1 | 0 | $-1$ | $-2$ | $-3$ | $-1$ | 1 | $\frac{3}{2}$ | 2 | $\frac{5}{2}$ | 3 | 2 | 1 | 0 |
| 0 | 0 | 0 | 0 | 0 | 0 | 0 | 0 | 0 | 0 | 0 | 0 | 0 | 0 |
| $i$ : | 0 | 1 | 2 | 3 | 4 | 5 | 6 | 7 | 8 | 9 | 10 | 11 | 12 |

- Complete tables of $W_n(i/n, j/n)$ and $W_n^0(i/n, j/n)$, $i, j = 0, 1, ..., n$, are given in Figure 2.7.a and Figure 2.7.b. The corresponding values of the process $(**)$ are given in Figure 2.7.c for comparison.

- We use (9) to (12) with the bandwidth $a = 0.40$ in order to evaluate the value

$$s = S(\vec{r}_1, \vec{r}_2) = 6.9705$$

of the omnibus statistic $S$.

- Since ties are present, there is no table with the corresponding exact conditional critical value $k_\alpha(3, 2, 4, 2, 1; 2, 2, 3, 1, 2, 2)$ but we utilize formula (15) in order to evaluate the actual (conditional) $p$-values by Monte Carlo simulation. With a Monte Carlo sample size of 10,000 we get

$$P_{\mathcal{H}_0^i}\left\{ S > 6.9705 \mid \tau = (3, 2, 4, 2, 1; 2, 2, 3, 1, 2, 2) \right\} = 0.157,$$

$$P_{\mathcal{H}_0^i}\left\{ S \geq 6.9705 \mid \tau = (3, 2, 4, 2, 1; 2, 2, 3, 1, 2, 2) \right\} = 0.157.$$

- This means that the omnibus test (13) will reject the null hypothesis of independence $\mathcal{H}_0^i$ in favor of the omnibus alternative $\mathcal{A}$ exactly for those levels $\alpha$ which are larger than or equal to the observed $p$-value

0.157 . Thus, the present data will **not** reject the null hypothesis for any reasonable level $\alpha$ , e.g. for $\alpha = 0.10$ .

- Using (16) to (18) with the bandwidth $a = 0.40$ we evaluate the value $s^0$ of the one-sided statistic $S^0$ as

$$s^0 = S^0(\vec{r}_1, \vec{r}_2) = 5.3399 .$$

- Since ties are present, there is no table with the corresponding exact conditional critical value $k_\alpha^0(3, 2, 4, 2, 1; 2, 2, 3, 1, 2, 2)$ but we utilize formula (15) in order to evaluate the actual (conditional) $p$–values by Monte Carlo simulation. With a Monte Carlo sample size of 10,000 we get

$$P_{\mathcal{H}_0^i}\left\{ S^0 > 5.3399 \mid \tau = (3, 2, 4, 2, 1;\, 2, 2, 3, 1, 2, 2) \right\} = 0.099 ,$$

$$P_{\mathcal{H}_0^i}\left\{ S^0 \geq 5.3399 \mid \tau = (3, 2, 4, 2, 1;\, 2, 2, 3, 1, 2, 2) \right\} = 0.099 .$$

- This means that the one-sided test will reject the null hypothesis of independence $\mathcal{H}_0^i$ in favor of the one-sided alternative $\mathcal{A}^0$ of positive quadrant dependence exactly for those levels $\alpha$ which are larger than or equal to the observed $p$–value 0.099 . Thus, the present data will reject the null hypothesis at level $\alpha = 0.10$ .

**The Projection Rank Test:**

Here we propose a second nonlinear rank statistic $S^\pi$ for testing the null hypothesis of independence $\mathcal{H}_0^i$ versus the **one-sided alternative** $\mathcal{A}^0$ of positive quadrant dependence.

The basic idea is to replace the single score function of a classical linear rank statistic by a finite number of sensibly chosen score functions $h_1, ..., h_r$ on $[0, 1]^2$ . As discussed in Section 6.3 we propose $r = 4$ and

$$h_1 = b_1 \times b_1 , \quad h_2 = b_1 \times b_3 , \quad h_3 = b_3 \times b_1 , \quad h_4 = b_3 \times b_3 , \qquad (21)$$

where $b_i \times b_j(s, t) = b_i(s)\, b_j(t)$ and where $b_1$ and $b_3$ are defined as in formula (18) of Section 2.1, i.e.

$$b_3(u) = -b_1(1 - u) = u(3u - 2), \qquad 0 \leq u \leq 1. \qquad (22)$$

Defining $B_j(t) = \int_0^t b_j(u)\, du$ , $0 \leq t \leq 1$ , $j = 1, 3$ , we get

$$B_1(t) = B_3(1 - t) = -t^3 + 2t^2 - t . \qquad (23)$$

**Figure 2.7.b**

*Numerical Example 2.7.1: The values of $W_n^0(i/n, j/n)$,   $i,j = 0,1,...,n$ .*

| $j$ | | | | | | | | | | | | | |
|---|---|---|---|---|---|---|---|---|---|---|---|---|---|
| 12 | 0 | 0 | 0 | 0 | 0 | 0 | 0 | 0 | 0 | 0 | 0 | 0 | 0 |
| 11 | 0 | 1 | 2 | 3 | 4 | 5 | 6 | 7 | 8 | 9 | 7 | 5 | 0 |
| 10 | 0 | 2 | 4 | 6 | 8 | 10 | 12 | 14 | 16 | 18 | 14 | 10 | 0 |
| 9 | 0 | 1 | 2 | 3 | 6 | 9 | $\frac{21}{2}$ | 12 | $\frac{27}{2}$ | 15 | 12 | 9 | 0 |
| 8 | 0 | 0 | 0 | 0 | 4 | 8 | 9 | 10 | 11 | 12 | 10 | 8 | 0 |
| 7 | 0 | 1 | 2 | 3 | 8 | 13 | 15 | 17 | 19 | 21 | 14 | 7 | 0 |
| 6 | 0 | $\frac{2}{3}$ | $\frac{4}{3}$ | 2 | 6 | 10 | 12 | 14 | 16 | 18 | 12 | 6 | 0 |
| 5 | 0 | $\frac{1}{3}$ | $\frac{2}{3}$ | 1 | 4 | 7 | 9 | 11 | 13 | 15 | 10 | 5 | 0 |
| 4 | 0 | 0 | 0 | 0 | 2 | 4 | 6 | 8 | 10 | 12 | 8 | 4 | 0 |
| 3 | 0 | 0 | 0 | 0 | $\frac{3}{2}$ | 3 | $\frac{9}{2}$ | 6 | $\frac{15}{2}$ | 9 | 6 | 3 | 0 |
| 2 | 0 | 0 | 0 | 0 | 1 | 2 | 3 | 4 | 5 | 6 | 4 | 2 | 0 |
| 1 | 0 | 0 | 0 | 0 | $\frac{1}{2}$ | 1 | $\frac{3}{2}$ | 2 | $\frac{5}{2}$ | 3 | 2 | 1 | 0 |
| 0 | 0 | 0 | 0 | 0 | 0 | 0 | 0 | 0 | 0 | 0 | 0 | 0 | 0 |
| $i$ : | 0 | 1 | 2 | 3 | 4 | 5 | 6 | 7 | 8 | 9 | 10 | 11 | 12 |

**Figure 2.7.c**

*Numerical Example 2.7.1: The values of $\max\big(0,\ W_n(i/n, j/n)\big)$ .*

| $j$ | | | | | | | | | | | | | |
|---|---|---|---|---|---|---|---|---|---|---|---|---|---|
| 12 | 0 | 0 | 0 | 0 | 0 | 0 | 0 | 0 | 0 | 0 | 0 | 0 | 0 |
| 11 | 0 | 1 | 2 | 3 | 4 | 5 | 6 | 7 | 8 | 9 | 7 | 5 | 0 |
| 10 | 0 | 2 | 4 | 6 | 8 | 10 | 12 | 14 | 16 | 18 | 14 | 10 | 0 |
| 9 | 0 | 1 | 2 | 3 | 6 | 9 | $\frac{21}{2}$ | 12 | $\frac{27}{2}$ | 15 | 12 | 9 | 0 |
| 8 | 0 | 0 | 0 | 0 | 4 | 8 | 9 | 10 | 11 | 12 | 10 | 8 | 0 |
| 7 | 0 | 1 | 2 | 3 | 8 | 13 | 15 | 17 | 19 | 21 | 14 | 7 | 0 |
| 6 | 0 | 0 | 0 | 0 | 4 | 10 | 12 | 14 | 16 | 18 | 12 | 6 | 0 |
| 5 | 0 | 0 | 0 | 0 | 0 | 7 | 9 | 11 | 13 | 15 | 10 | 5 | 0 |
| 4 | 0 | 0 | 0 | 0 | 0 | 4 | 6 | 8 | 10 | 12 | 8 | 4 | 0 |
| 3 | 0 | 0 | 0 | 0 | 0 | 3 | $\frac{9}{2}$ | 6 | $\frac{15}{2}$ | 9 | 6 | 3 | 0 |
| 2 | 0 | 0 | 0 | 0 | 0 | 2 | 3 | 4 | 5 | 6 | 4 | 2 | 0 |
| 1 | 0 | 0 | 0 | 0 | 0 | 1 | $\frac{3}{2}$ | 2 | $\frac{5}{2}$ | 3 | 2 | 1 | 0 |
| 0 | 0 | 0 | 0 | 0 | 0 | 0 | 0 | 0 | 0 | 0 | 0 | 0 | 0 |
| $i$ : | 0 | 1 | 2 | 3 | 4 | 5 | 6 | 7 | 8 | 9 | 10 | 11 | 12 |

With $0 = T_{10} < T_{11} < \cdots < T_{1d_1}$, $0 = T_{20} < T_{21} < \cdots < T_{2d_2}$, $\tau_{1q} = T_{1q} - T_{1(q-1)}$, $\tau_{2r} = T_{2r} - T_{2(r-1)}$, and $s(q,r)$, $q = 1, ..., d_1$, $r = 1, ..., d_2$, as in (4) and (8) we define the vector $\vec{S} = (S_1, S_2, S_3, S_4)^T$ of averaged scores linear rank statistics $S_\varrho$ according to

$$S_1 = S_n^\tau(b_1 \times b_1), \qquad S_2 = S_n^\tau(b_1 \times b_3),$$
$$S_3 = S_n^\tau(b_3 \times b_1), \qquad S_4 = S_n^\tau(b_3 \times b_3), \tag{24}$$

where, cf. (6.4.37) and (6.4.41),

$$S_n^\tau(b_j \times b_k) = \frac{1}{\sqrt{n}} \sum_{i=1}^n (b_j \times b_k)_n^\tau(R_{1i}, R_{2i}) \tag{25}$$

$$= n^{3/2} \sum_{q=1}^{d_1} \sum_{r=1}^{d_2} \left( B_j(\frac{T_{1q}}{n}) - B_j(\frac{T_{1(q-1)}}{n}) \right) \left( B_k(\frac{T_{2r}}{n}) - B_k(\frac{T_{2(r-1)}}{n}) \right) \frac{s(q,r)}{\tau_{1q}\,\tau_{2r}}$$

In the next step we'll define the $4 \times 4$ –matrix

$$\Gamma = \left( \gamma_{\varrho\sigma} \right)_{\varrho,\sigma=1,...,4} \tag{26}$$

according to

$$\Gamma = \begin{pmatrix} c_1(1,1) & c_1(1,3) \\ c_1(1,3) & c_1(3,3) \end{pmatrix} \otimes \begin{pmatrix} c_2(1,1) & c_2(1,3) \\ c_2(1,3) & c_2(3,3) \end{pmatrix} \tag{27}$$

$$= \begin{pmatrix} c_1(1,1)c_2(1,1) & c_1(1,1)c_2(1,3) & c_1(1,3)c_2(1,1) & c_1(1,3)c_2(1,3) \\ c_1(1,1)c_2(1,3) & c_1(1,1)c_2(3,3) & c_1(1,3)c_2(1,3) & c_1(1,3)c_2(3,3) \\ c_1(1,3)c_2(1,1) & c_1(1,3)c_2(1,3) & c_1(3,3)c_2(1,1) & c_1(3,3)c_2(1,3) \\ c_1(1,3)c_2(1,3) & c_1(1,3)c_2(3,3) & c_1(3,3)c_2(1,3) & c_1(3,3)c_2(3,3) \end{pmatrix}$$

where, for $i = 1, 3$ and $j = 1, 3$,

$$c_1(i,j) = \sum_{q=1}^{d_1} \left( B_i(\frac{T_{1q}}{n}) - B_i(\frac{T_{1(q-1)}}{n}) \right) \left( B_j(\frac{T_{1q}}{n}) - B_j(\frac{T_{1(q-1)}}{n}) \right) \frac{n}{\tau_{1q}},$$

$$\tag{28}$$

$$c_2(i,j) = \sum_{r=1}^{d_2} \left( B_i(\frac{T_{2r}}{n}) - B_i(\frac{T_{2(r-1)}}{n}) \right) \left( B_j(\frac{T_{2r}}{n}) - B_j(\frac{T_{2(r-1)}}{n}) \right) \frac{n}{\tau_{2r}},$$

For each subset $J$ of $\{1,2,3,4\}$ such that $J \neq \emptyset$ we put

$$\vec{S}_J = (S_\varrho : \varrho \in J)^T \qquad \text{and} \qquad \Gamma_J = \left( \gamma_{\varrho,\sigma} \right)_{\varrho,\sigma \in J}, \tag{29}$$

where the elements of $J$ are arranged in increasing order. Then the proposed *projection rank statistic* $S^\pi$ is defined by

$$S^\pi = \max\left\{ \vec{S}_J^T (\Gamma_J)^{-1} \vec{S}_J : (\Gamma_J)^{-1} \vec{S}_J \geq 0, \emptyset \neq J \subset \{1,2,3,4\} \right\} \qquad (30)$$

with $S^\pi := 0$ if $(\Gamma_J)^{-1}\vec{S}_J \not\geq 0$ $\forall J \subset \{1,2,3,4\}$, $J \neq \emptyset$. Of course the definition (30) requires a nonsingular $\Gamma$, otherwise $S^\pi$ is not defined.

The projection rank test will reject the null hypothesis $\mathcal{H}_0^i : F = G \times H$ at level $\alpha$ in favor of the one-sided alternative $\mathcal{A}^0 : F \geq G \times H, F \neq G \times H$, if the value of the test statistic $S^\pi$ is larger than the upper $\alpha$-quantile $k_\alpha^\pi(\tau)$ of the conditional distribution of $S^\pi$ under $\mathcal{H}_0^i$ given the tie-lengths vector $\tau = (\vec{r}_1; \vec{r}_2)$ of the $Y$'s and the $Z$'s, i.e. the projection rank test may be written in the form

$$\psi^\pi = 1\left( S^\pi > k_\alpha^\pi(\tau) \right) \qquad (31)$$

where $k_\alpha^\pi(\tau)$ is defined by

$$P_{\mathcal{H}_0^i}\left\{ S^\pi > k_\alpha^\pi(\tau) \mid \tau \right\} \leq \alpha < P_{\mathcal{H}_0^i}\left\{ S^\pi \geq k_\alpha^\pi(\tau) \mid \tau \right\}. \qquad (32)$$

The upper $\alpha$-quantiles $k_\alpha^\pi$ for the continuous model are given in Table 2.7.C of the Appendix.

In the Numerical Example 2.7.1 we evaluate the value $s^\pi$ of the projection rank statistic $S^\pi$ as

$$s^\pi = S^\pi(\vec{r}_1, \vec{r}_2) = 7.1969.$$

Since ties are present, there is no table with the exact conditional critical value $k_\alpha^\pi(3,2,4,2,1;2,2,3,1,2,2)$ but similar to the evaluation of the conditional $p$-values of the statistic (9) we may evaluate the actual (conditional) $p$-values of $S^\pi$ by Monte Carlo simulation. With a Monte Carlo sample size of 10,000 we get

$$P_{\mathcal{H}_0^i}\left\{ S^\pi > 7.1969 \mid \tau = (3,2,4,2,1;2,2,3,1,2,2) \right\} = 0.023,$$

$$P_{\mathcal{H}_0^i}\left\{ S^\pi \geq 7.1969 \mid \tau = (3,2,4,2,1;2,2,3,1,2,2) \right\} = 0.023.$$

This means that the one-sided test will reject the null hypothesis of independence $\mathcal{H}_0^i : F = G \times H$ in favor of the one-sided alternative $\mathcal{A}^0 : F \geq G \times H, F \neq G \times H$ exactly for those levels $\alpha$ which are larger than or equal to the observed $p$-value. Since the observed $p$-value of the example is $0.023$, the data will reject the null hypothesis for reasonable levels $\alpha$, e.g. for $\alpha = 0.05$.

Another approximate asymptotic computation of the $p$ -values is provided by the following formula:

For $t \geq 0$  we get from (6.3.15) to (6.3.20)

$$P_{\mathcal{H}_0^i}\left\{ S^\pi > t \mid \tau \right\} \tag{33}$$

$$\approx w_1 P\{\chi_1^2 > t\} + w_2 P\{\chi_2^2 > t\} + w_3 P\{\chi_3^2 > t\} + w_4 P\{\chi_4^2 > t\},$$

where $\chi_\varrho^2$ , $\varrho = 1, 2, 3, 4$ , denote random variables with respective $\chi_\varrho^2$ –distributions and where $w_\varrho$ , $\varrho = 1, 2, 3, 4$ , are computed according to formulae (6.3.16) to (6.3.19) but with the $\Gamma$ from (6.3.13) substituted by the $\tau$ – dependent $\Gamma$ defined in formula (27).

In the above numerical example we get

$$w_1 = 0.3461 , \qquad w_2 = 0.3499 , \qquad w_3 = 0.1539 , \qquad w_4 = 0.0247 .$$

Therefore the approximate observed (conditional) $p$ –value computed from (33) with $t = s^\pi = 7.1969$  is

$$P_{\mathcal{H}_0^i}\left\{ S^\pi > 7.1969 \mid \tau = (3, 2, 4, 2, 1; 2, 2, 3, 1, 2, 2) \right\} \approx 0.025 ,$$

which is a good approximation of the simulated value $0.023$  of the exact conditional $p$ –value.

# Part II

# Mathematical Foundation

# Chapter 3

# Two samples differing in location

In this chapter we'll develop the mathematical foundation and the asymptotic properties of suitable rank tests with estimated scores for the two-sample problem. We restrict the discussion to the two-sample model in order to present the basic ideas in the simplest form. The corresponding $k$-sample model will be treated in Section 4.2.

The general motivation has been given in Chapter 1. Especially, in Section 1.3 we've presented the nonparametric two-sample model for testing the null hypothesis of randomness $\mathcal{H}_0^r$ versus the alternative $\mathcal{A}_2^0$, which means that the distribution of the first sample is stochastically larger than the distribution of the second sample. This model will be the basis of the present chapter, i.e. as in Section 1.3 we assume the two sample model with the respective samples $X_1, ..., X_m$ and $X_{m+1}, ..., X_{m+n}$ and the respective underlying distribution functions $F$ and $G$ on the real line $\mathbb{R}$. All components of the pooled sample $X = (X_1, ..., X_N)$, $N = m + n$, are asssumed to be stochastically independent. For simplicity reasons we start with the assumption of *continuous* underlying distribution functions $F$ and $G$, but in Section 3.3 we'll give a complete theory *without the assumption of continuity*.

Under the continuity assumption

$$F, G \in \mathcal{F}_1^c = \{ H : H \text{ any continuous distribution function on } \mathbb{R} \} \quad (3.0.1)$$

we've reparametrized the parameter space $\mathcal{F}_1^c \times \mathcal{F}_1^c$ of the model in the form (1.3.24), i.e.

$$\mathcal{F}_1^c \times \mathcal{F}_1^c = \{ (F_{B,H}^N, G_{B,H}^N) : B \in \mathcal{B}_N^0, H \in \mathcal{F}_1^c \}, \quad (3.0.2)$$

where

$$F_{B,H}^{N} := H + \frac{1}{m}\sqrt{\frac{mn}{N}}\, B \circ H = H + c_{N1} B \circ H,$$

$$(3.0.3)$$

$$G_{B,H}^{N} := H - \frac{1}{n}\sqrt{\frac{mn}{N}}\, B \circ H = H + c_{NN} B \circ H,$$

and where the new parameter space $\mathcal{B}_N^0$ has been defined in (1.3.25) as the set of all absolutely continuous functions $B : [0,1] \to \mathbb{R}$ such that $B(0) = B(1) = 0$ and $-\sqrt{m/n} \le B'/\sqrt{N} \le \sqrt{n/m}$ $[\lambda - a.e.]$. Here and in the sequel we'll use the convenient two-sample coefficients

$$c_{Ni} = \sqrt{\frac{mn}{N}}\begin{cases} \frac{1}{m}, & \text{if}\quad 1 \le i \le m, \\ -\frac{1}{n}, & \text{if}\quad m+1 \le i \le N. \end{cases}$$

$$(3.0.4)$$

From formulae (3.0.2) and (1.3.26) we have the following representations, cf. (1.3.27) and (1.3.28), of the *null hypothesis of randomness* $\mathcal{H}_0^r$,

$$\mathcal{H}_0^r = \{ (F,G) \in \mathcal{F}_1^c \times \mathcal{F}_1^c : F = G \}$$

$$= \{ (F_{B,H}^{N}, G_{B,H}^{N}) : B \in \mathcal{B}_N^0,\ H \in \mathcal{F}_1^c,\ B = 0 \},$$

$$(3.0.5)$$

of the one-sided *alternative of stochastic ordering* $\mathcal{A}_2^0$,

$$\mathcal{A}_2^0 = \{ (F,G) \in \mathcal{F}_1^c \times \mathcal{F}_1^c : F \le G,\ F \ne G \}$$

$$= \{ (F_{B,H}^{N}, G_{B,H}^{N}) : B \in \mathcal{B}_N^0,\ H \in \mathcal{F}_1^c,\ B \le 0,\ B \ne 0 \},$$

$$(3.0.6)$$

and of the *omnibus alternative* $\mathcal{A}_2$,

$$\mathcal{A}_2 = \{ (F,G) \in \mathcal{F}_1^c \times \mathcal{F}_1^c : F \ne G \}$$

$$= \{ (F_{B,H}^{N}, G_{B,H}^{N}) : B \in \mathcal{B}_N^0,\ H \in \mathcal{F}_1^c,\ B \ne 0 \}.$$

$$(3.0.7)$$

Notice, the new parameter space $\mathcal{B}_N^0$ depends on the sample sizes $m$ and $n$. In order not to have both sample sizes $m$ and $n$ as indices in our asymptotics we assume that $m = m(N)$ and $n = n(N)$ are defined by the joint sample size $N$. Additionally, **all asymptotics will be under the assumption**

$$N \to \infty \quad \text{and} \quad \eta_N := m/N \to \eta$$

$$(3.0.8)$$

for some arbitrary $0 < \eta < 1$, without explicitly mention it in the respective theorems.

If we take any sequence of local asymptotic alternatives (1.3.30), i.e.

$$0 < \varrho \le 1,\ H \in \mathcal{F}_1^c \quad \text{and} \quad (F_{\varrho B_N, H}^{N}, G_{\varrho B_N, H}^{N}),\ N \ge 1,$$

$$(3.0.9)$$

where the directions $B_N \in \mathcal{B}_N^0$ are *almost fixed*, i.e. for the derivative $b_N = B_N'$ of $B_N$ we assume convergence in quadratic mean to some square integrable function $b : (0,1) \to \mathbb{R}$,

$$\int_0^1 (b_N - b)^2 \, dx \stackrel{N \to \infty}{\longrightarrow} 0, \tag{3.0.10}$$

then the following theorem proves the corresponding linear rank test $\psi_{N\alpha}(b) = 1(\, S_N(b) \geq u_\alpha \|b\| \,)$ defined in (1.3.29) to (1.3.32) to be asymptotically optimal for testing $\mathcal{H}_0^r$ versus $(F_{\varrho B_N, H}^N, G_{\varrho B_N, H}^N)$.

**Therefore $b_N$ is the optimal nonparametric score function to be estimated in the subsequent sections.**

In order to have a convenient notation we'll use the following terminology throughout the volume:

### 3.0.0 Notation

$\lambda$ *is the Lebesgue measure on the unit interval* $(0,1)$.

$\int_0^1 f \, d\lambda$ *or* $\int_0^1 f \, dx$ *is the Lebesgue integral of the measurable and integrable function* $f : (0,1) \to \mathbb{R}$.

$L_2(0,1) := \{f : \int_0^1 f^2 \, d\lambda < \infty\}$ *is the space of all measurable and square integrable functions on the unit interval with the usual inner product* $\langle f, g \rangle := \int_0^1 fg \, d\lambda$ *and the norm* $\|f\| := \sqrt{\langle f, f \rangle}$.

$L_2^0(0,1) := \{f \in L_2(0,1) : \int_0^1 f \, d\lambda = 0\}$ *is the set of all elements of* $L_2(0,1)$ *which are orthogonal to the constant function 1.*

Especially we have $b_N \in L_2^0(0,1)$ for all $B_N \in \mathcal{B}_N^0$, and thus $b \in L_2^0(0,1)$ if condition (3.0.10) holds true.

### 3.0.1 Theorem

*Assume* $b \in L_2^0(0,1)$ *such that* $\|b\| > 0$ *and let* $\psi_{N\alpha}(b)$, $N \geq 1$, *denote the linear rank tests defined in (1.3.29) to (1.3.32).*

*a) The sequence of linear rank tests* $\psi_{N\alpha}(b)$ *is asymptotically uniformly most powerful in the class of all asymptotic level* $\alpha$ *tests for testing the null hypothesis* $\mathcal{H}_0^r$ *versus the sequence of alternatives* $(F_{\varrho B_N, H}^N, G_{\varrho B_N, H}^N)$ *defined in (3.0.9) and (3.0.10) for the given* $b$.

*b) For any* $b_1 \in L_2^0(0,1)$ *and any sequence of directions* $B_{1N} \in \mathcal{B}_N^0$ *such that* $\|B_{1N}' - b_1\| \to 0$, *and any sequence of nuisance parameters* $H_N \in \mathcal{F}_1^c$ *the asymptotic power of* $\psi_{N\alpha}(b)$ *under the corresponding sequence of local* asymptotic alternatives $(H_N + c_{N1}\varrho B_{1N} \circ H_N,\ H_N + c_{NN}\varrho B_{1N} \circ H_N)$  is given by

$$\lim_{N\to\infty} \mathrm{Pr}_{(\varrho B_{1N},H_N)}\{\ S_N(b) \geq u_\alpha\|b\|\ \} = 1 - \Phi\left(u_\alpha - \varrho\,\frac{\langle b,b_1\rangle}{\|b\|}\right). \qquad (3.0.11)$$

*Especially we have*

$$\mathcal{L}\big[S_N(b)\,\big|\,(\varrho B_{1N},H_N)\big] \ \xrightarrow{\ \mathcal{L}\ }\ \mathcal{N}\big(\varrho\,\langle b,b_1\rangle,\ \|b\|^2\big). \qquad (3.0.12)$$


*The proof,* which follows the lines of Hájek and Šidák (1967), is given in Section 7.1 (Appendix).

As discussed in Section 1.3 we will construct various estimators $\hat{b}_N$ of the unknown underlying optimal nonparametric score function $b_N$ and reveal the respective asymptotic properties of the resulting nonlinear rank statistics $S_N(\hat{b}_N)$ .

## 3.1 Kernel estimators of the score function

If we consider the single alternative $(F, G)$, $F \neq G$, and construct the corresponding "least favourable" null hypothesis $(H_N, H_N)$ at sample size $N$ according to (1.3.8), i.e.

$$H_N = \eta_N F + (1 - \eta_N) G, \tag{3.1.1}$$

then the corresponding optimal score function $b_N$ for testing $(H_N, H_N)$ versus $(F, G)$ is given by, cf. (1.3.15) and (1.3.20),

$$b_N = B'_N, \quad B_N = \sqrt{\frac{m\,n}{N}}\,(F - G) \circ H_N^{-1}. \tag{3.1.2}$$

If the transformation $k : \mathbb{R} \to \mathbb{R}$ is bijective and strictly increasing the score function corresponding to the transformed observations $k(X_i)$, $1 \leq i \leq N$, is the same as for the original observations $X_i$, $1 \leq i \leq N$. Since the vector of ranks $(R_1, ..., R_N)$ is maximal invariant under the group of such transformations $k$, we'll try to estimate $b_N$ on the basis of the rank vector $(R_1, ..., R_N)$.

### A) The proposed test statistics

According to the formulae (1.3.10) and (1.3.11) the transformed random variables $H_N(X_1), \cdots, H_N(X_m)$ are i.i.d. with distribution function $F \circ H_N^{-1}$ and the transformed random variables $H_N(X_{m+1}), \cdots, H_N(X_n)$ are i.i.d. with distribution function $G \circ H_N^{-1}$. If these random variables could be observed, the empirical process

$$\sqrt{\frac{m\,n}{N}}\left(\frac{1}{m}\sum_{i=1}^{m} 1\big(H_N(X_i) \leq t\big) - \frac{1}{n}\sum_{j=m+1}^{N} 1\big(H_N(X_j) \leq t\big)\right), \tag{3.1.3}$$

$0 \leq t \leq 1$, would be the usual estimator of $B_N$. In real applications $F$ and $G$ and thus $H_N$ are unknown, i. e. the transformed random variables $H_N(X_i) = \eta_N F(X_i) + (1 - \eta_N) G(X_i)$, $1 \leq i \leq N$, are unobservable but we may substitute each $H_N(X_i)$ by its natural estimator $\hat{H}_N(X_i)$, where $\hat{H}_N = \eta_N \hat{F}_N + (1 - \eta_N)\hat{G}_N$ and where $\hat{F}_N$ and $\hat{G}_N$ are the respective empirical distribution functions of the first sample $X_1, \cdots, X_m$ and of the second sample $X_{m+1}, \cdots, X_N$. Since $\hat{H}_N$ obviously is the empirical distribution function of the pooled sample $X_1, \cdots, X_N$ we get, for each $i = 1, \cdots, N$, the equality $\hat{H}_N(X_i) = R_i/N$. Therefore the substitution yields the following basic rank estimator $\hat{B}_{N1}$ of $B_N$,

$$\hat{B}_{N1}(t) = \sqrt{\frac{m\,n}{N}}\left(\frac{1}{m}\sum_{i=1}^{m} 1\big(\frac{R_i}{N} \leq t\big) - \frac{1}{n}\sum_{j=m+1}^{N} 1\big(\frac{R_j}{N} \leq t\big)\right)$$

$$= \sum_{i=1}^{N} c_{Ni} \, 1\left(\frac{R_i}{N} \le t\right), \qquad 0 \le t \le 1, \tag{3.1.4}$$

where $c_{Ni}$ are the two-sample coefficients defined in formula (3.0.4). For technical reasons it will be more suitable to start with the piecewise linearized version $\hat{B}_{N2}$ of $\hat{B}_{N1}$ given by the following linear interpolation of $\hat{B}_{N1}$,

$$\hat{B}_{N2}(\frac{i}{N}) := \hat{B}_{N1}(\frac{i}{N}), \quad i = 0, 1, ..., N, \text{ and}$$

$$\hat{B}_{N2} \text{ linear on } [\frac{i-1}{N}, \, \frac{i}{N}] \text{ for any } 1 \le i \le N. \tag{3.1.5}$$

Since the score function $b_N$ is a derivative of $B_N$ we may take $\hat{b}_{N2}$ , the right continuous version of the derivative of $\hat{B}_{N2}$ , as a primitive estimator of $b_N$ , i.e. for $i = 1, ..., N$ we have

$$\hat{b}_{N2}(t) \; = \; N\left(\hat{B}_{N1}(\frac{i}{N}) - \hat{B}_{N1}(\frac{i-1}{N})\right) \quad \text{if } \frac{i-1}{N} \le t < \frac{i}{N}, \tag{3.1.6}$$

This primitive estimator will be modified and smoothed in order to get better estimators of the underlying score function $b_N$ .

In case of the *one-sided* (stochastically larger) alternative $\mathcal{A}_2^0$ the function $B_N$ has the property $B_N \le 0$ . In this case we start with the "projection"

$$\hat{B}_{N2}^0 \; := \; \hat{B}_{N2} \, 1(\, \hat{B}_{N2} < 0 \,) \tag{3.1.7}$$

of $\hat{B}_{N2}$ onto the cone of nonpositive functions on the interval $[0, 1]$ and take the special version

$$\hat{b}_{N2}^0 \; = \; \hat{b}_{N2} \, 1(\, \hat{B}_{N2} < 0 \,) \tag{3.1.8}$$

of the derivative of the absolutely continuous function $\hat{B}_{N2}^0$ as the corresponding primitive estimator of $b_N$ . (Other projections are utilized in Section 3.2.)

Using $\hat{B}_{N2}$ and its derivative $\hat{b}_{N2}$ , the optimal linear rank statistic $S_N(b_N)$ with scores

$$b_N(i) \; = \; N \int_{(i-1)/N}^{i/N} b_N \, d\lambda \tag{3.1.9}$$

has the following simple form,

$$S_N(b_N) \; = \; \sum_{i=1}^{N} c_{Ni} \, b_N(R_i) \; = \; \int_0^1 b_N \, d\hat{B}_{N2} \; = \; \langle b_N, \hat{b}_{N2} \rangle , \tag{3.1.10}$$

where $\langle \cdot, \cdot \rangle$ is the usual inner product in $L_2(0, 1)$ . Our aim will be the construction of suitable estimators $\hat{b}_N$ of $b_N$ and the discussion of the asymptotic properties of the resulting new test statistics $S_N(\hat{b}_N) \; = \; \langle \hat{b}_N, \hat{b}_{N2} \rangle$ .

Because of $b_N = \sqrt{mn/N}\,(f_N - g_N)$, where $f_N$ and $g_N$ are the respective $\lambda$-densities of $H_N(X_1)$ and $H_N(X_N)$, the problem of estimating the score function $b_N$ is closely related to the problem of density estimation. One of the most popular methods of density estimation is the kernel method, which has its root in classical results on singular integrals of the following type.

**Theorem**

*Let $\{R_N\}$ be a sequence of nonnegative functions on $[0,1]^2$ having the following three properties for any fixed $t \in (0,1)$,*

$$\int_0^1 R_N(s,t)\,ds = 1 \quad \forall\, N\,, \tag{3.1.11}$$

$$\begin{aligned}
R_N(s,t) \text{ is increasing for } 0 < s < t \\
\text{and decreasing for } t < s < 1\,,
\end{aligned} \tag{3.1.12}$$

$$\lim_{n \to \infty} R_N(s,t) = 0 \quad \forall\, s \neq t\,. \tag{3.1.13}$$

*Then, for any integrable function $f : [0,1] \to \mathbb{R}$ which is continuous at the point $t \in (0,1)$, we get*

$$\lim_{N \to \infty} \int_0^1 f(s)\,R_N(s,t)\,ds = f(t). \tag{3.1.14}$$

There are many refinements of this theorem, see e.g. the book of Dunford and Schwartz (1958), Vol I, Theorem III.12.11. They are used for density estimation in the following way:

Let $Z_1, ..., Z_N$ be i.i.d. random variables with values in $[0,1]$ and density $f$. Let $F$ be the corresponding distribution function and let $\hat{F}_{N1}$ denote the empirical distribution function of the $Z_i$'s. Then $f(t)$ is estimated by an estimator of the approximating integral $\int_0^1 f(s)\,R_N(s,t)\,ds = \int_0^1 R_N(s,t)\,dF(s)$ of formula (3.1.14). A natural estimator obviously is

$$\int_0^1 R_N(s,t)\,d\hat{F}_{N1}(s) = \frac{1}{N}\sum_{i=1}^N R_N(Z_i,t)\,,$$

which is the usual form of a kernel density estimator of $f(t)$, cf. the paper of Rosenblatt (1971). Instead of the empirical jump distribution function $\hat{F}_{N1}$ we may as well use the corresponding piecewise linearized version $\hat{F}_{N2}$, which is defined as the linear interpolation of the points $(0,0)$, $(Z_i, \hat{F}_{N1}(Z_i))$, $1 \leq i \leq N$. In this case the corresponding estimator of $f(t)$ is

$$\int_0^1 R_N(s,t)\,d\hat{F}_{N2}(s) = \int_0^1 R_N(s,t)\,\hat{f}_{N2}(s)\,ds\,, \tag{3.1.15}$$

where $\hat{f}_{N2}$ is the right continuous version of the derivative of $\hat{F}_{N2}$ . Thus the latter estimator of $f(t)$ is obtained by computing the *weighted average* of the *primitive* estimator $\hat{f}_{N2}$ with respect to the weight function $R_N(\cdot,t)$ .

Now let's go back to the problem of estimating the optimal score function $b_N$ defined in (3.1.2). In principle the procedure is the same as in formula (3.1.15) where the primitive estimators $\hat{b}_{N2}$ or $\hat{b}^0_{N2}$ will be used in place of $\hat{f}_{N2}$ . In order to be definite we will use a *kernel* $K : \mathbb{R} \to \mathbb{R}$ with the following three basic properties,

$$\int_{-\infty}^{+\infty} K(x) \, dx = 1 , \tag{3.1.16}$$

$$K(x) = K(-x) \quad \forall \, x \in \mathbb{R} , \tag{3.1.17}$$

$$K(x) = 0 \qquad \text{if } |x| \geq 1 . \tag{3.1.18}$$

Usually $K$ will be nonnegative, i.e. a probability density on $\mathbb{R}$ which is symmetric about zero and concentrates all its mass onto the interval $[-1, 1]$ . Sometimes it will be convenient to admit negative values of $K$ , cf. the Dirichlet kernel of Example 3.1.4. Given the kernel $K$ and some *bandwidth* $a \in (0, 1]$ we define the corresponding *convolution kernel* $K_a : [0, 1]^2 \to \mathbb{R}$ according to

$$K_a(s,t) = \frac{1}{a} \left( K(\frac{t+s}{a}) + K(\frac{t-s}{a}) + K(\frac{t+s-2}{a}) \right) , \tag{3.1.19}$$

for all $0 \leq s,t \leq 1$ . The kernel $K_a$ will substitute the function $R_N$ of the above theorem.

Usually a convolution kernel corresponding to the kernel $K$ and the bandwidth $a$ would be defined by $a^{-1}K(a^{-1}(t-s))$ . Since $b_N$ and its primitive estimators are defined on the compact interval $[0, 1]$ we want the convolution kernel to concentrate all its mass onto this interval. If $t$ approaches $0$ or $1$ , however, we obviously don't get $\int_0^1 a^{-1} K(a^{-1}(t-s)) \, ds = 1$ . In order to avoid such "disappearing of mass" from the interval $[0, 1]$ for the convolution kernel $K_a$ , we have added $a^{-1}K(a^{-1}(t+s))$ , which puts the "disappearing" mass back to the interval $[0, 1]$ if $t$ is near $0$ , and similarly $a^{-1}K(a^{-1}(t+s-2))$ ,which puts the "disappearing" mass back to the interval $[0, 1]$ if $t$ is near $1$ , cf. Figure 3.1.a. Therefore (3.1.16) and (3.1.18) imply

$$\int_0^1 K_a(s,t) \, ds = 1 \quad \forall \, 0 \leq t \leq 1, \quad \forall \, 0 < a \leq 1 . \tag{3.1.20}$$

### Figure 3.1a

*The usual kernel* $K\big((t-s)/a\big)/a$ $(\cdots)$ *and the modified kernel* $K_a(s,t)$ $(\text{***})$
*with bandwidth* $a = 0.2$ *for* $t = 0.05$ *(left),* $t = 0.50$ *(middle), and* $t =$
$0.98$ *(right).*

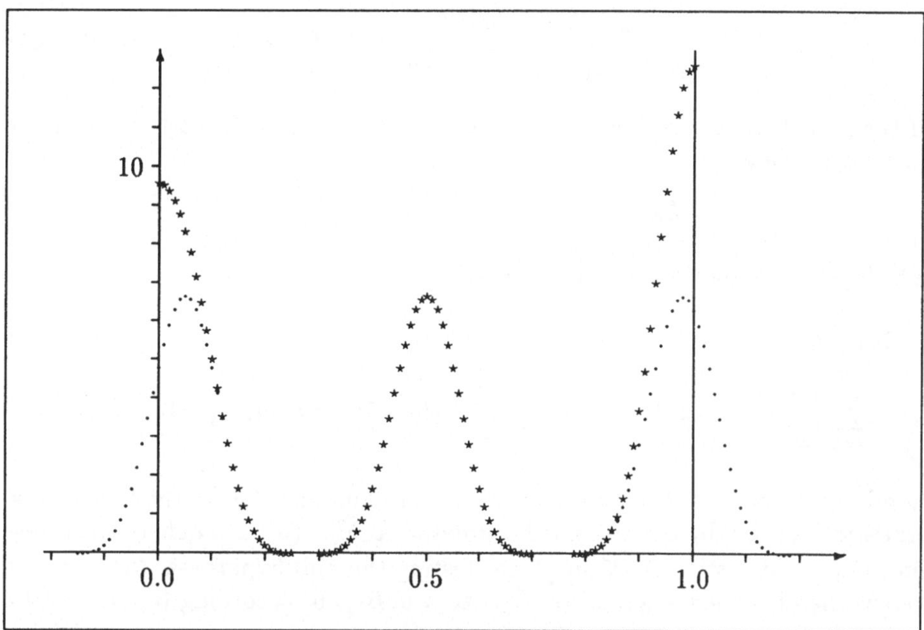

In accordance with (3.1.15) we define the estimator of $b_N$ in *case of the om-*
*nibus alternative* $\mathcal{A}_2 : B \neq 0$ as the convolution $\mathcal{K}_a \hat{b}_{N2}$ , i.e.

$$\mathcal{K}_a \hat{b}_{N2}(t) = \int_0^1 K_a(s,t)\, \hat{b}_{N2}(s)\, ds\ , \quad 0 \le t \le 1, \tag{3.1.21}$$

and in *case of the one-sided alternative* $\mathcal{A}_2^0 : B \le 0, B \neq 0$ as the convolution
$\mathcal{K}_a \hat{b}_{N2}^0$ , i.e.

$$\mathcal{K}_a \hat{b}_{N2}^0(t) = \int_0^1 K_a(s,t)\, \hat{b}_{N2}^0(s)\, ds\ , \quad 0 \le t \le 1. \tag{3.1.22}$$

Now the substitution of the optimal score function $b_N$ in the linear rank
statistic $S_N(b_N)$ by the estimator $\mathcal{K}_a \hat{b}_{N2}$ defines the (nonlinear) rank statistic $S_N(\mathcal{K}_a \hat{b}_{N2})$ for testing the null hypothesis $\mathcal{H}_0^r : B = 0$ versus the omnibus
alternative $\mathcal{A}_2 : B \neq 0$ according to, cf. (3.1.6),

$$S_N(\mathcal{K}_a \hat{b}_{N2}) = \int_0^1 ds\, \hat{b}_{N2}(s) \int_0^1 dt\, \hat{b}_{N2}(t)\, K_a(s,t) \tag{3.1.23}$$

$$= \sum_{i=1}^{N}\sum_{j=1}^{N} \left( \hat{B}_{N2}(\tfrac{i}{N}) - \hat{B}_{N2}(\tfrac{i-1}{N}) \right) \left( \hat{B}_{N2}(\tfrac{j}{N}) - \hat{B}_{N2}(\tfrac{j-1}{N}) \right) \bar{k}_{Na}(i,j),$$

where the weights $\bar{k}_{Na}(i,j)$ are defined by

$$\bar{k}_{Na}(i,j) \;=\; N^2 \int_{(i-1)/N}^{i/N} \int_{(j-1)/N}^{j/N} K_a(s,t)\,ds\,dt \;. \tag{3.1.24}$$

If the kernel $K$ is smooth we use instead of the weights (3.1.24) the simpler approximations

$$k_{Na}(i,j) \;=\; K_a(\frac{i-1/2}{N}, \frac{j-1/2}{N}) \;,$$

which yield the (nonlinear) rank statistic

$$S_N(a,K) \tag{3.1.25}$$
$$= \sum_{i=1}^{N}\sum_{j=1}^{N} \left( \hat{B}_{N2}(\tfrac{i}{N}) - \hat{B}_{N2}(\tfrac{i-1}{N}) \right) \left( \hat{B}_{N2}(\tfrac{j}{N}) - \hat{B}_{N2}(\tfrac{j-1}{N}) \right) k_{Na}(i,j).$$

Similarly the substitution of the optimal score function $b_N$ in the linear rank statistic $S_N(b_N)$ by the one-sided estimator $K_a \hat{b}_{N2}^0$ (3.1.22) defines the (nonlinear) rank statistic $S_N(K_a \hat{b}_{N2}^0)$ for testing the null hypothesis $\mathcal{H}_0^r : B = 0$ versus the one-sided alternative $\mathcal{A}_2^0 : B \le 0, B \ne 0$ according to, cf. (3.1.8) and (3.1.6),

$$S_N(K_a \hat{b}_{N2}^0) \;=\; \int_0^1 ds\, \hat{b}_{N2}^0(s) \int_0^1 dt\, \hat{b}_{N2}(t)\, K_a(s,t) \tag{3.1.26}$$
$$= \sum_{i=1}^{N}\sum_{j=1}^{N} \left( \hat{B}_{N2}(\tfrac{i}{N}) - \hat{B}_{N2}(\tfrac{i-1}{N}) \right) \left( \hat{B}_{N2}(\tfrac{j}{N}) - \hat{B}_{N2}(\tfrac{j-1}{N}) \right) \hat{k}_{Na}(i,j),$$

where the (random) one-sided weights $\hat{k}_{Na}(i,j)$ are defined by

$$\hat{k}_{Na}(i,j) \;=\; N^2 \int_{(i-1)/N}^{i/N} \int_{(j-1)/N}^{j/N} K_a(s,t)\, 1(\,\hat{B}_{N2}(s) < 0\,)\, ds\,dt \;. \tag{3.1.27}$$

If the kernel $K$ is smooth we use instead of the weights (3.1.27) the simpler approximations

$$K_a(\frac{i-1/2}{N}, \frac{j-1/2}{N})\, N \int_{(i-1)/N}^{i/N} 1(\,\hat{B}_{N2}(s) < 0\,)\, ds \;. \tag{3.1.28}$$

Since (3.1.6) and (3.1.8) imply the equality

$$N \left( \hat{B}_{N2}(\frac{i}{N}) - \hat{B}_{N2}(\frac{i-1}{N}) \right) \int_{(i-1)/N}^{i/N} 1( \hat{B}_{N2}(s) < 0 ) \, ds$$

$$= \int_{(i-1)/N}^{i/N} \hat{b}_{N2}^0(s) \, ds = \hat{B}_{N2}^0(\frac{i}{N}) - \hat{B}_{N2}^0(\frac{i-1}{N}) \tag{3.1.29}$$

for each $1 \leq i \leq N$, the substitution of $\hat{k}_{Na}(i,j)$ in (3.1.26) by its approximation (3.1.28) yields the (nonlinear) rank statistics

$$S_N^0(a, K) \tag{3.1.30}$$

$$= \sum_{i=1}^{N} \sum_{j=1}^{N} \left( \hat{B}_{N2}^0(\frac{i}{N}) - \hat{B}_{N2}^0(\frac{i-1}{N}) \right) \left( \hat{B}_{N2}(\frac{j}{N}) - \hat{B}_{N2}(\frac{j-1}{N}) \right) k_{Na}(i,j)$$

for testing the null hypothesis $\mathcal{H}_0^r : B = 0$ versus the one-sided alternative $\mathcal{A}_2^0 : B \leq 0, B \neq 0$.

**B) Equivalent representations and examples**

In addition to (3.1.16), (3.1.17), and (3.1.18) let us assume

$$\int_{-1}^{1} K^2(x) \, dx < \infty . \tag{3.1.31}$$

Then the convolution kernel (3.1.19) has the property

$$\int_0^1 \int_0^1 K_a^2(s,t) \, ds \, dt < \infty \tag{3.1.32}$$

and hence defines a linear operator $\mathcal{K}_a : L_2(0,1) \to L_2(0,1)$ of Hilbert-Schmidt type by putting, cf. (3.1.21) and (3.1.22),

$$\mathcal{K}_a g := \int_0^1 K_a(s, \cdot) \, g(s) \, \lambda(ds) \quad \forall \, g \in L_2(0,1) , \tag{3.1.33}$$

cf. Dunford and Schwartz (1963), Vol.2, XI.6. The symmetry assumption (3.1.17) implies the symmetry of $K_a$, i.e. $K_a(s,t) = K_a(t,s) \quad \forall \, s,t \in [0,1]$. Therefore the operator $\mathcal{K}_a$ is *selfadjoint*, i.e.

$$\langle \mathcal{K}_a f, g \rangle = \langle f, \mathcal{K}_a g \rangle \quad \forall \, f, g \in L_2(0,1) . \tag{3.1.34}$$

Using the operator $\mathcal{K}_a$ the rank statistic (3.1.23) may be written as, cf. (3.1.10),

$$S_N(\mathcal{K}_a \hat{b}_{N2}) = \langle \mathcal{K}_a \hat{b}_{N2}, \hat{b}_{N2} \rangle . \tag{3.1.35}$$

Similarly the rank statistic (3.1.26) has the representation

$$S_N(\mathcal{K}_a \hat{b}^0_{N2}) = \langle \mathcal{K}_a \hat{b}^0_{N2}, \hat{b}_{N2} \rangle .\qquad(3.1.36)$$

In order to get a better understanding of the performance of the proposed rank statistics (3.1.35) and (3.1.36) we shall utilize the spectral representation of the operator $\mathcal{K}_a$ .

### 3.1.1 Lemma

*For each $\kappa = 1, 2, \ldots$ we define $\psi_\kappa \in L_2^0(0,1)$ by*

$$\psi_\kappa(t) = -\sqrt{2}\,\cos(\pi \kappa t)\,,\qquad 0 < t < 1,\qquad(3.1.37)$$

*and for each $0 < a \le 1$ we put*

$$\lambda_\kappa(a) = \int_{-1}^{1} K(t)\,\cos(\pi \kappa a t)\,dt\,.\qquad(3.1.38)$$

*Then the following identity holds true,*

$$\mathcal{K}_a b = \langle b, 1 \rangle + \sum_{\kappa=1}^{\infty} \lambda_\kappa(a)\,\langle \psi_\kappa, b \rangle\,\psi_\kappa \qquad \forall\, b \in L_2(0,1),\qquad(3.1.39)$$

*where the convergence of the infinite series in (3.1.39) holds with respect to $L_2(0,1)$ .*

*Proof:*  From (3.1.20) we get $\mathcal{K}_a 1 = 1$ .  Since the system $\{1, \psi_1, \psi_2, \ldots\}$ is a complete orthonormal basis in $L_2(0,1)$ it suffices to prove $\mathcal{K}_a \psi_\kappa = \lambda_\kappa(a)\psi_\kappa\ \ \forall\,\kappa \ge 1$ .  Because of $0 < a \le 1$ and (3.1.17) to (3.1.19) we have for each $0 \le t \le 1$ the following chain of equations,

$$\int_0^1 K_a(s,t)\,\cos(\pi \kappa s)\,ds$$

$$= \int_{t/a}^{(t+1)/a} K(x)\,\cos(\pi \kappa(ax - t))\,dx + \int_{t/a}^{(t-1)/a} K(x)\,\cos(\pi \kappa(t - ax))\,dx$$

$$+ \int_{(t-2)/a}^{(t-1)/a} K(x)\,\cos(\pi \kappa(ax + 2 - t))\,dx$$

$$= \int_{(t-2)/a}^{(t+1)/a} K(x)\,\cos(\pi \kappa(ax - t))\,dx = \int_{-1}^{1} K(x)\,\cos(\pi \kappa(ax - t))\,dx$$

$$= \int_{-1}^{1} K(x)\,(\,\cos(\pi \kappa a x)\cos(\pi \kappa t) + \sin(\pi \kappa a x)\sin(\pi \kappa t)\,)\,dx$$

$$= \cos(\pi \kappa t) \int_{-1}^{1} K(x)\,\cos(\pi \kappa a x)\,dx = -\frac{1}{\sqrt{2}}\,\psi_\kappa(t)\,\lambda_\kappa(a)\,.$$

This concludes the proof, cf. definition (3.1.33).  □

Obviously the functions $1, \psi_1, \psi_2, \ldots$ are the *eigenfunctions* of $\mathcal{K}_a$ and the corresponding *eigenvalues* are $1, \lambda_1(a), \lambda_2(a), \ldots$ . Notice the remarkable feature that the operator $\mathcal{K}_a$ has the same system of eigenfunctions for all kernels (3.1.16) to (3.1.18) and (3.1.31) and for all bandwidths $a \in (0, 1]$ . Only the system of corresponding eigenvalues depends on the different kernels and on the different bandwidths. Using the equation (3.1.39) the following representations of the rank statistic (3.1.35) are obtained,

$$
\begin{aligned}
S_N&(\mathcal{K}_a \hat{b}_{N2}) \\
&= \langle \mathcal{K}_a \hat{b}_{N2}, \hat{b}_{N2} \rangle = \sum_{\kappa=1}^{\infty} \lambda_\kappa(a) \, \langle \psi_\kappa, \hat{b}_{N2} \rangle \, \langle \psi_\kappa, \hat{b}_{N2} \rangle \\
&= \sum_{\kappa=1}^{\infty} \lambda_\kappa(a) \, S_N^2(\psi_\kappa) = \sum_{\kappa=1}^{\infty} \lambda_\kappa(a) \, \langle \psi_\kappa', \hat{B}_{N2} \rangle^2,
\end{aligned} \tag{3.1.40}
$$

where the last equality is true since $\langle \hat{b}_{N2}, 1 \rangle = 0$ and partial integration yield $\langle \psi_\kappa', \hat{B}_{N2} \rangle = -\langle \psi_\kappa, \hat{b}_{N2} \rangle$ where $\psi_\kappa'(t) = d\psi_\kappa(t)/dt$ . Similarly we obtain the following representations of the one-sided rank statistic (3.1.36),

$$
\begin{aligned}
S_N(\mathcal{K}_a \hat{b}_{N2}^0) &= \langle \mathcal{K}_a \hat{b}_{N2}^0, \hat{b}_{N2} \rangle \\
&= \sum_{\kappa=1}^{\infty} \lambda_\kappa(a) \, \langle \psi_\kappa, \hat{b}_{N2}^0 \rangle \, \langle \psi_\kappa, \hat{b}_{N2} \rangle \\
&= \sum_{\kappa=1}^{\infty} \lambda_\kappa(a) \, \langle \psi_\kappa', \hat{B}_{N2}^0 \rangle \, \langle \psi_\kappa', \hat{B}_{N2} \rangle.
\end{aligned} \tag{3.1.41}
$$

From Theorem 3.0.1 we know the asymptotic optimality of the linear rank statistics $S_N(\psi_\kappa) = \langle \psi_\kappa, \hat{b}_{N2} \rangle$ for the null hypothesis $\mathcal{H}_0^r$ versus the corresponding sequences of local asymptotic alternatives $(F_{\varrho \Psi_\kappa, H}^N, G_{\varrho \Psi_\kappa, H}^N)$ defined in (3.0.9) with directions

$$
\Psi_\kappa = \int_0^t \psi_\kappa(s) \, ds = -(\pi\kappa)^{-1}\sqrt{2} \, \sin(\pi\kappa t) , \quad 0 \le t \le 1. \tag{3.1.42}
$$

Since the statistic $S_N(\mathcal{K}_a \hat{b}_{N2})$ is a quadratic form in these linear rank statistics $S_N(\psi_\kappa), \kappa \ge 1$ , the corresponding test $1(S_N(\mathcal{K}_a \hat{b}_{N2}) \ge k)$ should be sensitive with respect to a broad range of alternatives, where the actual power is governed by the weights $\lambda_\kappa(a), \kappa \ge 1$ . In general some of the eigenvalues $\lambda_\kappa(a)$ may become negative. This forces the corresponding test to be biased, since we shall prove that the test based on $\langle \mathcal{K}_a \hat{b}_{N2}, \hat{b}_{N2} \rangle$ is asymptotically unbiased, if and only if all the $\lambda_\kappa(a)$ 's are nonnegative, cf. (3.1.90) and (3.1.91).

Therefore we propose kernels K such that $\lambda_\kappa(a) \ge 0 \ \forall \ \kappa \ge 1 \ \forall \ a \in (0, 1]$ , cf. Example 3.1.3. For a given kernel K the shape of the sequence

$\{\lambda_\kappa(a), \kappa \geq 1\}$ is completely determined by the choice of the bandwidth $a \in (0,1)$. In general a small bandwidth causes a slow decrease of the eigenvalues $\lambda_\kappa(a)$ for $\kappa = 1, 2, ...$, while a large bandwidth causes a rapid decrease of the corresponding eigenvalues, see for example Figure 3.1.c. In principle the same remarks apply with respect to the one-sided test based on the rank statistic $S_N(\mathcal{K}_a \hat{b}^0_{N2})$ given in (3.1.41), though the single one-sided terms $\langle \psi_\kappa, \hat{b}^0_{N2} \rangle \langle \psi_\kappa, \hat{b}_{N2} \rangle$ don't allow as simple an interpretation as the corresponding omnibus terms $\langle \psi_\kappa, \hat{b}_{N2} \rangle \langle \psi_\kappa, \hat{b}_{N2} \rangle = S^2_N(\psi_\kappa)$ in (3.1.40). It seems plausible to propose rank statistics of the form $\sum \lambda_j \, S^2_N(b_j)$, $\lambda_j > 0$, for testing the null hypothesis $\mathcal{H}^r_0 : B = 0$ versus the omnibus alternative $\mathcal{A}_2 : B \neq 0$, whereas a comparable direct proposition for testing $\mathcal{H}^r_0$ versus the one-sided alternative $\mathcal{A}^0_2 : B \leq 0, B \neq 0$ seems to be much more difficult. In case of (3.1.41) the motivation is via the original representation $S_N(\mathcal{K}_a \hat{b}^0_{N2})$, i.e. the one-sided estimator $\mathcal{K}_a \hat{b}^0_{N2}$ of the optimal score function $b_N$ is used instead of the unknown $b_N$.

Special proposals and some connections to well-known statistics will be discussed in the following examples.

### 3.1.2 Example

Let us define the kernel $K : \mathbb{R} \to \mathbb{R}$ by

$$K(t) = \left( 2/3 - |t|/2 + |t|^2/4 \right) 1\left( |t| < 1 \right), \quad t \in \mathbb{R}. \tag{3.1.43}$$

By evaluation of (3.1.38) we get for any $0 < a \leq 1$ and any $\kappa \geq 1$

$$\lambda_\kappa(a) = \frac{\pi\kappa a - \sin(\pi\kappa a)}{(\pi\kappa a)^3} + \frac{5 \sin(\pi\kappa a)}{6(\pi\kappa a)}. \tag{3.1.44}$$

Putting $a = 1$ yields $\lambda_\kappa(1) = (\pi\kappa)^{-2} \; \forall \, \kappa \geq 1$. Therefore the corresponding omnibus statistic (3.1.40) has the form

$$S_N(\mathcal{K}_1 \hat{b}_{N2}) = \langle \mathcal{K}_1 \hat{b}_{N2}, \hat{b}_{N2} \rangle = \sum_{\kappa=1}^{\infty} \lambda_\kappa(1) \, \langle \psi'_\kappa, \hat{B}_{N2} \rangle^2$$

$$= \sum_{\kappa=1}^{\infty} \langle \sqrt{2} \, \sin(\pi\kappa \cdot), \hat{B}_{N2} \rangle^2 = \| \hat{B}_{N2} \|^2. \tag{3.1.45}$$

Similarly the corresponding one-sided statistic (3.1.41) has the form

$$S_N(\mathcal{K}_1 \hat{b}^0_{N2})$$

$$= \sum_{\kappa=1}^{\infty} \langle \sqrt{2} \, \sin(\pi\kappa \cdot), \hat{B}^0_{N2} \rangle \, \langle \sqrt{2} \, \sin(\pi\kappa \cdot), \hat{B}_{N2} \rangle \tag{3.1.46}$$

$$= \langle \hat{B}^0_{N2}, \hat{B}_{N2} \rangle = \| \hat{B}^0_{N2} \|^2,$$

where the last equality is immediate from the definition $\hat{B}_{N2} 1(\hat{B}_{N2} < 0)$ of $\hat{B}_{N2}^0$ . If we substitute the piecewise linear estimators $\hat{B}_{N2}$ and $\hat{B}_{N2}^0$ of $B_N$ in formulae (3.1.45) and (3.1.46) by their respective jump versions $\hat{B}_{N1}$ and $\hat{B}_{N1} 1(\hat{B}_{N1} < 0)$, then the corresponding statistics have the representations

$$\|\hat{B}_{N1}\|^2 = \frac{mn}{N} \int (\hat{F}_N - \hat{G}_N)^2 \, d\hat{H}_N \qquad (3.1.47)$$

and

$$\|\hat{B}_{N1} \, 1(\hat{B}_{N1} < 0)\|^2 = \frac{mn}{N} \int (\hat{F}_N - \hat{G}_N)^2 \, 1(\hat{F}_N < \hat{G}_N) \, d\hat{H}_N. \qquad (3.1.48)$$

The statistic (3.1.47) is the usual Cramér-von Mises statistic for testing $\mathcal{H}_0^r$ : $F = G$ versus $\mathcal{A}_2 : F \neq G$ , whereas the statistic (3.1.48) is the one-sided Cramér-von Mises statistic for testing $\mathcal{H}_0^r$ versus $\mathcal{A}_2^0 : F \leq G, F \neq G$ . Thus the Cramér-von Mises statistics essentially are linear rank statistics with estimated scores, where the estimators of the optimal score function are based on the rather artificial kernel (3.1.43) and the very large bandwidth $a = 1$ . The large bandwidth induces the rapid decrease of the eigenvalues $\lambda_\kappa(1) = (\pi\kappa)^{-2}$ for $\kappa = 1, 2, \dots$ . In practice only the first eigenvalues may be regarded as significantly different from zero, i.e. the omnibus Cramér-von Mises statistic (3.1.45) may be approximated by the rank statistic $\pi^{-2}(S_N^2(\psi_1) + S_N^2(\psi_2)/4)$ . In the spirit of Durbin and Knott (1972) the linear rank statistics $S_N(\psi_\kappa)$ in (3.1.40) should be called *components* of $S_N(\mathcal{K}_a \hat{b}_{N2})$ .

In the preceding example some of the eigenvalues $\lambda_\kappa(a), \kappa \geq 1$ , may become negative if $0 < a < 1$ . In the next example the kernel K will have the usual bellshaped form and all eigenvalues $\lambda_\kappa(a)$ will be nonnegative.

### 3.1.3 Example

Let us define the kernel $K : \mathbb{R} \to \mathbb{R}$ as the density of the random variable $(V_1 + V_2 + V_3 + V_4)/4$ , where $V_1, \dots, V_4$ are i.i.d. random variables with uniform distribution on the interval $(-1, 1)$ . By easy calculation we find

$$K(t) = \begin{cases} 4/3 - 8|t|^2 + 8|t|^3 , & \text{if } |t| \leq 1/2 , \\ 8(1 - |t|)^3/3 , & \text{if } 1/2 \leq |t| < 1 , \\ 0 , & \text{if } |t| \geq 1 . \end{cases} \qquad (3.1.49)$$

This kernel is known as *Parzen-2 kernel*, cf. Figure 3.1.b.

Since the characteristic function of each $V_j$ is given by $x^{-1}\sin(x)$ , the characteristic function $g$ of the probability density $K$ of $(V_1 + \dots + V_4)/4$ obviously is

$$g(x) = \left( \frac{\sin(x/4)}{x/4} \right)^4 , \qquad x \in \mathbb{R}. \qquad (3.1.50)$$

**Figure 3.1b**

*Graph of the Parzen-2 kernel* (3.1.49)

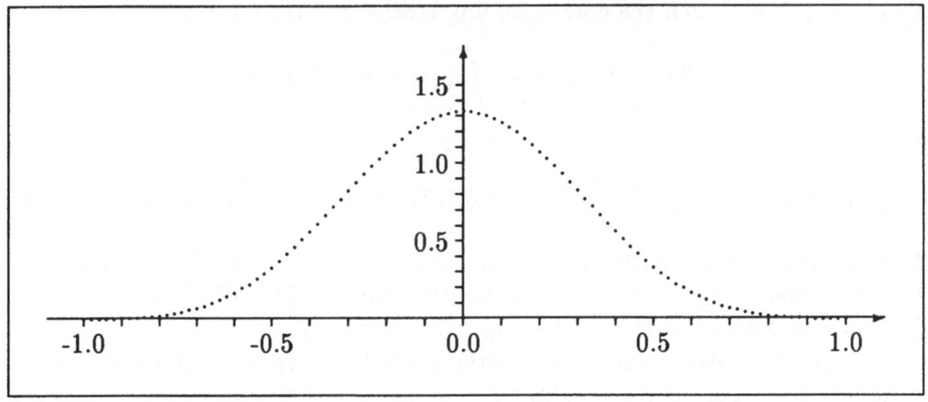

Therefore we may evaluate the $\lambda_\kappa(a)$ 's as

$$
\begin{aligned}
\lambda_\kappa(a) &= \int_{-1}^{1} \cos(\pi\kappa a t)\, K(t)\, dt \\[2mm]
&= \int_{-1}^{1} \exp(i\,\pi\kappa a t)\, K(t)\, dt \; = \; g(\pi\kappa a)\ .
\end{aligned}
\tag{3.1.51}
$$

The evaluation (3.1.51) is quite general, if K is a symmetric probability density with corresponding characteristic function g. The Parzen-2 kernel is well-known in time series analysis as a so-called lag window, see e.g. Chap. 8 of Koopmans (1974). Since the Parzen-2 kernel yields nonnegative eigenvalues $\lambda_\kappa(a)$ and since there is no essential difference between various bellshaped kernels, cf. Example 3.1.12, the Parzen-2 kernel is proposed as the basic kernel for the procedures of Chapter 2.

The kernels $K$ of Example 3.1.2 and Example 3.1.3 are probability densities on the interval $(-1, 1)$. If we allow negative values of $K$, the sequence of the corresponding eigenvalues $\{\lambda_\kappa(a), \kappa \geq 1\}$ may contain only a finite number of elements such that $\lambda_\kappa(a) \neq 0$, cf. the Dirichlet kernel of the next example.

### 3.1.4 Example

For each $r = 1, 2, \ldots$ let us define the kernel $K_r : \mathbb{R} \to \mathbb{R}$ by

$$
K_r(t) \; = \; \frac{\sin(\pi(r + 1/2)t)}{2\,\sin(\pi t/2)}\, 1\big(|t| < 1\big)\ , \quad t \in \mathbb{R}\ ,
\tag{3.1.52}
$$

## Figure 3.1c

*The graphs of the eigenvalues $\lambda_\kappa(a)$, $\kappa = 1, ..., 6$, with respect to the Parzen-2 kernel (3.1.49) as functions of the bandwidth $a \in (0, 1]$ .*

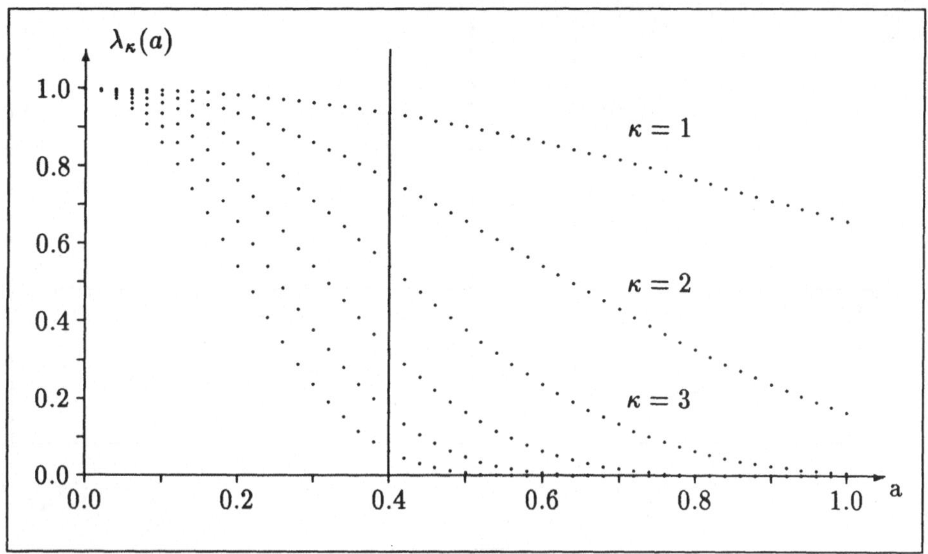


which is called *Dirichlet kernel* , cf. Figure 3.1.d.

In this case we define the convolution kernel (3.1.19) only for $a = 1$ , i.e. $K_r(s,t) = K_r(t+s) + K_r(t-s) + K_r(t+s-2)$ , and $r^{-1}$ plays the role of the bandwidth. If $\mathcal{K}_r$ denotes the operator according to (3.1.33) the corresponding eigenvalues $\lambda_\kappa(r) = \int K_r(t) \cos(\pi \kappa t) dt$ have the values

$$\lambda_\kappa(r) = 1 \quad \forall\, 1 \leq \kappa \leq r \quad \text{and} \quad \lambda_\kappa(r) = 0 \quad \forall\, \kappa > r \,. \tag{3.1.53}$$

Therefore the representation (3.1.39) becomes

$$\mathcal{K}_r b \;=\; \langle b, 1 \rangle + \sum_{\kappa=1}^{r} \langle \psi_\kappa, b \rangle\, \psi_\kappa \qquad \forall\, b \in L_2(0, 1), \tag{3.1.54}$$

which is a finite Fourier expansion of b in $L_2(0, 1)$ . If the linear subspace of $L_2(0, 1)$ generated by $1, \psi_1, ..., \psi_r$ is denoted by $V_r = [1, \psi_1, ..., \psi_r]$ then the operator $\mathcal{K}_r$ simply is the orthogonal projection onto $V_r$ , i.e. $\mathcal{K}_r = \Pi_{V_r} : L_2(0, 1) \to V_r$ . In case of this example the respective statistics (3.1.40) and (3.1.41) have the simple representations

$$S_N(\mathcal{K}_r \hat{b}_{N2}) \;=\; \sum_{\kappa=1}^{r} S_N^2(\psi_\kappa) \tag{3.1.55}$$

**Figure 3.1d**

*Graph of the Dirichlet kernel $K_r$  for  $r = 1$*

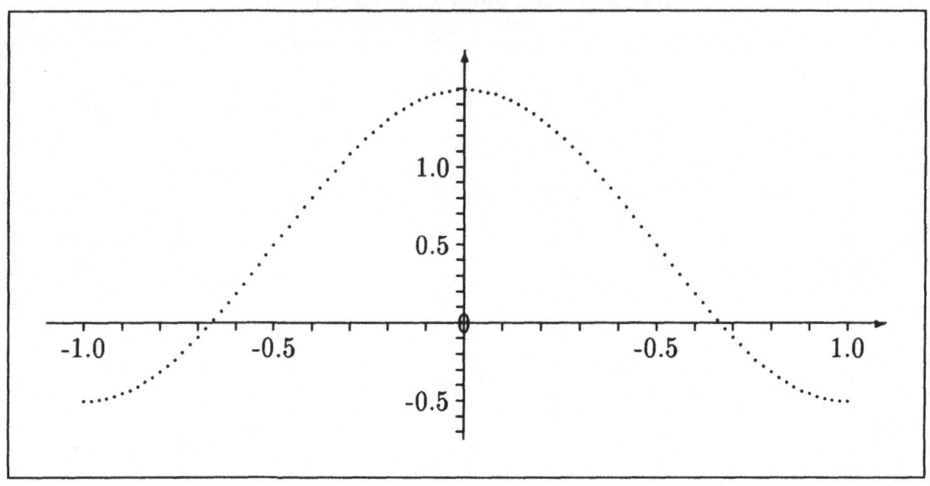

and

$$S_N(\mathcal{K}_r \hat{b}_{N2}^0) \; = \; \sum_{\kappa=1}^{r} \langle \psi_\kappa, \hat{b}_{N2}^0 \rangle \, S_N(\psi_\kappa). \tag{3.1.56}$$

In Section 3.2 we'll extend the situation of Example 3.1.4 by considering statistics of the form  $S_N(\Pi_V \hat{b}_{N2})$ , where  $\Pi_V$  is a suitable projection onto a suitable subspace or cone of  $L_2(0,1)$ .


## C) Asymptotic distribution under local asymptotic alternatives

It's quite easy to derive the limiting distribution of the two rank statistics  $S_N(\mathcal{K}_a \hat{b}_{N2})$  and  $S_N(\mathcal{K}_a \hat{b}_{N2}^0)$  if we use the respective expansions (3.1.40) and (3.1.41) and if the bandwidth  $a \in (0,1]$  is kept fixed while N tends to infinity. At first sight you would expect an asymptotic theory where  $a = a_N \to o$  with suitable speed while  $N \to \infty$ , since the estimation of  $b_N$  is similar to usual density estimation.  In fact you find such asymptotic theory in Behnen and Huskova (1984) under *fixed*  alternatives (F,G).It turns out that this theory collapses in case of sequences of alternatives  $(F_N, G_N)$  which are *contiguous* to the null hypothesis  $H_0 : F = G$ . Additionally, the asymptotic power results from local asymptotic theory with fixed bandwidth  $a \in (0,1]$  fit much better to our extensive power simulations under various types of alternatives than the respective power results from asymptotic theory with fixed alternatives and  $a = a_N \to 0$ . The deeper reason is that for fixed  $a \in (0,1]$  the smoothing operator $\mathcal{K}_a$ reduces the set of all possible score functions $\{B' : B \in \mathcal{B}_N^0\}$, cf.(1.3.25), to the much smaller (relatively compact) set $\{\mathcal{K}_a B' : B \in \mathcal{B}_N^0\}$. The generalizations of (3.1.14) may be utilized in order to prove $\mathcal{K}_a b \to b$ in $L_2(0,1)$ as $a \to 0$. Therefore the set $\{\mathcal{K}_a B' : B \in \mathcal{B}_N^0\}$ would tend to the huge set $L_2^0(0,1)$ if $a = a_N \to 0$ and $N \to \infty$. With respect to $L_2^0(0,1)$ there is no hope to find a uniformly good (data based) approximation of the unknown score function.

The limit results are based on a well-known functional limit theorem for the processes $\hat{B}_{N2} = (\hat{B}_{N2}(t) : 0 \le t \le 1), N \ge 1$, with values in the space $C[0,1]$ of continuous functions on the compact interval $[0,1]$ endowed with the sup-norm $\|\cdot\|_\infty$ -topology which generates the Borel $\sigma$ -algebra $\mathcal{B}(C[0,1])$. In the sequel let $W_0 = (W_0(t) : 0 \le t \le 1)$ denote the *Brownian-bridge* on $C[0,1]$, i.e. $W_0$ is a centered Gaussian process with covariances $EW_0(s)W_0(t) = s(1-t) \,\forall\, 0 \le s \le t \le 1$, and having continuous paths. The processes $\hat{B}_{N2}$ and $W_0$ may be regarded as random variables with values in $(C[0,1], \mathcal{B}(C[0,1]))$.

### 3.1.8 Theorem

*Assume $b \in L_2^0(0,1)$ and a corresponding sequence of directions $B_N \in \mathcal{B}_N^0$ such that $\|B_N' - b\| \to 0$ as $N \to \infty$. Similar to (3.0.9) define the sequences of local asymptotic alternatives*

$$(H_N + c_{N1}\varrho B_N \circ H_N, \; H_N + c_{NN}\varrho B_N \circ H_N), \quad N \ge 1, \quad 0 \le \varrho \le 1. \quad (3.1.57)$$

*Then we have the limiting law*

$$\mathcal{L}[\hat{B}_{N2} \mid (\varrho B_N, H_N)] \xrightarrow{\mathcal{L}} \mathcal{L}[W_0 + \varrho B] \; \text{ in } (C[0,1], \mathcal{B}(C[0,1])), \quad (3.1.58)$$

*where $B(t) = \int_0^t b(s)\, ds \,\forall\, 0 \le t \le 1$ and where $\xrightarrow{\mathcal{L}}$ means convergence in distribution, i.e. for any bounded continuous function $g : C[0,1] \to \mathbb{R}$ we have*

$$E[g(\hat{B}_{N2}) \mid (\varrho B_N, H_N)] \xrightarrow{N \to \infty} E[g(W_0 + \varrho B)]. \quad (3.1.59)$$

*Proof:* Theorem 3.1.8 is an immediate consequence of the (k–sample) Theorem 4.2.2. $\square$

This theorem will be utilized in order to prove convergence in distribution for the rank statistics $S_N(\mathcal{K}_a \hat{b}_{N2})$ and $S_N(\mathcal{K}_a \hat{b}_{N2}^0)$ under local asymptotic alternatives (3.1.57).

### 3.1.9 Theorem

*Let $0 < a \le 1$ be a fixed bandwidth and let $\lambda_\kappa = \lambda_\kappa(a)$ denote the eigenvalues (3.1.38) of the operator $\mathcal{K}_a$ corresponding to a kernel $K$ with the* *properties (3.1.16) to (3.1.18), (3.1.31), and*

$$\sum_{\kappa=1}^{\infty} |\lambda_\kappa| < \infty . \tag{3.1.60}$$

*Then, under the same assumptions and notations as in Theorem 3.1.8, we have the limiting law*

$$\mathcal{L}[\, S_N(\mathcal{K}_a \hat{b}_{N2}) \mid (\varrho B_N, H_N)] \xrightarrow{\mathcal{L}} \mathcal{L}[\sum_{\kappa=1}^{\infty} \lambda_\kappa \, (Z_\kappa + \varrho\langle \psi_\kappa, b\rangle)^2 ], \tag{3.1.61}$$

*where*

$$Z_\kappa := -\langle \psi'_\kappa, W_0\rangle , \quad \kappa = 1, 2, \cdots , \tag{3.1.62}$$

*are i.i.d. random variables with standard normal distribution.*

*Under the more stringent condition*

$$\sum_{\kappa=1}^{\infty} \kappa \, |\lambda_\kappa| < \infty \tag{3.1.63}$$

*we have the additional limiting law*

$$\mathcal{L}[\, S_N(\mathcal{K}_a \hat{b}_{N2}^0) \mid (\varrho B_N, H_N)] \xrightarrow{\mathcal{L}} \tag{3.1.64}$$

$$\mathcal{L}[\sum_{\kappa=1}^{\infty} \lambda_\kappa \, \langle \psi'_\kappa, W_0 + \varrho B\rangle \, \langle \psi'_\kappa, (W_0 + \varrho B) \, 1(W_0 + \varrho B < 0)\rangle ].$$

*Proof:*    As a first step let's prove that the random variables (3.1.62) are i.i.d. with standard normal distribution:
If $\{G_\kappa : \kappa \geq 1\}$ is a sequence of absolutely continuous functions $G_\kappa : [0, 1] \to$ IR  with derivatives $g_\kappa = G'_\kappa \in L_2(0, 1)$ then it's a wellknown fact that the stochastic integrals $\langle g_\kappa, W_0\rangle$, $\kappa \geq 1$ , are real-valued random variables with (joint) normal distribution, zero expectation, and covariances according to

$$\begin{aligned}
E[\, & \langle g_\kappa, W_0\rangle\langle g_\tau, W_0\rangle ] \\
&= \int_0^1 \int_0^1 g_\kappa(s) \, g_\tau(t) \, (\min(s, t) - st) \, \lambda(ds) \, \lambda(dt) \\
&= \int_0^1 G_\kappa \, G_\tau \, d\lambda - \int_0^1 G_\kappa \, d\lambda \int_0^1 G_\tau \, d\lambda,
\end{aligned} \tag{3.1.65}$$

where the first equality in (3.1.65) is a consequence of Theorem 1.3.10 in Ash and Gardener (1975), while the second equality in (3.1.65) is proved by partial integration.

Putting $G_\kappa = \psi_\kappa \; \forall \; \kappa \geq 1$ the orthonormality of the $\psi_\kappa$'s implies that the random variables (3.1.62) indeed are i.i.d. with $Z_\kappa \sim \mathcal{N}(0,1)$.

In a second step we'll consider the right hand sides of (3.1.61) and (3.1.64): Since for each $1 \leq \kappa < \infty$ the function $(\langle \psi_\kappa', f \rangle, \; f \in C[0,1])$ and the function $(\langle \psi_\kappa', f1(f < 0) \rangle, \; f \in C[0,1])$ are continuous functions from $(C[0,1], \|\cdot\|_\infty)$ to $\mathbb{R}$, we may define real-valued random variables $U_\kappa$ and $V_\kappa$ by putting

$$U_\kappa := \langle \psi_\kappa', W_0 + \varrho B \rangle \, ,$$
$$V_\kappa := \langle \psi_\kappa', (W_0 + \varrho B) \, 1(W_0 + \varrho B < 0) \rangle \, . \qquad (3.1.66)$$

Therefore

$$S_\infty := \sum_{\kappa=1}^{\infty} |\lambda_\kappa| \, U_\kappa^2 \quad \text{and} \quad T_\infty := \sum_{\kappa=1}^{\infty} |\lambda_\kappa| \, |U_\kappa \, V_\kappa| \qquad (3.1.67)$$

are well-defined $\overline{\mathbb{R}}$–valued random variables and Fatou's lemma implies

$$0 \leq E[S_\infty] \leq \liminf_{k \to \infty} \sum_{\kappa=1}^{k} |\lambda_\kappa| \, E[U_\kappa^2] \, ,$$
$$0 \leq E[T_\infty] \leq \liminf_{k \to \infty} \sum_{\kappa=1}^{k} |\lambda_\kappa| \, \sqrt{E[U_\kappa^2]} \, \sqrt{E[V_\kappa^2]} \, . \qquad (3.1.68)$$

On one hand from $\|\psi_\kappa\|^2 = 1$, (3.1.65), and $\langle \psi_\kappa', B \rangle = -\langle \psi_\kappa, b \rangle$ we get the inequality

$$E(U_\kappa^2) \leq 2E(\langle \psi_\kappa', W_0 \rangle^2) + 2\varrho^2 \langle \psi_\kappa', B \rangle^2$$
$$= 2 + 2\varrho^2 \langle \psi_\kappa, b \rangle^2 \leq 2 + 2\varrho^2 \|\psi_\kappa\|^2 \|b\|^2$$
$$= 2 + 2\varrho^2 \|b\|^2,$$

and therefore from (3.1.68) and assumption (3.1.60) the result

$$0 \leq E[S_\infty] \leq (2 + 2\varrho^2 \|b\|^2) \sum_{\kappa=1}^{\infty} |\lambda_\kappa| < \infty \, , \qquad (3.1.69)$$

on the other hand from $\|\psi_\kappa'\| = \pi\kappa$ we get the inequality

$$E(V_\kappa^2) \leq \|\psi_\kappa'\|^2 E(\|W_0 + \varrho B\|^2) = (\pi\kappa)^2 E(\|W_0 + \varrho B\|^2) < \infty,$$

and therefore from (3.1.68) and assumption (3.1.63) the result

$$0 \leq E[T_\infty] \leq \sqrt{(2 + 2\varrho^2 \|b\|^2) \, \pi^2 E[\|W_0 + \varrho B\|^2]} \sum_{\kappa=1}^{\infty} \kappa \, |\lambda_\kappa| < \infty. \qquad (3.1.70)$$

The inequalities (3.1.69) and (3.1.70) imply the random variables (3.1.67) to be finite [a.s.]. Therefore

$$X(a,\varrho,b) := \sum_{\kappa=1}^{\infty} \lambda_\kappa\, U_\kappa^2 = \sum_{\kappa=1}^{\infty} \lambda_\kappa\, (Z_\kappa + \varrho\langle\psi_\kappa,b\rangle)^2 \qquad (3.1.71)$$

and

$$Y(a,\varrho,b) := \sum_{\kappa=1}^{\infty} \lambda_\kappa\, U_\kappa\, V_\kappa$$
$$= \sum_{\kappa=1}^{\infty} \lambda_\kappa\, \langle\psi'_\kappa, W_0 + \varrho B\rangle\, \langle\psi'_\kappa, (W_0 + \varrho B)\, 1(W_0 + \varrho B < 0)\rangle \qquad (3.1.72)$$

are well-defined [a.s.] real random variables, and

$$X_k(a,\varrho,b) := \sum_{\kappa=1}^{k} \lambda_\kappa\, U_\kappa^2 \overset{k\to\infty}{\longrightarrow} X(a,\varrho,b) \quad [a.s.],$$
$$Y_k(a,\varrho,b) := \sum_{\kappa=1}^{k} \lambda_\kappa\, U_\kappa\, V_\kappa \overset{k\to\infty}{\longrightarrow} Y(a,\varrho,b) \quad [a.s.]. \qquad (3.1.73)$$

Now let's consider the left hand sides of (3.1.61) and (3.1.64):
For each $1 \le k < \infty$ the functions

$$\mathcal{G}_k(f) := \sum_{\kappa=1}^{k} \lambda_\kappa\, \langle\psi'_\kappa, f\rangle^2 \quad \forall\, f \in C[0,1]\,,$$
$$\mathcal{H}_k(f) := \sum_{\kappa=1}^{k} \lambda_\kappa\, \langle\psi'_\kappa, f\rangle\, \langle\psi'_\kappa, f1(f < 0)\rangle \quad \forall\, f \in C[0,1]\,, \qquad (3.1.74)$$

are continuous functions from $(\,C[0,1],\ \|\cdot\|_\infty\,)$ to $\mathbb{R}$, and (3.1.40) and (3.1.41) imply the representations $(\,\forall\, k \ge 1)\,$,

$$S_N(\mathcal{K}_a\hat{b}_{N2}) = \mathcal{G}_k(\hat{B}_{N2}) + \sum_{\kappa=k+1}^{\infty} \lambda_\kappa\, S_N^2(\psi_\kappa),$$
$$S_N(\mathcal{K}_a\hat{b}_{N2}^0) = \mathcal{H}_k(\hat{B}_{N2}) + \sum_{\kappa=k+1}^{\infty} \lambda_\kappa\, \langle\psi'_\kappa, \hat{B}_{N2}^0\rangle\, S_N(\psi_\kappa). \qquad (3.1.75)$$

Thus, on one hand for each fixed $k \ge 1$ and $N \to \infty$ we get from Theorem 3.1.8 the results

$$\mathcal{L}[\mathcal{G}_k(\hat{B}_{N2} \mid (\varrho B_N, H_N)] \overset{\mathcal{L}}{\longrightarrow} \mathcal{L}[\mathcal{G}_k(W_0 + \varrho B)] = \mathcal{L}[X_k(a,\varrho,b)] \qquad (3.1.76)$$

and

$$\mathcal{L}[\mathcal{H}_k(\hat{B}_{N2}\mid(\varrho B_N, H_N))] \xrightarrow{\mathcal{L}} \mathcal{L}[\mathcal{H}_k(W_0 + \varrho B)]] = \mathcal{L}[Y_k(a,\varrho,b)], \quad (3.1.77)$$

on the other hand we shall prove ( $\forall\, \varepsilon > 0$ )

$$\lim_{k\to\infty} \limsup_{N\to\infty} P_{(\varrho B_N, H_N)}\{|S_N(\mathcal{K}_a\hat{b}_{N2}) - \mathcal{G}_k(\hat{B}_{N2})| \ge \varepsilon\} = 0 \qquad (3.1.78)$$

and

$$\lim_{k\to\infty} \limsup_{N\to\infty} P_{(\varrho B_N, H_N)}\{|S_N(\mathcal{K}_a\hat{b}_{N2}^0) - \mathcal{H}_k(\hat{B}_{N2})| \ge \varepsilon\} = 0. \qquad (3.1.79)$$

This will complete the proof since (3.1.71), (3.1.73), (3.1.76), (3.1.78), and Theorem 4.2 of Billingsley (1967) prove the assertion (3.1.61), whereas the formulae (3.1.72), (3.1.73), (3.1.77), (3.1.79), and the same theorem prove the assertion (3.1.64). Notice, the assumption (3.1.63) has been used in proving the second part of (3.1.73).

The final proofs of (3.1.78) and (3.1.79) will be based on a contiguity argument. Under the null hypothesis $H_0 : F = G$ we get the following chain of inequalities (for any $k \ge 1$ and $N \ge 2$), cf. (3.1.75),

$$
\begin{aligned}
E_{H_0}|S_N(\mathcal{K}_a\hat{b}_{N2}) - \mathcal{G}_k(\hat{B}_{N2})| &\le \sum_{\kappa=k+1}^{\infty} |\lambda_\kappa|\, E_{H_0} S_N^2(\psi_\kappa) \\
&= \sum_{\kappa=k+1}^{\infty} |\lambda_\kappa|\, \frac{1}{N-1} \sum_{i=1}^{N} (N \int_{(i-1)/N}^{i/N} \psi_\kappa(x)\, dx)^2 \\
&\le \sum_{\kappa=k+1}^{\infty} |\lambda_\kappa|\, \frac{N}{N-1} \sum_{i=1}^{N} \int_{(i-1)/N}^{i/N} \psi_\kappa^2(x)\, dx) \\
&\le 2 \sum_{\kappa=k+1}^{\infty} |\lambda_\kappa|\, \|\psi_\kappa\|^2 = 2 \sum_{\kappa=k+1}^{\infty} |\lambda_\kappa|. \qquad (3.1.80)
\end{aligned}
$$

Therefore assumption (3.1.60) and Markov's inequality yield

$$\lim_{k\to\infty} \limsup_{N\to\infty} P_{H_0}\{|S_N(\mathcal{K}_a\hat{b}_{N2}) - \mathcal{G}_k(\hat{B}_{N2})| \ge \varepsilon\} = 0 \ \forall\, \varepsilon > 0. \qquad (3.1.81)$$

Since the definition (3.1.57) of local asymptotic alternatives forces the sequence of distributions $\{\, \mathcal{L}[(X_1,...,X_N)\mid(\varrho B_N, H_N)],\ N \ge 1\,\}$ to be contiguous to the null hypothesis sequence $\{\, \mathcal{L}[(X_1,...,X_N)\mid(0, H_N)],\ N \ge 1\,\}$, formula (3.1.81) implies (3.1.78), cf. Section 7.1 (Appendix).

The proof of (3.1.79) is completely similar, since $\|\psi'_\kappa\| = \pi\kappa$ , (3.1.75), and (3.1.80) imply the following chain of inequalities (for any $k \geq 1$ and $N \geq 2$ ),

$$E_{H_0}|S_N(\mathcal{K}_a \hat{b}^0_{N2}) - \mathcal{H}_k(\hat{B}_{N2})|$$

$$\leq \sum_{\kappa=k+1}^{\infty} |\lambda_\kappa| \sqrt{E_{H_0}\langle \psi'_\kappa, \hat{B}^0_{N2}\rangle^2} \sqrt{E_{H_0} S^2_N(\psi_\kappa)}$$

$$\leq \sqrt{2} \sum_{\kappa=k+1}^{\infty} |\lambda_\kappa| \sqrt{\|\psi'_\kappa\|^2 E_{H_0}\|\hat{B}_{N2}\|^2}$$

$$\leq \pi\sqrt{2E_{H_0}\|\hat{B}_{N2}\|^2} \sum_{\kappa=k+1}^{\infty} \kappa |\lambda_\kappa|$$

$$\leq \pi \sum_{\kappa=k+1}^{\infty} \kappa |\lambda_\kappa|, \tag{3.1.82}$$

where the last inequality holds true, since for each $0 \leq t \leq 1$ we have the representation, cf. (3.1.4) and (3.1.5),

$$\hat{B}_{N2}(t) = \sum_{i=1}^{N} c_{Ni} \left(1(R_i \leq [Nt]) + (Nt - [Nt])1([Nt] < R_i \leq [Nt]+1)\right),$$

$$\tag{3.1.83}$$

which implies $E_{H_0}\hat{B}_{N2}(t) = 0$ and

$$E_{H_0}\hat{B}^2_{N2}(t) = \frac{[Nt] + (Nt - [Nt])^2 - Nt^2}{N-1} \leq \frac{1}{4} \frac{N}{N-1}. \tag{3.1.84}$$

Now use assumption (3.1.63) instead of (3.1.60) and the contiguity argument in order to prove (3.1.79). $\square$

Because of (3.1.51) and (3.1.53) Theorem 3.1.9 applies to the Parzen-2 kernel (3.1.49) for each bandwidth $0 < a \leq 1$ and to the Dirichlet kernel (3.1.52) for each $1 \leq r < \infty$ . Under the null hypothesis $H_0 : B = 0$ the limiting distribution of the corresponding omnibus statistic $S_N(\mathcal{K}_a \hat{b}_{N2})$ is $\mathcal{L}\left[\sum_{\kappa=1}^{\infty} \lambda_\kappa(a) Z^2_\kappa\right]$ , where $(Z^2_\kappa, \kappa \geq 1)$ are independent random variables with $\chi^2_1$ -distribution. Nowadays there are standard methods for the numerical computation of such distributions as well as for the computation of the limiting distribution $\mathcal{L}[X(a, \varrho, b)]$ , cf. (3.1.71) and (3.1.61), under local asymptotic alternatives.

Even under the null hypothesis $H_0 : B = 0$ the limiting distribution of the one-sided statistic $S_N(\mathcal{K}_a \hat{b}^0_{N2})$ is much more complicated, cf. (3.1.64). At the time being a method for numerical computation of the distribution is not available since the random variables $\langle \psi'_\kappa, W_0 1(W_0 < 0)\rangle$ , $\kappa \geq 1$ , are neither normally distributed nor stochastically independent. Therefore we have to use tables of the exact critical values or Monte Carlo simulation.

In Section 2.1 we have proposed the approximation $S_N^0(a, K)$ of the statistic $S_N(\mathcal{K}_a \hat{b}_{N2}^0)$, cf. definition (3.1.30), based on the Parzen-2 kernel (3.1.49) and on the corresponding convolution kernel (3.1.19) with bandwidth $a = 0.40$. The limiting distribution of $S_N^0(a, K)$ is the same as the limiting distribution of $S_N(\mathcal{K}_a \hat{b}_{N2}^0)$, cf. Corollary 3.1.10, and the discussion of the final subsection of the present section will reveal the attractive power properties of the corresponding rank test.

For $m, n \in \{10, 20, 30, 40, 50\}$ the (simulated) critical values of the statistic $S_N = S_N^0(0.40, \text{Parzen-2})$ are given in Table 2.1.B. For other values of $m, n$ and $a$ we use a Monte Carlo simulation program for the evaluation of the actual p-values

$$P_{H_0}\{S_N^0(a, \text{Parzen-2}) > s_0\} \quad \text{and} \quad P_{H_0}\{S_N^0(a, \text{Parzen-2}) \geq s_0\}$$

for any observed value $s_0$ of the statistic $S_N^0(a, \text{Parzen-2})$.

### 3.1.10 Corollary

*Assume local asymptotic alternatives of the form (3.1.57) corresponding to $b \in L_2^0(0, 1)$. Assume $K$ to be a kernel with the properties (3.1.16) to (3.1.18) and (3.1.31). Let $0 < a \leq 1$ be a fixed bandwidth and let $\lambda_\kappa = \lambda_\kappa(a)$ denote the eigenvalues (3.1.38) of the operator $\mathcal{K}_a$ corresponding to $K$ and $a$.*

*a) If condition (3.1.60) holds true, then the statistic $S_N(a, K)$ defined in formula (3.1.25) has the same limiting law as the statistic $S_N(\mathcal{K}_a \hat{b}_{N2})$, cf. (3.1.61).*

*b) If condition (3.1.63) holds true, then the statistic $S_N^0(a, K)$ defined in formula (3.1.30) has the same limiting law as the statistic $S_N(\mathcal{K}_a \hat{b}_{N2}^0)$, cf. (3.1.64).*

*Proof:* In order to prove the assertions we define a modified empirical (jump) rank process $\tilde{B}_{N1}$ according to

$$\tilde{B}_{N1}(t) = \hat{B}_{N1}(t + \frac{1}{2N}) = \sum_{i=1}^{N} c_{Ni}\, 1(\frac{R_i - 1/2}{N} \leq t),\ 0 \leq t \leq 1. \qquad (3.1.85)$$

Then $\hat{B}_{N2}(i/N) = \hat{B}_{N1}(i/N) = \tilde{B}_{N1}((i - 1/2)/N)$, (3.1.25), and (3.1.39) imply

$$S_N(a, K) = \sum_{i=1}^{N}\sum_{j=1}^{N} K_a(\frac{i - 1/2}{N}, \frac{j - 1/2}{N})\, \Delta\hat{B}_{N1}(\frac{i}{N})\, \Delta\hat{B}_{N1}(\frac{j}{N})$$

$$= \int_0^1 \int_0^1 K_a(s,t)\, \tilde{B}_{N1}(ds)\, \tilde{B}_{N1}(dt) = \sum_{\kappa=1}^{\infty} \lambda_\kappa \left( \int_0^1 \psi_\kappa\, d\tilde{B}_{N1} \right)^2$$

$$= \sum_{\kappa=1}^{\infty} \lambda_\kappa \, \langle \psi'_\kappa, \tilde{B}_{N1} \rangle^2 \,, \tag{3.1.86}$$

where

$$\Delta \hat{B}_{N1}\Big(\frac{i}{N}\Big) \;=\; \hat{B}_{N1}\Big(\frac{i}{N}\Big) - \hat{B}_{N1}\Big(\frac{i-1}{N}\Big).$$

Here the last equality is implied by Fubini's theorem using the representation
$\psi_\kappa(t) = - \int_0^1 \psi'_\kappa(s) 1(t \le s)\, ds + \psi_\kappa(1)$ . In order to get a similar representation
of $S_N^0(a, K)$ we use the abbreviations

$$\delta_N(i) = N \int_{(i-1)/N}^{i/N} 1(\hat{B}_{N2}(s) < 0)\, ds \quad \text{and}$$
$$\tilde{B}_{N1}^0 = \tilde{B}_{N1} 1(\tilde{B}_{N1} < 0). \tag{3.1.87}$$

As a consequence we get the equality

$$\delta_N(i)\Big(\hat{B}_{N1}\Big(\frac{i}{N}\Big) - \hat{B}_{N1}\Big(\frac{i-1}{N}\Big)\Big)$$
$$= \Big(\hat{B}_{N1}\Big(\frac{i}{N}\Big) \wedge 0\Big) - \Big(\hat{B}_{N1}\Big(\frac{i-1}{N}\Big) \wedge 0\Big) \tag{3.1.88}$$
$$= \tilde{B}_{N1}^0\Big(\frac{i-1/2}{N}\Big) - \tilde{B}_{N1}^0\Big(\frac{i-1/2-1}{N}\Big).$$

Therefore the formulae (3.1.29), (3.1.30), and (3.1.39) imply

$$S_N^0(a, K) = \sum_{i=1}^{N} \sum_{j=1}^{N} K_a\Big(\frac{i-1/2}{N}, \frac{j-1/2}{N}\Big)\, \delta_N(i)\, \Big(\hat{B}_{N1}\Big(\frac{i}{N}\Big) - \hat{B}_{N1}\Big(\frac{i-1}{N}\Big)\Big)$$
$$\Big(\hat{B}_{N1}\Big(\frac{j}{N}\Big) - \hat{B}_{N1}\Big(\frac{j-1}{N}\Big)\Big)$$
$$= \int_0^1 \int_0^1 K_a(s,t)\, \tilde{B}_{N1}^0(ds)\, \tilde{B}_{N1}(dt)$$
$$= \sum_{\kappa=1}^{\infty} \lambda_\kappa \Big( \int_0^1 \psi_\kappa\, d\tilde{B}_{N1}^0 \Big) \Big( \int_0^1 \psi_\kappa\, d\tilde{B}_{N1} \Big)$$
$$= \sum_{\kappa=1}^{\infty} \lambda_\kappa \, \langle \psi'_\kappa, \tilde{B}_{N1}^0 \rangle \, \langle \psi'_\kappa, \tilde{B}_{N1} \rangle. \tag{3.1.89}$$

Comparing the representation (3.1.40) of the statistic $S_N(K_a \hat{b}_{N2})$ and the representation (3.1.41) of the statistic $S_N(K_a \hat{b}_{N2}^0)$ with the respective representations (3.1.86) of $S_N(a, K)$ and (3.1.89) of $S_N^0(a, K)$ we notice only one difference, the rank processes $\hat{B}_{N2}$ and $\hat{B}_{N2}^0 = \hat{B}_{N2} 1(\hat{B}_{N2} < 0)$ have been substituted by the respective rank processes $\tilde{B}_{N1}$ and $\tilde{B}_{N1}^0 = \tilde{B}_{N1} 1(\tilde{B}_{N1} < 0)$ .

From the definition of $\tilde{B}_{N1}$ and from the representation (3.1.83) of $\hat{B}_{N2}$ we get the inequality

$$\|\tilde{B}_{N1} - \hat{B}_{N2}\|_\infty \leq 2 \max_{1 \leq i \leq N} |c_{Ni}| \leq \max(\frac{1}{\sqrt{m}}, \frac{1}{\sqrt{n}}) \overset{N \to \infty}{\longrightarrow} 0.$$

Using the notations of the proof of Theorem 3.1.9 this implies, for each fixed $k \geq 1$ and $N \to \infty$,

$$\mathcal{L}[\mathcal{G}_k(\tilde{B}_{N1}) \mid (\varrho B_N, H_N)] \overset{\mathcal{L}}{\longrightarrow} \mathcal{L}[X_k(a, \varrho, b)]$$

and

$$\mathcal{L}[\mathcal{H}_k(\tilde{B}_{N1}) \mid (\varrho B_N, H_N)] \overset{\mathcal{L}}{\longrightarrow} \mathcal{L}[Y_k(a, \varrho, b)].$$

The remaining parts of the proof are essentially the same as before since $\int_0^1 \psi_\kappa \, d\tilde{B}_{N1}$ may be written as a linear rank statistic,

$$\int_0^1 \psi_\kappa \, d\tilde{B}_{N1} = \sum_{i=1}^N \psi_\kappa(\frac{i - 1/2}{N}) \, (\hat{B}_{N1}(\frac{i}{N}) - \hat{B}_{N1}(\frac{i-1}{N}))$$

$$= \sum_{i=1}^N c_{Ni} \, \psi_\kappa(\frac{R_i - 1/2}{N}).$$

Therefore the proof is concluded. $\square$

## D) Bandwidth and asymptotic power

We restrict the discussion of asymptotic power properties to the (simpler) omnibus case. The case of one-sided alternatives $\mathcal{A}_2^0 : B \leq 0, B \neq 0$ is more difficult but the results are similar. Under the null hypothesis $\mathcal{H}_0^r : B = 0$ the limiting distribution of the omnibus statistics $S_N(\mathcal{K}_a \hat{b}_{N2})$ and $S_N(a, K)$ is

$$\mathcal{L}[X(a, 0, 0)], \quad \text{with} \quad X(a, 0, 0) = \sum_{\kappa=1}^\infty \lambda_\kappa(a) \, Z_\kappa^2,$$

where $(Z_\kappa, \kappa \geq 1)$ are i.i.d. random variables with standard normal distribution $\mathcal{N}(0, 1)$. Let's fix some level $0 < \alpha < 1$ and let $c_\alpha(a)$ denote the upper $\alpha$-quantile of the limiting distribution,

$$P\{X(a, 0, 0) > c_\alpha(a)\} = \alpha.$$

Then the sequence of tests $\varphi_{Na} = 1(S_N(a, K) > c_\alpha(a))$ has the asymptotic level $\alpha$ for testing the null hypothesis $\mathcal{H}_0^r$ versus the omnibus alternative $\mathcal{A}_2 : B \neq 0$.

Let's take any $k \geq 1$ and apply Theorem 3.1.9 to the local asymptotic alternatives corresponding to $b = \psi_k$ . Obviously the limiting power of the test $\varphi_{Na}$ is

$$P\{ \sum_{\kappa=1,\kappa\neq k}^{\infty} \lambda_\kappa(a)\, Z_\kappa^2 + \lambda_k(a)(Z_k + \varrho)^2 \; > \; c_\alpha(a) \; \}. \qquad (3.1.90)$$

Therefore the asymptotic level $\alpha$ test $(\varphi_{Na}, N \geq 1)$ is asymptotically unbiased for all local asymptotic alternatives of the form (3.1.57) if and only if

$$\lambda_\kappa(a) \geq 0 \quad \forall\, \kappa \geq 1. \qquad (3.1.91)$$

Additionally the sequence of tests $(\varphi_{Na}, N \geq 1)$ may be proved to be asymptotically admissible if the condition (3.1.91) is fulfilled, cf. Neuhaus (1987). Therefore the test corresponding to the Parzen-2 kernel (3.1.49) is asymptotically unbiased and asymptotically admissible for any bandwidth $0 < a \leq 1$ , cf. Example 3.1.3. This means that this test cannot be improved for some special direction without diminishing the asymptotic power for other directions. In this sense the asymptotic test $(\varphi_{Na}, N \geq 1)$ puts its power in an optimal way onto the *principal directions* $b = \psi_\kappa, \; \kappa = 1, 2, \dots$ , if condition (3.1.91) is fulfilled. The influence of each principal direction $\psi_\kappa$ is weighted by the corresponding eigenvalue $\lambda_\kappa(a)$ . Because of the representation (3.1.86),

$$S_N(a, K) \; = \; \sum_{\kappa=1}^{\infty} \lambda_\kappa(a)\, \langle \psi_\kappa', \tilde{B}_{N1} \rangle^2,$$

the test $\varphi_{Na}$ is completely determined by the sequence $(\lambda_\kappa(a), \kappa \geq 1)$ .

In order to apply the test $\varphi_{Na}$ in statistical practice we have to select the kernel $K$ and the bandwidth $0 < a \leq 1$ . Similarly to density estimation the choice of the bandwidth is much more important than the choice of the special kernel, as long as the kernels are bellshaped. The following theorem will illustrate this point.

### 3.1.11 Theorem

*Let $K$ be a bounded and symmetric probability density on $\mathbb{R}$ with the support contained in $[-1, 1]$ , i.e., especially the conditions (3.1.16) to (3.1.18) and (3.1.31) are fulfilled. Then, if $a \to 0$ , the following limiting law holds true,*

$$\mathcal{L}[\sqrt{a} \sum_{\kappa=1}^{\infty} \lambda_\kappa(a)\, (Z_\kappa^2 - 1)] \xrightarrow{\mathcal{L}} \mathcal{N}(0, 2 \int_{-1}^{1} K^2(x)\, dx), \qquad (3.1.92)$$

*where $(Z_\kappa, \kappa \geq 1)$ are i.i.d. $\mathcal{N}(0,1)$ random variables, cf. Theorem 3.1.9.*

Before giving the proof let's discuss the consequences. According to (3.1.92) the two kernels $K_1$ and $K_2$ lead (approximately) to the same limiting distribution if the corresponding bandwidths $a_1$ and $a_2$ fulfil the condition

$$\frac{1}{a_1} \int_{-1}^{1} K_1^2(x)\, dx \; = \; \frac{1}{a_2} \int_{-1}^{1} K_2^2(x)\, dx. \qquad (3.1.93)$$

Now take two of the typical bellshaped kernels $K_1, K_2$ and choose the bandwidths $a_1$ and $a_2$ such that condition (3.1.93) is fulfilled. At least in case of the usual examples we've found that this implies the corresponding sequences of eigenvalues $(\lambda_\kappa(a_1, K_1), \kappa \geq 1)$ and $(\lambda_\kappa(a_2, K_2), \kappa \geq 1)$ to be almost identical, cf. Example 3.1.12. Therefore the corresponding tests are almost identical, too.

*Proof of Theorem 3.1.11:* In order to prove (3.1.92) it suffices to prove the convergence of any moment of the left-hand side to the corresponding moment of the $\mathcal{N}\big(0,\, 2\int K^2(x)dx\,\big)$ –distribution, if $0 < a \to 0$. For this proof we utilize the following well-known equality, cf. Smirnow (1966), Chap. I §21–24,

$$\int_0^1 K_a^{(i)}(x,x)dx \; = \; 1 + \sum_{\kappa=1}^{\infty} \lambda_\kappa^i(a) \quad \text{if } i \geq 2, \qquad (3.1.94)$$

where $K_a^{(i)}$ is the i-th iterated kernel corresponding to $K_a$ ,

$$K_a^{(i+1)}(s,t) = \int_0^1 K_a^{(i)}(s,x)\, K_a(x,t)\, dx, \qquad K_a^{(1)} \;=\; K_a. \qquad (3.1.95)$$

Especially for each $0 < a \leq 1$ we get the property

$$\sum_{\kappa=1}^{\infty} |\lambda_\kappa(a)|^i < \infty \quad \text{if } i \geq 2. \qquad (3.1.96)$$

Now let's define an approximating sequence according to

$$X_k(a) \; := \; \sum_{\kappa=1}^{k} \lambda_\kappa(a)(Z_\kappa^2 - 1), \quad k \geq 1. \qquad (3.1.97)$$

Obviously $EX_k(a) \; = \; 0$, and for any $p = 2, 3, \ldots$ we get the following representation of the p-th moment,

$$EX_k^p(a) = \sum_{i_1=1}^{k} \cdots \sum_{i_p=1}^{k} \lambda_{i_1}(a) \cdots \lambda_{i_p}(a) \, E[\,(Z_{i_1}^2 - 1)\ldots(Z_{i_p}^2 - 1)\,]$$

$$= \sum_{r=1}^{[p/2]} \sum_{\mathcal{P}_r} \prod_{j=1}^{r} (E[\,(Z_1^2 - 1)^{|I_j|}\,] \sum_{\kappa=1}^{k} \lambda_\kappa^{|I_j|}(a)), \qquad (3.1.98)$$

where the summation with respect to $\mathcal{P}_r$ means the summation with respect to all partitions $\{I_1, ..., I_r\}$ of the set $\{1, ..., p\}$ such that $|I_j| \geq 2 \;\; \forall\, j$. Since (3.1.96) and (3.1.98) obviously imply

$$\limsup_{k \to \infty} |EX_k^p(a)| < \infty \quad \forall\, p \geq 2 \quad \forall\, 0 < a \leq 1, \tag{3.1.99}$$

the sequence $(X_k^p(a),\; k \geq 1)$ is uniformly integrable for each $p \geq 1$ and each $0 < a \leq 1$. Therefore (3.1.98) and (3.1.94) imply the representation

$$EX_\infty^p(a) = \lim_{k \to \infty} EX_k^p(a) \tag{3.1.100}$$

$$= \sum_{r=1}^{[p/2]} \sum_{\mathcal{P}_r} \prod_{j=1}^{r} (E[(Z_1^2 - 1)^{|I_j|}] \sum_{\kappa=1}^{\infty} \lambda_\kappa^{|I_j|}(a))$$

$$= \sum_{r=1}^{[p/2]} \sum_{\mathcal{P}_r} \prod_{j=1}^{r} (E[(Z_1^2 - 1)^{|I_j|}] \, (\int_0^1 K_a^{(|I_j|)}(x, x)\, dx - 1))$$

for each $p \geq 2$. In addition the iteration (3.1.95) implies the following inequality ($\forall\, i \geq 1$)

$$\int_0^1 K_a^{(i+1)}(x, x)\, dx \leq \|K_a\|_\infty \int_0^1 \int_0^1 K_a^{(i)}(x, y)\, dx\, dy,$$

and

$$\int_0^1 \int_0^1 K_a^{(i)}(x, y)\, dx\, dy = 1 \quad \forall\, i \geq 1.$$

Therefore we get

$$\int_0^1 K_a^{(i+1)}(x, x)\, dx \leq \frac{3}{a} \|K\|_\infty \quad \forall\, i \geq 1 \quad \forall\, 0 < a \leq 1. \tag{3.1.101}$$

If $p = 2q + 1$ for some integer $q \geq 1$ then $[p/2] = q$. Thus (3.1.100) and (3.1.101) imply

$$E(\sqrt{a}X_\infty(a))^p = a^{q+1/2} O(a^{-q}) = O(\sqrt{a}) \xrightarrow{a \to 0} 0, \tag{3.1.102}$$

which concludes the proof for the convergence of the odd moments.

Now assume $p = 2q$ for some integer $q \geq 1$. Because of $[p/2] = q$ the formulae (3.1.100) and (3.1.101) in this case imply

$$E(\sqrt{a}X_\infty(a))^p$$

$$= a^q \Big( O(a^{1-q}) + \sum_{\mathcal{P}_q} \prod_{j=1}^{q} E(Z_1^2 - 1)^2 (\int_0^1 K_a^{(2)}(x, x)\, dx - 1) \Big)$$

$$= O(a) + |\mathcal{P}_q| \, (E(Z_1^2 - 1)^2)^q \, a^q \Big( \int_0^1 K_a^{(2)}(x, x)\, dx - 1 \Big)^q.$$

As the next step we'll prove the equality

$$\int_0^1 K_a^{(2)}(x,x)\,dx \;=\; \frac{1}{a}\int_{-1}^1 K^2(x)\,dx + \frac{1}{2}\,, \qquad (3.1.103)$$

which in turn implies

$$E(\sqrt{a}X_\infty(a))^p \;=\; O(a) + |\mathcal{P}_q|\,2^q\left(\int_{-1}^1 K^2(x)\,dx - \frac{a}{2}\right)^q. \qquad (3.1.104)$$

For the proof of (3.1.103) we use $0 < a \le 1$ and the definition (3.1.38) of $\lambda_\kappa(a)$ in order to get the Fourier expansion

$$\frac{1}{a}K(\frac{t}{a}) \;=\; \frac{1}{2} + \sum_{\kappa=1}^\infty \lambda_\kappa(a)\,\cos(\pi\kappa t) \qquad \text{in } L_2(-1,1). \qquad (3.1.105)$$

Then Parseval's identity and again $0 < a \le 1$ imply

$$\frac{1}{2} + \sum_{\kappa=1}^\infty \lambda_\kappa^2(a) = \int_{-1}^1 \left(\frac{1}{a}K(\frac{t}{a})\right)^2 dt = \frac{1}{a}\int_{-1}^1 K^2(x)\,dx.$$

Therefore formula (3.1.94) finally proves

$$\int_0^1 K_a^{(2)}(x,x)\,dx \;=\; 1 + \sum_{\kappa=1}^\infty \lambda_\kappa^2(a) \;=\; \frac{1}{2} + \frac{1}{a}\int_{-1}^1 K^2(x)\,dx.$$

Because of $p = 2q$ we have $|\mathcal{P}_q| = (2q)!\,/\,(q!\,2^q)$. Therefore formula (3.1.104) implies

$$E(\sqrt{a}X_\infty(a))^{2q} \;=\; O(a) + \frac{(2q)!}{q!\,2^q}\left(2\int_{-1}^1 K^2(x)\,dx - a\right)^q. \qquad (3.1.106)$$

Since the $(2q)$–th moment of the $\mathcal{N}(0,\sigma^2)$–distribution is $(2q)!\,\sigma^{2q}/(q!\,2^q)$ the proof of (3.1.92) is complete. $\square$

The result of Theorem 3.1.11 is in accordance with Theorem 2.2 of Behnen and Huskova (1984). Under additional smoothness assumptions on $K$ they prove

$$\mathcal{L}_{H_0}\left[\sqrt{a}\left(S_N(a,K) - \frac{1}{a}K(0)\right)\right] \xrightarrow{\mathcal{L}} \mathcal{N}(0, 2\int K^2(x)\,dx),$$

$$E_{H_0}S_N(a,K) \;=\; \frac{1}{a}K(0) - \frac{1}{2} + o(1),$$

if $N \to \infty$ and $a = a_N \to 0$, $Na_N \to \infty$. On the other hand the additional smoothness of $K$ implies, cf. (3.1.105),

$$\frac{1}{a}K(0) - \frac{1}{2} = \sum_{\kappa=1}^\infty \lambda_\kappa(a).$$

Therefore in this case the limiting law (3.1.92) has the form

$$\mathcal{L}\Big[\sqrt{a}\,\Big(\sum_{\kappa=1}^{\infty}\lambda_\kappa(a)\,Z_\kappa^2-\frac{1}{a}K(0))\Big)\Big]\ \xrightarrow{\mathcal{L}}\ \mathcal{N}\Big(0,2\int K^2(x)\,dx\Big)\quad\text{as}\ \ a\to 0.$$

If we use asymptotic critical values then we should utilize the limiting law (3.1.62) where the bandwidth is fixed while $N$ tends to infinity, since this approximation of the exact critical values seems to be better than the normal approximation for $a=a_N\to 0$, cf. Neuhaus (1987).

### 3.1.12 Example

Let $K_1$ be the Parzen-2 kernel (3.1.49) and let $K_2$ be the quartic kernel

$$K_2(x)\ =\ \frac{15}{16}\,(1-x^2)^2\,1(\,|x|\le 1\,),\qquad x\in\mathbb{R}.\tag{3.1.107}$$

Both of the kernels are bellshaped and fulfil the conditions of Theorem 3.1.11. By easy calculation we get

$$\int_{-1}^{1}K_1^2(x)\,dx\ =\ \frac{302}{315}\qquad\text{and}\qquad\int_{-1}^{1}K_2^2(x)\,dx\ =\ \frac{5}{7}.$$

Therefore the condition (3.1.93) means

$$a_2\ =\ \frac{225}{302}\,a_1,\tag{3.1.108}$$

i.e., if $0<a_1\le 1$ is a given bandwidth for the Parzen-2 kernel $K_1$ then we have to take the bandwidth $a_2=225a_1/302$ for the quartic kernel $K_2$ in order to get (approximately) the same limiting distribution of the corresponding rank statistics $S_N(a_1,K_1)$ and $S_N(a_2,K_2)$. But in fact the condition (3.1.108) has a much bigger effect: The resulting sequences of eigenvalues $(\lambda_\kappa(a_1,K_1),\kappa\ge 1)$ and $(\lambda_\kappa(a_2,K_2),\kappa\ge 1)$ are almost identical, cf. the subsequent table, which means that the corresponding tests are almost identical.

In order to evaluate the eigenvalues we use formulae (3.1.50) and (3.1.51) for the Parzen-2 kernel $K_1$, i.e.

$$\lambda_\kappa(a_1,K_1)\ =\ g_1(\pi\kappa a_1)\qquad\text{where}\qquad g_1(x)\ =\ \Big(\frac{\sin(x/4)}{x/4}\Big)^4,$$

and the corresponding formulae for the quartic kernel $K_2$,

$$\lambda_\kappa(a_2,K_2)=g_2(\pi\kappa a_2)\quad\text{where}\quad g_2(x)=\frac{15}{8}\,\frac{(24-8x^2)\sin(x)-24x\cos(x)}{x^5}.$$

The following table shows the first eight eigenvalues $\lambda_\kappa(a_1,K_1)$ and the corresponding eigenvalues $\lambda_\kappa(225a_1/302,K_2)$ for the three bandwidths

$$a_1\ =\ 0.3,\quad a_1\ =\ 0.4,\quad a_1\ =\ 0.5.$$

*Table of eigenvalues*
*corresponding to the kernels* $K_1$ *and* $K_2$
*for some equivalent bandwidths* $a_1$ *and* $a_2 = 225a_1/302$ .

| $\kappa$ : | 1 | 2 | 3 | 4 | 5 | 6 | 7 | 8 |
|---|---|---|---|---|---|---|---|---|
| $\lambda_\kappa(0.30, K_1)$ : | .96 | .86 | .71 | .54 | .38 | .24 | .13 | .06 |
| $\lambda_\kappa(0.22, K_2)$ : | .96 | .87 | .72 | .55 | .37 | .21 | .09 | .01 |
| $\lambda_\kappa(0.40, K_1)$ : | .94 | .77 | .54 | .33 | .16 | .01 | .00 | .00 |
| $\lambda_\kappa(0.30, K_2)$ : | .94 | .77 | .55 | .32 | .13 | .01 | −.04 | −.04 |
| $\lambda_\kappa(0.50, K_1)$ : | .90 | .66 | .38 | .16 | .05 | .01 | .00 | .00 |
| $\lambda_\kappa(0.37, K_2)$ : | .90 | .66 | .37 | .13 | −.01 | −.04 | −.02 | .00 |

The corresponding equivalent bandwidths for $K_2$ are

$$a_2 = 0.22, \quad a_2 = 0.30, \quad a_2 = 0.37.$$

Obviously there is no essential difference between the corresponding sets of eigenvalues. Therefore we propose to use a fixed kernel, namely the Parzen-2 kernel. The choice of the special bandwidth $a_1 = 0.40$ is based on extensive power simulations for different types of generalized shift alternatives and sample sizes ranging from $m = n = 10$ to $m = n = 40$ .

Having fixed the Parzen-2 kernel as the standard kernel we have to specify a suitable bandwidth $0 < a \leq 1$ . From (3.1.51), (3.1.50) or from Figure 3.1.c it may be seen that a small bandwidth makes quite a number of the eigenvalues $\lambda_\kappa(a)$ substantially different from zero whereas a bandwidth close to one only makes the first two eigenvalues substantially different from zero. As a result the test corresponding to a small bandwidth has rather low power but this power is kept on a huge set of alternatives. On the other hand the test corresponding to a bandwidth near one is almost a linear rank test and therefore almost optimal for just one direction, whereas the power for other directions may decrease to the level $\alpha$ . The compromise will be some bandwidth of medium size. In order to get a better insight into the situation of plausible models we have done some systematic computation of the expansion

$$b_N = \sum_{K=1}^{\infty} \langle b_N, \psi_\kappa \rangle \psi_\kappa, \tag{3.1.109}$$

where $b_N = B_N'$, $B_N = \sqrt{mn/N} \, (F - G) \circ H_N^{-1}$ , cf. formula (3.1.2). Especially we have done the computation for (F,G) corresponding to the generalized shift models (3.1.110), (3.1.111), with underlying normal distribution, logistic distribution, and Cauchy distribution, respectively. In all cases considered the first three or four of the weights $\langle b_N, \psi_\kappa \rangle$ dominate in the expansion (3.1.109). Therefore the bandwidth $a = 0.40$ (together with the Parzen-2 kernel) seems to be a good compromise, since the first eigenvalues $\lambda_\kappa(0.40)$, $\kappa = 1, \ldots 6$ have the values,

$$0.94, \quad 0.77, \quad 0.54, \quad 0.33, \quad 0.16, \quad 0.01 \,,$$

cf. the table in Example 3.1.12. The asymptotic discussion has been confirmed by extensive Monte Carlo power simulation under various types of generalized shift alternatives

$$F_{\vartheta D}(x) \;=\; G\big(x - \vartheta D(x)\big) \quad \forall\, x \in \mathbb{R}, \tag{3.1.110}$$

cf. Section 1.2. Especially the following types of shift functions $D$ have been considered,

$$
\begin{aligned}
&\text{1) lower shift:} && D(x) \;=\; 1 - G(x) \quad \forall\, x \in \mathbb{R}, \\[2mm]
&\text{2) central shift:} && D(x) \;=\; 4G(x)(1 - G(x)) \quad \forall\, x \in \mathbb{R}, \\[2mm]
&\text{3) upper shift:} && D(x) \;=\; G(x) \quad \forall\, x \in \mathbb{R}, \\[2mm]
&\text{4) exact shift:} && D(x) \;=\; 1 \quad \forall\, x \in \mathbb{R},
\end{aligned}
\tag{3.1.111}
$$

for the following cases of the underlying (symmetric) distribution function $G$ ,

$$
\begin{aligned}
&\text{1) Normal:} && G(x) = \Phi(x) = \frac{1}{\sqrt{2\pi}} \int_{-\infty}^{x} \exp(-\tfrac{1}{2}\, y^2)\, dy, \quad x \in \mathbb{R}, \\[3mm]
&\text{2) Logistic:} && G(x) = \frac{\exp(x)}{1 + \exp(x)}, \quad x \in \mathbb{R}, \\[3mm]
&\text{3) Cauchy:} && G(x) = \frac{1}{2} + \frac{1}{\pi}\, \arctan(x), \quad x \in \mathbb{R}.
\end{aligned}
\tag{3.1.112}
$$

Some of the results are shown in the graphs of Figure 2.1.c to Figure 2.1.h. For comparison and reference we have included the Wilcoxon rank test, which is a competitor in practical application.

The asymptotic power discussion has been concentrated to the omnibus situation, since here the interpretation of the corresponding asymptotic formulae is simpler than in the one-sided case. Since the one-sided problem is more important in practice, we have included the results of the Monte Carlo simulation for this case as a counterbalance, cf. Section 2.1. Additional results, also for the omnibus tests, may be found in Neuhaus (1987). Competitors based on projection methods will be developed in the next section.

## 3.2 Projection estimators of the score function

### A) Introduction

In Example 3.1.4 it has been shown that the Dirichlet kernel (3.1.52) defines a convolution operator $K_r$ on $L_2^0(0,1)$ which is equal to orthogonal projection $\Pi_{V_r}$ onto the linear subspace $V_r = [\psi_1, .., \psi_r]$ of $L_2^0(0,1)$. In this case the corresponding kernel estimator $K_r \hat{b}_{N2}$ of the underlying score function $b_N$ is $\Pi_{V_r} \hat{b}_{N2}$ and the corresponding (omnibus) rank statistic has the form $S_N(\Pi_{V_r} \hat{b}_{N2}) = < \Pi_{V_r} \hat{b}_{N2}, \hat{b}_{N2} >$, cf. (3.1.10). Obviously, $\Pi_{V_r} \hat{b}_{N2}$ may be regarded as that score function from the subspace $V_r$ which fits best the primitive estimator $\hat{b}_{N2}$ of $b_N$. Additionally, under generalized shift models of the form (3.1.110) to (3.1.112) the expansion (3.1.109) of $b_N$ may be approximated by the corresponding finite expansion of order three or four. Therefore we will generalize the idea of approximating $b_N$ by suitable projections of $\hat{b}_{N2}$ to general linear subspaces $V$ of $L_2^0(0,1)$.

In the omnibus problem $\mathcal{H}_0^r : B = 0$ versus $\mathcal{A}_2 : B \neq 0$ there is no special restriction for the corresponding score functions $b = B'$, whereas the one-sided problem $\mathcal{H}_0^r : B = 0$ versus $\mathcal{A}_2^0 : B \leq 0, B \neq 0$ obviously leads to the restriction $B \leq 0$ for the corresponding score functions $b = B'$. Therefore we will also consider projections of $\hat{b}_{N2}$ onto a closed convex cone in $L_2^0(0,1)$, i.e. onto a subset $V \subset L_2^0(0,1)$ which is closed with respect to the $\| \cdot \|$ –topology and which fulfils the following conditions,

$$v_1, v_2 \in V, \ \alpha \in (0,1) \ \implies \ \alpha v_1 + (1-\alpha)v_2 \in V, \tag{3.2.1}$$

$$v \in V, \ \alpha \geq 0 \ \implies \ \alpha v \in V. \tag{3.2.2}$$

In many examples the linear subspace will be finite dimensional, i.e. for some fixed $b_1, ..., b_r \in L_2^0(0,1)$ we have

$$V = [b_1, ..., b_r] = \{\sum_{i=1}^r \vartheta_i \, b_i \ : \ \vartheta_i \in \mathbb{R} \ \forall \ i\}, \tag{3.2.3}$$

or the closed convex cone will have the special form

$$V = [b_1, ..., b_r]^+ = \{\sum_{i=1}^r \vartheta_i \, b_i \ : \ \vartheta_i \geq 0 \ \forall \ i\}. \tag{3.2.4}$$

The latter cone will be a subcone of the closed convex cone

$$V_0 = \{b \in L_2^0(0,1) : \ B(t) = \int_0^t b(x) \, dx \leq 0 \ \ \forall \ 0 \leq t \leq 1\}, \tag{3.2.5}$$

if

$$B_i(t) \; = \; \int_0^t b_i(x)\, dx \le 0 \quad \forall\, 0 \le t \le 1 \quad \forall\, i = 1, ..., r. \tag{3.2.6}$$

Obviously $V_0$ corresponds to the one-sided testing problem $\mathcal{H}_0^r : B = 0$ versus $\mathcal{A}_2^0 : B \le 0, B \ne 0$ .

In the sequel we'll use the following general result.

### 3.2.1 Proposition

*Assume $\mathcal{H}$ to be a real Hilbert space with inner product $< \cdot, \cdot >$ and norm $\| \cdot \|$ .*

*a) If $V$ is a closed convex subset of $\mathcal{H}$ , then for any $b \in \mathcal{H}$ there exists an unique element $\Pi_V b \in V$ such that*

$$\|b - \Pi_V b\| \; = \; \inf_{v \in V} \|b - v\|. \tag{3.2.7}$$

*Therefore $\Pi_V b$ is called the projection of $b$ onto $V$ .*

*b) If $V \in \mathcal{H}$ is a closed convex cone, then the projection $\Pi_V b \in V$ of $b \in \mathcal{H}$ is uniquely determined by the following two conditions,*

$$< \Pi_V b, b > \; = \; \|\Pi_V b\|^2, \tag{3.2.8}$$

$$< b, v > \; \le \; < \Pi_V b, v > \quad \forall\, v \in V. \tag{3.2.9}$$

The *proof* is given in Section 7.2 (Appendix).

As  a consequence of Proposition 3.2.1 we get

$$\begin{aligned}
< \Pi_V b, b > \; &= \; \|\Pi_V b\|^2 \; = \; \|b\|^2 - \|b - \Pi_V b\|^2 \\
&= \; \|b\|^2 - \inf_{v \in V}(\|b\|^2 - 2 < b, v > + \|v\|^2) \\
&= \; \sup_{v \in V}(2 < b, v > - \|v\|^2) \\
&= \; \sup_{\lambda \ge 0}(2\lambda T_V(b) - \lambda^2),
\end{aligned} \tag{3.2.10}$$

where

$$T_V(b) \; = \; \sup\{< b, v >: \; v \in V, \; \|v\| = 1\}. \tag{3.2.11}$$

Obviously (3.2.10) implies the equality

$$< \Pi_V b, b > \; = \; \|\Pi_V b\|^2 \; = \; (\max(T_V(b), 0))^2 \; = \; (T_V^+(b))^2. \tag{3.2.12}$$

Applying these results to the case $\mathcal{H} = L_2(0,1)$ we get the following representation of any test statistic of the form $S_N(\Pi_V \hat{b}_{N2})$ , where $V$ is any convex cone in $L_2(0,1)$ ,

$$S_N(\Pi_V \hat{b}_{N2}) \;=\; < \Pi_V \hat{b}_{N2}, \hat{b}_{N2} > \;=\; (T_V^+(\hat{b}_{N2}))^2 . \tag{3.2.13}$$

Therefore the rank statistics $S_N(\Pi_V \hat{b}_{N2})$ and $T_V^+(\hat{b}_{N2}) = \max(T_V(\hat{b}_{N2}), 0)$ are equivalent statistics for testing $\mathcal{H}_0^r$ versus $\mathcal{A}_2^0$ or $\mathcal{A}_2$ . The subsequent examples will specify the respective cones and evaluate the corresponding rank statistics.

## B) Projecting onto infinite dimensional cones: Concave majorants, Galton's statistic, convex minorants

### 3.2.2 Example

As mentioned before, the closed convex cone $V_0 \subset L_2^0(0,1)$ defined in (3.2.5) corresponds to the one-sided testing problem $\mathcal{H}_0^r : B = 0$ versus $\mathcal{A}_2^0 : B \leq 0, B \neq 0$ . Therefore the rank statistic $S_N(\Pi_{V_0} \hat{b}_{N2})$ may be viewed as a test statistic which tries hard to detect *any* special alternative from $\mathcal{A}_2^0$ . Since $V_0$ is a very big cone it may be expected that the resulting power can't be very large. But let's try to compute the projection $\Pi_{V_0} \hat{b}_{N2}$ .

Let $\hat{B}_{N2}^{cc}$ denote the *concave majorant* of the piecewise linear rank process $\hat{B}_{N2}$ , i.e. for each given set of ranks $\hat{B}_{N2}^{cc}$ is the smallest concave function on the interval $[0,1]$ such that $\hat{B}_{N2}^{cc}(t) \geq \hat{B}_{N2}(t) \; \forall \; 0 \leq t \leq 1$ . Obviously $\hat{B}_{N2}^{cc}$ also is piecewise linear. Therefore the derivative

$$\hat{b}_{N2}^{cc}(t) \;=\; d\hat{B}_{N2}^{cc}(t)/dt \tag{3.2.14}$$

is well defined for almost all $0 \leq t \leq 1$ . (In order to be definite we may take the right continuous version.)

Using the derivative of the concave majorant of $\hat{B}_{N2}$ we'll prove the following equality,

$$\Pi_{V_0} \hat{b}_{N2} \;=\; \hat{b}_{N2} - \hat{b}_{N2}^{cc}. \tag{3.2.15}$$

Since $\hat{B}_{N2}^{cc}(0) = \hat{B}_{N2}^{cc}(1) = 0$ and $\hat{B}_{N2} - \hat{B}_{N2}^{cc} \leq 0$ are immediate consequences of the definition of $\hat{B}_{N2}^{cc}$ , we get $\hat{b}_{N2} - \hat{b}_{N2}^{cc} \in V_0$ . Because of the second part of Proposition 3.2.1 it suffices to prove the two statements

$$< \hat{b}_{N2} - \hat{b}_{N2}^{cc}, \hat{b}_{N2} > \;=\; \|\hat{b}_{N2} - \hat{b}_{N2}^{cc}\|^2 , \tag{3.2.16}$$

$$< \hat{b}_{N2}, b > \;\leq\; < \hat{b}_{N2} - \hat{b}_{N2}^{cc}, b > \quad \forall \, b \in V_0 . \tag{3.2.17}$$

In Section 7.3 (Appendix) it's proved that any function $g \in L_2(0,1)$ fulfils the condition $< g, b > \; \leq 0 \quad \forall \, b \in V_0$ if and only if the function $g$ is $\lambda - a.s.$ nonincreasing.

**Figure 3.2.a**

*The graphs of* $\hat{B}_{N2}$ *(* —— *),* $\hat{B}_{N2}^{cc}$ *(* ∘∘∘ *), and* $\hat{B}_{N2} - \hat{B}_{N2}^{cc}$ *(* ⋆⋆⋆ *).*

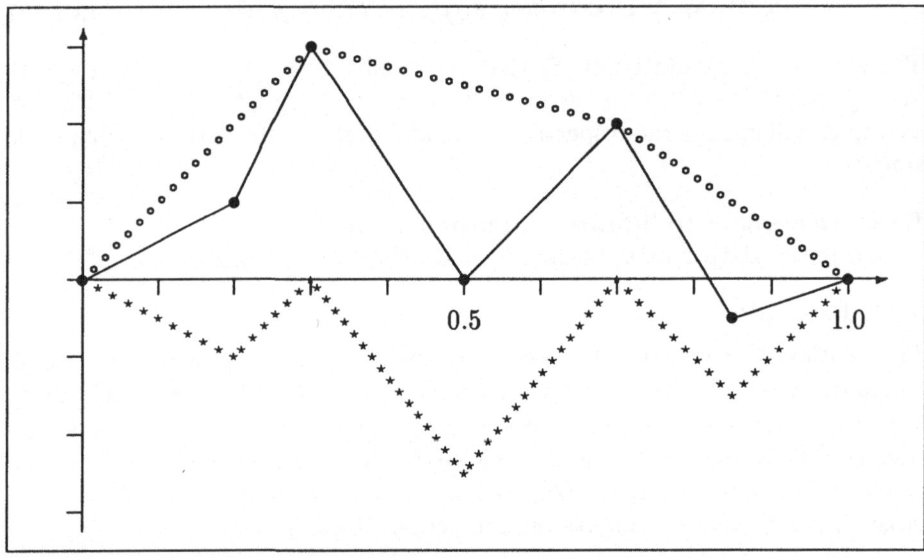


Since $\hat{B}_{N2}^{cc}$ is concave the derivative $\hat{b}_{N2}^{cc}$ is nonincreasing, which proves (3.2.17). For the proof of (3.2.16) we notice that $\hat{B}_{N2}^{cc}$ and $\hat{B}_{N2}$ coincide at the vertices of the graph of $\hat{B}_{N2}^{cc}$, cf. the illustration given in Figure 3.2.a. Since $\hat{b}_{N2}^{cc}$ is constant between two adjoining vertices, this implies the equality

$$< \hat{b}_{N2}^{cc}, \hat{b}_{N2} > = \ \|\hat{b}_{N2}^{cc}\|^2 \qquad (3.2.18)$$

which is equivalent to (3.2.16).

Because of (3.2.15) and (3.2.18) we get the one-sided rank statistic

$$\begin{aligned} S_N(\Pi_{V_0}\hat{b}_{N2}) \ &= \ < \Pi_{V_0}\hat{b}_{N2}, \hat{b}_{N2} > \\ &= \ \|\hat{b}_{N2}\|^2 - \|\hat{b}_{N2}^{cc}\|^2 \ = \ N - \|\hat{b}_{N2}^{cc}\|^2, \end{aligned} \qquad (3.2.19)$$

where the last equality holds true since

$$\int_0^1 \hat{b}_{N2}^2(x) \, dx \ = \ N \sum_{i=1}^N c_{Ni}^2 \ = \ N.$$

If the rank process $\hat{B}_{N2}$ is concave (for the given set of data), then the definition of the process $\hat{B}_{N2}^{cc}$ implies the equality $\hat{B}_{N2}^{cc} = \hat{B}_{N2}$ and hence the equalities $\Pi_{V_0}\hat{b}_{N2} = 0$ and $S_N(\Pi_{V_0}\hat{b}_{N2}) = 0$. Since the concavity of $\hat{B}_{N2}$ especially implies the inequality $\hat{B}_{N2}(t) \geq 0 \quad \forall\, 0 \leq t \leq 1$, the small value of the test statistic is in accordance with intuition. If $\hat{B}_{N2}(t) \leq 0 \quad \forall\, 0 \leq t \leq 1$ then $\hat{B}_{N2}^{cc} = 0$ and $\Pi_{V_0}\hat{b}_{N2} = \hat{b}_{N2}$. Again the maximal value $S_N(\Pi_{V_0}\hat{b}_{N2}) = N$ is in accordance with intuition.

Instead of the upper tests

$$1(S_N(\Pi_{V_0}\hat{b}_{N2}) > c) \;=\; 1(\|\hat{b}_{N2}^{cc}\|^2 < N - c) \qquad (3.2.20)$$

we may as well use the lower tests of the form

$$1(\|\hat{b}_{N2}^{cc}\| < \tilde{c}). \qquad (3.2.21)$$

In Groeneboom and Pyke (1983) the following limiting law has been proved,

$$\mathcal{L}_{\mathcal{H}_0^r}\Big[\frac{\|\hat{b}_{N2}^{cc}\|^2 - \log N}{\sqrt{3\log N}}\Big] \;\xrightarrow{\mathcal{L}}\; \mathcal{N}(0,1) \quad \text{as} \quad N \to \infty. \qquad (3.2.22)$$

From formula (3.2.22) it's immediate that $\|\hat{b}_{N2}^{cc}\|^2 / \log N$ converges to 1 in $\mathcal{H}_0^r$–probability. Therefore the representation

$$\frac{\|\hat{b}_{N2}^{cc}\|^2 - \log N}{\sqrt{3\log N}} \;=\; \frac{(\|\hat{b}_{N2}^{cc}\| - \sqrt{\log N})}{\sqrt{3}}\, \frac{(\|\hat{b}_{N2}^{cc}\| + \sqrt{\log N})}{\sqrt{\log N}}$$

and (3.2.22) prove the limiting law

$$\mathcal{L}_{\mathcal{H}_0^r}\Big[\frac{2}{\sqrt{3}}(\|\hat{b}_{N2}^{cc}\| - \sqrt{\log N})\Big] \;\xrightarrow{\mathcal{L}}\; \mathcal{N}(0,1). \qquad (3.2.23)$$

Groeneboom and Pyke (1983) have some comments about the speed of convergence of (3.2.22) and (3.2.23). The convergence in (3.2.22) seems to be extremely slow, while the speed of convergence in (3.2.23) seems to be somewhat better. For practical purposes we better rely on critical values or p-values derived by Monte Carlo simulation.

### 3.2.3 Galton's rank statistic

In Example 3.2.2 we have utilized the projection of the primitive estimator $\hat{b}_{N2}$ of $b_N$ onto the big cone $V_0$ which contains any $b_N$ corresponding to the one-sided alternative $\mathcal{A}_2^0 : B \leq 0, B \neq 0$. Here we'll try directly the projection of the rank process $\hat{B}_{N2}$ onto the $\mathcal{A}_2^0$–cone of all absolutely continuous functions $B$ on $[0,1]$ such that $B(0) = B(1) = 0$ and $B \leq 0$. It's easily seen that this projection is nothing but the rank process $\hat{B}_{N2}^0 = \hat{B}_{N2}1(\hat{B}_{N2} < 0)$, which has been defined in (3.1.7) as the natural one-sided version of the two-sample rank process. As before let's use the version $\hat{b}_{N2}^0 = \hat{b}_{N2}1(\hat{B}_{N2} < 0)$

of the derivative of $\hat{B}^0_{N2}$ . Clearly, $\hat{b}^0_{N2}$ is an element of the cone $V_0$ but in general $\hat{b}^0_{N2}$ is not the projection of $\hat{b}_{N2}$ onto $V_0$ , cf. Example 3.2.2. (The $\hat{b}^0_{N2}$ is the derivative of a projection instead of the projection of a derivative.) Nevertheless it does make sense to consider the corresponding one-sided rank statistic $S_N(\hat{b}^0_{N2}) = <\hat{b}^0_{N2}, \hat{b}_{N2}>$ .

On one hand we get from (3.1.6) and (3.1.4) the equalities

$$
\begin{aligned}
S_N(\hat{b}^0_{N2}) &= \sum_{i=1}^{N} \int_{(R_i-1)/N}^{R_i/N} 1(\hat{B}_{N2}(x) < 0)\hat{b}^2_{N2}(x)\, dx \\
&= N^2 \sum_{i=1}^{N} c^2_{Ni} \int_{(R_i-1)/N}^{R_i/N} 1(\hat{B}_{N2}(x) < 0)\, dx \\
&= N(\frac{n}{m} \sum_{i=1}^{m} \int_{(R_i-1)/N}^{R_i/N} 1(\hat{B}_{N2}(x) < 0)\, dx \\
&\quad + \frac{m}{n} \sum_{i=m+1}^{N} \int_{(R_i-1)/N}^{R_i/N} 1(\hat{B}_{N2}(x) < 0)\, dx), \qquad (3.2.24)
\end{aligned}
$$

on the other hand the slope of $\hat{B}_{N2}$ is equal to $Nc_{N1}$ on each of the intervals $[(R_i-1)/N,\ R_i/N)$, $i = 1,...,m$ and the slope of $\hat{B}_{N2}$ is equal to $Nc_{NN}$ on each of the intervals $[(R_i-1)/N, R_i/N)$, $i = m+1,...,N$ which implies

$$
\begin{aligned}
|c_{N1}| \sum_{i=1}^{m} & \int_{(R_i-1)/N}^{R_i/N} 1(\hat{B}_{N2}(x) < 0)\, dx \\
& = |c_{NN}| \sum_{i=m+1}^{N} \int_{(R_i-1)/N}^{R_i/N} 1(B_{N2}(x) < 0)\, dx.
\end{aligned}
\qquad (3.2.25)
$$

Because of $|c_{N1}|/|c_{NN}| = n/m$ the combination of (3.2.24) and (3.2.25) yields

$$
\begin{aligned}
S_N(\hat{b}^0_{N2}) &= N \sum_{i=1}^{N} \int_{(R_i-1)/N}^{R_i/N} 1(\hat{B}_{N2}(x) < 0)\, dx \\
&= N \int_0^1 1(\hat{B}_{N2}(x) < 0)\, dx = N\lambda\{x \in [0,1] : \hat{B}_{N2}(x) < 0\},
\end{aligned}
\qquad (3.2.26)
$$

which is the (generalized) *Galton rank statistic* , cf. Behnen and Neuhaus (1983). In the case of equal sample sizes $m = n$ the usual form of Galton's rank statistic is

$$
S^G_N = |\{i \in \{1,...,m\} : R^{(i)}_2 < R^{(i)}_1\}|, \qquad (3.2.27)
$$

where $R_1^{(1)} < \cdots < R_1^{(m)}$ and $R_2^{(1)} < \cdots < R_2^{(m)}$ are the ordered ranks of the first sample and the ordered ranks of the second sample, respectively. Plotting the graph of $\hat{B}_{N2}$ easily reveals the equality of (3.2.26) and (3.2.27). The exact null distribution of $S_N^G$ is given in Feller (1968), p.94,

$$P_{\mathcal{H}_0^\tau}\{S_N^G = i\} = \frac{1}{m+1}, \quad i = 0, 1, ..., m. \tag{3.2.28}$$

Since

$$\Lambda(f) = \lambda\{x \in [0,1] : f(x) < 0\}, \quad f \in C[0,1], \tag{3.2.29}$$

defines a function $\Lambda : C[0,1] \longrightarrow \mathbb{R}$ which is Borel measurable and continuous (with respect to $\|\cdot\|_\infty$ –norm) except on a set of $W_0$ –measure zero, cf. Billingsley (1968), pp. 231-232, Theorem 3.1.8 implies the following limiting law under local alternatives of the form (3.1.57),

$$\mathcal{L}[\frac{1}{N}S_N(\hat{b}_{N2}^0)|(\varrho B_N, H_N)] \xrightarrow{\mathcal{L}} \mathcal{L}[\Lambda(W_0 + \varrho B)]. \tag{3.2.30}$$

According to Billingsley (1968), pp.85-86, the random variable $\Lambda(W_0) = \lambda\{x \in [0,1] : W_0(x) < 0\}$ has a uniform distribution on the interval $[0,1]$. Therefore (3.2.30) implies the following approximate null distribution,

$$\lim_{N \to \infty} P_{\mathcal{H}_0^\tau}\{\frac{1}{N}S_N(\hat{b}_{N2}^0) > 1 - \alpha\} = \alpha \quad \forall\, 0 < \alpha < 1. \tag{3.2.31}$$

The (generalized) Galton test has some remarkable features: Since $\hat{B}_{N2}$ has been projected onto the very large cone of all absolutely continuous functions $B$ on $[0,1]$ such that $B(0) = B(1) = 0$ and $B \leq 0$ (which contains any direction $B$ corresponding to the alternative $\mathcal{A}_2^0$ ), it's not surprising that the power of Galton's test for any special direction is rather small, cf. Table 3 in Behnen and Neuhaus (1983). Galton's test, however, takes extreme precaution against acceptation of the alternative $\mathcal{A}_2^0 : B \leq 0, B \neq 0$ , if $\mathcal{A}_2^0$ in fact is not true. (At least approximately Galton's test may be viewed as a level $\alpha$ test for testing the null hypothesis "not $\mathcal{A}_2^0$ " versus the alternative $\mathcal{A}_2^0$ .) In order to specify this point let's assume $m/N \to \eta$ for some $0 < \eta < 1$ as $N \to \infty$ and let's take any pair $(F, G) \in \mathcal{F}^c \times \mathcal{F}^c$ . Putting $H_N = \frac{m}{N}F + \frac{n}{N}G$ and $H = \eta F + (1 - \eta)G$ the following properties can be proved, cf. Section 7.4 (Appendix),

$$\|\sqrt{\frac{N}{mn}}\hat{B}_{N2} - (F - G) \circ H_N^{-1}\|_\infty \xrightarrow{N \to \infty} 0 \quad \text{in (F,G)-probability}, \tag{3.2.32}$$

$$\|(F - G) \circ H_N^{-1} - (F - G) \circ H^{-1}\|_\infty \leq \frac{|m/N - \eta|}{\eta(1 - \eta)} \xrightarrow{N \to \infty} 0. \tag{3.2.33}$$

If $\Lambda$ is continuous (with respect to $\|\cdot\|_\infty$ –norm) at the point $(F-G)\circ H^{-1} \in C[0,1]$, then (3.2.32) and (3.2.33) obviously imply

$$\Lambda(\hat{B}_{N2}) \; = \; \Lambda(\sqrt{\frac{N}{mn}}\hat{B}_{N2}) \; \overset{N\to\infty}{\Longrightarrow} \; \Lambda((F-G)\circ H^{-1}) \qquad (3.2.34)$$

in $(F,G)$ –probability. Now take $(F,G) \notin \mathcal{A}_2^0 = \{(F,G) : F \leq G, F \neq G\}$ and assume

$$\Lambda((F-G)\circ H^{-1}) = \lambda\{x \in [0,1] : (F-G)\circ H^{-1}(x) < 0\} \; < \; 1-\alpha \quad (3.2.35)$$

and

$$|\{x \in [0,1] : (F-G)\circ H^{-1}(x) = 0\}| < \infty. \qquad (3.2.36)$$

Since (3.2.36) implies the continuity of $\Lambda$ at the point $(F-G)\circ H^{-1}$, we get from (3.2.34) the result

$$P_{(F,G)}\{\frac{1}{N}S_N(\hat{b}_{N2}^0) > 1-\alpha\} \; = \; P_{(F,G)}\{\Lambda(\hat{B}_{N2}) > 1-\alpha\} \overset{N\to\infty}{\Longrightarrow} 0. \quad (3.2.37)$$

This means especially that the power of Galton's test asymptotically stays below the level $\alpha$ for any fixed $(F,G)$ which deviates from the alternative $\mathcal{A}_2^0$ more than $\alpha$, where the measure of deviation is defined under the side condition (3.2.36) as $\Lambda\{x \in [0,1] : (F-G)\circ H^{-1}(x) \geq 0\}$. Obviously the condition (3.2.36) may be weakened substantially.

Additionally it's clear from (3.2.30) that Galton's test is asymptotically unbiased for testing the null hypothesis $\mathcal{H}_0 : B \geq 0$ versus the alternative $\mathcal{A}_2^0 : B \leq 0, B \neq 0$. Moreover, similar to the proof of (3.2.37) the consistency of Galton's test may be proved for any fixed alternative

$$(F,G) \in \mathcal{A}_2^0 \text{ with } \Lambda\{x \in [0,1] : (F-G)\circ H^{-1}(x) < 0\} > 1-\alpha.$$

### 3.2.4 Example

It's well-known that a linear rank test of the form $1(S_N(\psi) > c)$ is unbiased for testing the null hypothesis $\mathcal{H}_0 : F \geq G$ versus the alternative $\mathcal{A}_2^0 : F \leq G, F \neq G$, if the score function $\psi$ is nondecreasing $[\lambda - a.e.]$. Therefore we may consider the corresponding closed convex cone in $L_2(0,1)$, namely

$$V_* = \{b \in L_2^0(0,1) : \; b \text{ nondecreasing } [\lambda - a.e.]\}. \qquad (3.2.38)$$

Obviously $V_*$ is a subcone of $V_0$ and we'll prove

$$\Pi_{V_*}\hat{b}_{N2} = \hat{b}_{N2}^{cv}, \qquad (3.2.39)$$

where $\hat{b}_{N2}^{cv}$ is the (right continuous) derivative of the *convex minorant* $\hat{B}_{N2}^{cv}$ of the piecewise linear rank process $\hat{B}_{N2}$ , i.e. for each given set of ranks $\hat{B}_{N2}^{cv}$ is the largest convex function on the interval $[0,1]$ such that $\hat{B}_{N2}^{cv}(t) \leq \hat{B}_{N2}(t)$ $\forall\ 0 \leq t \leq 1$. Obviously $\hat{B}_{N2}$ is piecewise linear and $\hat{B}_{N2}^{cv}$ and $\hat{B}_{N2}$ coincide at the vertices of the graph of $\hat{B}_{N2}^{cv}$ . Since $\hat{B}_{N2}^{cv}$ is convex by definition, we get $\hat{b}_{N2}^{cv} \in V_*$ . Completely similar to the proof of (3.2.18) we get the equality

$$< \hat{b}_{N2}^{cv}, \hat{b}_{N2} > = \ \|\hat{b}_{N2}^{cv}\|^2. \tag{3.2.40}$$

Because of Proposition 3.2.1 the proof of (3.2.39) is concluded by proving the assertion

$$< \hat{b}_{N2}, b > \ \leq \ < \hat{b}_{N2}^{cv}, b > \quad \forall\ b \in V_*. \tag{3.2.41}$$

For the proof of (3.2.41) we use the abbreviations

$$b_{Ni} := \int_{(i-1)/N}^{i/N} b(x)\ dx \quad \text{and} \quad D_N := \hat{B}_{N2} - \hat{B}_{N2}^{cv}.$$

From $b \in V_*$ we get $b_{N1} \leq b_{N2} \leq \ldots \leq b_{NN}$ and the definition of $\hat{B}_{N2}^{cv}$ implies $D_N(0) = D_N(1) = 0$ and $D_N(t) \geq 0$ $\forall\ 0 \leq t \leq 1$. Since $\hat{b}_{N2}$ and $\hat{b}_{N2}^{cv}$ are constant on each of the intervals $[(i-1)/N, i/N)$ , we get

$$\begin{aligned}
< \hat{b}_{N2} - \hat{b}_{N2}^{cv}, b > \ &= \ \sum_{i=1}^{N} N\left(D_N(\frac{i}{N}) - D_N(\frac{i-1}{N})\right) b_{Ni} \\
&= \ N \sum_{i=1}^{N-1} D_N(\frac{i}{N})\, b_{Ni} \ - \ N \sum_{i=2}^{N} D_N(\frac{i-1}{N})\, b_{Ni} \\
&= \ -N \sum_{i=1}^{N-1} D_N(\frac{i}{N})\, (b_{Ni+1} - b_{Ni}) \ \leq \ 0,
\end{aligned}$$

which proves (3.2.41). Because of (3.2.39) and (3.2.40) the resulting one-sided rank statistic $S_N(\Pi_{V_*} \hat{b}_{N2})$ has the form

$$S_N(\Pi_{V_*} \hat{b}_{N2}) \ = \ < \Pi_{V_*} \hat{b}_{N2}, \hat{b}_{N2} > \ = \ \|\hat{b}_{N2}^{cv}\|^2. \tag{3.2.42}$$

We may as well use the equivalent statistic $\|\hat{b}_{N2}^{cv}\|$ . From (3.2.13) and (3.2.11) we get the representation

$$\|\hat{b}_{N2}^{cv}\| \ = \ S_N^{1/2}(\Pi_{V_*} \hat{b}_{N2}) \ = \ T_{V_*}(\hat{b}_{N2}) \vee 0, \tag{3.2.43}$$

where

$$T_{V_*}(\hat{b}_{N2}) \ = \ \sup\{< b, \hat{b}_{N2} >:\ b \in V_*, \|b\| = 1\}.$$

The statistic (3.2.43) is closely related to the statistic

$$S_N \;=\; \sup\{S_N(b): \; b \in V_*^N, \|b\| = 1\}, \tag{3.2.44}$$

where

$$V_*^N \;:=\; \{b \in V_* : b \text{ constant on } [\frac{i-1}{N}, \frac{i}{N}) \;\; \forall \, 1 \le i \le N\}.$$

This statistic has been proposed in Behnen (1975) and we'll prove the following equality,

$$S_N^+ \;=\; S_N \vee 0 \;=\; T_{V_*}(\hat{b}_{N2}) \vee 0 \;=\; \|\hat{b}_{N2}^{cv}\|. \tag{3.2.45}$$

Because of $S_N(b) = <b, \hat{b}_{N2}>$ and $V_*^N \subset V_*$ we get $S_N \le T_{V_*}(\hat{b}_{N2})$ and therefore $S_N^+ \le \|\hat{b}_{N2}^{cv}\|$. If $T_{V_*}(\hat{b}_{N2}) \le 0$ then the inequality $S_N^+ \ge 0 = \|\hat{b}_{N2}^{cv}\|$ is obvious from (3.2.43). Therefore the proof of (3.2.45) is concluded, if we prove the following implication

$$b \in V_*, \; \|b\| = 1, \; 0 < \int_0^1 \hat{b}_{N2} \, b \, dx \;\; \Longrightarrow \;\; S_N \ge \; <b, \hat{b}_{N2}> .$$

For the proof we use the abbreviations

$$\bar{b}_N[i] \;=\; N \int_{(i-1)/N}^{i/N} b(x) \, dx \text{ and } b_N(x) \;=\; \sum_{i=1}^{N} \bar{b}_N[i] \, 1(\frac{i-1}{N} \le x < \frac{i}{N})$$

In a first step the inequality

$$0 < \int_0^1 \hat{b}_{N2} \, b \, dx \;=\; \sum_{i=1}^{N} c_{Ni} \bar{b}_N[R_i]$$

proves $\|b_N\| > 0$. In a second step we get from $b \in V_*$ the monotonicity of $b_N$ and $\int_0^1 b_N \, dx = 0$. Finally $\|b\| = 1$ implies $\|b_N\| \le \|b\| = 1$. Thus $\tilde{b}_N := b_N/\|b_N\| \in V_*^N$ and

$$<\tilde{b}_N, \hat{b}_{N2}> \;=\; \frac{1}{\|b_N\|} <b_N, \hat{b}_{N2}> \;=\; \frac{1}{\|b_N\|} <b, \hat{b}_{N2}> \ge <b, \hat{b}_{N2}>$$

which concludes the proof.

For any critical value $c_N \ge 0$ the tests

$$1(S_N > c_N) \quad \text{and} \quad 1(\|\hat{b}_{N2}^{cv}\| > c_N) \tag{3.2.46}$$

are identical. In addition we get

$$P_{\mathcal{H}_0^r}\{S_N \le 0\} \;=\; P_{\mathcal{H}_0^r}\{\|\hat{b}_{N2}^{cv}\| = 0\}$$

$$=\; P_{\mathcal{H}_0^r}\{\hat{B}_{N2} \ge 0\} \;=\; P_{\mathcal{H}_0^r}\{\Lambda(\hat{B}_{N2}) = 0\} \stackrel{N \to \infty}{\longrightarrow} 0.$$

A symmetry argument proves

$$\mathcal{L}_{\mathcal{H}_0^r}(\|\hat{b}_{N2}^{cc}\|) \;=\; \mathcal{L}_{\mathcal{H}_0^r}(\|\hat{b}_{N2}^{cv}\|).$$

Therefore (3.2.23) implies the limiting law

$$\mathcal{L}_{\mathcal{H}_0^r}[\frac{2}{\sqrt{3}}(\|\hat{b}_{N2}^{cv}\| - \sqrt{\log N})] \xrightarrow{\mathcal{L}} \mathcal{N}(0,1). \tag{3.2.47}$$

According to the above remark the limiting law (3.2.47) also holds true for $S_N$ or $S_N^+$ instead of $\|\hat{b}_{N2}^{cv}\|$ . The Monte Carlo results of Behnen (1975) reveal the rather good power properties of the one-sided $S_N$ –test.


## C) Projecting onto finite dimensional cones

In Subsection B we have evaluated and discussed some projections onto infinite dimensional cones which are closely related to the general properties of the one-sided alternative $\mathcal{A}_2^0 : B \leq 0, B \neq 0$ . Since the cones are very large, the resulting tests have rather low power for large classes of alternatives. At the beginning of the section we have mentioned that for many relevant types of alternatives the underlying score functions $b_N$ can be approximated by a given finite set of special score functions. Thus, for practical purposes it may be better to derive the estimated score function as a projection of $\hat{b}_{N2}$ onto a suitable finite dimensional cone.


## a) Computation of the test statistic

Let the integer $r \geq 1$ be given and assume $b_1, ..., b_r \in L_2^0(0,1)$ to be special score functions corresponding to the one-sided testing problem $\mathcal{H}_0^r : B = 0$ versus $\mathcal{A}_2^0 : B \leq 0, B \neq 0$ , i.e. for each $i = 1, ..., r$ we assume

$$B_i(t) := \int_0^t b_i(x)\, dx \;\leq\; 0 \quad \forall\, 0 \leq t \leq 1. \tag{3.2.48}$$

In order to have a minimal set of score functions let's assume that $b_1, ..., b_r$ are linearly independent in $L_2(0,1)$ . Obviously this is equivalent to the condition

$$\Gamma \;=\; (<b_i, b_j>)_{\substack{i=1,...,r \\ j=1,...,r}} \qquad \text{positive definite .} \tag{3.2.49}$$

Finally we define the $r$ –dimensional cone $V \subset L_2^0(0,1)$ according to (3.2.4),

$$V := [b_1, ..., b_r]^+ \;=\; \{\sum_{i=1}^r \vartheta_i\, b_i : \vartheta_i \geq 0 \;\forall\, i\}. \tag{3.2.50}$$

Because of (3.2.48) this is a closed convex subcone of the cone $V_0$ which represents the one-sided testing problem. In the sequel let

$$\vec{S}_N = (< b_1, \hat{b}_{N2} >, ..., < b_r, \hat{b}_{N2} >)^T \qquad (3.2.51)$$

denote the column vector of the linear rank statsitics $S_N(b_i) = < b_i, \hat{b}_{N2} >$, $i = 1, ..., r$. In addition we use the abbreviations $\vartheta = (\vartheta_1, ..., \vartheta_r)^T \in \mathbb{R}^r$ and $\vartheta \geq 0$ iff $\vartheta_i \geq 0 \ \forall \ i = 1, ..., r$. Then, according to (3.2.10), the test statistic $S_N(\Pi_V \hat{b}_{N2})$ may be written in the form

$$\begin{aligned} S_N(\Pi_V \hat{b}_{N2}) &= \|\Pi_V \hat{b}_{N2}\|^2 = \sup_{b \in V}(2 < b, \hat{b}_{N2} > -\|b\|^2) \\ &= \sup_{\vartheta \geq 0}(2\vartheta^T \vec{S}_N - \vartheta^T \Gamma \vartheta) = f_0(\vec{S}_N), \end{aligned} \qquad (3.2.52)$$

where the function $f_0 : \mathbb{R}^r \to [0, \infty)$ is defined by

$$f_0(x) = \sup_{\vartheta \geq 0}(2\vartheta^T x - \vartheta^T \Gamma \vartheta), \quad x = (x_1, ..., x_r)^T \in \mathbb{R}^r. \qquad (3.2.53)$$

In a first step we simplify the evaluation of the test statistic $f_0(\vec{S}_N)$, in a second step we'll prove a limiting law of $S_N(\Pi_V \hat{b}_{N2})$ under the null hypothesis and under local alternatives, where the limiting null distribution will be given in an explicit form.

In order to evaluate $f_0(\vec{S}_N)$ let's define the $\Gamma$–scalar-product $< \cdot, \cdot >_\Gamma$ on $\mathbb{R}^r$ with corresponding norm $|\cdot|_\Gamma$ according to

$$< x, y >_\Gamma = x^T \Gamma y \quad \forall \ x, y \in \mathbb{R}^r. \qquad (3.2.54)$$

Remember, $\Gamma$ is a symmetric positive definite $r \times r$–matrix. Given any $x \in \mathbb{R}^r$ we get from (3.2.54)

$$f_0(x) = \sup_{\vartheta \geq 0}(2 < \vartheta, \Gamma^{-1}x >_\Gamma -|\vartheta|_\Gamma^2). \qquad (3.2.55)$$

Now let $\Pi_\Gamma : \mathbb{R}^r \to \{\vartheta \in \mathbb{R}^r : \vartheta \geq 0\}$ denote the (unique) projection onto the closed convex cone $\{\vartheta \in \mathbb{R}^r : \vartheta \geq 0\}$ with respect to the $\Gamma$–scalar-product (3.2.54). Then (3.2.10) and Proposition 3.2.1 imply

$$f_0(x) = < \Pi_\Gamma(\Gamma^{-1}x), \Gamma^{-1}x >_\Gamma = |\Pi_\Gamma(\Gamma^{-1}x)|_\Gamma^2. \qquad (3.2.56)$$

Especially we get from (3.2.52) the equalities

$$S_N(\Pi_V \hat{b}_{N2}) = \|\Pi_V \hat{b}_{N2}\|^2 = f_0(\vec{S}_N) = |\Pi_\Gamma(\Gamma^{-1}\vec{S}_N)|_\Gamma^2, \qquad (3.2.57)$$

where $\vec{S}_N$ is the vector of linear rank statistics defined in formula (3.2.51).

In order to evaluate the projection $\Pi_\Gamma(\Gamma^{-1}x)$ we need the following notation:
For $J \subset \{1,...,r\} =: R$ we put

$$\mathbb{R}^r_J := \{\vartheta = (\vartheta_1,...,\vartheta_r)^T \in \mathbb{R}^r : \vartheta_j = 0 \; \forall \, j \in R \setminus J\}$$

$$\mathbb{R}^{r+}_J := \{\vartheta \in \mathbb{R}^r_J : \vartheta_j > 0 \; \forall \, j \in J\}. \qquad (3.2.58)$$

Clearly, the definitions (3.2.58) imply

$$\mathbb{R}^r_\emptyset = \mathbb{R}^{r+}_\emptyset = \{0\}, \quad \mathbb{R}^r_R = \mathbb{R}^r, \quad \mathbb{R}^{r+}_R = \{\vartheta \in \mathbb{R}^r : \vartheta_j > 0 \; \forall \, j\},$$

$$\{\vartheta \in \mathbb{R}^r : \vartheta \geq 0\} = \sum_{J \subset R} \mathbb{R}^{r+}_J, \qquad (3.2.59)$$

where the sum-sign stands for disjoint union. Additionally, $\mathbb{R}^r_J$ is a $|J|-$
dimensional linear subspace generated by $\mathbb{R}^{r+}_J$ .

Using these notations the following lemma will facilitate the computation of
the projection $\Pi_\Gamma$ onto $\{\vartheta \in \mathbb{R}^r : \vartheta \geq 0\}$ .

### 3.2.5 Lemma

*For each $J \subset R = \{1,...,r\}$ let $\Pi_\Gamma^J : \mathbb{R}^r \to \mathbb{R}^r_J$ denote the projection with
respect to $< \cdot, \cdot >_\Gamma$ of $\mathbb{R}^r$ onto its linear subspace $\mathbb{R}^r_J$ .*

*a) For each $z \in \mathbb{R}^r$ and each $J \subset R$ we have the implication*

$$\Pi_\Gamma z \in \mathbb{R}^{r+}_J \quad \Longrightarrow \quad \Pi_\Gamma z = \Pi_\Gamma^J z. \qquad (3.2.60)$$

*b) For each $z \in \mathbb{R}^r$ we have the equalities*

$$f_0(\Gamma z) = |\Pi_\Gamma z|^2_\Gamma = \max\{|\Pi_\Gamma^J z|^2_\Gamma : \Pi_\Gamma^J z \geq 0, J \subset R\}. \qquad (3.2.61)$$

*Especially we get*

$$S_N(\Pi_V \hat{b}_{N2}) = \|\Pi_V \hat{b}_{N2}\|^2$$

$$= \max\{|\Pi_\Gamma^J(\Gamma^{-1}\vec{S}_N)|^2_\Gamma : \Pi_\Gamma^J(\Gamma^{-1}\vec{S}_N) \geq 0, J \subset R\}. \qquad (3.2.62)$$

*Proof:* Assume $z \in \mathbb{R}^r$ and put $\hat{z} := \Pi_\Gamma z$ .

a) For $J = \emptyset$ we have $\mathbb{R}^{r+}_\emptyset = \mathbb{R}^r_\emptyset = \{0\}$. In this case the implication (3.2.60)
is obvious. Therefore let's assume $\emptyset \neq J \subset R$ and $\hat{z} \in \mathbb{R}^{r+}_J$ . Then the
definition of $\hat{z}$ and the inclusion $\mathbb{R}^{r+}_J \subset \{\vartheta \in \mathbb{R}^r : \vartheta \geq 0\}$ imply

$$|\hat{z} - z|_\Gamma = \inf\{|\vartheta - z|_\Gamma : \vartheta \in \mathbb{R}^r, \vartheta \geq 0\} = \inf\{|\vartheta - z|_\Gamma : \vartheta \in \mathbb{R}^{r+}_J\}.$$

In order to prove $\Pi_\Gamma^J z = \hat{z}$ let's assume the contrary, i.e., $\tilde{z} := \Pi_\Gamma^J z \neq \hat{z}$ .
This assumption implies $\tilde{z} \in \mathbb{R}_J^r$ and $|\tilde{z} - z|_\Gamma < |\hat{z} - z|_\Gamma$ . Since $\hat{z} \in \mathbb{R}_J^{r+}$
implies $\hat{z}_j > 0 \; \forall \, j \in J$ , and since $\tilde{z}_j = \hat{z}_j = 0 \; \forall \, j \in R \setminus J$ , there exists some
(small) $0 < \varepsilon < 1$ such that $z^* := \varepsilon\tilde{z} + (1 - \varepsilon)\hat{z} \in \mathbb{R}_J^{r+}$ . This implies the
contradiction $|z^* - z|_\Gamma < |\hat{z} - z|_\Gamma$ , since we'll prove the function

$$g(t) \;:=\; |t\tilde{z} + (1 - t)\hat{z} - z|_\Gamma^2, \quad 0 \leq t \leq 1,$$

to be strictly convex and since

$$\inf_{0 \leq t \leq 1} g^2(t) \;=\; g(1) \;=\; |\tilde{z} - z|_\Gamma^2 \;<\; |\hat{z} - z|_\Gamma^2 \;=\; g(0).$$

The strict convexity of $g$ is obvious, since the representation

$$g(t) \;=\; t^2 |\tilde{z} - \hat{z}|_\Gamma^2 + 2t < \tilde{z} - \hat{z}, \hat{z} - z >_\Gamma + |\hat{z} - z|_\Gamma^2$$

implies

$$g"(t) \;=\; 2|\tilde{z} - \hat{z}|_\Gamma^2 \;>\; 0 \quad \forall \, 0 < t < 1.$$

b) Because of (3.2.59) there is a $J \subset R$ such that $\Pi_\Gamma z \in \mathbb{R}_J^{r+}$ . Therefore
(3.2.60) implies

$$|\Pi_\Gamma z|_\Gamma^2 \;\leq\; \max\{|\Pi_\Gamma^J z|_\Gamma^2 : \; \Pi_\Gamma^J z \geq 0, J \subset R\}.$$

On the other hand the assumption $\Pi_\Gamma^J z \geq 0$ and the definition of $\Pi_\Gamma$ imply
$|z - \Pi_\Gamma z|_\Gamma^2 \leq |z - \Pi_\Gamma^J z|_\Gamma^2$ . Additionally from formula (3.2.10) we get the equal-
ities $|z - \Pi_\Gamma z|_\Gamma^2 = |z|_\Gamma^2 - |\Pi_\Gamma z|_\Gamma^2$ and $|z - \Pi_\Gamma^J z|_\Gamma^2 = |z|_\Gamma^2 - |\Pi_\Gamma^J z|_\Gamma^2$ , which proves
$|\Pi_\Gamma^J z|_\Gamma^2 \leq |\Pi_\Gamma z|_\Gamma^2$ . This concludes the proof of (3.2.61). Formula (3.2.62) is an
immediate consequence of (3.2.57) and (3.2.61). $\square$


According to formula (3.2.62) the statistic $S_N(\Pi_V \hat{b}_{N2})$ may be evaluated by
evaluating suitable orthogonal projections $\Pi_\Gamma^J$ with respect to the $\Gamma$ –scalar-
product onto the respective linear subspaces $\mathbb{R}_J^r$ . Therefore we'll have a closer
look at the projections $\Pi_\Gamma^J$ :

By definition $\Pi_\Gamma^J z$ is the (unique) element in $\mathbb{R}_J^r$ such that

$$(z - \Pi_\Gamma^J z)^T \Gamma(z - \Pi_\Gamma^J z) \;=\; \inf\{(z - \vartheta)^T \Gamma(z - \vartheta) : \; \vartheta \in \mathbb{R}_J^r\}.$$

Therefore, if $\Gamma^{1/2}$ denotes the root of the symmetric positive definite $r \times r$ –
matrix $\Gamma$ , the vector $\Gamma^{1/2}\Pi_\Gamma^J z$ is the usual projection with respect to the
*Euclidean norm* $|\cdot|$ of the vector $\Gamma^{1/2} z$ onto the subspace $\Gamma^{1/2}\mathbb{R}_J^r$ , i.e.

$$\Pi_\Gamma^J \;=\; \Gamma^{-1/2} P_{J,\Gamma} \Gamma^{1/2}, \tag{3.2.63}$$

where $P_{J,\Gamma} : \mathbb{R}^r \rightarrow \Gamma^{1/2}\mathbb{R}^r_J$ denotes the $|\cdot|$–projection onto the linear subspace $\Gamma^{1/2}\mathbb{R}^r_J$. For the evalution of (3.2.63) we use the well-known fact from the theory of linear models that for any $r \times r$–matrix A the $|\cdot|$–projection onto $im(A) := \{Ax : x \in \mathbb{R}^r\}$ is nothing but $A(A^TA)^-A^T$, where $(A^TA)^-$ is a generalized inverse, i.e. the characterizing property $B = BB^-B$ is fulfilled for $B = (A^TA)$. In order to utilize this result for the evaluation of (3.2.63) we need some additional **notation:**

If $A = (a_{ij})_{i,j=1,...,r}$ is a $r \times r$–matrix and if $I, J \subset \{1,...,r\} = R$, then $A_{I \times J}$ denotes the $|I| \times |J|$–submatrix of $A$ according to

$$A_{I \times J} = (a_{ij})_{i \in I, j \in J} \tag{3.2.64}$$

with natural ordering of $I$ and $J$. In addition let $P_J$ denote the $|\cdot|$–projection of $\mathbb{R}^r$ onto $\mathbb{R}^r_J$, i.e., $P_J$ is a $r \times r$–diagonal-matrix where the diagonal elements corresponding to $J$ are equal to 1 and all other elements are equal to 0. Finally the $r \times r$–unit-matrix is denoted by E.

Since the definition of $P_J$ implies $\Gamma^{1/2}\mathbb{R}^r_J = im(\Gamma^{1/2}P_J)$, we get the representation

$$P_{J,\Gamma} = \Gamma^{1/2}P_J(P_J\Gamma P_J)^- P_J\Gamma^{1/2} \tag{3.2.65}$$

and because of (3.2.63)

$$\Pi^J_\Gamma = P_J(P_J\Gamma P_J)^- P_J\Gamma. \tag{3.2.66}$$

If $J = \emptyset$ then $\Pi^J_\Gamma \equiv 0$. Therefore we may assume $J \neq \emptyset$. Using the notation (3.2.64) we get

$$P_J\Gamma P_J = E_{R \times J}\Gamma_{J \times J}E_{J \times R}.$$

Since $\Gamma$ is positive definite, $\Gamma_{J \times J}$ is positive definite, too. Especially the inverse $(\Gamma_{J \times J})^{-1}$ exists. An easy verification of the defining equality $B = BB^-B$ of a generalized inverse proves

$$(P_J\Gamma P_J)^- = E_{R \times J}(\Gamma_{J \times J})^{-1}E_{J \times R}.$$

Therefore (3.2.66) implies

$$\Pi^J_\Gamma = P_J E_{R \times J}(\Gamma_{J \times J})^{-1}E_{J \times R}P_J\Gamma = E_{R \times J}(\Gamma_{J \times J})^{-1}\Gamma_{J \times R} \tag{3.2.67}$$

and (3.2.65) implies

$$P_{J,\Gamma} = \Gamma^{1/2}E_{R \times J}(\Gamma_{J \times J})^{-1}E_{J \times R}\Gamma^{1/2}. \tag{3.2.68}$$

As an immediate consequence we get ( $\forall\, x \in \mathbb{R}^r$ )

$$\Pi^J_\Gamma(\Gamma^{-1}x) = E_{R \times J}(\Gamma_{J \times J})^{-1}E_{J \times R}x \tag{3.2.69}$$

and

$$|\Pi_\Gamma^J(\Gamma^{-1}x)|_\Gamma^2 = (E_{J\times R}x)^T(\Gamma_{J\times J})^{-1}(E_{J\times R}x).$$ (3.2.70)

Because of

$$x_J := (x_i, i \in J) = E_{J\times R}x \quad \forall\, x = (x_1, ..., x_r)^T \in \mathbb{R}^r$$ (3.2.71)

Lemma 3.2.5, (3.2.69), and (3.2.70) imply the equalities

$$f_0(x) = \sup_{\vartheta \geq 0}(2\vartheta^T x - \vartheta^T\Gamma\vartheta)$$ (3.2.72)
$$= \max\{x_J^T(\Gamma_{J\times J})^{-1}x_J : (\Gamma_{J\times J})^{-1}x_J \geq 0, \emptyset \neq J \subset R\}$$

and

$$S_N(\Pi_V\hat{b}_{N2}) = \|\Pi_V\hat{b}_{N2}\|^2$$ (3.2.73)
$$= \max\{(\vec{S}_N)_J^T(\Gamma_{J\times J})^{-1}(\vec{S}_N)_J : (\Gamma_{J\times J})^{-1}(\vec{S}_N)_J \geq 0, \emptyset \neq J \subset R\},$$

where the convention $\max \emptyset = 0$ is used. For small values of $r$ the representation (3.2.73) provides an easy method for the evaluation of the test statistic $S_N(\Pi_V\hat{b}_{N2})$. An example with $r = 3$ is given in Subsection c.

## b) Asymptotic distribution under local alternatives

Because of (3.2.52) and (3.2.53) it's quite easy to derive a limiting law for $S_N(\Pi_V\hat{b}_{N2}) = f_0(\vec{S}_N)$.

### 3.2.6 Corollary

Let $X$ denote a $\mathbb{R}^r$ -valued random variable with the $r$ -variate normal distribution $\mathcal{N}(0, \Gamma) = \mathcal{L}(X)$. Then, for any $a \in L_2^0(0,1)$, any sequence of directions $A_N \in \mathcal{B}_N^0$ such that $\|A_N' - a\| \to 0$, and any sequence $H_N \in \mathcal{F}_1^c$, we have the limiting law (as $N \to \infty$)

$$\mathcal{L}[S_N(\Pi_V\hat{b}_{N2}) \mid (A_N, H_N)] \xrightarrow{\mathcal{L}} \mathcal{L}[f_0(X + <\vec{b}, a >)],$$ (3.2.74)

where $\vec{b} := (b_1, ..., b_r)^T$ and $<\vec{b}, a >:= (<b_1, a >, ..., <b_r, a >)^T$.

*Especially we have*

$$\mathcal{L}_{\mathcal{H}_0^c}[S_N(\Pi_V\hat{b}_{N2})] \xrightarrow{\mathcal{L}} \mathcal{L}[f_0(X)].$$ (3.2.75)

*Proof:*  The Cramer-Wold-device and Theorem 3.0.1 imply

$$\mathcal{L}[\vec{S}_N \mid (A_N, H_N)] = \mathcal{L}[< \vec{b}, \hat{b}_{N2} > \mid (A_N, H_N)]$$
$$\xrightarrow{\mathcal{L}} \mathcal{N}(< \vec{b}, a >, \Gamma) = \mathcal{L}[X + < \vec{b}, a >]. \tag{3.2.76}$$

According to (3.2.10) and (3.2.12) the function $f_0$ may be written as

$$f_0(x) \;=\; (\sup\{\vartheta^T x : \; \vartheta \ge 0, \vartheta^T \Gamma \vartheta = 1\} \vee 0)^2, \quad x \in \mathbb{R}^r. \tag{3.2.77}$$

Since the supremum is taken over a compact set in $\mathbb{R}^r$ , the function $f_0$ is continuous. Therefore (3.2.76) and (3.2.52) imply the assertion (3.2.74). $\square$

The next theorem will utilize the limiting law (3.2.75) and the representation (3.2.72) in order to give an explicit form of the limiting null distribution $\mathcal{L}[\,f_0(X)\,]$ .

### 3.2.7 Theorem

*Let  $X = (X_1, ..., X_r)^T$  be a random vector with nonsingular  $r$ – variate normal distribution  $\mathcal{N}(0, \Gamma)$ . For each  $J \subset R = \{1, ..., r\}, J \ne \emptyset$ , let  $Y_J$ and  $\tilde{Y}_J$  be  $\mathbb{R}^{|J|}$ –dimensional random variables with the respective normal distributions  $\mathcal{N}(0, (\Gamma_{J \times J})^{-1})$  and  $\mathcal{N}(0, ((\Gamma^{-1})_{J \times J})^{-1})$ . For  $J = \emptyset$  put  $Y_J = \tilde{Y}_J = 0$ .*

*Then, for each  $t \ge 0$ , the following equality holds true,*

$$P\{f_0(X) > t\} = \sum_{\emptyset \ne J \subset R} P\{Y_J \ge 0\}\, P\{\tilde{Y}_{R \setminus J} \le 0\}\, P\{\chi^2_{|J|} > t\}, \tag{3.2.78}$$

*where  $\chi^2_k$  denotes a random variable with  $\chi^2_k$  – distribution.*

*Proof:*  Using the abbreviation  $Z := \Gamma^{-1} X$  and the assumption  $t \ge 0$  we get from (3.2.59), (3.2.56), and (3.2.60) the equality

$$\{f_0(X) > t\} \;=\; \bigcup_{\emptyset \ne J \subset R} \{|\Pi_\Gamma Z|_\Gamma^2 > t, \; \Pi_\Gamma Z = \Pi_\Gamma^J Z\}.$$

Applying Proposition 3.2.1 to the projections  $\Pi_\Gamma$  and  $\Pi_\Gamma^J$  implies the following chain of equalities,

$$\{\Pi_\Gamma Z = \Pi_\Gamma^J Z\} = \{\Pi_\Gamma^J Z \ge 0, < Z, \vartheta >_\Gamma \le \; < \Pi_\Gamma^J Z, \vartheta >_\Gamma \quad \forall\, \vartheta \ge 0\}$$
$$= \{\Pi_\Gamma^J Z \ge 0, (\Gamma Z)^T \vartheta \le (\Gamma \Pi_\Gamma^J Z)^T \vartheta \quad \forall\, \vartheta \ge 0\}$$
$$= \{\Pi_\Gamma^J Z \ge 0, \Gamma Z \le \Gamma \Pi_\Gamma^J Z\}.$$

Combining the results yields

$$\{f_0(X) > t\} = \bigcup_{\emptyset \neq J \subset R} \{|\Pi_\Gamma^J Z|_\Gamma^2 > t, \Pi_\Gamma^J Z \geq 0, (\Gamma - \Gamma \Pi_\Gamma^J) Z \leq 0\}.$$

Now we use (3.2.63) and the abbreviation $\tilde{Z} := \Gamma^{-1/2} X$ in order to get

$$\Pi_\Gamma^J Z = \Gamma^{-1/2} P_{J,\Gamma} \tilde{Z} \quad \text{and} \quad |\Pi_\Gamma^J Z|_\Gamma^2 = |P_{J,\Gamma} \tilde{Z}|^2.$$

Thus,

$$\begin{aligned}
&\{f_0(X) > t\} \\
&= \bigcup_{\emptyset \neq J \subset R} \{|P_{J,\Gamma} \tilde{Z}|^2 > t, \Gamma^{-1/2} P_{J,\Gamma} \tilde{Z} \geq 0, \Gamma^{1/2}(E - P_{J,\Gamma}) \tilde{Z} \leq 0\}
\end{aligned} \tag{3.2.79}$$

and

$$\mathcal{L}(\tilde{Z}) = \mathcal{N}(0, E). \tag{3.2.80}$$

Since

$$Cov(P_{J,\Gamma} \tilde{Z}, (E - P_{J,\Gamma}) \tilde{Z}) = P_{J,\Gamma} Cov(\tilde{Z})(E - P_{J,\Gamma}) = P_{J,\Gamma}(E - P_{J,\Gamma}) = 0$$

implies the independence of the normally distributed random variables $P_{J,\Gamma} \tilde{Z}$ and $(E - P_{J,\Gamma}) \tilde{Z}$ , and since (3.2.80) implies

$$P\{P_{J_1,\Gamma} \tilde{Z} = P_{J_2,\Gamma} \tilde{Z}\} = 0 \quad \forall\, J_1 \neq J_2,$$

we get from (3.2.79) the equality

$$P\{f_0(X) > t\} \tag{3.2.81}$$
$$= \sum_{\emptyset \neq J \subset R} P\{|P_{J,\Gamma} \tilde{Z}|^2 > t, \Gamma^{-1/2} P_{J,\Gamma} \tilde{Z} \geq 0\}\, P\{\Gamma^{1/2}(E - P_{J,\Gamma}) \tilde{Z} \leq 0\}.$$

As the next step we'll evaluate the covariance-matrices of the centered normal random variables $P_{J,\Gamma} \tilde{Z}$ and $(E - P_{J,\Gamma}) \tilde{Z}$ . On one hand formula (3.2.68) immediately implies

$$Cov(P_{J,\Gamma} \tilde{Z}) = \Gamma^{1/2} E_{R \times J} (\Gamma_{J \times J})^{-1} E_{J \times R} \Gamma^{1/2}$$

and therefore

$$\mathcal{L}[\Gamma^{-1/2} P_{J,\Gamma} \tilde{Z}] = \mathcal{N}(0, E_{R \times J}(\Gamma_{J \times J})^{-1} E_{J \times R}). \tag{3.2.82}$$

On the other hand $E - P_{J,\Gamma}$ is the usual orthogonal projection onto the orthogonal complement

$$(\Gamma^{1/2} \mathbb{R}_J^r)^\perp := \{x \in \mathbb{R}^r : x^T \Gamma^{1/2} y = 0 \quad \forall\, y \in \mathbb{R}_J^r\}.$$

Since an easy computation proves

$$(\Gamma^{1/2}\mathrm{I\!R}^r_J)^{\perp} = \Gamma^{-1/2}\mathrm{I\!R}^r_{R\setminus J},$$

we get the representation

$$E - P_{J,\Gamma} = P_{R\setminus J,\Gamma^{-1}}. \tag{3.2.83}$$

Therefore formula (3.2.68) with $\Gamma$ substituted by $\Gamma^{-1}$ yields

$$Cov((E - P_{J,\Gamma})\tilde{Z}) = \Gamma^{-1/2}E_{R\times(R\setminus J)}((\Gamma^{-1})_{(R\setminus J)\times(R\setminus J)})^{-1}E_{(R\setminus J)\times R}\Gamma^{-1/2}$$

and therefore

$$\begin{aligned}
\mathcal{L}[\Gamma^{1/2}&(E - P_{J,\Gamma})\tilde{Z}] \\
&= \mathcal{N}(0,, E_{R\times(R\setminus J)}((\Gamma^{-1})_{(R\setminus J)\times(R\setminus J)})^{-1}E_{(R\setminus J)\times R}).
\end{aligned} \tag{3.2.84}$$

Obviously (3.2.82) and (3.2.84) imply

$$P\{\Gamma^{-1/2}P_{J,\Gamma}\tilde{Z} \geq 0\} = P\{Y_J \geq 0\}$$

and

$$P\{\Gamma^{1/2}(E - P_{J,\Gamma})\tilde{Z} \leq 0\} = P\{\tilde{Y}_{R\setminus J} \leq 0\} = P\{\tilde{Y}_{R\setminus J} \geq 0\}.$$

Therefore (3.2.81) concludes the proof, if we prove the independence of the two events $\{|P_{J,\Gamma}\tilde{Z}|^2 > t\}$ and $\{\Gamma^{-1/2}P_{J,\Gamma}\tilde{Z} \geq 0\}$, and if we prove the equality $P\{|P_{J,\Gamma}\tilde{Z}|^2 > t\} = P\{\chi^2_{|J|} > t\}$ .

For the final proofs fix $J \subset R$ such that $k := |J| > 0$ . Assume $(e_j, j \in J)$ to be an orthonormal basis of the linear space $\Gamma^{1/2}\mathrm{I\!R}^r_J = im(P_{J,\Gamma})$ . Putting $\zeta_j = <e_j, \tilde{Z}>$ and $\eta_j = \Gamma^{-1/2}e_j$ we get

$$P_{J,\Gamma}\tilde{Z} = \sum_{j \in J} \zeta_j\, e_j, \quad |P_{J,\Gamma}\tilde{Z}|^2 = \sum_{j \in J} \zeta_j^2, \quad \Gamma^{-1/2}P_{J,\Gamma}\tilde{Z} = \sum_{j \in J} \zeta_j\, \eta_j.$$

Since the random variables $(\zeta_j, j \in J)$ are i.i.d. with standard normal distribution, we have

$$\mathcal{L}[\,|P_{J,\Gamma}\tilde{Z}|^2\,] = \chi^2_k.$$

For the independence proof we may assume w.l.o.g. $J = \{1, ..., k\}$ . Making a polar transformation to the random variables $\varrho, \vartheta_1, ..., \vartheta_{k-1}$ according to

$$\begin{aligned}
\zeta_1 &= \sqrt{\varrho}\cos\vartheta_1 \cdots \cos\vartheta_{k-1}, \\
\zeta_j &= \sqrt{\varrho}\cos\vartheta_1 \cdots \cos\vartheta_{k-j}\sin\vartheta_{k-j+1}, \quad j = 2, ..., k - 1, \\
\zeta_k &= \sqrt{\varrho}\sin\vartheta_1,
\end{aligned}$$

we get  $\varrho = |P_{J,\Gamma}\tilde{Z}|^2$  and also the independence of  $\varrho$  and  $(\vartheta_1, ..., \vartheta_{k-1})$ , cf. e.g. Kendall and Stuart (1969), Chapter 11.1. Since the event

$$\{\Gamma^{-1/2}P_{J,\Gamma}\tilde{Z} \geq 0\} \;=\; \{\sum_{j\in J}\zeta_j\eta_j \geq 0\}$$

does not depend on  $\varrho$ , the proof of the independence is concluded, too.  $\square$

### 3.2.8 Corollary

*If  $\Gamma$  is the  $r \times r$ –unit-matrix  $E$ , then formula (3.2.78) has the special form*

$$P\{f_0(X) > t\} = \sum_{k=1}^{r} \binom{r}{k} (\tfrac{1}{2})^r \, P\{\chi_k^2 > t\} \quad \forall\, t \geq 0. \tag{3.2.85}$$

*Proof:*  The assumption  $\Gamma = E$  implies that the components of the random vector  $Y_J = (Y_j, j \in J)$  are i.i.d. with standard normal distribution and that the components of  $\check{Y}_{R\setminus J} = (\bar{Y}_i, i \in R \setminus J)$  are i.i.d. with standard normal distribution, too. Therefore (3.2.78) implies

$$P\{f_0(X) > t\} \;=\; \sum_{\emptyset \neq J \subset R} (\tfrac{1}{2})^{|J|}\,(\tfrac{1}{2})^{r-|J|}\, P\{\chi_{|J|}^2 > t\},$$

and  $|\{J \subset R :\; |J| = k\}| = \binom{r}{k}$  concludes the proof.  $\square$

For general  $\Gamma$  and large values of  $r$  the computation of the weights  $P\{Y_J \geq 0\}P\{\check{Y}_{R\setminus J} \leq 0\}$  is difficult but in case of  $1 \leq r \leq 3$  an explicit evaluation is possible. In the following subsection we'll discuss an example with  $r = 3$ . Finally, let's remark that there is no easy way for extending Theorem 3.2.7 to the evaluation of the limiting law  $\mathcal{L}[f_0(X+ <\vec{b}, a >)]$  under  $<\vec{b}, a > \neq 0$ , cf. Corollary 3.2.6.

## c) Three representative score functions and the corresponding nonlinear rank test

In Section 1.2 we have discussed the concept of generalized shift alternatives in order to get a more suitable description of the underlying reality. Given the distribution function  $F_0$  and the bounded shift function  $D : \mathbb{R} \to [0, 1)$  with bounded derivative  $d : \mathbb{R} \to \mathbb{R}$  the corresponding generalized shift alternative  $\mathcal{A}_2^0(F_0, D)$  has been defined in (1.2.2) according to

$$\mathcal{A}_2^0(F_0, D) : \quad \begin{cases} F, G \in \{F_{0,D,\vartheta} :\; \vartheta \in \mathbb{R},\; \vartheta\|d\|_\infty \leq 1\}, \\ F \leq G,\; F \neq G, \end{cases}$$

where $F_{0,D,\vartheta}$ is the $D$–shifted distribution function

$$F_{0,D,\vartheta}(x) = F_0(x - \vartheta D(x)), \quad x \in \mathbb{R}.$$

According to the discussion around formula (1.2.4) the asymptotically optimal score function for testing $\mathcal{H}_0^r$ versus $\mathcal{A}_2^0(F_0, D)$ is

$$\varphi(u, f_0, D) = \varphi(u, f_0)D(F_0^{-1}(u)) - d(F_0^{-1}(u)), \quad 0 < u < 1,$$

if $F_0$ has an absolutely continuous density $f_0$ with finite Fisher-information $I(f_0) = \int (f_0'/f_0)^2 f_0 \, dx$. Therefore it's reasonable to use the $r$–dimensional cone $V = [b_1, ..., b_r]^+$ defined in (3.2.50), where the functions $b_i$ are chosen as $\varphi(\cdot, f_0, D)$ for suitable $f_0$'s and $D$'s.

From the variety of all possible generalized shifts we'll choose three special types according to the following considerations: If $F_0$ is the distribution function of the standard treatment and if the new treatment mainly shows a positive reaction for values of $x$ where $F_0(x)$ is small, then we call this a *lower shift* situation. (The new treatment shows a positive effect for the bad cases.) If the new treatment mainly shows a positive reaction for values of $x$ where $F_0(x)$ is near $1/2$, then we call this a *central shift* situation. (The new treatment shows a positive effect for the average cases.) If the new treatment mainly shows a positive reaction for values of $x$ where $F_0(x)$ is large, then we call this a *upper shift* situation. (The new treatment shows a positive effect for the good cases.) Assuming $F_0(0) = 1/2$ we formalize these notions by the following selections of the shift function $D$,

$$
\begin{aligned}
D_1 &:= 1 - F_0 &&\text{(lower shift)}, \\
D_2 &:= 4F_0(1 - F_0) &&\text{(central shift)}, \qquad (3.2.86) \\
D_3 &:= F_0 &&\text{(upper shift)}.
\end{aligned}
$$

Obviously this implies $D_1(x) \approx 1$ for small values of $x$ and $D_1(x) \approx 0$ for large values of $x$. The behavior of $D_3$ is dual to $D_1$, whereas $D_2(0) = 1$ and $D_2(x) \approx 0$ for large values of $|x|$.

Notice, the score function $\varphi(\cdot, f_0)$, which is optimal for the exact shift model $D = 1$, is contained in the cone, cf. (3.2.50),

$$V = [\varphi(\cdot, f_0, D_i), i = 1, 2, 3]^+ = \{ \sum_{i=1}^{3} \vartheta_i \, \varphi(\cdot, f_0, D_i) : \vartheta_i \geq 0 \, \forall \, i\}, \quad (3.2.87)$$

since the definition (3.2.86) implies the identity

$$\varphi(u, f_0) \equiv \varphi(u, f_0, D_1) + \varphi(u, f_0, D_3).$$

In the Monte Carlo simulations of Section 3.1 we have used the shift functions (3.2.86) and $D_4 = 1$ for $F_0$ corresponding to the normal distribution, the logistic distribution, and the Cauchy distribution, respectively. Because of simplicity and since the corresponding cone $V$ contains suitable approximations of the other score functions, we'll restrict the discussion to the case of underlying logistic distribution, i.e. we choose

$$F_0(x) = \frac{\exp(x)}{1 + \exp(x)}, \quad x \in \mathbb{R}. \tag{3.2.88}$$

By elementary evaluation, cf. Table 1.1, we get the following representation of the corresponding three score functions $(0 \le u \le 1)$,

$$\begin{aligned}
b_1(u) &:= \varphi(u, f_0, D_1) = (1 - u)(3u - 1) = -b_3(1 - u),\\
b_2(u) &:= \varphi(u, f_0, D_2) = 8u(1 - u)(2u - 1),\\
b_3(u) &:= \varphi(u, f_0, D_3) = u(3u - 2).
\end{aligned} \tag{3.2.89}$$

We'll use $b_1, b_2, b_3$ as a representative set of score functions, which take care of potential lower shift, central shift, upper shift, and exact shift, too. Because of $b_1(u) + b_3(u) \equiv 2u - 1$ the Wilcoxon score function $2u - 1$ is contained in the cone $V = [b_1, b_2, b_3]^+$ . The covariance matrix $\Gamma = ( <b_i, b_j> )$ and its inverse may be evaluated as

$$\Gamma = \frac{1}{210} \begin{pmatrix} 28 & 28 & 7 \\ 28 & 64 & 28 \\ 7 & 28 & 28 \end{pmatrix} \tag{3.2.90}$$

and

$$\Gamma^{-1} = \frac{1}{16} \begin{pmatrix} 240 & -140 & 80 \\ -140 & 175 & -140 \\ 80 & -140 & 240 \end{pmatrix}. \tag{3.2.91}$$

Now assume that the observed value of

$$\vec{S}_N = (<b_1, \hat{b}_{N2}>, <b_2, \hat{b}_{N2}>, <b_3, \hat{b}_{N2}>)^T = (S_N(b_1), S_N(b_2), S_N(b_3))^T$$

is $x = (x_1, x_2, x_3)^T$ . In order to compute the corresponding value of the test statistic $f_0(\vec{S}_N) = f_0(x)$ we use (3.2.72),

$$f_0(x) = \max\{x_J^T(\Gamma_{J \times J})^{-1}x_J : (\Gamma_{J \times J})^{-1}x_J \ge 0, \emptyset \ne J \subset \{1, 2, 3\} \},$$

with the convention $\max \emptyset = 0$ . Therefore we have to evaluate $(\Gamma_{J \times J})^{-1}$ for each $J \subset \{1, 2, 3\}$ . In case of $J = \{1, 2, 3\}$ the inverse $(\Gamma_{J \times J})^{-1} = \Gamma^{-1}$ is given in (3.2.91). In case of $|J| = 2$ we use the formula

$$\begin{pmatrix} a & b \\ b & c \end{pmatrix}^{-1} = \frac{1}{ac - b^2} \begin{pmatrix} c & -b \\ -b & a \end{pmatrix}.$$

**Figure 3.2.b**

*The graphs of the score functions (3.2.89):*
$b_1$ ( $\ast\ast\ast$ ), $b_2$ ( $\cdot\,\cdot\,\cdot$ ), $b_3$ ( $\circ\,\circ\,\circ$ ).

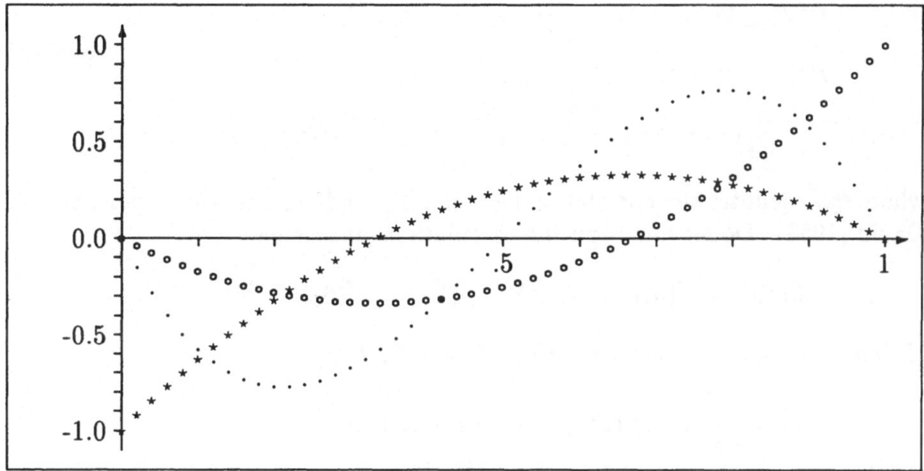

Finally, in case of $J = \{j\}$, $j = 1, 2, 3$, we have $(\Gamma_{J \times J})^{-1} = 1/\gamma_{jj}$.

In practical applications it's more convenient to substitute the components of $\vec{S}_N = (S_N(b_1), S_N(b_2), S_N(b_3))^T$ by the asymptotically equivalent linear rank statistics

$$\tilde{S}_N(b_j) \; = \; \sum_{i=1}^{N} c_{Ni} \, b_j\left(\frac{R_i - 1/2}{N}\right), \quad j = 1, 2, 3, \tag{3.2.92}$$

i.e. we evaluate $f_0(x)$ on the basis of the observed value $x = (x_1, x_2, x_3)^T$ of $(\tilde{S}_N(b_1), \tilde{S}_N(b_2), \tilde{S}_N(b_3))^T$.

Since the approximation of the limiting law (3.2.75) is quite good, we use formula (3.2.78) in order to evaluate the (approximate) $p$–value corresponding to the observed value $f_0(x)$,

$$P\{f_0(X) > f_0(x)\} \; = \; \sum_{k=1}^{3} w_k \, P\{\chi_k^2 > f_0(x)\}, \tag{3.2.93}$$

where the weights $w_k$ have the form

$$w_k \; = \; \sum_{J \subset R, \, |J|=k} P\{Y_J \geq 0\} \, P\{\tilde{Y}_{R \setminus J} \geq 0\}, \quad R = \{1, 2, 3\}. \tag{3.2.94}$$

In the present case $(|R| = 3)$ an explicit evaluation of the $w_k$ is possible, since for trivariate normal random variables $(Z_1, Z_2, Z_3)$ we have

$$P\{Z_i \geq 0\} = 1/2,$$

$$P\{Z_i \geq 0, Z_j \geq 0\} = \frac{1}{2\pi}(\pi - \arccos(\varrho_{ij})),$$

$$P\{Z_1 \geq 0, Z_2 \geq 0, Z_3 \geq 0\} =$$

$$\frac{1}{4\pi}(2\pi - \arccos(\varrho_{12}) - \arccos(\varrho_{13}) - \arccos(\varrho_{23})),$$

$$(3.2.95)$$

where $\varrho_{ij}$ denotes the correlation between $Z_i$ and $Z_j$, cf. the paper by F.N. David (1953). Doing some routine calculations and using

$$\mathcal{L}(Y_J) = \mathcal{N}(0, (\Gamma_{J\times J})^{-1}), \quad \mathcal{L}(\tilde{Y}_J) = \mathcal{N}(0, ((\Gamma^{-1})_{J\times J})^{-1}),$$

cf. Theorem 3.2.7, we get from (3.2.94) and (3.2.95),

$$w_1 = \frac{1}{4\pi}\left(\arccos(\varrho_{12}^*) + \arccos(\varrho_{13}^*) + \arccos(\varrho_{23}^*)\right),$$

$$w_2 = \frac{1}{4\pi}\left(\arccos(\varrho_{12}) + \arccos(\varrho_{13}) + \arccos(\varrho_{23})\right),$$

$$w_3 = \frac{1}{2} - w_1,$$

$$(3.2.96)$$

where now $(\varrho_{ij})_{i,j=1,2,3}$ is the correlation matrix corresponding to $\Gamma$ and $(\varrho_{ij}^*)_{i,j=1,2,3}$ is the correlation matrix corresponding to $\Gamma^{-1}$. Using the special $3 \times 3$ –matrices $\Gamma$ and $\Gamma^{-1}$ given in (3.2.90) and (3.2.91) we get the weights

$$w_1 = 0.46765, \qquad w_2 = 0.23987, \qquad w_3 = 0.03235. \qquad (3.2.97)$$

Therefore (3.2.93) implies the following explicit representation of the (approximate) $p$ –value of the observed value $t = f_0((\tilde{S}_N(b_1), \tilde{S}_N(b_2), \tilde{S}_N(b_3))^T)$,

$$P\{f_0(X) > t\}$$
$$= 0.46765\, P\{\chi_1^2 > t\} + 0.23987\, P\{\chi_2^2 > t\} + 0.03235\, P\{\chi_3^2 > t\}.$$

$$(3.2.98)$$

Upper critical values of $f_0\big(\tilde{S}_N(b_1), \tilde{S}_N(b_2), \tilde{S}_N(b_3)\big)$ are given in Table 2.1.C of the Appendix. The graphs of the three representative score functions (3.2.89) are shown in Figure 3.2.b.

## 3.3 Treatment of ties

At the beginning of this chapter we have assumed the underlying distribution functions $F$ and $G$ to be *continuous* , cf. (3.0.1). But either by rounding or by the properties of the underlying stochastic mechanism the distribution functions $F$ and $G$ may actually have discontinuities. As a consequence a considerable amount of *ties* may show up in the data. In the present section we therefore generalize the asymptotic optimality of linear rank tests to arbitrary underlying distribution functions $F$ and $G$ . Then, using the same motivation as in the continuous model of Section 3.1, we'll estimate the underlying optimal score function in order to get suitable tests without knowing the underlying direction of the alternative.

The proposed tests will be conditional rank tests given the vector of the tie-lengths of the pooled sample, i.e. they will be conditionally distribution-free under the null hypothesis $F = G$ .

Using the abbreviation

$$\mathcal{F}_1 = \big\{ H : H \text{ is an arbitrary distribution function on } \mathbb{R} \big\} \qquad (3.3.1)$$

the obvious parameter space of this general model is

$$\mathcal{M} = \big\{ (F,G) : F, G \in \mathcal{F}_1 \big\}, \qquad (3.3.2)$$

and the *general hypothesis of randomness* may be written in the form

$$\mathcal{H}_0^r = \big\{ (F,G) \in \mathcal{M} : F = G \big\} = \big\{ (H,H) : H \in \mathcal{F}_1 \big\}. \qquad (3.3.3)$$

The aim of the present section is the construction of suitable rank tests for testing $\mathcal{H}_0^r$ versus the *general one-sided alternative*

$$\mathcal{A}_2^0 = \big\{ (F,G) \in \mathcal{M} : F \leq G, F \neq G \big\}, \qquad (3.3.4)$$

and also for testing $\mathcal{H}_0^r$ versus the *general omnibus alternative*

$$\mathcal{A}_2 = \big\{ (F,G) \in \mathcal{M} : F \neq G \big\}. \qquad (3.3.5)$$

In a first step we'll construct a partition of the general model $\mathcal{M}$ such that the continuous submodel is just one part of this partition and all the other parts show some basic similarities to the continuous case. Given the special part we'll construct a reparametrization which is completely similar to the reparametrization (3.0.2) and which is used in order to identify the optimal score function for testing the corresponding part of $\mathcal{H}_0^r$ versus any given alternative $(F,G)$ .

In a second step we'll estimate the underlying optimal score function in order to construct suitable (nonlinear) rank tests for testing $\mathcal{H}_0^r$ versus $\mathcal{A}_2^0$ or versus $\mathcal{A}_2$ . If no ties are present the proposed tests coincide with the tests of Section 3.1.

Before stating the partition and the new parametrization we have to introduce some additional definitions.

### 3.3.1 Definition

1. *For each distribution function $F \in \mathcal{F}_1$ we define $\mathcal{I}(F)$ to be the family of all halfopen subintervals of the unit interval $[0,1]$ which correspond to points of discontinuity of $F$ ,*

$$\mathcal{I}(F) = \Big\{ \, (F(x_-), F(x)] : \; x \in \mathbb{R}, \; F(x_-) < F(x) \, \Big\}, \qquad (3.3.6)$$

*where $F(x_-) = \sup\{ F(y) : y \in \mathbb{R}, y < x \}$ is the left-sided limit of $F$ at the point $x$ .*

2. *Let $\mathcal{J}_{0,1}^*$ denote the system of all families $\mathcal{J}$ of pairwise disjoint halfopen intervals $(s,t] \subset [0,1]$ . Then definition (3.3.6) implies the equality*

$$\mathcal{J}_{0,1}^* = \{\mathcal{I}(F) : F \in \mathcal{F}_1\}. \qquad (3.3.7)$$

3. *For each $\mathcal{J} \in \mathcal{J}_{0,1}^*$ let $\mathcal{F}_1(\mathcal{J})$ denote the set of all $F \in \mathcal{F}_1$ such that $\mathcal{I}(F)$ coincides with $\mathcal{J}$ , i.e.*

$$\mathcal{F}_1(\mathcal{J}) = \{ F \in \mathcal{F}_1 : \mathcal{I}(F) = \mathcal{J} \} \subset \mathcal{F}_1. \qquad (3.3.8)$$

*Defining $H_{\mathcal{J}} : \mathbb{R} \to \mathbb{R}$ according to*

$$H_{\mathcal{J}}(x) = \begin{cases} 0, & \text{if } x < 0, \\[2mm] x, & \text{if } x \in [0,1] \setminus \cup\{ (s,t) : (s,t] \in \mathcal{J} \}, \\[2mm] s, & \text{if } x \in (s,t) \;\; \text{for some } (s,t] \in \mathcal{J}, \\[2mm] 1, & \text{if } x > 1, \end{cases} \qquad (3.3.9)$$

*we obviously have*

$$H_{\mathcal{J}} \in \mathcal{F}_1(\mathcal{J}) \quad \text{and} \quad H_{\mathcal{J}}(\mathbb{R}) = [0,1] \setminus \cup\{ (s,t) : (s,t] \in \mathcal{J} \}. \qquad (3.3.10)$$

*In addition formula (3.3.7) induces the following disjoint partition of $\mathcal{F}_1$ ,*

$$\mathcal{F}_1 = \sum_{\mathcal{J} \in \mathcal{J}_{0,1}^*} \mathcal{F}_1(\mathcal{J}), \qquad (3.3.11)$$

*where the summation-sign indicates disjoint union.*

4. *Finally, for each* $J \in J_{0,1}^*$ *we define*

$$\mathcal{B}_N^0(J) \;=\; \{\, B \in \mathcal{B}_N^0 : B \text{ is linear on each } (s,t] \in J \,\}, \qquad (3.3.12)$$

*where* $\mathcal{B}_N^0$ *has been defined in formula (1.3.25) as the set of all absolutely continuous functions* $B : [0,1] \to \mathbb{R}$ *such that* $B(0) = B(1) = 0$ *and* $-\sqrt{m/n} \le B'/\sqrt{N} \le \sqrt{n/m}$ $[\lambda - a.e.]$.

### 3.3.2 Remark

1. A distribution function $F$ on $\mathbb{R}$ is continuous, iff $\mathcal{I}(F) = \emptyset$. Therefore the continuous model $\mathcal{F}_1^c$ corresponds to $J = \emptyset$, i.e. $\mathcal{F}_1(\emptyset) = \mathcal{F}_1^c$, the set of all continuous distribution functions on $\mathbb{R}$.

2. Each family $J \in J_{0,1}^*$ consists of countably many elements (subintervals of the unit interval). This property is used in the proof of (3.3.7) in order to construct a distribution function $H_J$ on $\mathbb{R}$ such that $\mathcal{I}(H_J) = J$, cf. (3.3.9) and (3.3.10).

3. Defining the $J$ –part of the general model $\mathcal{M}$ at sample size $N$ as

$$\mathcal{M}_N(J) \;:=\; \Big\{\, (F,G) \in \mathcal{M} : \frac{m}{N} F + \frac{n}{N} G \in \mathcal{F}_1(J) \,\Big\}, \qquad (3.3.13)$$

the partition (3.3.11) of $\mathcal{F}_1$ obviously induces a disjoint partition of $\mathcal{M}$ according to

$$\mathcal{M} \;=\; \sum_{J \in J_{0,1}^*} \mathcal{M}_N(J). \qquad (3.3.14)$$

The corresponding parts of the induced disjoint partitions of the hypothesis $\mathcal{H}_0^r$ and of the alternatives $\mathcal{A}_2^0$ and $\mathcal{A}_2$ are

$$\mathcal{H}_{0N}^r(J) \;:=\; \{\, (F,G) \in \mathcal{M}_N(J) : \; F = G \,\}$$

$$=\; \{\, (H,H) : \; H \in \mathcal{F}_1(J) \,\}, \qquad (3.3.15)$$

$$\mathcal{A}_{2N}^0(J) \;:=\; \{\, (F,G) \in \mathcal{M}_N(J) : \; F \le G, \, F \ne G \,\}, \qquad (3.3.16)$$

$$\mathcal{A}_{2N}(J) \;:=\; \{\, (F,G) \in \mathcal{M}_N(J) : \; F \ne G \,\}. \qquad (3.3.17)$$

Since $J = \emptyset$ represents the continuous case, formulae (3.0.5) to (3.0.7) will coincide with the respective reparametrizations of the continuous hypotheses $\mathcal{H}_{0N}^r(\emptyset)$, $\mathcal{A}_{2N}^0(\emptyset)$, and $\mathcal{A}_{2N}(\emptyset)$.

The next proposition will prove a completely similar reparametrization for any given $\mathcal{J} \in \mathcal{J}_{0,1}^*$ .

### 3.3.3 Proposition

*Given any sample size $N$ and any $\mathcal{J} \in \mathcal{J}_{0,1}^*$ and using the abbreviations, cf. (3.0.3) and (3.0.4),*

$$F_{B,H}^N = H + c_{N1} B \circ H, \qquad G_{B,H}^N = H + c_{NN} B \circ H, \qquad (3.3.18)$$

*we have the following equalities,*

$$\mathcal{M}_N(\mathcal{J}) = \{ (F_{B,H}^N, G_{B,H}^N) :\ B \in \mathcal{B}_N^0(\mathcal{J}),\ H \in \mathcal{F}_1(\mathcal{J}) \}, \quad (3.3.19)$$

$$\mathcal{H}_{0N}^r(\mathcal{J}) = \{ (F_{B,H}^N, G_{B,H}^N) \in \mathcal{M}_N(\mathcal{J}) :\ B = 0 \}, \qquad (3.3.20)$$

$$\mathcal{A}_{2N}^0(\mathcal{J}) = \{ (F_{B,H}^N, G_{B,H}^N) \in \mathcal{M}_N(\mathcal{J}) :\ B \leq 0,\ B \neq 0 \}, \quad (3.3.21)$$

$$\mathcal{A}_{2N}(\mathcal{J}) = \{ (F_{B,H}^N, G_{B,H}^N) \in \mathcal{M}_N(\mathcal{J}) :\ B \neq 0 \}. \qquad (3.3.22)$$

*The above $(B, H)$ –parametrization is identifiable, cf. Corollary 3.3.4.*

*Proof* : In a first step we'll prove that the right-hand sides are subsets of the left-hand sides. The second step will prove the reverse inclusions.

a) Assume $\mathcal{J} \in \mathcal{J}_{0,1}^*$ , $B \in \mathcal{B}_N^0(\mathcal{J})$ , and $H \in \mathcal{F}_1(\mathcal{J})$ . Put

$$F := (I + c_{N1} B) \circ H, \qquad G := (I + c_{NN} B) \circ H,$$

where $I : [0,1] \to [0,1]$ denotes the identity. From assumption $B \in \mathcal{B}_N^0(\mathcal{J})$ we get $[\lambda - a.e.]$ the inequalities

$$(I + c_{N1} B)' = 1 + c_{N1} B' \geq 0, \qquad (I + c_{NN} B)' = 1 + c_{NN} B' \geq 0.$$

Therefore $F$ and $G$ are rightcontinuous nondecreasing functions. Additionally, the assumptions $B(0) = B(1) = 0$ and $H \in \mathcal{F}_1(\mathcal{J})$ imply $F$ and $G$ to be distribution functions on $\mathbb{R}$ . Finally the definition of the $c_{Ni}$ 's implies $m\, c_{N1} + n\, c_{NN} = 0$ and therefore

$$\frac{m}{N} F + \frac{n}{N} G = H \in \mathcal{F}_1(\mathcal{J}).$$

Thus we've proved

$$(F, G) \in \mathcal{M}_N(\mathcal{J}).$$

Moreover, the property $F - G = (c_{N1} - c_{NN})B \circ H = \sqrt{N/(mn)}\, B \circ H$ obviously implies the following implications,

$$B = 0 \quad \Longrightarrow \quad F = G,$$
$$B \geq 0 \quad \Longrightarrow \quad F \geq G,$$
$$B \leq 0 \quad \Longrightarrow \quad F \leq G.$$

Now assume $F = G$. This implies $B = 0$ on the set $H(\mathbb{R})$ and thus $B = 0$ on the set $\overline{H(\mathbb{R})}$, too. Since $B$ is linear on each of the intervals $(s,t] \in \mathcal{J} = \mathcal{I}(H)$ we get $B = 0$ on $[0,1]$. This proves the final implication

$$B \neq 0 \quad \Longrightarrow \quad F \neq G.$$

b) Now let's assume $\mathcal{J} \in \mathcal{J}_{0,1}^{*}$ and $(F,G) \in \mathcal{M}_N(\mathcal{J})$. Since $H = (mF + nG)/N$ dominates $F$ and $G$ we may define the functions

$$f^* := \frac{dF}{dH} \circ H^{-1} \quad \text{and} \quad g^* := \frac{dG}{dH} \circ H^{-1}. \tag{3.3.23}$$

Obviously $f^*$ and $g^*$ are nonnegative measurable functions on the interval $[0,1]$ and

$$\frac{m}{N}f^* + \frac{n}{N}g^* = 1 \quad [\lambda - a.e.], \tag{3.3.24}$$

where $\lambda$ denotes the Lebesgue measure on $[0,1]$. Since the assumption $H \in \mathcal{F}_1(\mathcal{J})$ implies $H^{-1}$ to be constant on $(s,t]$ for any $(s,t] \in \mathcal{J}$, we get in addition

$$f^* \text{ and } g^* \text{ are constant on } (s,t] \text{ for any } (s,t] \in \mathcal{J}. \tag{3.3.25}$$

Now let's define

$$F^*(t) := \int_0^t f^* \, d\lambda, \quad G^*(t) := \int_0^t g^* \, d\lambda, \quad 0 \leq t \leq 1, \tag{3.3.26}$$

and

$$B := \sqrt{\frac{mn}{N}}(F^* - G^*). \tag{3.3.27}$$

Because of (3.3.24) and (3.3.26) the function $B$ is absolutely continuous on $[0,1]$ with the property

$$-\sqrt{m/n} \leq B'/\sqrt{N} \leq \sqrt{n/m} \quad [\lambda - a.e.].$$

In addition, (3.3.25) implies $B$ to be linear on each $(s,t] \in \mathcal{J}$. Therefore $B \in \mathcal{B}_N^0(\mathcal{J})$ is proved, if we verify $B(0) = B(1) = 0$.

While $B(0) = 0$ is a direct consequence of the definitions (3.3.26) and (3.3.27), we'll prove $B(1) = 0$ by showing

$$F^*(1) \; = \; G^*(1) \; = 1. \tag{3.3.28}$$

For the proof of (3.3.28) take any $x \in \mathbb{R}$ . Then

$$F^*\big(H(x)\big) = \int 1\big(s \le H(x)\big) \, f^*(s) \, d\lambda(s)$$

$$= \int 1\big(H^{-1}(s) \le x\big) \, \frac{dF}{dH}\big(H^{-1}(s)\big) \, d\lambda(s)$$

$$= \int 1(y \le x) \, \frac{dF}{dH}(y) \, dH(y) \; = \; F(x)$$

and similarly $G^*\big(H(x)\big) = G(x)$ . Thus we've proved

$$F = F^* \circ H \quad \text{and} \quad G = G^* \circ H, \tag{3.3.29}$$

which especially proves $F^*$ and $G^*$ to be absolutely continuous distribution functions on $[0, 1]$ with the respective densities $f^*$ and $g^*$ . Especially the proof of (3.3.28) and $B \in \mathcal{B}^0_N(\mathcal{J})$ is concluded.

Since (3.3.24) implies $(mF^* \circ H + nG^* \circ H)/N = H$ , formula (3.3.29) implies

$$\begin{aligned}
F &= H + \frac{n}{N}(F^* - G^*) \circ H \; = \; H + c_{N1} B \circ H, \\
G &= H - \frac{m}{N}(F^* - G^*) \circ H \; = \; H + c_{NN} B \circ H.
\end{aligned} \tag{3.3.30}$$

Therefore the proof is complete, if we prove the implications

$$\begin{aligned}
F = G &\implies B = 0, \\
F \ge G &\implies B \ge 0, \\
F \le G &\implies B \le 0, \\
F \ne G &\implies B \ne 0.
\end{aligned}$$

Since (3.3.30) implies

$$F - G = \sqrt{\frac{N}{mn}} \, B \circ H, \tag{3.3.31}$$

the last implication is obvious, whereas the respective assumptions $F = G$ , $F \ge G$ , $F \le G$ , and the continuity of $B$ imply the respective conclusions $B = 0$ , $B \ge 0$ , $B \le 0$ for the function $B$ restricted to the set $\overline{H(\mathbb{R})} \subset [0, 1]$ . Therefore the linearity of $B$ on each of the intervals $(s, t] \in \mathcal{J} = \mathcal{I}(H)$ concludes the proof. $\square$

The following corollary proves that the $(B, H)$ –parametrization is injective.

### 3.3.4 Corollary

*Assume $\mathcal{J} \in \mathcal{J}_{0,1}^*$ , $B_1, B_2 \in \mathcal{B}_N^0(\mathcal{J})$ , and $H_1, H_2 \in \mathcal{F}_1(\mathcal{J})$ . Then we have the implication*

$$(B_1, H_1) \neq (B_2, H_2) \Longrightarrow (F_{B_1,H_1}^N, G_{B_1,H_1}^N) \neq (F_{B_2,H_2}^N, G_{B_2,H_2}^N). \qquad (3.3.32)$$

*Proof* : Assume $F := H_1 + c_{N1} B_1 \circ H_1 = H_2 + c_{N1} B_2 \circ H_2$ and $G := H_1 + c_{NN} B_1 \circ H_1 = H_2 + c_{NN} B_2 \circ H_2$ . Then we get on one hand

$$H_1 = \frac{m}{N} F + \frac{n}{N} G = H_2,$$

and therefore on the other hand

$$B_1 \circ H_1 = \sqrt{\frac{m\,n}{N}} (F - G) = B_2 \circ H_2 = B_2 \circ H_1,$$

i.e. continuity of $B_1$ and $B_2$ implies the equality $B_1 = B_2$ on the closure of $H_1(\mathbb{R}) \subset [0, 1]$ . Since $B_1$ and $B_2$ are linear on each interval $(s, t] \in \mathcal{J}$ and since $H_1 \in \mathcal{F}_1(\mathcal{J})$ implies

$$\overline{H_1(\mathbb{R})} = [0, 1] \setminus \bigcup_{(s,t] \in \mathcal{J}} (s, t), \qquad (3.3.33)$$

we get $B_1 = B_2$ on $[0, 1]$ . $\square$

The following propositions will demonstrate that for arbitrary $\mathcal{J} \in \mathcal{J}_{0,1}^*$ there is a complete analogy between the $\mathcal{J}$ –part $\mathcal{M}_N(\mathcal{J})$ of the general model $\mathcal{M}$ and the special continuous model $\mathcal{F}_1^c \times \mathcal{F}_1^c = \mathcal{M}_N(\emptyset)$ .

The first proposition will prove the usual vector of ranks $(R_1, ..., R_N)$ ,

$$R_i = \sum_{j=1}^{N} 1(X_j \leq X_i), \qquad i = 1, ..., N, \qquad (3.3.34)$$

to have a distribution which is independent of the underlying $H \in \mathcal{F}_1(\mathcal{J})$ . Under the null hypothesis $B = 0$ this is an extension of the well-known fact that the ranks are distribution- free under the continuous null hypothesis $F = G \in \mathcal{F}_1^c$ .

### 3.3.5 Proposition

*For any given $\mathcal{J} \in \mathcal{J}_{0,1}^*$ , $B \in \mathcal{B}_N^0(\mathcal{J})$ , and $H_1, H_2 \in \mathcal{F}_1(\mathcal{J})$ we have*

$$\mathcal{L}\big[(R_1, ..., R_N) \mid (B, H_1)\big] = \mathcal{L}\big[(R_1, ..., R_N) \mid (B, H_2)\big]. \qquad (3.3.35)$$

*Proof* :  Assume  $J \in \mathcal{J}_{0,1}^*$ ,  $B \in \mathcal{B}_N(J)$ , and  $H \in \mathcal{F}_1(J)$ .  Since the  $X_1, ..., X_N$  are independent and since for each  $i = 1, ..., N$  we have

$$R_i = \sum_{j=1}^N 1(X_j \leq X_i) = \sum_{j=1}^N 1(H(X_j) \leq H(X_i)) \qquad [(B,H) - a.e.]$$

it suffices to prove the following assertion:

$$\mathcal{L}\big[ H(X_i) \,|\, (B,H) \big] \;=\; Q_i \qquad \forall\, H \in \mathcal{F}_1(J), \tag{3.3.36}$$

where  $i \in \{1, ..., N\}$ , and  $Q_i$  is some suitable distribution on  $[0,1]$ .

In order to prove the assertion (3.3.36) put  $S := \cup\{\, [s,t] : (s,t] \in J\}$  and  $T := (0,1) \setminus S$ .  Since the assumption  $H \in \mathcal{F}_1(J)$  means  $\mathcal{I}(H) = J$ , we get the inclusion  $T \subset \{H(x) : x \in \mathbb{R}\} = H(\mathbb{R})$ .  In a first step let's assume  $u \in T$ , which implies  $u \in H(\mathbb{R})$ .  Since  $H + c_{Ni} B \circ H$  is the distribution function of  $X_i$  we get

$$P_{(B,H)}\{H(X_i) \leq u\} = P_{(B,H)}\{X_i \leq H^{-1}(u)\} = u + c_{Ni} B(u).$$

Obviously the right-hand side does not depend on  $H$ .  In a second step we assume  $s < u < t$  for some  $(s,t] \in J$ .  Since  $J(H) = J$ , this assumption implies the existence of some discontinuity point  $x$  of  $H$  such that  $H(x_-) = s$  and  $H(x) = t$ .  Therefore the continuity of  $B$  implies

$$\begin{aligned} P_{(B,H)}\{H(X_i) \leq u\} = P_{(B,H)}\{X_i < x\} &= H(x_-) + c_{Ni} B(H(x_-)) \\ &= s + c_{Ni} B(s), \end{aligned}$$

which again does not depend on  $H$ .

Since the set  $\{s,t : (s,t] \in J\}$  is a countable subset of  $[0,1]$  the combination of the two steps proves (3.3.36). $\square$


The following proposition proves invariance properties of the ranks which are similar to the continuous case.  In order to formulate the result we call a mapping  $T : \mathbb{R} \to \mathbb{R}$  a *transformation of the measurement scale* , if  $T$  is strictly increasing and continuous such that  $T(\mathbb{R}) = \mathbb{R}$ .


### 3.3.6 Proposition

*a) For each  $J \in \mathcal{J}_{0,1}^*$  the hypotheses  $\mathcal{H}_{0N}^r(J)$ ,  $\mathcal{A}_{2N}^0(J)$ , and  $\mathcal{A}_{2N}(J)$  are invariant under the group of transformations of the measurement scale.*

*b) The vector of ranks $R = (R_1, ..., R_N)$ is maximal invariant with respect to the group of transformations of the measurement scale.*

*Proof* : a) Assume $\mathcal{J} \in \mathcal{J}_{0,1}^*$ and take $(F, G) \in \mathcal{M}_N(\mathcal{J})$ . Let $T$ be a transformation of the measurement scale. Then the distribution functions of the transformed random variables $T(X_i)$ are $F \circ T^{-1}$ and $G \circ T^{-1}$ , respectively. Therefore, on one hand we have to prove $H \circ T^{-1} \in \mathcal{F}_1(\mathcal{J})$ , on the other hand we have to prove the following implications

$$
\begin{aligned}
F = G &\implies F \circ T^{-1} = G \circ T^{-1}, \\
F \geq G &\implies F \circ T^{-1} \geq G \circ T^{-1}, \\
F \leq G &\implies F \circ T^{-1} \leq G \circ T^{-1}, \\
F \neq G &\implies F \circ T^{-1} \neq G \circ T^{-1}.
\end{aligned}
$$

Since $T$ and $T^{-1}$ are continuous, we get the equality

$$
D(H \circ T^{-1}) = T(D(H))
$$

where $D(f)$ denotes the set of discontinuity points of the function $f$ . Now take any $y \in D(H \circ T^{-1})$ . Then there is a corresponding $x \in D(H)$ such that $y = T(X)$ and $(H \circ T^{-1}(y_-), H \circ T^{-1}(y)] = (H(x_-), H(x)]$ . Therefore $H \in \mathcal{F}_1(\mathcal{J})$ implies $H \circ T^{-1} \in \mathcal{F}_1(\mathcal{J})$ . Finally, the proof of the first three implications is trivial, whereas the proof of the fourth implication obviously follows from continuity of $T^{-1}$ and $T(\mathbb{R}) = \mathbb{R}$ .

b) The proof of part b) is identical with the usual proof since $R_i = R_j$ and $X_i = X_j$ are equivalent, cf. (3.3.34). $\square$


For the moment let's assume that the parameter $\mathcal{J} \in \mathcal{J}_{0,1}^*$ is given. Then the original testing problem $\mathcal{H}_0^r$ versus $\mathcal{A}_2^0$ [ or $\mathcal{H}_0^r$ versus $\mathcal{A}_2$ ] reduces to the problem of testing

$$
\mathcal{H}_{0N}^r(\mathcal{J}) \quad \text{versus} \quad \mathcal{A}_{2N}^0(\mathcal{J}) \quad [\text{ or } \quad \mathcal{H}_{0N}^r(\mathcal{J}) \quad \text{versus} \quad \mathcal{A}_{2N}(\mathcal{J})\,].
$$

Because of Proposition 3.3.6 it's reasonable to use a rank test in such a situation. Moreover, in a local asymptotic model which corresponds to the continuous situation (3.0.9) we get an optimality result which is completely similar to Theorem 3.0.1.

As a starting point we consider linear rank statistics $S_N(b)$ as defined in (1.3.31) and (1.3.32), i.e. for any given score function $b \in L_2^0(0, 1)$ we define the corresponding scores $b_N(i)$ as

$$
b_N(i) = N \int_{(i-1)/N}^{i/N} b(x)\, dx = N \left( B(\tfrac{i}{N}) - B(\tfrac{i-1}{N}) \right), \quad 1 \leq i \leq N, \quad (3.3.37)
$$

where $B(t) = \int_0^t b(x)\, dx$ , $0 \le t \le 1$ .

If ties (equal observations) are present, it's not suitable to use the original form (1.3.31) of linear rank statistics. Instead we'll use *averaged scores linear rank statistics* in order to get asymptotic optimality, and we use *randomized linear rank statistics* as a convenient tool.

In a first step we define *randomized ranks* $R^* = (R_1^*, ..., R_N^*)$ corresponding to the original ranks $R = (R_1, ..., R_N)$ defined in formula (3.3.34):

Assume $U_1, ..., U_N$ to be i.i.d. random variables with uniform distribution $\mathcal{R}(0,1)$ on the unit interval $(0,1)$ such that the (randomization) vector $U = (U_1, ..., U_N)$ and the observation vector $X = (X_1, ..., X_N)$ are stochastically independent. Then we define the *randomized ranks* $R_i^* = R_i^*(R, U)$ according to

$$R_i^* \;=\; \sum_{j=1}^{N} 1\big( R_j + U_j \le R_i + U_i \big), \qquad 1 \le i \le N, \qquad (3.3.38)$$

and *randomized linear rank statistics* $S_N^*(b)$ according to

$$S_N^*(b) \;=\; \sum_{i=1}^{N} c_{Ni}\, b_N(R_i^*), \qquad\qquad (3.3.39)$$

i.e. the randomized linear rank statistic $S_N^*(b)$ is the usual linear rank statistic $S_N(b)$ with $R$ substituted by $R^*$ .

In a second step we define *averaged scores* $b_N^\tau(i)$ and the corresponding *averaged scores linear rank statistic*

$$S_N^\tau(b) \;=\; \sum_{i=1}^{N} c_{Ni}\, b_N^\tau(R_i) \qquad\qquad (3.3.40)$$

in the following way:

Let $d$ denote the number of different values among the ranks $R_1, ..., R_N$ , which is the same as the number of different values among the original observations $X_1, ..., X_N$ . In the continuous model (no ties) we have $d = N$ . In the general model (ties are possible) $d$ is a random quantity.

Let $T_1 < \cdots < T_d$ denote the ordered values of the different values in $R_1, ..., R_N$ . If no ties are present we have $d = N$ and $T_i = i$ for $i = 1, ..., N$ . If all observations are tied, we have $d = 1$ and $T_1 = N$ . In any case we have $1 \le d \le N$ and $T_d = N$ .

If $X^{(1)} \le X^{(2)} \le ... \le X^{(N)}$ is the ordered pooled sample, then

$$X^{(1)} = ... = X^{(T_1)} \; < \; X^{(T_1+1)} = ... = X^{(T_2)} \; < \; ...$$

$$... < \; X^{(T_{d-1}+1)} = ... = X^{(T_d)}. \qquad (3.3.41)$$

Therefore

$$\tau_i := T_i - T_{i-1}, \qquad 1 \le i \le d, \tag{3.3.42}$$

denote the *lengths of the ties* in the ordered pooled sample, where $T_0 := 0$.

Using $\tau = (\tau_1, ..., \tau_d)$ and $T_0, T_1, ..., T_d$ we define the *averaged scores* $b_N^\tau(i)$ according to $(\forall \, 1 \le i \le N)$,

$$b_N^\tau(i) = \frac{1}{\tau_k} \sum_{j=T_{k-1}+1}^{T_k} b_N(j) \qquad \text{if } T_{k-1} < i \le T_k. \tag{3.3.43}$$

Since the averaged scores depend on the structure of the ties in the observations, the averaged scores are random variables: The original scores $b_N(i)$ are averaged with respect to the lengths of ties in the ordered pooled observations. Using the definition (3.3.37) of the scores $b_N(i)$ we get the representation $(\forall \, 1 \le i \le N)$,

$$b_N^\tau(i) = \frac{N}{\tau_k} \int_{T_{k-1}/N}^{T_k/N} b(x) \, dx, \qquad \text{if } T_{k-1} < i \le T_k. \tag{3.3.44}$$

Additionally, if $R_i = T_k$ then the definition (3.3.38) of $R_i^*$ implies the inequality $T_{k-1} < R_i^* \le T_k$. Therefore we have

$$b_N^\tau(R_i) = b_N^\tau(R_i^*), \qquad 1 \le i \le N, \tag{3.3.45}$$

and thus

$$S_N^\tau(b) = \sum_{i=1}^{N} c_{Ni} \, b_N^\tau(R_i) = \sum_{i=1}^{N} c_{Ni} \, b_N^\tau(R_i^*). \tag{3.3.46}$$

Since the randomization of ranks is a very convenient tool for proving asymptotic properties without the assumption of underlying continuous distribution functions, the relation (3.3.46) will be used in order to prove the asymptotics of averaged scores linear rank statistics $S_N^\tau(b)$ by utilizing the asymptotics of randomized linear rank statistics $S_N^*(b)$.

### 3.3.7 Theorem

*Assume* $b \in L_2^0(0, 1)$.

*a) Under the general null hypothesis* $\mathcal{H}_0^\tau = \{ (H, H) : H \in \mathcal{F}_1 \}$ *the randomized linear rank statistic (3.3.39) has the limiting law*

$$\mathcal{L}\big[ S_N^*(b) \,|\, \mathcal{H}_0^\tau \big] \xrightarrow{\mathcal{L}} \mathcal{N}(0, \|b\|^2). \tag{3.3.47}$$

*b) Under the* $\mathcal{J}$ *-part* $\mathcal{H}_{0N}^\tau(\mathcal{J}) = \{ (H, H) : H \in \mathcal{F}_1(\mathcal{J}) \}$ *of the general null hypothesis* $\mathcal{H}_0^\tau$ *the averaged scores linear rank statistic (3.3.40) has the limiting law*

$$\mathcal{L}\big[ S_N^\tau(b) \,|\, \mathcal{H}_{0N}^\tau(\mathcal{J}) \big] \xrightarrow{\mathcal{L}} \mathcal{N}(0, \|L_{\mathcal{J}} b\|^2), \tag{3.3.48}$$

where the linear mapping $L_{\mathcal{J}} : L_2(0,1) \to L_2(0,1)$ is defined by

$$
(L_{\mathcal{J}} f)(u) = \begin{cases} f(u), & \text{if } u \in (0,1) \setminus \cup\{(s,t] \in \mathcal{J}\}, \\[2mm] \frac{1}{t-s} \int_s^t f \, d\lambda, & \text{if } s < u \le t \text{ for some } (s,t] \in \mathcal{J}, \end{cases} \tag{3.3.49}
$$

for all $f \in L_2(0,1)$. Obviously $b \in L_2^0(0,1)$ implies $L_{\mathcal{J}} b \in L_2^0(0,1)$. Additionally we have the following convergence in $\mathcal{H}_{0N}^{\tau}(\mathcal{J})$ –probability,

$$
\frac{1}{N-1} \sum_{i=1}^{N} \left( b_N^{\tau}(i) \right)^2 \quad \longrightarrow \quad \|L_{\mathcal{J}} b\|^2. \tag{3.3.50}
$$

*Proof of a)* : Since the randomized ranks may be viewed as usual ranks with respect to randomized random variables with underlying continuous distribution functions $F^*$ and $G^*$ , cf. Lemma 3.3.9, the limiting law (3.3.47) is the well-known limiting law of linear rank statistics under the continuous null hypothesis. $\square$

Before proving part b) let's give some remarks:

Obviously the vector $\tau = (\tau_1, ..., \tau_d)$ of the tie-lengths of the ordered pooled sample defines an element $\mathcal{J}_N^{\tau}$ of $\mathcal{J}_{0,1}^*$ according to

$$
\mathcal{J}_N^{\tau} := \left\{ (T_0/N, T_1/N], ..., (T_{d-1}/N, T_d/N] \right\}. \tag{3.3.51}
$$

Using $\mathcal{J}_N^{\tau}$ and the definition (3.3.49) we get from (3.3.44) and (3.3.46) the equalities

$$
S_N^{\tau}(b) = S_N^*(L_{\mathcal{J}_N^{\tau}} b) \tag{3.3.52}
$$

and

$$
\|L_{\mathcal{J}_N^{\tau}} b\|^2 = \frac{1}{N} \sum_{i=1}^{N} \left( b_N^{\tau}(i) \right)^2 . \tag{3.3.53}
$$

If $\hat{H}_N$ denotes the empirical distribution function of the pooled sample and if $\mathcal{I}(\hat{H}_N) = \{ (\hat{H}_N(x_-), \hat{H}_N(x)] : x \in \mathbb{R} \}$ is the family of jump-intervals of $\hat{H}_N$ , then we have in addition

$$
\mathcal{I}(\hat{H}_N) = \mathcal{J}_N^{\tau} , \tag{3.3.54}
$$

i.e. the family $\mathcal{I}(\hat{H}_N) \in \mathcal{J}_{0,1}^*$ only depends on $\tau$ .

*Proof of part b)* : Assume $F = G = H \in \mathcal{F}_1(\mathcal{J})$ , i.e. $\mathcal{I}(H) = \mathcal{J}$ , and let $\hat{H}_N$ be the empirical distribution function of the pooled sample. Then (3.3.54), $\|\hat{H}_N - H\|_\infty \to 0$ in probability, and Lemma 7.5.5 imply

$$
\|L_{\mathcal{J}_N^{\tau}} b - L_{\mathcal{J}} b\| \xrightarrow{N \to \infty} 0 \quad \text{in } H - \text{probability.} \tag{3.3.55}
$$

Because of (3.3.53) and since the right-hand side of (3.3.53) is distribution-free under $\mathcal{H}_{0N}^r(\mathcal{J})$ , cf. Proposition 3.3.5, the proof of (3.3.50) is concluded.

For the proof of (3.3.48) we prove

$$S_N^r(b) - S_N^*(L_{\mathcal{J}}b) \overset{N\to\infty}{\longrightarrow} 0 \quad \text{in} \quad \mathcal{H}_{0N}^r(\mathcal{J}) - \text{probability}, \tag{3.3.56}$$

and apply part a) with $L_{\mathcal{J}}b$ instead of $b$ .

For the final proof of (3.3.56) we apply (3.3.52) and the independence of $R^*$ and $\tau$ under $\mathcal{H}_{0N}^r$ in order to get

$$E_0\!\left[\left(S_N^r(b) - S_N^*(L_{\mathcal{J}}b)\right)^2 \mid \tau\right] = E_0\!\left[\left(S_N^*(L_{\mathcal{J}_N^r}b - L_{\mathcal{J}}b)\right)^2 \mid \tau\right]$$

$$= \frac{1}{N-1} \sum_{i=1}^{N} \left(N \int_{(i-1)/N}^{i/N} (L_{\mathcal{J}_N^r}b - L_{\mathcal{J}}b)\, d\lambda\right)^2$$

$$\leq \frac{N}{N-1} \sum_{i=1}^{N} \int_{(i-1)/N}^{i/N} (L_{\mathcal{J}_N^r}b - L_{\mathcal{J}}b)^2\, d\lambda$$

$$= \frac{N}{N-1}\, \|L_{\mathcal{J}_N^r}b - L_{\mathcal{J}}b\|^2.$$

Therefore (3.3.55) implies (3.3.56). $\square$

For any given level $0 < \alpha < 1$ Theorem 3.3.7 implies the *randomized linear rank test*

$$\psi_N^*(b) = 1\!\left(S_N^*(b) \geq u_\alpha \|b\|\right) \tag{3.3.57}$$

and the *averaged scores linear rank test*

$$\psi_N^r(b) = 1\!\left(S_N^r(b) \geq u_\alpha \sqrt{\frac{1}{N-1}\sum_{i=1}^{N} \left(b_N^r(i)\right)^2}\ \right) \tag{3.3.58}$$

to be asymptotically of level $\alpha$ for testing the null hypothesis $\mathcal{H}_{0N}^r(\mathcal{J})$ , if the condition

$$\|L_{\mathcal{J}}b\|^2 = \int_0^1 (L_{\mathcal{J}}b)^2\, d\lambda > 0 \tag{3.3.59}$$

holds true.

**Remark:** *In Theorem 3.3.7 and in the definition of the asymptotic tests (3.3.57) and (3.3.58) the score function $b \in L_2^0(0,1)$ may be substituted by any $b_N \in L_2^0(0,1)$ such that $\|b_N - b\| \to 0$ as $N \to \infty$ .*

The next theorem will state the asymptotic optimality and the asymptotic power of the averaged scores linear rank test (3.3.58) for suitable local asymptotic alternatives.

**The Local Asymptotic Model**

For any given $\mathcal{J} \in \mathcal{J}_{0,1}^*$ the local asymptotic model is defined in the following way, cf. (3.0.9) and (3.0.10):

Using the notation (3.3.18) we take sequences $(N \geq 2,\ N \to \infty)$ of alternatives

$$
\begin{aligned}
&\left( F^N_{\varrho B_N, H_N},\ G^N_{\varrho B_N, H_N} \right) \\
&= \left( H_N + c_{N1}\varrho B_N \circ H_N,\ H_N + c_{NN}\varrho B_N \circ H_N \right) \in \mathcal{M}_N(\mathcal{J}),
\end{aligned}
\tag{3.3.60}
$$

where $H_N \in \mathcal{F}_1(\mathcal{J})$ is arbitrary and where the directions $B_N \in \mathcal{B}^0_N(\mathcal{J})$ are *almost fixed*, i.e. for the derivative $b_N = B'_N$ of $B_N$ we assume convergence in quadratic mean to some function $b \in L_2(0,1)$,

$$
\|b_N - b\| \overset{N \to \infty}{\longrightarrow} 0.
\tag{3.3.61}
$$

Because of $B_N \in \mathcal{B}^0_N(\mathcal{J})$ the limiting $b$ must be constant $[\lambda - a.e.]$ on each of the intervals $(s,t] \in \mathcal{J}$, i.e. $b = L_{\mathcal{J}} b \ [\lambda - a.e.]$, and $\int_0^1 b\,d\lambda = 0$ is necessary because of $b_N \in L_2^0(0,1)$. For the corresponding pair of underlying distribution functions

$$
(F_N, G_N) = (F^N_{\varrho B_N, H_N},\ G^N_{\varrho B_N, H_N})
$$

we have the property

$$
F_N - G_N \;=\; \varrho\,\sqrt{\frac{N}{mn}}\; B_N \circ H_N,
\tag{3.3.62}
$$

which especially implies the contiguity of the sequence $(F_N^{(m)} \times G_N^{(n)},\ N \geq 2)$ to the (null hypothesis) sequence $(H_N^{(N)},\ N \geq 2)$.

### 3.3.8 Theorem

*Assume* $\mathcal{J} \in \mathcal{J}_{0,1}^*$ *and let* $(F^N_{\varrho B_N, H_N}, G^N_{\varrho B_N, H_N})$ *be any sequence of local asymptotic alternatives as defined in (3.3.60) and (3.3.61), which especially implies* $\|B'_N - b\| \to 0$, $b = L_{\mathcal{J}} b \ [\lambda - a.e.]$, *and* $b \in L_2^0(0,1)$.

*Let* $h_N \in L_2^0(0,1)$, $N \geq 1$, *be any sequence of score functions such that* $\|h_N - h\| \to 0$ *(as* $n \to \infty$*) for some* $h \in L_2^0(0,1)$.

*a) Then we have the limiting laws*

$$
\mathcal{L}[S_N^*(h_N) \mid (F^N_{\varrho B_N, H_N}, G^N_{\varrho B_N, H_N})] \overset{\mathcal{L}}{\longrightarrow} \mathcal{N}(\varrho \langle L_{\mathcal{J}} h, b \rangle, \|h\|^2)
\tag{3.3.63}
$$

and

$$\mathcal{L}[S_N^\tau(h_N) \mid (F_{\varrho B_N,H_N}^N, G_{\varrho B_N,H_N}^N)] \xrightarrow{\mathcal{L}} \mathcal{N}(\varrho\langle L_{\mathcal{J}}h, b\rangle, \|L_{\mathcal{J}}h\|^2). \quad (3.3.64)$$

b) If $\|b\|^2 > 0$ and $L_{\mathcal{J}}h = b \, [\lambda - a.e.]$, then the averaged scores linear rank test $\psi_N^\tau(h_N)$ defined in (3.3.58) is asymptotically optimal at the level $\alpha$ for testing $\mathcal{H}_{0N}^\tau(\mathcal{J})$ versus $(F_{\varrho B_N,H_N}^N, G_{\varrho B_N,H_N}^N)$.

c) If $\|L_{\mathcal{J}}h\| > 0$ holds true, the asymptotic power of the averaged scores linear rank test $\psi_N^\tau(h_N)$ is

$$\lim_{N\to\infty} E\left[\psi_N^\tau(h_N) \mid (F_{\varrho B_N,H_N}^N, G_{\varrho B_N,H_N}^N)\right] = 1 - \Phi(u_\alpha - \frac{\varrho\langle L_{\mathcal{J}}h, b\rangle}{\|L_{\mathcal{J}}h\|}) \quad (3.3.65)$$

and the asymptotic power of the randomized linear rank test $\psi_N^*(h_N)$ is

$$\lim_{N\to\infty} E\left[\psi_N^*(h_N) \mid (F_{\varrho B_N,H_N}^N, G_{\varrho B_N,H_N}^N)\right] = 1 - \Phi(u_\alpha - \frac{\varrho\langle L_{\mathcal{J}}h, b\rangle}{\|h\|}). \quad (3.3.66)$$

d) If the function $h$ is nondecreasing such that $\|L_{\mathcal{J}}h\| > 0$ and if the asymptotic direction $b$ corresponds to the one-sided alternative $\mathcal{A}_2^0 : F \le G, F \ne G$, i.e. $B(t) = \int_0^t b \, d\lambda \le 0 \ \ \forall \, 0 \le t \le 1$, $B(0) = B(1) = 0$, and $B \ne 0$, then we have the inequalities $\langle L_{\mathcal{J}}h, b\rangle \ge 0$ and

$$\alpha \le \lim_{N\to\infty} E\left[\psi_N^*(h_N) \mid (F_{\varrho B_N,H_N}^N, G_{\varrho B_N,H_N}^N)\right]$$

$$\le \lim_{N\to\infty} E\left[\psi_N^\tau(h_N) \mid (F_{\varrho B_N,H_N}^N, G_{\varrho B_N,H_N}^N)\right]. \quad (3.3.67)$$

*Proof:* Because of Proposition 3.3.5 the proofs of a) and b) are direct consequences of Behnen (1976). Part c) is immediate from a) and Theorem 3.3.7. For the proof of d) we apply the result of Section 7.3 (Appendix) in order to get the inequality $\langle L_{\mathcal{J}}h, b\rangle \ge 0$, since $L_{\mathcal{J}}h$ is nondecreasing if $h$ is nondecreasing. Then (3.3.65), (3.3.66), and $0 < \|L_{\mathcal{J}}h\| \le \|h\|$ imply the assertion (3.3.67). $\square$

At least approximately the optimality result of Theorem 3.3.8 assumes the structure of discontinuity $\mathcal{J} \in \mathcal{J}_{0,1}^*$ and the direction $B_N \in \mathcal{B}_N^0(\mathcal{J})$ to be given. In reality neither the underlying family $\mathcal{J}$ nor the corresponding underlying direction $B_N \in \mathcal{B}_N^0(\mathcal{J})$ are given quantities. In accordance with Section 3.1 and Section 3.2 we therefore try to find estimators of the underlying optimal score function $B_N'$ in order to construct suitable tests without knowing the special type of the alternative. As a preparatory step we'll exemplify the randomization of the underlying $X_1, ..., X_N$ , which has been used in the proofs of Theorem 3.3.7 and Theorem 3.3.8.

### 3.3.9 Lemma

*Let $X_1, ..., X_N, U_1, ..., U_N$ be independent real random variables on some probability space $(\Omega, \mathcal{A}, P)$ . For each $1 \leq i \leq N$ assume $\mathcal{L}(U_i) = \mathcal{R}(0, 1)$ and let $F_i \in \mathcal{F}_1$ denote the (arbitrary) distribution function of $\mathcal{L}(X_i)$ . Define the mixture $H \in \mathcal{F}_1$ according to $H = p_1 F_1 + p_2 F_2 + ... + p_N F_N$ for given weights $0 < p_i < 1$ such that $p_1 + p_2 + ... + p_N = 1$ . Finally we define the randomized random variables $X_1^*, ..., X_N^*$ according to*

$$X_i^* = T(X_i, U_i), \quad i = 1, ..., N, \tag{3.3.68}$$

*where $T : \mathbb{R} \times [0, 1] \longrightarrow [0, 1]$ is defined by*

$$T(x, u) = H(x_-) + u\big(H(x) - H(x_-)\big). \tag{3.3.69}$$

*Then the following assertions hold true:*

*a) $X_1^*, ..., X_N^*$ are independent random variables with values in $[0, 1]$ .*

*b) For each $1 \leq i \leq N$ we have with probability 1*

$$\sum_{j=1}^{N} 1(X_j^* \leq X_i^*) = \sum_{j=1}^{N} 1(R_j + U_j \leq R_i + U_i) = R_i^*. \tag{3.3.70}$$

*This means that the randomized ranks (3.3.38) may be viewed as the usual ranks of the randomized random variables $X_1^*, ..., X_N^*$ .*

*c) For each $1 \leq i \leq N$ the distribution function $F_i^*$ of $\mathcal{L}(X_i^*)$ is absolutely continuous with $\lambda$ –density*

$$f_i^* = \frac{dF_i}{dH} \circ H^{-1}, \tag{3.3.71}$$

*which is constant on each of the intervals $(s, t] \in \mathcal{I}(H)$ .*

*d) For each $1 \leq i \leq N$ the conditional distribution of $R_i^*$ given the ranks $R = (R_1, ..., R_N)$ is the uniform distribution on $\{R_i^0 + 1, ..., R_i\}$ , where*

$$R_i^0 = \sum_{j=1}^{N} 1(R_j < R_i). \tag{3.3.72}$$

**Remark:** Using $T_0 = 0$ and $T_1, ..., T_d$ as defined in (3.3.41) we have

$$\{R_1, ..., R_N\} = \{T_1, ..., T_d\} \tag{3.3.73}$$

and

$$\left[ R_i = T_j \iff R_i^0 = T_{j-1} \right] \qquad \forall\, 1 \le i \le N,\ 1 \le j \le d. \tag{3.3.74}$$

*Proof:* Assertion a) is obvious from (3.3.68) and (3.3.69). For the proof of b) we notice that the following chain of equalities holds true with probability 1,

$$\sum_{j=1}^{N} 1(X_j^* \le X_i^*) = \sum_{j=1}^{N} 1(X_j < X_i) + \sum_{j=1}^{N} 1(X_j = X_i)\, 1(U_j \le U_i)$$

$$= \sum_{j=1}^{N} 1(R_j < R_i) + \sum_{j=1}^{N} 1(R_j = R_i)\, 1(U_j \le U_i) = \sum_{j=1}^{N} 1(R_j + U_j \le R_i + U_i).$$

c) In a first step we get for each $x \in \mathbb{R}$

$$F_i^*(H(x)) = P\{\, X_i^* \le H(x)\, \}$$

$$= P\{\, H(X_i) \le H(x), X_i \in \mathbb{R} \setminus D(H)\, \}$$

$$+ \sum_{y \in D(H)} P\{\, H(y_-) + U_i(H(y) - H(y_-)) \le H(x), X_i = y\, \}$$

$$= P\{\, H(X_i) \le H(x)\, \}$$

$$+ \sum_{y \in D(H)} P\{X_i = y\} \left( P\{\, U_i \le \frac{H(x) - H(y_-)}{H(y) - H(y_-)}\, \} - 1(\, H(y) \le H(x)\, ) \right),$$

$$= P\{\, H(X_i) \le H(x)\, \}, \quad \text{since } H(x) < H(y) \text{ implies } H(x) \le H(y_-),$$

$$= P\{\, X_i \le x\, \}, \quad \text{since } \mathcal{L}(X_i) \text{ is dominated by } H,$$

$$= F_i(x).$$

In a second step we assume $H(x_-) < u < H(x)$. Then we get

$$F_i^*(u) = P\{\, X_i^* \le u\, \} = P\{\, X_i^* \le H(x)\, \} - P\{\, u < X_i^* \le H(x)\, \}$$

$$= F_i(x) - P\{\, u < H(x_-) + U_i(H(x) - H(x_-)) \le H(x),\ X_i = x\, \}$$

$$= F_i(x) - \left( F_i(x) - F_i(x_-) \right) \left( 1 - P\{U_i \le (u - H(x_-))/(H(x) - H(x_-))\} \right)$$

$$= F_i(x_-) + \left( u - H(x_-) \right) \left( F_i(x) - F_i(x_-) \right) / \left( H(x) - H(x_-) \right).$$

Therefore the second part of the proof of Proposition 3.3.3 proves the equality

$$\int_0^t f_i^* \, d\lambda \; = \; F_i^*(t) \quad \forall \, 0 \le t \le 1.$$

d) Given $R$, i.e. $R_i = T_q$ for some $q \in \{1, ..., d\}$, we get from (3.3.38) and $0 < U_i < 1$ the inequality $R_i^0 \le R_i^* \le R_i$ and

$$R_i^* - R_i^0 \; = \; \sum_{j=1}^{N} 1(R_j = T_q) \, 1(U_j < U_i). \tag{3.3.75}$$

Since $R$ and $U = (U_1, ..., U_N)$ are independent, the conditional distribution of $R_i^* - R_i^0$ given $R$ is equal to the distribution of any rank in an i.i.d. sample of size $T_q - T_{q-1} = R_i - R_i^0$ from Uniform(0,1)-distribution. $\square$

**Approximating the optimal score function by kernel estimators in the presence of ties**

Now let's go back to the two-sample case with given $\mathcal{J} \in \mathcal{J}_{0,1}^*$ , i.e.,

$X_1, ..., X_m$ are i.i.d. with distribution function $F \in \mathcal{F}_1$,

$X_{m+1}, ..., X_N$ are i.i.d. with distribution function $G \in \mathcal{F}_1$,

and $H_N = (mF + nG)/N \in \mathcal{F}_1(\mathcal{J})$.

Then Theorem 3.3.8 implies the averaged scores linear rank statistic $S_N^\tau(b)$ to be approximately optimal for testing $\mathcal{H}_{0N}^\tau(\mathcal{J})$ versus $(F, G)$, if the score function $b$ is given by

$$b \; = \; \sqrt{\frac{mn}{N}} \left( \frac{dF}{dH_N} \circ H_N^{-1} - \frac{dG}{dH_N} \circ H_N^{-1} \right) \; = \; \sqrt{\frac{mn}{N}} (f_N^* - g_N^*). \tag{3.3.76}$$

Since (3.3.43), part d) of Lemma 3.3.9, and (3.3.74) imply

$$E\big[b_N(R_i^*) \,|\, R\big] \; = \; \frac{1}{R_i - R_i^0} \sum_{j=1}^{N} b_N(j) \, 1(R_i^0 < j \le R_i) \; = \; b_N^\tau(R_i), \tag{3.3.77}$$

with scores $b_N(i)$ according to (3.3.37), the definitions (3.3.39) and (3.3.40) yield the equality

$$S_N^\tau(b) \; = \; E\big[S_N^*(b) \,|\, R\big]. \tag{3.3.78}$$

The randomized rank statistic $S_N^*(b)$ is a linear rank statistic without ties. Therefore formula (3.1.10) implies the representation

$$S_N^*(b) \; = \; \int_0^1 b \, d\hat{B}_{N2}^* \; = \; \langle b, \hat{b}_{N2}^* \rangle, \tag{3.3.79}$$

where $\hat{B}^*_{N2}$ is the piecewise linearized version, cf. (3.1.5), of the *randomized rank process*

$$\hat{B}^*_{N1}(t) \;=\; \sum_{i=1}^{N} c_{Ni}\, 1\!\left(\frac{R^*_i}{N} \le t\right), \qquad 0 \le t \le 1, \qquad (3.3.80)$$

and $\hat{b}^*_{N2}$ is the rightcontinuous version of the derivative of $\hat{B}^*_{N2}$ , cf. (3.1.6),

$$\hat{b}^*_{N2} = N\left(\hat{B}^*_{N1}(\frac{i}{N}) - \hat{B}^*_{N1}(\frac{i-1}{N})\right) \qquad \text{if } \frac{i-1}{N} \le t < \frac{i}{N}. \qquad (3.3.81)$$

Using (3.3.79) and defining the *averaged two-sample rank process* $\hat{B}_{N2}$ according to

$$\hat{B}_{N2}(t) \;=\; E\big[\hat{B}^*_{N2}(t) \mid R\big], \qquad 0 \le t \le 1, \qquad (3.3.82)$$

we get from (3.3.78) the representation

$$S^\tau_N(b) \;=\; \langle b, \hat{b}_{N2}\rangle, \qquad (3.3.83)$$

where $\hat{b}_{N2}$ is the rightcontinuous version of the derivative of $\hat{B}_{N2}$ ,

$$\hat{b}_{N2}(t) = N\left(\hat{B}_{N2}(\frac{i}{N}) - \hat{B}_{N2}(\frac{i-1}{N})\right) \qquad \text{if } \frac{i-1}{N} \le t < \frac{i}{N}. \qquad (3.3.84)$$

Notice, since $\hat{B}^*_{N2}(\cdot)$ is linear on each of the intervals $[(i-1)/N, i/N]$, $i = 1, ..., N$ , the same is true for $\hat{B}_{N2}(\cdot)$ . Therefore $\hat{B}_{N2}$ is completely determined by $\hat{B}_{N2}(i/N)$ , $i = 0, 1, ..., N$ .

Obviously the definition (3.3.82) implies $\hat{B}_{N2}(0) = \hat{B}_{N2}(1) = 0$ and ( $\forall\, i = 1, ..., N$)

$$\begin{aligned}
\Delta\hat{B}_{N2}(\frac{i}{N}) &:= \hat{B}_{N2}(\frac{i}{N}) - \hat{B}_{N2}(\frac{i-1}{N}) \\[2mm]
&= \sum_{j=1}^{N} c_{Nj}\, P\big[R^*_j = i \mid R\big] \\[2mm]
&= \sum_{j=1}^{N} c_{Nj}\, 1\!\left(R^0_j < i \le R_j\right)/(R_j - R^0_j). \qquad (3.3.85)
\end{aligned}$$

Especially, the averaged two-sample rank process $\hat{B}_{N2}$ takes values in $C[0,1]$ and fulfils the side condition $\hat{B}_{N2}(0) = \hat{B}_{N2}(1) = 0$ . If no ties are present the general definition (3.3.82) of $\hat{B}_{N2}$ coincides with the (continuous) definition (3.1.5).

With the same argument as in the continuous model we'll estimate the unknown optimal score function $b$ defined in (3.3.76) and substitute the $b$ in (3.3.83) by the estimator in order to get a suitable test statistic:

According to Lemma 3.3.9 (with $H = H_N$ ) we have

$X_1^*, ..., X_m^*$ are i.i.d. with $\lambda$–density $f_N^*$ ,

$X_{m+1}^*, ..., X_N^*$ are i.i.d. with $\lambda$–density $g_N^*$ ,

and $\quad h_N^* = (mf_N^* + ng_N^*)/N = 1 \quad [\lambda - a.e.]$.

Since $\hat{B}_{N2}^*$ is a natural estimator of the underlying direction

$$B(t) = \int_0^t b\, d\lambda = \sqrt{\frac{mn}{N}} \int_0^t (f_N^* - g_N^*)\, d\lambda, \qquad 0 \le t \le 1, \qquad (3.3.86)$$

the derivative $b = B'$ should be estimated by a derivative of a suitably smoothed version of $\hat{B}_{N2}^*$ , cf. Section 3.1. The disadvantage of such estimators is their dependence on external randomization. But fortunately the function $B$ is linear on each of the intervals $(s,t] \in \mathcal{J}$ . Therefore the conditional expectation $\hat{B}_{N2} = E[\hat{B}_{N2}^* \mid R]$ will even do better than $\hat{B}_{N2}^*$ , cf. Theorem 3.3.11 and Theorem 3.3.13.

This means that we may adapt the procedures of Section 3.1 and Section 3.2 to the general case of this section by substituting the rank process $\hat{B}_{N2}$ defined in (3.1.4) and (3.1.5) by the averaged rank process $\hat{B}_{N2}$ defined in (3.3.82) or (3.3.85). Notice, there is no problem in using the same symbol in both cases, since in the special continuous case we have with probability 1 the identity $(R_1^*, ..., R_N^*) = (R_1, ..., R_N)$ which implies the coincidence of the definitions in this case. If ties are present, we have to use the general definition of this section.

**The Omnibus Test Statistic**

In case of the omnibus alterrnative $\mathcal{A}_2 : F \ne G$ the kernel estimator of $b$ is

$$\mathcal{K}_a\hat{b}_{N2}(t) = \int_0^1 K_a(s,t)\, \hat{b}_{N2}(s)\, ds, \qquad 0 \le t \le 1. \qquad (3.3.87)$$

Therefore the corresponding rank statistic $S_N^\tau(\mathcal{K}_a\hat{b}_{N2})$ has the form, cf. formulae (3.3.83) and (3.1.23),

$$S_N^\tau(\mathcal{K}_a\hat{b}_{N2}) = \langle \mathcal{K}_a\hat{b}_{N2}, \hat{b}_{N2}\rangle \qquad\qquad (3.3.88)$$

$$= \sum_{i=1}^N \sum_{j=1}^N \left( \hat{B}_{N2}(\frac{i}{N}) - \hat{B}_{N2}(\frac{i-1}{N}) \right) \left( \hat{B}_{N2}(\frac{j}{N}) - \hat{B}_{N2}(\frac{j-1}{N}) \right) \bar{k}_{Na}(i,j),$$

where the weights $\bar{k}_{Na}(i,j)$ are defined in (3.1.24). With the same argument as in Section 3.1 we may use the simpler weights

$$k_{Na}(i,j) = K_a(\frac{i-1/2}{N}, \frac{j-1/2}{N}) \tag{3.3.89}$$

if the underlying kernel $K$ is smooth. This yields the (nonlinear) rank statistic $S_N(a,K)$, cf. (3.1.25), for testing the null hypothesis $\mathcal{H}_0^{\tau} : F = G$ versus the omnibus alternative $\mathcal{A}_2 : F \neq G$.

From the construction it's clear that the test should reject the null hypothesis if the statistic $S_N(a,K)$ is sufficiently large.

**The One-Sided Test Statistic**

In case of the one-sided alternative $\mathcal{A}_2^0 : F \leq G, F \neq G$ we use the corresponding projection

$$\hat{B}_{N2}^0 = \hat{B}_{N2}\, 1(\hat{B}_{N2} < 0) \tag{3.3.90}$$

and its derivative

$$\hat{b}_{N2}^0 = \hat{b}_{N2}\, 1(\hat{B}_{N2} < 0) \tag{3.3.91}$$

instead of $\hat{B}_{N2}$ and $\hat{b}_{N2}$, cf. Section 3.1. Therefore the corresponding one-sided rank statistic $S_N^{\tau}(\mathcal{K}_a \hat{b}_{N2}^0)$ has the form, cf. (3.3.83) and (3.1.26),

$$S_N^{\tau}(\mathcal{K}_a \hat{b}_{N2}^0) = \langle \mathcal{K}_a \hat{b}_{N2}^0, \hat{b}_{N2} \rangle \tag{3.3.92}$$

$$= \sum_{i=1}^{N} \sum_{j=1}^{N} \left( \hat{B}_{N2}(\frac{i}{N}) - \hat{B}_{N2}(\frac{i-1}{N}) \right) \left( \hat{B}_{N2}(\frac{j}{N}) - \hat{B}_{N2}(\frac{j-1}{N}) \right) \hat{k}_{Na}(i,j),$$

where the (random) one-sided weights $\hat{k}_{Na}(i,j)$ are defined by

$$\hat{k}_{Na}(i,j) = N^2 \int_{(i-1)/N}^{i/N} \int_{(j-1)/N}^{j/N} K_a(s,t)\, 1(\hat{B}_{N2}(s) < 0)\, ds\, dt. \tag{3.3.93}$$

Again, if the underlying kernel $K$ is smooth, we may use instead of (3.3.93) the simpler weights

$$k_{Na}(i,j)\Delta_N(i), \tag{3.3.94}$$

with $k_{Na}(i,j)$ given in (3.3.89) and

$$\Delta_N(i) = N \int_{(i-1)/N}^{i/N} 1(\hat{B}_{N2}(s) < 0)\, ds. \tag{3.3.95}$$

Completely similar to (3.1.29) we get the equality

$$\left( \hat{B}_{N2}(\frac{i}{N}) - \hat{B}_{N2}(\frac{i-1}{N}) \right) \Delta_N(i) = \hat{B}_{N2}^0(\frac{i}{N}) - \hat{B}_{N2}^0(\frac{i-1}{N}). \tag{3.3.96}$$

Therefore the substitution of the weights $\hat{k}_{Na}(i,j)$ in (3.3.92) by their approximations (3.3.94) yields the (nonlinear) rank statistic

$$S_N^0(a,K) \tag{3.3.97}$$

$$= \sum_{i=1}^{N}\sum_{j=1}^{N} \left( \hat{B}_{N2}^0(\frac{i}{N}) - \hat{B}_{N2}^0(\frac{i-1}{N}) \right) \left( \hat{B}_{N2}(\frac{j}{N}) - \hat{B}_{N2}(\frac{j-1}{N}) \right) k_{Na}(i,j)$$

for testing the null hypothesis $\mathcal{H}_0^r : F = G$ versus the one-sided alternative $\mathcal{A}_2^0 : F \leq G, F \neq G$ .

From the construction it's clear that the test should reject the null hypothesis if the statistic $S_N^0(a,K)$ is sfficiently large.

**Evaluation of the Tests**

It's clear that in case of discontinuous underlying distribution functions the ranks $R = (R_1, ..., R_N)$ are not distribution-free under the general null hypothesis $\mathcal{H}_0^r : F = G$ , cf. Proposition 3.3.5. But according to Lemma 3.3.9 the randomized ranks $R^* = (R_1^*, ..., R_N^*)$ are distribution-free under $\mathcal{H}_0^r$ . Therefore the conditional distribution of the ranks $R$ given the lengths of the ties $\tau = (\tau_1, ..., \tau_d)$ in the ordered pooled sample will be distribution-free under the general null hypothesis $\mathcal{H}_0^r$ , too, cf. Lemma 3.3.10. As a consequence we'll use the conditional tests (given $\tau$ ) based on the nonlinear rank statistics $S_N(a,K)$ (in the omnibus case) and $S_N^0(a,K)$ (in the one-sided case), i.e.

$$\psi_{N\alpha} = 1\Big( S_N(a,K) > q_\alpha(\tau) \Big),$$

$$\psi_{N\alpha}^0 = 1\Big( S_N^0(a,K) > q_\alpha^0(\tau) \Big). \tag{3.3.98}$$

### 3.3.10 Lemma

Let $X_1, ..., X_N$ be i.i.d. real random variables with arbitrary underlying distribution function $F \in \mathcal{F}_1$ . Let $R = (R_1, ...R_N)$ and $\tau = (\tau_1, ..., \tau_d)$ denote the ranks and the lengths of ties in the ordered $X$ –sample, cf. (3.3.42). Then $\mathcal{L}_F[R \mid \tau]$ , the conditional distribution of $R$ given $\tau$ , is the same for any $F \in \mathcal{F}_1$ .

*Proof*: Define the randomized ranks $R^* = (R_1^*, ..., R_N^*)$ according to (3.3.38). Then the well-known Theorem 29A of Hájek (1969) proves the stochastic independence of $R^*$ and $\tau$ , and the uniform distribution of $R^*$ on the $N!$ permutations of $(1, ..., N)$ , independently of the underlying $F \in \mathcal{F}_1$ . Since the sets of values $\{T_1, ..., T_d\}$ and $\{R_1, ..., R_N\}$ coincide and since $R_i = T_j$

implies $R_i^0 = T_{j-1}$ , formula (3.3.75) implies the following representation $R_i = g(R_i^*, \tau)$ of the ranks in terms of the randomized ranks and $\tau$ ,

$$R_i = \sum_{j=1}^{d} T_j \, 1(T_{j-1} < R_i^* \le T_j), \qquad 1 \le i \le N. \tag{3.3.99}$$

For any $F \in \mathcal{F}_1$ and any given $\tau = \tau_0$ this proves

$$\mathcal{L}_F\big[ R \mid \tau = \tau_0 \big] = \mathcal{L}_0\big[ g(R_1^*, \tau_0), ..., g(R_N^*, \tau_0) \big], \tag{3.3.100}$$

which concludes the proof. $\square$

As a special consequence we get the conditional distribution of $R_i$ given $\tau$ under $\mathcal{H}_0^r$ in the form

$$P_{\mathcal{H}_0^r}\big[ R_i \in \{T_1, ..., T_d\} \mid \tau \big] = 1 \qquad \text{and}$$

$$P_{\mathcal{H}_0^r}\big[ R_i = T_j \mid \tau \big] = \tau_j/N, \qquad j = 1, ..., d. \tag{3.3.101}$$

Because of (3.3.99) and (3.3.100) there is no principal problem to evaluate the exact conditional critical values $q_\alpha(\tau)$ and $q_\alpha^0(\tau)$ or the conditional $p$-values of the respective test statistics $S_N(a, K)$ and $S_N^0(a, K)$ . But the evaluation may become very time-consuming for larger sample sizes. Starting with an observed $\tau_0$ and using a $\mathcal{R}(0, 1)$ random generator we utilize the representation (3.3.99) in order to simulate the conditional $p$-values of $S_N^0(a, K)$ for the Parzen-2 kernel (3.1.49) and the bandwidth $a = 0.40$ , cf. Section 2.1.

For the practical evaluation of the two-sample rank process $\hat{B}_{N2}$ the following representation seems to be most convenient:

Using $0 = T_0 < T_1 < \cdots < T_d = N$ and $\{T_1, ..., T_d\} = \{R_1, ..., R_N\}$ we get from (3.3.85) and (3.3.74) for any $i \in \{1, ..., N\}$ such that $T_{q-1} < i \le T_q$ the following equalities,

$$\Delta \hat{B}_{N2}(\frac{i}{N}) = \sum_{j=1}^{N} c_{Nj} \, 1(R_j = T_q)/\tau_q$$

$$= \frac{1}{\tau_q} \sqrt{\frac{mn}{N}} \left( \frac{1}{m} \sum_{j=1}^{m} 1(R_j = T_q) - \frac{1}{n} \sum_{j=m+1}^{N} 1(R_j = T_q) \right)$$

$$= \frac{1}{\sqrt{mnN}} \frac{n\tau_{1q} - m\tau_{2q}}{\tau_q} , \tag{3.3.102}$$

where $( \ \forall \ q = 1, ..., d)$

$$\tau_{1q} := \sum_{j=1}^{m} 1(R_j = T_q) = \sum_{j=1}^{m} 1\left( X_j = X^{(T_q)} \right) \qquad (3.3.103)$$

is the *contribution of the first sample* to the length of the $q$-th tie, and

$$\tau_{2q} := \sum_{j=m+1}^{N} 1(R_j = T_q) = \sum_{j=m+1}^{N} 1\left( X_j = X^{(T_q)} \right) \qquad (3.3.104)$$

is the *contribution of the second sample* to the length of the $q$-th tie. Especially we have $\tau_{1q} + \tau_{2q} = \tau_q \ \ \forall \ q = 1, ..., d$.

An illustration of the practical evalution of $\hat{B}_{N2}$ via (3.3.102) to (3.3.104) is given in Example 2.1.1.

**The Asymptotic Distribution**

The asymptotics of the various statistics is based on the asymptotics of the process $\hat{B}_{N2}$, cf. Section 3.1 and Section 3.2. The asymptotics of $\hat{B}_{N2}$ is derived from the asymptotics of the randomized rank process $\hat{B}_{N2}^*$.

**3.3.11 Theorem**

*Assume $\mathcal{J} \in \mathcal{J}_{0,1}^*$ and let $(F_{\varrho B_N, H_N}^N, G_{\varrho B_N, H_N}^N)$ be a sequence of local asymptotic alternatives as defined in (3.3.60) and (3.3.61), which especially implies $\|B_N' - b\| \to 0$, $b = L_{\mathcal{J}} b \ [\lambda - a.e.]$, and $b \in L_2^0(0,1)$.*

*Then we have the following limiting law in $\left( C[0,1], \mathcal{B}(C[0,1]) \right)$,*

$$\mathcal{L}\left[ \hat{B}_{N2}^* \mid (F_{\varrho B_N, H_N}^N, G_{\varrho B_N, H_N}^N) \right] \overset{\mathcal{L}}{\longrightarrow} \mathcal{L}[W_0 + \varrho B], \qquad (3.3.105)$$

*where $B(t) = \int_0^t b \, d\lambda \ \forall \ 0 \le t \le 1$ and $W_0$ is the Brownian-bridge.*

*Since $b = L_{\mathcal{J}} b \ [\lambda - a.e.]$ the limiting direction $B$ is an absolutely continuous function on $[0,1]$ such that $B$ is linear on each of the intervals $(s,t] \in \mathcal{J}$ and $B(0) = B(1) = 0$.*

*Proof*

: Given $(F_N, G_N) = (F_{\varrho B_N, H_N}^N, G_{\varrho B_N, H_N}^N)$ and $H = (mF_N + nG_N)/N = H_N$ let $X_{N1}^*, ..., X_{NN}^*$ denote the randomized random variables defined in Lemma 3.3.9. Then the random variables $X_{N1}^*, ..., X_{Nm}^*$ are i.i.d. with $\lambda$-density $f_N^* = 1 + c_{N1} \varrho b_N$ and the random variables $X_{Nm+1}^*, ..., X_{NN}^*$ are i.i.d. with $\lambda$-density $g_N^* = 1 + c_{NN} \varrho b_N$. Since $R_1^*, ..., R_N^*$ are the ranks of $X_{N1}^*, ..., X_{NN}^*$ the assertion (3.3.105) is a direct consequence of Theorem 3.1.8. $\square$

For the proof of the limiting law of $\hat{B}_{N2}$ it's helpful to have a suitable representation of $\hat{B}_{N2}$ in terms of $\hat{B}_{N2}^{*}$ . For that purpose let $\hat{H}_N$ denote the empirical distribution function of the pooled sample $X_1, ..., X_N$ and let $\mathcal{I}(\hat{H}_N) = \{(\hat{H}_N(x_-), \hat{H}_N(x)] : x \in \mathbb{R}\}$ be the family of jump-intervals of $\hat{H}_N$ . Using the definition $\mathcal{J}_N^{\tau} = \{(T_{i-1}/N, T_i/N] : i = 1, ..., d\}$ , cf. (3.3.51), we have $\mathcal{I}(\hat{H}_N) = \mathcal{J}_N^{\tau}$ , cf. (3.3.54).

In addition we define for each family $\mathcal{J} \in \mathcal{J}_{0,1}^{*}$ of halfopen subintervals of the unit interval $[0, 1]$ the mapping $T_{\mathcal{J}} : C[0, 1] \to C[0, 1]$ by ( $\forall f \in C[0, 1]$ )

$$(T_{\mathcal{J}} f)(t) = \begin{cases} f(t), & \text{if } t \in [0, 1] \setminus \cup \{J \in \mathcal{J}\}, \\ \frac{y-t}{y-x} f(x) + \frac{t-x}{y-x} f(y), & \text{if } t \in (x, y] \in \mathcal{J}. \end{cases} \qquad (3.3.106)$$

### 3.3.12 Proposition

*The averaged two-sample rank process $\hat{B}_{N2}$ defined in (3.3.82) has the following representation:*

$$\hat{B}_{N2} \text{ is linear on } \left[\frac{T_{i-1}}{N}, \frac{T_i}{N}\right] \qquad \forall\, i = 1, ..., d, \qquad (3.3.107)$$

*and*

$$\hat{B}_{N2}\left(\frac{T_i}{N}\right) = \hat{B}_{N2}^{*}\left(\frac{T_i}{N}\right) \qquad \forall\, i = 1, ..., d. \qquad (3.3.108)$$

*In other words,*

$$\hat{B}_{N2} = T_{\mathcal{I}(\hat{H}_N)} \hat{B}_{N2}^{*} = T_{\mathcal{J}_N^{\tau}} \hat{B}_{N2}^{*} . \qquad (3.3.109)$$

*Proof*: According to formula (3.3.102) the increment $\Delta \hat{B}_{N2}(i/N)$ is constant if $T_{q-1} < i \leq T_q$ . Since $\hat{B}_{N2}(\cdot)$ is linear on each interval $[(i-1)/N, i/N]$ this proves (3.3.107).

From formula (3.3.75) we get the equalities

$$1(R_j^{*} \leq T_i) = 1(R_j \leq T_i) \qquad \forall\, 1 \leq j \leq N \quad \forall\, 0 \leq i \leq d,$$

which obviously imply

$$\hat{B}_{N2}(T_i/N) = E[\hat{B}_{N2}^{*}(T_i/N) \mid R] = \sum_{j=1}^{N} c_{Nj}\, 1(R_j \leq T_i) = \hat{B}_{N2}^{*}(T_i/N).$$

This concludes the proof. $\square$

Notice, the family $\mathcal{I}(\hat{H}_N) = \mathcal{J}_N^\tau$ only depends on $\tau$ while $\hat{B}_N^*$ only depends on $R^* = (R_1^*, ..., R_N^*)$. Since under the null hypothesis $\mathcal{H}_0^\tau$ the random variables $\tau$ and $R^*$ are stochastically independent we have, under the condition $\tau = \tau_0$, the equality

$$\mathcal{L}_{\mathcal{H}_0^\tau}\big[\,\hat{B}_{N2} \mid \tau = \tau_0\,\big] \;=\; \mathcal{L}_0\big[\,T_{\mathcal{J}_N^{\tau_0}}\hat{B}_{N2}^*\,\big], \qquad (3.3.110)$$

where $\mathcal{J}_N^{\tau_0}$ is the family of jump-intervals corresponding to $\tau_0$.


### 3.3.13 Theorem

Assume $\mathcal{J} \in \mathcal{J}_{0,1}^*$ and let $(F_{\varrho B_N, H_N}^N, G_{\varrho B_N, H_N}^N)$ be any sequence of local asymptotic alternatives as defined in (3.3.60) and (3.3.61), which especially implies $\|B_N' - b\| \to 0$, $b = L_{\mathcal{J}}b\,[\lambda - a.e.]$, and $b \in L_2^0(0,1)$.

Then we have the following limiting law in $\big(C[0,1], \mathcal{B}(C[0,1])\big)$,

$$\mathcal{L}\big[\,\hat{B}_{N2} \mid (F_{\varrho B_N, H_N}^N, G_{\varrho B_N, H_N}^N)\,\big] \;\xrightarrow{\mathcal{L}}\; \mathcal{L}\big[\,T_{\mathcal{J}}W_0 + \varrho B\,\big]. \qquad (3.3.111)$$

Under the special null hypothesis $\mathcal{H}_{0N}^\tau(\mathcal{J})$ we have in $\big(C[0,1], \mathcal{B}(C[0,1])\big)$,

$$\mathcal{L}_{\mathcal{H}_{0N}^\tau(\mathcal{J})}\big[\,\hat{B}_{N2} \mid \tau\,\big] \;\xrightarrow{\mathcal{L}}\; \mathcal{L}\big[\,T_{\mathcal{J}}W_0\,\big] \quad a.s.. \qquad (3.3.112)$$

The process $T_{\mathcal{J}}W_0$ is a $C[0,1]$–valued Gaussian process with zero expectation, $(T_{\mathcal{J}}W_0)(0) = (T_{\mathcal{J}}W_0)(1) = 0$, and covariance function $K_{\mathcal{J}}$ given by $(\forall\, 0 \le s \le t \le 1)$

$$K_{\mathcal{J}}(s,t) \;=\; \begin{cases} x + \frac{t-x}{y-x} - st, & \text{if } s,t \in (x,y] \in \mathcal{J}, \\[2mm] s - st, & \text{elsewhere}. \end{cases} \qquad (3.3.113)$$


*Proof*: Because of formula (3.3.94), $\mathcal{I}(H_N) = \mathcal{J}$, and Lemma 7.5.4 we have the inequality

$$P\{\|\hat{B}_{N2} - T_{\mathcal{J}}\hat{B}_{N2}^*\|_\infty \ge \eta\}$$
$$(3.3.114)$$
$$\le\; P\{\omega(\hat{B}_{N2}^*, \varepsilon) \ge \eta/23\} + P\{\|\hat{H}_N - H_N\|_\infty \ge \varepsilon/2\}$$

for any $\eta > 0$ and any $\varepsilon \in (0,1)$. Now the continuity of $f \to \omega(f,\varepsilon)$, the continuity of $W_0 + \varrho B$, and $\|\hat{H}_N - H_N\|_\infty \to 0$ in probability prove $\|\hat{B}_{N2} - T_{\mathcal{J}}\hat{B}_{N2}^*\|_\infty \to 0$ in probability. Hence (3.3.111) is implied by (3.3.105) and the continuity of $T_{\mathcal{J}}$.

Under $F = G \in \mathcal{F}_1(\mathcal{J})$ we have $H_N = F = G$ and therefore $\|\hat{H}_N - F\|_\infty \to 0$ [a.s.]. For any fixed sequence $(\hat{H}_N, N \ge 1)$ with $\|\hat{H}_N - F\|_\infty \to 0$ let

$(\mathcal{J}_N := \mathcal{I}(\hat{H}_N), N \geq 1)$ and $(\hat{\tau}_N, N \geq 1)$ be the corresponding (non-random) sequences of jump-interval-systems and tie-lengths-vectors. Then, according to (3.3.110), $\mathcal{I}(F) = \mathcal{J}$ , Lemma 7.5.4, and (3.3.90) we get

$$\mathcal{L}_{(F,F)}[\hat{B}_{N2} \mid \tau = \hat{\tau}_N] = \mathcal{L}_0[T_{\mathcal{J}_N}\hat{B}^*_{N2}] \xrightarrow{\mathcal{L}} \mathcal{L}[T_{\mathcal{J}}W_0].$$

The covariance function (3.3.113) follows from the definition (3.3.106) of $T_{\mathcal{J}}$ by simple direct computation using $EW_0(s)W_0(t) = \min(s,t) - st$ . $\square$

**Approximating the optimal score function by projections onto finite dimensional cones when ties are present.**

Our aim is to extend the (continuous) results of part C) of Section 3.2 to the case when ties are present. In accordance with the continuous case let $b_1, ..., b_r \in L_2^0(0,1)$ be special score functions which fulfil the side condition (3.2.48) of the one-sided model. Define

$$\vec{S}_N = \Big( \langle b_1, \hat{b}_{N2} \rangle, ..., \langle b_r, \hat{b}_{N2} \rangle \Big)^T , \tag{3.3.115}$$

as in (3.2.51), but with the general $\hat{b}_{N2}$ defined in (3.3.84). If no ties are present, $\hat{b}_{N2}$ coincides with its original definition (3.1.6). Using (3.3.83), (3.3.52), and (3.3.79) we get

$$\langle b_i, \hat{b}_{N2} \rangle = S_N^{\tau}(b_i) = S_N^*(L_{\mathcal{J}_N^{\tau}}b_i) = \langle L_{\mathcal{J}_N^{\tau}}b_i, \hat{b}^*_{N2} \rangle , \tag{3.3.116}$$

where $\mathcal{J}_N^{\tau} = \{ (T_0/N, T_1/N], ..., (T_{d-1}/N, T_d/N] \}$ , cf. (3.3.51), and $\tau = (\tau_1, ..., \tau_d)$ is the vector of tie-lengths.

Now we are in the position to prove the following convergence result.

### 3.3.14 Theorem

*Assume* $\mathcal{J} \in \mathcal{J}_{0,1}^*$ *and let* $(F_{\varrho B_N, H_N}^N, G_{\varrho B_N, H_N}^N)$ *be any sequence of local asymptotic alternatives as defined in (3.3.60) and (3.3.61), which especially implies* $\|B_N' - b\| \to 0$ , $b = L_{\mathcal{J}}b \, [\lambda - a.e.]$ , *and* $b \in L_2^0(0,1)$ .

*a) Under the above assumptions we have the limiting law*

$$\mathcal{L}[\vec{S}_N \mid (F_{\varrho B_N, H_N}^N, G_{\varrho B_N, H_N}^N)] \xrightarrow{\mathcal{L}} \mathcal{N}(\vec{\mu}_{\mathcal{J}}, \Gamma_{\mathcal{J}}) \tag{3.3.117}$$

*with mean vector* $\vec{\mu}_{\mathcal{J}} = \varrho(\langle L_{\mathcal{J}}b_1, b \rangle, ..., \langle L_{\mathcal{J}}b_r, b \rangle)^T$ *and covariance matrix* $\Gamma_{\mathcal{J}} = \Big( \langle L_{\mathcal{J}}b_i, L_{\mathcal{J}}b_j \rangle \Big)_{i,j=1,...,r}$ .

*b) Under the null hypothesis* $F = G \in \mathcal{F}_1(\mathcal{J})$ *we have the conditional limiting law*

$$\mathcal{L}_{(F,F)}[\vec{S}_N \mid \tau] \xrightarrow{\mathcal{L}} \mathcal{N}(0, \Gamma_{\mathcal{J}}) \quad [a.s.]. \tag{3.3.118}$$

c) Let $\hat{H}_N$ denote the empirical distribution function of the pooled sample $X_1, ..., X_N$ and let $\mathcal{I}(\hat{H}_N) = \mathcal{J}_N^\tau$ be the corresponding family of jump-intervals, cf. (3.3.54). Then, under the null hypothesis $F = G \in \mathcal{F}_1(\mathcal{J})$, we have the convergence

$$\Gamma_{\mathcal{J}_N^\tau} := \left( \langle L_{\mathcal{J}_N^\tau} b_i, L_{\mathcal{J}_N^\tau} b_j \rangle \right)_{i,j=1,...,r} \xrightarrow{a.s.} \Gamma_{\mathcal{J}}. \tag{3.3.119}$$

*Proof*: Because of (3.3.116) the proof of part a) is immediate from (3.3.64) and the Cramér-Wold device.

Under $F = G \in \mathcal{F}_1(\mathcal{J})$ we have $\|\hat{H}_N - F\|_\infty \to 0$ [a.s.]. For any fixed sequence $\{\hat{H}_N\}$ with $\|\hat{H}_N - F\|_\infty \to 0$ let $\{\tau_N^0\}$ be the corresponding (*non-random*) sequence of tie-lengths and $\mathcal{J}_N^0 := \mathcal{I}(\hat{H}_N)$. Then, because of $\mathcal{I}(F) = \mathcal{J}$, $\langle b_i, \hat{b}_{N2} \rangle = \langle L_{\mathcal{J}_N^0} b_i, \hat{b}_{N2}^* \rangle$ $\forall\, 1 \leq i \leq r$, Lemma 7.5.5, and the indepdendence of $\tau$ and $\hat{b}_{N2}^*$, we get

$$\mathcal{L}_{(F,F)}\big[\, \vec{S}_N \mid \tau = \tau_N^0 \,\big] = \mathcal{L}_0\Big[ \big( \langle L_{\mathcal{J}_N^0} b_i, \hat{b}_{N2}^* \rangle, i = 1, ..., r \big)^T \Big] \xrightarrow{\mathcal{L}} \mathcal{N}(0, \Gamma_{\mathcal{J}}),$$

where the convergence in law is implied by (3.3.63) and the Cramér-Wold device. Notice, Lemma 7.5.5 implies $L_{\mathcal{J}_N^0} b_i \to L_{\mathcal{J}} b_i$ in $L_2(0,1)$ as $N \to \infty$. This concludes the proof of part b).

Part c) is an immediate consequence of Lemma 7.5.5. $\square$

In order to emphasize the $\Gamma$–dependence of the function $f_0$ defined in formula (3.2.53), we explicitly write

$$f(x, \Gamma) = \sup_{\vartheta \geq 0} \big( 2\vartheta^T x - \vartheta^T \Gamma \vartheta \big), \qquad x = (x_1, ..., x_r)^T \in \mathbb{R}^r. \tag{3.3.120}$$

Then, according to (3.2.52), we'll use

$$S_N^\tau(b_1, ..., b_r) := f(\vec{S}_N, \Gamma_{\mathcal{J}_N^\tau}) \tag{3.3.121}$$

as the *projection rank statistic*, if ties are present.

Notice, even in the continuous case (no ties) the original statistic $f_0(\vec{S}_N) = f(\vec{S}_N, \Gamma)$ defined in (3.2.52) differs from the above statistic $f(\vec{S}_N, \Gamma_{\mathcal{J}_N^\tau})$, since the limiting $\Gamma$ is substituted by its approximation $\Gamma_{\mathcal{J}_N^\tau}$. In the continuous model the matrix $\Gamma_{\mathcal{J}_N^\tau}$ is non-random, since $d = N$ and $\tau_1 = ... = \tau_N = 1$ in this case. In the general model the matrix $\Gamma_{\mathcal{J}_N^\tau}$ will depend on the random vector of tie-lengths $\tau$.

According to Lemma 7.5.7 the function $f(x, \Gamma)$ is jointly continuous in $x$ and $\Gamma$. Therefore, under the null hypothesis $F = G \in \mathcal{F}_1(\mathcal{J})$ and if the matrix $\Gamma_{\mathcal{J}}$ is non-singular, Theorem 3.3.14 implies the limiting law

$$\mathcal{L}_{(F,F)}[f(\vec{S}_N, \Gamma_{\mathcal{J}_N^\tau})] \xrightarrow{\mathcal{L}} \mathcal{L}[f(X, \Gamma_{\mathcal{J}})] \qquad (3.3.122)$$

and also the conditional limiting law

$$\mathcal{L}_{(F,F)}[f(\vec{S}_N, \Gamma_{\mathcal{J}_N^\tau}) \mid \tau] \xrightarrow{\mathcal{L}} \mathcal{L}[f(X, \Gamma_{\mathcal{J}})] \quad [a.s.], \qquad (3.3.123)$$

where $X = (X_1, ..., X_r)^T$ is a random vector with $r$-variate normal distribution $\mathcal{N}(0, \Gamma_{\mathcal{J}})$.

The $p$-value of the limiting distribution may be computed (approximately) by using formula (3.2.78) with $\Gamma$ replaced by $\Gamma_{\mathcal{J}_N^\tau}$. The more ties are present the more questionable the above approximation will usually be.

In contrast, the computation of the $p$-value of the conditional distribution $\mathcal{L}_{(F,F)}[f(\vec{S}_N, \Gamma_{\mathcal{J}_N^\tau}) \mid \tau]$ given the observed $\tau$ is always possible by Monte-Carlo simulation.

# Chapter 4

# Randomness versus related alternatives

In Chapter 1 we have motivated the need of new testing procedures in the case of two treatments differing in location. In Chapter 3 the corresponding mathematical foundation and the local asymptotic properties of the proposed testing statistics have been given. In this chapter we will discuss related problems. Therefore the basic parts of the motivation and the exposition are quite similar to those of Chapter 1 and Chapter 3.

## 4.1  Two samples differing in scale

The general model and the null hypothesis of randomness $\mathcal{H}_0^r$ are the same as in Chapter 1, i.e. we assume $X_1, ..., X_m, X_{m+1}, ..., X_{m+n}$ to be independent real random variables with $X_i \sim F \ \ \forall \ i = 1, ..., m$ and $X_j \sim G \ \ \forall \ j = m + 1, ..., N$, $N = m + n$, where $F$ and $G$ are continuous distribution functions, and the *hypothesis of randomness* has the form

$$\mathcal{H}_0^r = \{ (F, G) \in \mathcal{F}_1^c \times \mathcal{F}_1^c : F = G \}. \tag{4.1.1}$$

Given any $\mu \in [0, 1]$ we call (the distribution) $F$ to be *more dispersed about $\mu$* than (the distribution) $G$ if the pair $(F, G)$ belongs to the *dispersion alternative* $\mathcal{A}_2^\mu$ which is defined by

$$\mathcal{A}_2^\mu = \left\{ \begin{array}{l} (F, G) \in \mathcal{F}_1^c \times \mathcal{F}_1^c \ \ \text{such that } F \neq G \ \ \text{ and} \\[1ex] \qquad\qquad F(x) \geq G(x), \text{ if } G(x) < \mu, \\[1ex] \qquad\qquad F(x) \leq G(x), \text{ if } G(x) \geq \mu \end{array} \right\} . \tag{4.1.2}$$

Obviously the assumption that $F$ is more dispersed about $\mu$ than $G$ means

$$F(y) - F(x) \leq G(y) - G(x) \qquad \forall \ (x,y] \ni G^{-1}(\mu),$$

i.e. the probability mass of $G$ is more concentrated about $G^{-1}(\mu)$ than the probability mass of $F$.

In the special case $\mu = 0$ the dispersion alternative $\mathcal{A}_2^0$ is nothing but the stochastically larger alternative $\mathcal{A}_2^0 : F \leq G, F \neq G$ which has been discussed in Chapter 3. In this case the phrase ' $F$ is more dispersed about $\mu = 0$ than $G$ ' seems to be artificial. Nevertheless, if all distributions are concentrated on the positive real axis, i.e. if $F(0) = G(0) = 0$, then the assumption $F \leq G$ is equivalent to $F(x) = G(x/\sigma(x)) \ \forall \ x \geq 0$ for some 'scale function' $\sigma(x) \geq 1$ $\forall \ x \geq 0$. Thus, in the so-called *positive case* $(F(0) = G(0) = 0)$ it's again quite natural to call $F$ more dispersed than $G$ if $(F,G) \in \mathcal{A}_2^0$. In the *positive case* all the testing procedures of Chapter 3 for testing $\mathcal{H}_0^r$ versus the one-sided alternative $\mathcal{A}_2^0$ may be used for testing $\mathcal{H}_0^r$ versus the special dispersion alternative $\mathcal{A}_2^0$.

Besides the dispersion alternative $\mathcal{A}_2^0$ for the positive case, the most interesting case in practice is the *dispersion about the median*, i.e. the dispersion alternative $\mathcal{A}_2^\mu$ for the special case $\mu = 1/2$. In this case definition (4.1.2) means that $F$ has less probability mass about the median than $G$. The dispersion alternative $\mathcal{A}_2^{1/2}$ is especially interesting if the distribution $G$ may be assumed to be (nearly) symmetric about the median.

**Exact scale:**

The classical forerunner of the dispersion alternative $\mathcal{A}_2^\mu$ is the exact *scale alternative* $\mathcal{A}_s(F_0)$ corresponding to some given continuous distribution function $F_0$,

$$\mathcal{A}_s(F_0) = \left\{ (F,G) = (F_{0,\vartheta_1}, F_{0,\vartheta_2}) : -\infty < \vartheta_2 < \vartheta_1 < \infty \right\}, \qquad (4.1.3)$$

where

$$F_{0,\vartheta}(x) = F_0(xe^{-\vartheta}), \qquad x \in \mathbb{R}, \quad \vartheta \in \mathbb{R}. \qquad (4.1.4)$$

Traditionally the equivalent parameter $\sigma = exp(\vartheta) \in (0,\infty)$ is used in this setting, but the parametrization (4.1.4) will be more suitable for our purposes.

Clearly the dispersion alternatives $\mathcal{A}_2^\mu$ are more general than scale alternatives since we have the inclusion

$$\mathcal{A}_s(F_0) \subset \mathcal{A}_2^\mu, \quad \text{if} \ \mu = F_0(0).$$

In case of the exact scale model let's assume that the underlying $F_0$ has an absolutely continuous (on all finite intervals) Lebesgue density $f_0$ with derivative $f_0'$. Additionally we assume the corresponding *Fisher-information* $I_1(f_0)$ to be finite, i.e.

$$I_1(f_0) = \int_{-\infty}^{+\infty} \left(-1 - x\,\frac{f_0'(x)}{f_0(x)}\right)^2 f_0(x)\,dx$$

$$= \int_0^1 \varphi_1^2(u, f_0)\,du = \|\varphi_1(\cdot, f_0)\|^2 < \infty, \tag{4.1.5}$$

where the function, cf. (1.1.8),

$$\varphi_1(u, f_0) = -1 - F_0^{-1}(u)\,\frac{f_0'(F_0^{-1}(u))}{f_0(F_0^{-1}(u))}$$

$$= F_0^{-1}(u)\,\varphi(u, f_0) - 1, \qquad 0 < u < 1, \tag{4.1.6}$$

is the *optimal score function* for testing $\mathcal{H}_0^r$ versus the scale alternative $\mathcal{A}_s(F_0)$, cf. Hájek and Šidák (1967), II.4.5 and VII.1.3. This means, cf. (1.1.11) to (1.1.13), that the sequence $\{\psi_N(f_0), N \geq 2\}$ of rank tests

$$\psi_N(f_0) = 1\left(S_N(f_0) \geq u_\alpha\,\sqrt{I_1(f_0)}\right) \tag{4.1.7}$$

with linear rank statistics $S_N(f_0)$ corresponding to $\varphi_1(\cdot, f_0)$, i.e.

$$S_N(f_0) = \sum_{i=1}^N c_{Ni}\,a_N(R_i) \tag{4.1.8}$$

with scores $a_N(i)$, $1 \leq i \leq N$, according to

$$\int_0^1 \left(a_N(1 + [Nu]) - \varphi_1(u, f_0)\right)^2 \lambda(du) \xrightarrow{N \to \infty} 0, \tag{4.1.9}$$

is asymptotically optimal at level $\alpha$ for testing $\mathcal{H}_0^r$ versus $\mathcal{A}_s(F_0)$.

The optimal (scale) score function $\varphi_1(\cdot, f_0)$ and the corresponding Fisher-information $I_1(f_0)$ may be found in Table 1.1 (Section 1.1) for the underlying density $f_0$ corresponding to the standard normal distribution, the logistic distribution, and the Cauchy distribution, respectively. The respective standardized optimal score functions $\varphi_1(\cdot, f_0)/\sqrt{I_1(f_0)}$ are plotted in Figure 4.1.a.

The score functions $\varphi_1(\cdot, f_0)$ corresponding to the normal density, the logistic density, and the Cauchy density, respectively, are very smooth. Therefore the simple scores $a_N(i) = \varphi_1(\frac{i}{N+1}, f_0)$ fulfil condition (4.1.9) in all three cases, and the corresponding asymptotically optimal statistics have the form

$$S_N(f_0) = \sum_{i=1}^N c_{Ni}\,\varphi_1\left(\frac{R_i}{N+1}, f_0\right) \tag{4.1.10}$$

### Figure 4.1a

*The standardized optimal score functions $\varphi_1(\cdot, f_0)/\sqrt{I_1(f_0)}$ for the normal scale model ( $\star\,\star\,\star$ ) , the logistic scale model ( $\cdot\;\cdot\;\cdot$ ) , and the Cauchy scale model ( $\circ\;\circ\;\circ$ ) .*

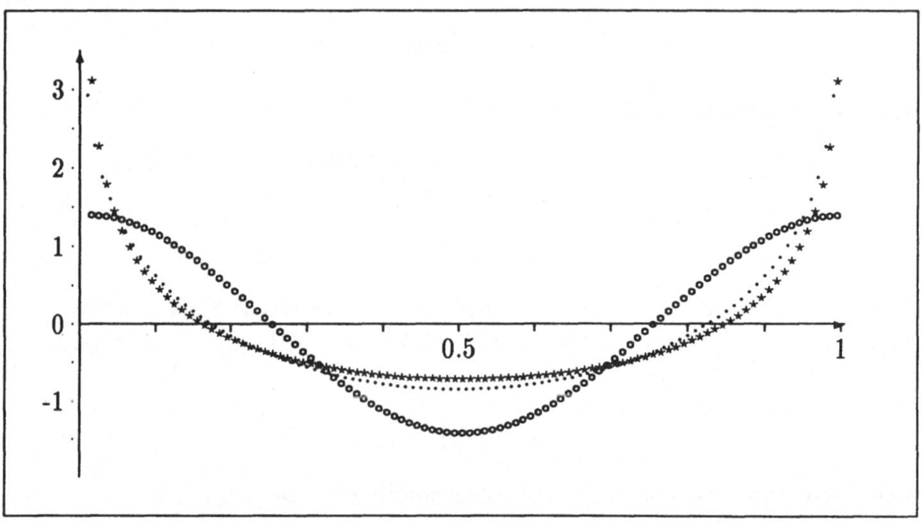

where the $c_{Ni}$ are the two-sample coefficients defined in formula (1.1.16).

In case of the *normal scale model* the optimal score function is $\varphi_1(u, f_0) = \left(\Phi^{-1}(u)\right)^2 - 1$ and the corresponding optimal statistic (4.1.10) is the *Klotz statistic*

$$S_N = \sum_{i=1}^{N} c_{Ni} \left(\Phi^{-1}(\frac{R_i}{N+1})\right)^2. \tag{4.1.11}$$

### Generalized scale:

Similar to Section 1.2 we may define *generalized scale  alternatives* in order to demonstrate how sensitive optimal score functions react to violations of the exact scale model:

Let $F_0$ be a given distribution function and—for easy representation—assume a bounded *scale function* $D : \mathbb{R} \rightarrow [0, \infty)$ with bounded derivative $d : \mathbb{R} \rightarrow \mathbb{R}$ such that $d_0 := \sup\{|x\,d(x)| : x \in \mathbb{R}\} < \infty$ . Then we define the generalized scale alternative $\mathcal{A}_s(F_0, D)$ according to

$$\mathcal{A}_s(F_0, D) = \left\{ \begin{array}{c} (F, G) \in \{ F_{0,D,\vartheta} : \vartheta \in \mathbb{R}, \ d_0|\vartheta| \leq 1 \}, \\ \vartheta_2 < \vartheta_1, \ F \neq G, \end{array} \right\} \tag{4.1.12}$$

where

$$F_{0,D,\vartheta}(x) = F_0\big( x\, e^{-\vartheta D(x)} \big), \qquad x \in \mathbb{R}. \tag{4.1.13}$$

It's easily proved that (4.1.13) defines a proper distribution function on $\mathbb{R}$ if $|\vartheta| d_0 \leq 1$. In case of exact scale $(D \equiv 1, d \equiv 0)$ there is no restriction on $\vartheta \in \mathbb{R}$.

Introducing general scale functions $D$ allows the modelling of alternatives where different parts of the distribution $F_0$ are dispersed at different rates. Also in the case of the generalized scale alternative (4.1.12) we have the inclusion

$$\mathcal{A}_s(F_0, D) \subset \mathcal{A}_2^\mu, \quad \text{if } \mu = F_0(0).$$

As in the exact scale model we assume that the given distribution function $F_0$ has an absolutely continuous density $f_0$ with derivative $f_0'$. Therefore the distribution function $F_{0,D,\vartheta}$ has a density $f_{0,D,\vartheta}$ according to

$$f_{0,D,\vartheta}(x) = e^{-\vartheta D(x)} \left( 1 - \vartheta\, x\, d(x) \right) f_0\big( x\, e^{-\vartheta D(x)} \big), \quad x \in \mathbb{R}, \tag{4.1.14}$$

and the optimal score function $\varphi_1(\cdot, f_0)$ defined in (4.1.6) has a natural analogue $\varphi_1(\cdot, f_0, D)$ for the generalized scale alternative $\mathcal{A}_s(F_0, D)$, namely

$$\varphi_1(u, f_0, D) = \partial \ln f_{0,D,\vartheta}\big(F_0^{-1}(u)\big)/\partial\vartheta\big|_{\vartheta=0} \tag{4.1.15}$$

$$= \varphi_1(u, f_0)D\big(F_0^{-1}(u)\big) - F_0^{-1}(u)d\big(F_0^{-1}(u)\big), \quad 0 < u < 1.$$

Similar to the above result of Hájek and Šidák (1967) we may prove the sequence of rank tests

$$\psi_N(f_0, D) = 1\Big( S_N(f_0, D) \geq u_\alpha \|\varphi_1(\cdot, f_0, D)\| \Big) \tag{4.1.16}$$

to be asymptotically optimal for testing $\mathcal{H}_0^\tau$ versus $\mathcal{A}_s(F_0, D)$ where the statistic $S_N(f_0, D)$ is the linear rank statistic (4.1.8) with scores $a_N(i)$ according to

$$\int_0^1 \Big( a_N(1 + [Nu]) - \varphi_1(u, f_0, D) \Big)^2 \lambda(du) \overset{N\to\infty}{\longrightarrow} 0. \tag{4.1.17}$$

Also in this generalized scale model the shape of the optimal score function $\varphi_1(\cdot, f_0, D)$ is influenced by *two independent quantities*, namely the density $f_0$ and the scale function $D$. The following lemma demonstrates how drastically $\varphi_1(\cdot, f_0, D)$ may vary if the usual scale function $D \equiv 1$ is replaced by some general scale function $D$.


### 4.1.1 Lemma

*Let $F$ and $G$ be distribution functions on $\mathbb{R}$ with $F(0) = G(0)$ and respective densities $f$ and $g$. Assume $f$ and $g$ to be strictly positive and* *absolutely continuous (on all finite intervals) with finite Fisher informations $I_1(f)$ and $I_1(g)$ , cf. (4.1.5). Define the scale function $D$ by*

$$D(x) = \frac{\int_0^{F(x)} \varphi_1(y,g)\, dy}{\int_0^{F(x)} \varphi_1(y,f)\, dy} = \frac{G^{-1}\big(F(x)\big)\; g \circ G^{-1}\big(F(x)\big)}{F^{-1}\big(F(x)\big)\; f \circ F^{-1}\big(F(x)\big)}. \qquad (4.1.18)$$

*Then the following equation holds true*

$$\varphi_1(\cdot, f, D) \;=\; \varphi_1(\cdot, g). \qquad (4.1.19)$$

The *proof* is immediate from the following two formulae,

$$\int_0^t \varphi_1(y,f)\, dy \;=\; -F^{-1}(t)\,\big(f \circ F^{-1}(t)\big), \qquad 0 < t < 1, \qquad (4.1.20)$$

$$\int_0^t \varphi_1(y,f,D)\, dy = -F^{-1}(t)\big(f \circ F^{-1}(t)\big)\big(D \circ F^{-1}(t)\big), \quad 0 < t < 1, \ (4.1.21)$$

both of which may be proved by differentiating both sides. The inequality $D \geq 0$ is implied by the assumption $F(0) = G(0)$ .

**Remark:** Because of $F(0) = G(0)$ an application of de l'Hospital's rule to $G^{-1}(F(x))/x$ yields $D(x) \to 1$ if $0 \neq x \to 0$. Using the results listed in Table 1.1 (Section 1.1) we may compute the following examples. The corresponding examples in case of generalized shift models have been given in Example 1.2.2.

### 4.1.2 Example

a) Let $f = \varphi$ be the normal density and let $g$ be the logistic density. Let $D_{NL}$ denote the scale function (4.1.18) in this case. Then we get

$$D_{NL}(x) = \ln\!\Big(\frac{\Phi(x)}{1 - \Phi(x)}\Big)\, \frac{\Phi(x)(1 - \Phi(x))}{x\varphi(x)}, \qquad x \neq 0. \qquad (4.1.22)$$

b) Let $f = \varphi$ be the normal density and let $g$ be the Cauchy density. Let $D_{NC}$ denote the scale function (4.1.18) in this case. The we get

$$D_{NC}(x) = \frac{\tan[\pi(\Phi(x) - \tfrac{1}{2})]}{x\varphi(x)} \; \frac{1}{\pi\Big(1 + \tan^2[\pi(\Phi(x) - \tfrac{1}{2})]\Big)}, \qquad x \neq 0. \qquad (4.1.23)$$

Let $f$ be the logistic density and let $g$ be the Cauchy density. Let $D_{LC}$ denote the scale function (4.1.18) in this case. Then we get

$$D_{LC}(x) = \frac{\tan[\pi(F(x) - \tfrac{1}{2})]}{x F(x)\big(1 - F(x)\big)\, \pi\,\Big(1 + \tan^2[\pi(F(x) - \tfrac{1}{2})]\Big)}, \qquad x \neq 0, \qquad (4.1.24)$$

**Figure 4.1b**

*The graphs of the scale functions* $D_{NL}$ $(\star\,\star\,\star)$, $D_{NC}$ $(\cdots)$, $D_{LC}$ $(\circ\,\circ\,\circ)$
*given in Example 4.1.2.*

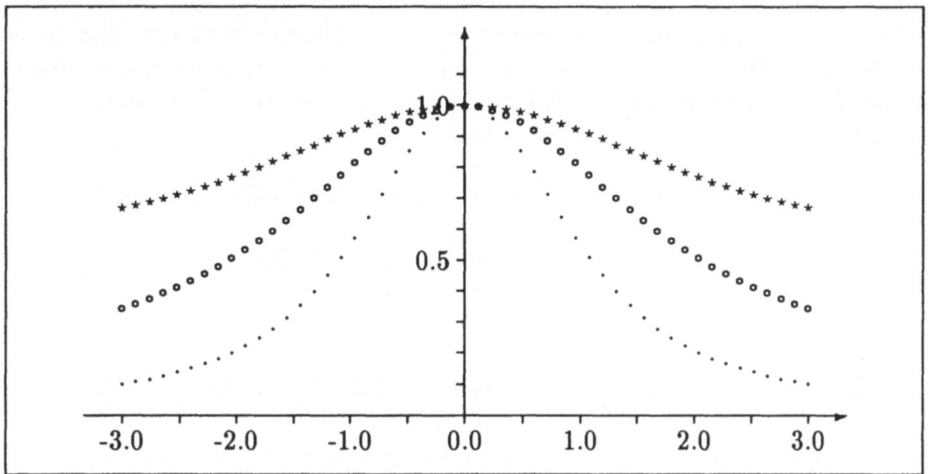

where $F(x) = \exp(x)/(1 - \exp(x))$ is the logistic distribution function.

Figure 4.1.b shows the graphs of the above scale functions $D_{NL}$, $D_{NC}$, and $D_{LC}$. Similar to the generalized shift model we notice that e.g. the optimal score function for *generalized normal scale* alternatives corresponding to the scale function $D_{NL}$ is the same as the optimal score function for *usual logistic scale* alternatives. Thus, instead of the Klotz test, which is optimal for usual normal scale alternatives, the test corresponding to the score function $\varphi_1(u, \text{logistic}) = (2u-1)\ln\big(u/(1-u)\big) - 1$ becomes optimal for the generalized normal scale alternatives defined by the scale function $D_{NL}$.

Again we recognize that the shape of the optimal score function has no connection with 'heavy' or 'light' tails of the underlying distribution if we take generalized scale alternatives into account.

As in the location case we will use the notion of generalized scale alternatives in order to select some representative score functions for the construction of suitable nonlinear rank tests in the general dispersion model (4.1.1), (4.1.2).

## General dispersion:

Now let's go back to the general dispersion model and let's use the reparametrization (3.0.2) of the (continuous) two sample model, i.e. the original parameter $(F, G) \in \mathcal{F}_1^c \times \mathcal{F}_1^c$ of the joint distribution of $X_1, ..., X_m, X_{m+1}, ..., X_N$,

$N = m + n$ , is substituted by the new parameter $(B, H) \in \mathcal{B}_N^0 \times \mathcal{F}_1^c$ via the correspondence

$$\begin{cases} F = H + c_{N1} B \circ H = F_{B,H}^N, \\ G = H + c_{NN} B \circ H = G_{B,H}^N, \end{cases} \qquad (4.1.25)$$

where $\mathcal{F}_1^c$ has been defined as the set of all continuous distribution functions on $\mathbb{R}$ and $\mathcal{B}_N^0$ has been defined as the set of all absolutely continuous functions $B : [0,1] \to \mathbb{R}$ which fulfil the side conditions $B(0) = B(1) = 0$ and $-\sqrt{m/n} \le B'/\sqrt{N} \le \sqrt{n/m}$ $[\lambda - a.e.]$ .

Since $G$ is continuous it's easily proved that the two defining conditions

$$\Big( F(x) \ge G(x) \text{ if } G(x) < \mu \Big) \quad \text{and} \quad \Big( F(x) \le G(x) \text{ if } G(x) \ge \mu \Big)$$

of $\mathcal{A}_2^\mu$ imply the equivalences

$$\Big( G(x) < \mu \text{ iff } H_N(x) < \mu \Big) \quad \text{and} \quad \Big( G(x) \ge \mu \text{ iff } H_N(x) \ge \mu \Big),$$

where $H_N = (mF + nG)/N$ . Thus, for any $0 \le \mu \le 1$ , the dispersion alternative (4.1.2) can be written in the form

$$\mathcal{A}_2^\mu = \left\{ \begin{array}{l} (F,G) \in \mathcal{F}_1^c \times \mathcal{F}_1^c \quad \text{such that } F \ne G \quad \text{and} \\[4pt] \qquad F(x) \ge G(x), \text{ if } \frac{m}{N} F(x) + \frac{n}{N} G(x) < \mu, \\[4pt] \qquad F(x) \le G(x), \text{ if } \frac{m}{N} F(x) + \frac{n}{N} G(x) \ge \mu \end{array} \right\} . \qquad (4.1.26)$$

Therefore (4.1.25) implies the following $(B, H)$ –representation of $\mathcal{A}_2^\mu$

$$\mathcal{A}_2^\mu = \left\{ \begin{array}{l} (F_{B,H}^N, G_{B,H}^N) : \ B \in \mathcal{B}_N^0, \ H \in \mathcal{F}_1^c, \ B \ne 0, \\[4pt] \qquad B(t) \ge 0 \ \forall \, t < \mu, \ B(t) \le 0 \ \forall \, t \ge \mu \end{array} \right\} . \qquad (4.1.27)$$

Thus, the only differences between the general dispersion model of this section and the general location model of Chapter 3 are the different conditions on the parameter $B \in \mathcal{B}_N^0$ in order to formulate the various alternatives. Here the $\mu$ -dispersion alternative is defined by the side conditions

$$B \ge 0 \text{ on } [0, \mu), \quad B \le 0 \text{ on } [\mu, 1], \quad B \ne 0, \qquad (4.1.28)$$

whereas e.g. the general one-sided location alternative is defined by the side conditions $B \le 0$ , $B \ne 0$ .

Especially the optimality result of Theorem 3.0.1 holds true for testing the null hypothesis $\mathcal{H}_0^r$ versus any sequence $(F_{\varrho B_N, H}^N, G_{\varrho B_N, H}^N)$ of local asymptotic alternatives as defined in (3.0.9) and (3.0.10). Since the optimal score function

$b_N = B'_N$ is unknown in practice, the construction of (rank) tests for testing $\mathcal{H}_0^r$ versus $\mathcal{A}_2^\mu$ is completely similar to Chapter 3, the only difference being the consideration of the new side conditions (4.1.28) instead of the old side conditions $B \leq 0$, $B \neq 0$.

## A) Approximating the optimal score function by kernel estimators

According to Section 3.1 the piecewise linear rank process $\hat{B}_{N2}$ is a natural unrestricted estimator of the underlying $B_N$. Under the null hypothesis $\mathcal{H}_0^r$ and under the alternative $\mathcal{A}_2^\mu$ the function $B_N$ fulfils the side conditions $B_N \geq 0$ on $[0, \mu)$ and $B_N \leq 0$ on $[\mu, 1]$. Therefore we should start with a suitable adaptation of $\hat{B}_{N2}$ with respect to these side conditions.

As mentioned before the case $\mu = 0$ corresponds to the problem of testing $\mathcal{H}_0^r : F = G \; [B = 0]$ versus $\mathcal{A}_2^0 : F \leq G, F \neq G \; [B \leq 0, B \neq 0]$, which has been discussed in Chapter 3. Similarly, the case $\mu = 1$ corresponds to the problem of testing the null hypothesis $\mathcal{H}_0^r : F = G \; [B = 0]$ versus the dual one-sided alternative $\mathcal{A}_2^{0d} : F \geq G, F \neq G \; [B \geq 0, B \neq 0]$. Obviously, this problem reduces to the case $\mu = 0$ if we interchange the role of the first and the second sample. Thus, for the rest of the present section we'll assume $0 < \mu < 1$, where the most interesting case $\mu = 1/2$ means dispersion about the median.

Given the $\mu \in (0, 1)$ we construct a 'projection' $\hat{B}_{N2}^\mu$ of the rank process $\hat{B}_{N2}$ onto the side conditions (4.1.28) in the following way:

$$
\hat{B}_{N2}^\mu(t) = \begin{cases} \hat{B}_{N2}^+(t), & \text{if } 0 \leq t \leq \mu_1, \\[2mm] \frac{\mu_2 - t}{\mu_2 - \mu_1} \hat{B}_{N2}^+(\mu_1) + \frac{t - \mu_1}{\mu_2 - \mu_1} \hat{B}_{N2}^0(\mu_2), & \text{if } \mu_1 < t \leq \mu_2, \\[2mm] \hat{B}_{N2}^0(t), & \text{if } \mu_2 < t \leq 1, \end{cases} \qquad (4.1.29)
$$

with $\hat{B}_{N2}^+(t) = \max(\hat{B}_{N2}(t), 0)$, $\hat{B}_{N2}^0(t) = \min(\hat{B}_{N2}(t), 0)$, and

$$
\begin{aligned}
\mu_1 &= \mu_{1N} = \max\{ i/N : i/N < \mu, i = 0, 1, ..., N \}, \\
\mu_2 &= \mu_{2N} = \min\{ i/N : i/N \geq \mu, i = 0, 1, ..., N \}.
\end{aligned} \qquad (4.1.30)
$$

Apparently the definition implies

$$\hat{B}_{N2}^\mu \in C[0, 1] \quad \text{and} \quad \hat{B}_{N2}^\mu(t) \geq 0 \; \forall\, t \leq \mu_1, \quad \hat{B}_{N2}^\mu(t) \leq 0 \; \forall\, t \geq \mu_2.$$

Since the score function $b_N$ is a derivative of $B_N$ we'll take a suitable derivative $\hat{b}_{N2}^\mu$ of $\hat{B}_{N2}^\mu$ as the corresponding primitive estimator of $b_N$, i.e. we use the special version

$$
\hat{b}^{\mu}_{N2} = \begin{cases} \hat{b}_{N2}(t)\,1\big(\hat{B}_{N2}(t) > 0\big), & \text{if } 0 \le t \le \mu_1, \\[2mm] \dfrac{\hat{B}^0_{N2}(\mu_2) - \hat{B}^+_{N2}(\mu_1)}{\mu_2 - \mu_1}, & \text{if } \mu_1 < t < \mu_2, \\[2mm] \hat{b}_{N2}(t)\,1\big(\hat{B}_{N2}(t) < 0\big), & \text{if } \mu_2 \le t \le 1. \end{cases} \tag{4.1.31}
$$

In accordance with (3.1.22) we define the final estimator of $b_N$ as the convolution $\mathcal{K}_a \hat{b}^{\mu}_{N2}$ , i.e.

$$
\mathcal{K}_a \hat{b}^{\mu}_{N2}(t) = \int_0^1 K_a(s,t)\,\hat{b}^{\mu}_{N2}(s)\,ds, \qquad 0 \le t \le 1, \tag{4.1.32}
$$

where the convolution kernel $K_a$ is defined in formulae (3.1.19) and (3.1.16) to (3.1.18). Substitution of the optimal score function $b_N$ in the linear rank statistic $S_N(b_N)$ by the estimator $\mathcal{K}_a \hat{b}^{\mu}_{N2}$ defines the (nonlinear) rank statistic $S_N(\mathcal{K}_a \hat{b}^{\mu}_{N2})$ for testing the null hypothesis of randomness $\mathcal{H}^r_0 : B = 0$ versus the dispersion alternative $\mathcal{A}^{\mu}_2 : B \ge 0$ on $[0,\mu]$, $B \le 0$ on $[\mu, 1]$, $B \ne 0$ . Because of (3.1.10) we have

$$
\begin{aligned}
S_N(\mathcal{K}_a \hat{b}^{\mu}_{N2}) &= \,< \mathcal{K}_a \hat{b}^{\mu}_{N2}, \hat{b}_{N2} > \\[3mm]
&= \int_0^1 \int_0^1 K_a(s,t)\,\hat{b}^{\mu}_{N2}(s)\,\hat{b}_{N2}(t)\,ds\,dt,
\end{aligned} \tag{4.1.33}
$$

and completely similar to (3.1.41) we get the representation

$$
\begin{aligned}
S_N(\mathcal{K}_a \hat{b}^{\mu}_{N2}) &= \sum_{\kappa=1}^{\infty} \lambda_\kappa(a)\, < \psi_\kappa, \hat{b}^{\mu}_{N2} > \, < \psi_\kappa, \hat{b}_{N2} > \\[3mm]
&= \sum_{\kappa=1}^{\infty} \lambda_\kappa(a)\, < \psi'_\kappa, \hat{B}^{\mu}_{N2} > \, < \psi'_\kappa, \hat{B}_{N2} >,
\end{aligned} \tag{4.1.34}
$$

where the last equality holds true since $\hat{B}^{\mu}_{N2}(0) = \hat{B}^{\mu}_{N2}(1) = 0$ and since partial integration yields the equalities $< \psi_\kappa, \hat{b}^{\mu}_{N2} > = - < \psi'_\kappa, \hat{B}^{\mu}_{N2} >$ and $< \psi_\kappa, \hat{b}_{N2} > = - < \psi'_\kappa, \hat{B}_{N2} >$ . Therefore, and since apparently

$$
|\hat{B}^{\mu}_{N2}(t)| \le |\hat{B}_{N2}(t)| \qquad \forall\, 0 \le t \le 1, \tag{4.1.35}
$$

the proof of the limiting law (3.1.64) can easily be extended in order to prove a corresponding limiting law for $S_N(\mathcal{K}_a \hat{b}^{\mu}_{N2})$ , i.e. if the condition (3.1.63) and the other assumptions of Theorem 3.1.9 are fulfilled, we have the limiting law

$$
\mathcal{L}\Big[ S_N(\mathcal{K}_a \hat{b}^{\mu}_{N2}) \mid (F^N_{\varrho B_N, H_N}, G^N_{\varrho B_N, H_N}) \Big] \xrightarrow{\mathcal{L}} \mathcal{L}\Big[ \sum_{\kappa=1}^{\infty} \lambda_\kappa(a)\, Z_\kappa \Big] \tag{4.1.36}
$$

with

$$Z_\kappa := \left( \int_0^\mu (W_0 + \varrho B)^+ \psi_\kappa' \, d\lambda + \int_\mu^1 (W_0 + \varrho B)^0 \psi_\kappa' \, d\lambda \right) \int_0^1 (W_0 + \varrho B) \, \psi_\kappa' \, d\lambda$$

and $f^+ := \max(f, 0)$ and $f^0 := \min(f, 0)$.

For practical purposes it's more convenient to use a smooth kernel $K$, e.g. the Parzen-2 kernel, in order to approximate $K_a(s, t)$ on each of the squares $\big((i-1)/N, i/N\big) \times \big((j-1)/N, j/N\big)$ by the constants

$$k_{Na}(i, j) \;=\; K_a(\frac{i - 1/2}{N}, \frac{j - 1/2}{N}), \qquad i, j = 1, ., .., N. \tag{4.1.37}$$

Using these approximations the representation (4.1.33) yields the simple approximate (nonlinear) rank statistic

$$S_N^\mu(a, K) \tag{4.1.38}$$

$$= \sum_{i=1}^N \sum_{j=1}^N \left( \hat{B}_{N2}^\mu(\frac{i}{N}) - \hat{B}_{N2}^\mu(\frac{i-1}{N}) \right) \left( \hat{B}_{N2}(\frac{j}{N}) - \hat{B}_{N2}(\frac{j-1}{N}) \right) k_{Na}(i, j)$$

for testing the null hypothesis of randomness $\mathcal{H}_0^r$ versus the dispersion alternative $\mathcal{A}_2^\mu$.

## B) Approximating the optimal score function by projections

The projection onto very large cones will produce rank tests which may have rather low power for large classes of alternatives. Therefore, and since many relevant types of alternatives can be approximated by a given finite set of score functions of the form (4.1.15), we'll restrict the discussion to the case of finite dimensional cones.

Let the integer $r \geq 1$ be given and assume $b_1, ..., b_r \in L_2^0(0, 1)$ to be special score functions corresponding to the testing problem $\mathcal{H}_0^r$ versus $\mathcal{A}_2^\mu$, i.e. for each $i = 1, ..., r$ we assume

$$B_i(t) \geq 0 \quad \forall \, t < \mu, \qquad B_i(t) \leq 0 \quad \forall \, t \geq \mu, \qquad B_i \neq 0, \tag{4.1.39}$$

where

$$B_i(t) := \int_0^t b_i(x) \, dx \qquad \forall \, 0 \leq t \leq 1. \tag{4.1.40}$$

Obviously any linear combination of the form

$$B = \sum_{i=1}^r \vartheta_i B_i \qquad \text{such that} \quad \vartheta_i \geq 0 \quad \forall \, i = 1, ..., r$$

has the property

$$B(t) \geq 0 \quad \forall\, t < \mu, \qquad B(t) \leq 0 \quad \forall\, t \geq \mu.$$

Therefore it's reasonable to define the cone $V \subset L_2^0(0,1)$ according to

$$V := \left[ b_1, ..., b_r \right]^+ = \Big\{ \sum_{i=1}^{r} \vartheta_i b_i : \vartheta_i \geq 0 \,\forall\, i \Big\}. \tag{4.1.41}$$

In order to have a $r$–dimensional cone let's assume that $b_1, ..., b_r$ are linearly independent in $L_2(0,1)$, i.e. the $r \times r$–matrix $\Gamma = (< b_i, b_j >)$ is positive definite. Thus we are in a setting which is completely similar to the situation of Subsection 3.2.C, the only difference being the different assumptions on the given $b_1, ..., b_r$. Especially all the results of Subsection 3.2.C are applicable.

**Dispersion about the median:**

As a representative example we take $r = 3$ and

$$b_i = \varphi_1(\cdot, f_0, D_i), \qquad i = 1, 2, 3, \tag{4.1.42}$$

the optimal score functions for the generalized scale alternatives $\mathcal{A}_s(F_0, D_i)$, where $F_0$ is the logistic distribution function (3.2.88) and the $D_i$ are the special scale functions corresponding to (3.2.86),

$$\begin{aligned}
&D_1 := 1 - F_0 && \text{(lower dispersion)}, \\
&D_2 := 4F_0(1 - F_0) && \text{(central dispersion)}, && \tag{4.1.43} \\
&D_3 := F_0 && \text{(upper dispersion)}.
\end{aligned}$$

Because of $F_0(0) = 1/2$ we have $\mathcal{A}_s(F_0, D_i) \subset \mathcal{A}_2^{1/2}$, $i = 1, 2, 3$, which means that the projection $\Pi_V \hat{b}_{N2}$ of $\hat{b}_{N2}$ onto the cone $V = [b_1, b_2, b_3]^+$ yields the (nonlinear) rank statistic $S_N(\Pi_V \hat{b}_{N2}) = \|\Pi_V \hat{b}_{N2}\|^2$ for testing the null hypothesis of randomness $\mathcal{H}_0^r$ versus the dispersion alternative $\mathcal{A}_2^{1/2}$ [*dispersion about the median*]. Notice, the optimal score function $\varphi_1(\cdot, f_0)$ for testing $\mathcal{H}_0^r$ versus the exact scale alternative $\mathcal{A}_s(F_0)$ is contained in the cone $[\varphi_1(\cdot, f_0, D_i), i = 1, 2, 3]^+$ since (4.1.15) immediately implies

$$\varphi_1(\cdot, f_0, 1 - F_0) + \varphi_1(\cdot, f_0, F_0) = \varphi_1(\cdot, f_0). \tag{4.1.44}$$

By elementary evaluation, cf. (4.1.15) and Table 1.1, we get the following

### Figure 4.1c

*The graphs of the three representative score functions $b_1$ , $b_2$ , $b_3$ defined in formula (4.1.45):  $b_1$  ( ⋆ ⋆ ⋆ ⋆ ⋆ ) , $b_2$  ( · · · · · ) , $b_3$  ( ∘ ∘ ∘ ∘ ∘ ) .*

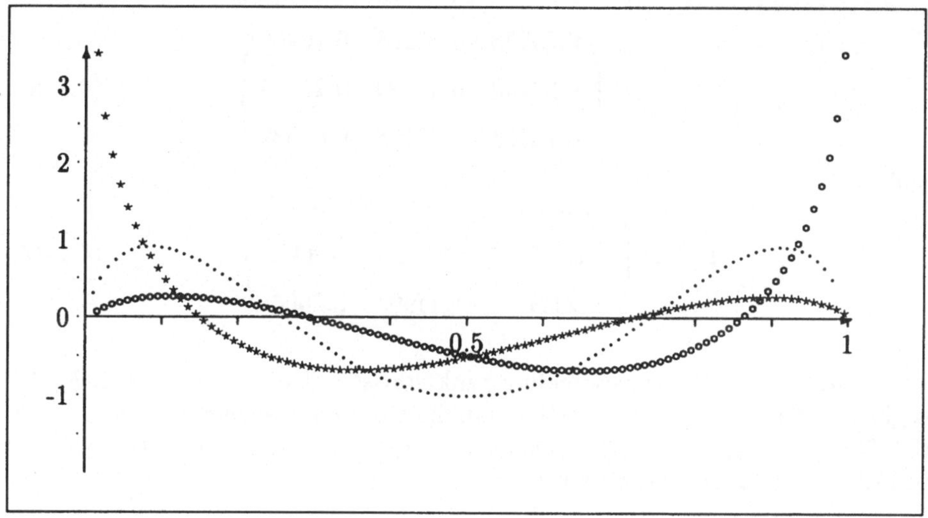

representation of the corresponding three score functions $(0 < u < 1)$ ,

$$b_1(u) = \varphi_1\big(u, f_0, 1 - F_0\big)$$
$$= (1-u)\big(3(1-u) - 2\big)\ln\big(\frac{1-u}{u}\big) - (1-u),$$

$$b_2(u) = \varphi_1\big(u, f_0, 4F_0(1-F_0)\big)$$
$$= 4u(1-u)\Big(2(2u-1)\ln\big(\frac{u}{1-u}\big) - 1\Big), \tag{4.1.45}$$

$$b_3(u) = \varphi_1\big(u, f_0, F_0\big)$$
$$= u(3u-2)\ln\big(\frac{u}{1-u}\big) - u = b_1(1-u).$$

The score functions $b_1, b_2, b_3$ are square-integrable and the integrals $B_1$, $B_2$, $B_3$ fulfil the side conditions (4.1.39) for $\mu = 1/2$ . The graphs of the $B_i$ 's are shown in Figure 4.1.c. The covariance matrix $\Gamma = (< b_i, b_j >)$ and its inverse $\Gamma^{-1}$ are evaluated as

$$\Gamma = \frac{1}{90} \begin{pmatrix} 4\pi^2 + 15 & 4\pi^2 - 15 & \pi^2 \\ 4\pi^2 - 15 & \frac{320}{35}\pi^2 - 48 & 4\pi^2 - 15 \\ \pi^2 & 4\pi^2 - 15 & 4\pi^2 + 15 \end{pmatrix}$$

$$= \begin{pmatrix} 0.60532 & 0.27198 & 0.10966 \\ 0.27198 & 0.46929 & 0.27198 \\ 0.10966 & 0.27198 & 0.60532 \end{pmatrix} \tag{4.1.46}$$

and

$$\Gamma^{-1} = \begin{pmatrix} 2.2596 & -1.4499 & 0.24211 \\ -1.4499 & 3.8115 & -1.4499 \\ 0.24211 & -1.4499 & 2.2596 \end{pmatrix}. \tag{4.1.47}$$

The evaluation of the corresponding rank statistic $S_N(\Pi_V \hat{b}_{N2}) = \|\Pi_V \hat{b}_{N2}\|^2$ is given in Section 3.2. The limiting null distribution is evaluated according to (3.2.93), (3.2.94), and (3.2.96). Doing the explicit calculations on the basis of (4.1.46) and (4.1.47) we get ( $\forall\, t > 0$ )

$$P_{\mathcal{H}_0^r}\{\, \|\Pi_V \hat{b}_{N2}\|^2 > t \,\} \overset{N \to \infty}{\longrightarrow} \tag{4.1.48}$$

$$0.44870\, P\{\chi_1^2 > t\} \;+\; 0.27527\, P\{\chi_2^2 > t\} \;+\; 0.05130\, P\{\chi_3^2 > t\}$$

and

$$P_{\mathcal{H}_0^r}\{\, \|\Pi_V \hat{b}_{N2}\|^2 = 0 \,\} \overset{N \to \infty}{\longrightarrow} 0.22473\,. \tag{4.1.49}$$

### C) Dispensing with continuity: Treatment of ties

For the remainder of this section we'll assume that the underlying distribution functions $F$ and $G$ may have discontinuities, cf. Section 3.3, i.e. the parameter of the general model is $(F, G) \in \mathcal{M} = \mathcal{F}_1 \times \mathcal{F}_1$ and the *general null hypothesis of randomness* has the form

$$\mathcal{H}_0^r \;=\; \{\, (F, G) \in \mathcal{M} : F = G \,\}. \tag{4.1.50}$$

Since $F$ and $G$ may have jumps at $G^{-1}(\mu)$, it's more suitable to use formula (4.1.26) instead of formula (4.1.2) as the general definition of dispersion alternatives. Thus, in accordance with the continuous case, we define the *general dispersion alternative* at sample sizes $(m, n)$ as

$$\mathcal{A}_{2N}^{\mu} = \left\{ \begin{array}{ll} (F, G) \in \mathcal{F}_1 \times \mathcal{F}_2 : & F(x) \geq G(x) \text{ if } H_N(x) < \mu, \\ \quad\;\; F \neq G, & F(x) \leq G(x) \text{ if } H_N(x) \geq \mu \end{array} \right\}. \tag{4.1.51}$$

Using the notion and the results of Section 3.3 we get the $(B, H)$ – reparametrization of $\mathcal{M}$ according to, cf. (3.3.14) and (3.3.19),

$$\mathcal{M} = \sum_{\mathcal{J} \in \mathcal{J}^*_{0,1}} \mathcal{M}_N(\mathcal{J}), \tag{4.1.52}$$

where the $\mathcal{J}$ –part of $\mathcal{M}$ has the form

$$\mathcal{M}_N(\mathcal{J}) = \left\{ (F^N_{B,H}, G^N_{B,H}) : \ B \in \mathcal{B}^0_N(\mathcal{J}), \ H \in \mathcal{F}_1(\mathcal{J}) \right\} \tag{4.1.53}$$

and

$$F^N_{B,H} = H + c_{N1} B \circ H, \qquad G^N_{B,H} = H + c_{NN} B \circ H. \tag{4.1.54}$$

The reparametrization (4.1.52), (4.1.53) of the underlying $(F, G) \in \mathcal{M}$ obviously induces a corresponding $(B, H)$ –parametrization of the hypotheses $\mathcal{H}^r_0$
and $\mathcal{A}^\mu_{2N}$ ,

$$\mathcal{H}^r_0 = \sum_{\mathcal{J} \in \mathcal{J}^*_{0,1}} \mathcal{H}^r_{0N}(\mathcal{J}), \qquad \mathcal{A}^\mu_{2N} = \sum_{\mathcal{J} \in \mathcal{J}^*_{0,1}} \mathcal{A}^\mu_{2N}(\mathcal{J}), \tag{4.1.55}$$

and

$$\mathcal{H}^r_{0N}(\mathcal{J}) = \left\{ (F^N_{B,H}, G^N_{B,H}) \in \mathcal{M}_N(\mathcal{J}) : \ B = 0 \right\}, \tag{4.1.56}$$

$$\mathcal{A}^\mu_{2N}(\mathcal{J}) = \left\{ \begin{array}{l} (F^N_{B,H}, G^N_{B,H}) \in \mathcal{M}_N(\mathcal{J}) : \ B(t) \geq 0 \ \text{if} \ t \leq \mu_1(\mathcal{J}), \\ \qquad\qquad B \neq 0, \quad B(t) \leq 0 \ \text{if} \ t \geq \mu_2(\mathcal{J}) \end{array} \right\}, \tag{4.1.57}$$

where

$$\begin{aligned} \mu_1(\mathcal{J}) &= \mu, & \mu_2(\mathcal{J}) &= \mu, & \text{if} \ \mu &\in (0, 1) \setminus \cup \{J \in \mathcal{J}\}, \\ \mu_1(\mathcal{J}) &= x, & \mu_2(\mathcal{J}) &= y, & \text{if} \ \mu &\in (x, y] \in \mathcal{J}. \end{aligned} \tag{4.1.58}$$

Given $\mathcal{J} \in \mathcal{J}^*_{0,1}$ the testing problem $\mathcal{H}^r_{0N}(\mathcal{J})$ versus $\mathcal{A}^\mu_{2N}(\mathcal{J})$ is quite similar
to the *continuous* testing problem $\mathcal{H}^r_0$ versus $\mathcal{A}^\mu_2$ . If we take any sequence
of directions $B_N \in \mathcal{B}^0_N(\mathcal{J})$ such that $b_N = B'_N \to b$ in $L_2(0, 1)$ for some
$b \in L^0_2(0, 1)$ and $\|b\| > 0$ then the *averaged scores linear rank test* $\psi^r_N(b_N)$
defined in (3.3.58) is *asymptotically optimal* at level $\alpha$ for testing $\mathcal{H}^r_{0N}(\mathcal{J})$
versus any sequence $(N \geq 2, N \to \infty)$ of corresponding local asymptotic
alternatives,

$$0 < \varrho \leq 1, \ H_N \in \mathcal{F}_1(\mathcal{J}) \quad \text{and} \quad (F^N_{\varrho B_N, H_N}, G^N_{\varrho B_N, H_N}), \tag{4.1.59}$$

cf. Theorem 3.3.8.

Obviously $(F^N_{\varrho B_N, H_N}, G^N_{\varrho B_N, H_N}) \in \mathcal{A}^\mu_{2N}(\mathcal{J})$ iff the conditions $B_N(t) \geq 0$ for $t \leq \mu_1(\mathcal{J})$, $B_N(t) \leq 0$ for $t \geq \mu_2(\mathcal{J})$, and $B_N \not\equiv 0$ hold true.

In practice $B_N$ is unknown. Therefore we substitute the score function $b_N$ of the optimal averaged scores rank statistic $S^\tau_N(b_N)$ by a suitable (rank) estimator $\hat{b}_N$ of $b_N$, which approximately fulfils the side conditions.

According to Section 3.3 the *averaged two sample rank process* $\hat{B}_{N2}$ defined in (3.3.82) is a natural estimator of $B_N$. Recall that $\hat{B}_{N2}$ is linear on each interval $[T_{i-1}/N, T_i/N]$ for $i = 1, ..., d$, c.f. Proposition 3.3.12. Therefore the 'projection' $\hat{B}^\mu_{N2}$ of $\hat{B}_{N2}$ is defined as in formula (4.1.29), but with $\mu_1 = \hat{\mu}_{1N}$ and $\mu_2 = \hat{\mu}_{2N}$ according to the definitions

$$\mu_1 = \hat{\mu}_{1N} = \max\{T_i/N : T_i/N < \mu, i = 0, 1, ..., N\},$$
$$\mu_2 = \hat{\mu}_{2N} = \min\{T_i/N : T_i/N \geq \mu, i = 0, 1, ..., N\}. \tag{4.1.60}$$

With this modification formulae (4.1.31) to (4.1.34) will be used, resp. hold true, when ties are present.

In the continuous case (no ties) we have $d = N$ and $T_i = i$ $\forall\, i = 1, ..., N$, which implies $\mu_{1N} = \hat{\mu}_{1N}$ and $\mu_{2N} = \hat{\mu}_{2N}$. If the underlying distribution functions $(F_N, G_N)$ are not continuous, $\hat{\mu}_{1N}$ and $\hat{\mu}_{2N}$ are random variables having the representation

$$\hat{\mu}_{1N} = \hat{H}_N\big(\hat{H}_N^{-1}(\mu)_-\big),$$
$$\hat{\mu}_{2N} = \hat{H}_N\big(\hat{H}_N^{-1}(\mu)\big), \tag{4.1.61}$$

where $\hat{H}_N$ is the empirical distribution function of the sample $X_1, ..., X_N$. Under the null hypothesis $F_N = G_N = H \in \mathcal{F}_1(\mathcal{J})$ it follows immediately from the Glivenko-Cantelli Theorem that

$$\lim_{N \to \infty} \hat{\mu}_{1N} = H\big(H^{-1}(\mu)_-\big) = \mu_1(\mathcal{J}) \quad [a.s.],$$
$$\lim_{N \to \infty} \hat{\mu}_{2N} = H\big(H^{-1}(\mu)\big) = \mu_2(\mathcal{J}) \quad [a.s.], \tag{4.1.62}$$

if $H^{-1}$ is continuous at $\mu$, i.e. if there is at most one point $z \in \mathbb{R}$ with $H(z) = \mu$. The numbers $\mu_1(\mathcal{J})$ and $\mu_2(\mathcal{J})$ are defined in formula (4.1.58). It is easily seen that the following generalization of the limiting law (4.1.36) holds true under (3.1.63), (4.1.62), and the assumptions of Theorem 3.3.13:

$$\mathcal{L}\big[S_N(\mathcal{K}_a \hat{b}^\mu_{N2}) \,|\, (F^N_{\varrho B_N, H_N}, G^N_{\varrho B_N, H_N})\big]$$
$$\xrightarrow{\mathcal{L}} \mathcal{L}\Big[\sum_{\kappa=1}^\infty \lambda_\kappa(a) < \psi'_\kappa, V^\mu_\varrho > < \psi'_\kappa, V_\varrho >\Big] \tag{4.1.63}$$

with $V_\varrho := T_{\mathcal{J}} W_0 + \varrho B$ , and $V_\varrho^\mu$ defined by formula (4.1.29) with $\hat{B}_{N2}$ replaced by $V_\varrho$ and $\mu_1 := \mu_1(\mathcal{J})$ , $\mu_2 := \mu_2(\mathcal{J})$ .

Likewise, under the special null hypothesis $\mathcal{H}_{0N}^r(J) : F_N = G_N = H \in \mathcal{F}_1(\mathcal{J})$ we have the conditional limiting law

$$\mathcal{L}\big[ S_N(\mathcal{K}_a \hat{b}_{N2}^\mu) \mid \tau \big] \xrightarrow{\mathcal{L}} \mathcal{L}\Big[ \sum_{\kappa=1}^{\infty} \lambda_\kappa(a) \; < \psi_\kappa', V_0^\mu > \; < \psi_\kappa', V_0 > \Big]. \qquad (4.1.64)$$

In practical applications we'll use a smooth kernel K, e.g. the Parzen-2 kernel. Therefore we may use the statistic $S_N^\mu(a, K)$ from (4.1.38) ( with $\hat{B}_{N2}^\mu$ given by (4.1.60) and (4.1.29) ) as an approximation of $S_N(\mathcal{K}_a \hat{b}_{N2}^\mu)$ for testing $\mathcal{H}_0^r$ versus $\mathcal{A}_2^\mu$ .

The considerations concerning the approximation of the unknown score function by projections onto finite dimensional cones (in the presence of ties) have already been treated in Section 3.3, cf. formulae (3.3.115) to (3.3.122). The only difference is that in the present dispersion problem the special score functions $b_1$ , $b_2$ , and $b_3$ from (4.1.45) are used instead of those from (3.2.89).

## 4.2  Several samples on the real line

In order to concentrate onto the essential points we have developed the basic ideas of rank tests with estimated scores in the well-known two sample case, cf. Chapter 1 and Chapter 3. In the present section we'll extend the theory to the case of several samples.

Let $k \geq 2$ denote the number of samples and let

$$X_{11}, ..., X_{1n_1}, X_{21}, ..., X_{2n_2}, ..., X_{k1}, ..., X_{kn_k} \tag{4.2.1}$$

be independent real random variables, where the i-th sample $X_{i1}, ..., X_{in_i}$ is i.i.d. with distribution function $F_i$ . Let $N = n_1 + ... + n_k$ denote the total number of observations and assume the partial numbers of observations $n_i$ to be given in terms of $N$ , i.e. $n_i = n_i(N) \quad \forall \, i = 1, ..., k$ . In a first step we'll assume the *continuous model*

$$F_i \in \mathcal{F}_1^c, \quad i = 1, ..., k, \tag{4.2.2}$$

and we'll consider the *null hypothesis of randomness*

$$\mathcal{H}_0^r : \; F_1 = F_2 = \cdots = F_k \tag{4.2.3}$$

versus the *omnibus alternative*

$$\mathcal{A}_k : \; F_i \neq F_j \quad \text{for some} \quad i \neq j, \; 1 \leq i, j \leq k, \tag{4.2.4}$$

or $\mathcal{H}_0^r$ versus the *one sided alternative of trend*

$$\mathcal{A}_k^0 : \; F_1 \leq F_2 \leq \cdots \leq F_k, \; F_1 \neq F_k. \tag{4.2.5}$$

Completely similar to the two sample case we identify the optimal score functions for testing $\mathcal{H}_0^r$ versus local asymptotic alternatives corresponding to a given alternative point $(F_1, ..., F_k)$ . Since in reality these score functions are unknown, we use instead suitable (rank) estimators and plug them into the optimal linear rank statistics.

In the sequel let $F_1, ..., F_k \in \mathcal{F}_1^c$ be fixed and define the mixture

$$H_N = \eta_{N1} F_1 + ... + \eta_{Nk} F_k, \quad \eta_{Ni} = n_i/N. \tag{4.2.6}$$

Then the random variable $H_N(X_{ij})$ has the distribution function $F_i \circ H_N^{-1}$ on $[0, 1]$ and the Lebesgue density

$$f_{Ni} := \frac{dF_i}{dH_N} \circ H_N^{-1} = \frac{d(F_i \circ H_N^{-1})}{d\lambda}, \tag{4.2.7}$$

where $\lambda$ is the Lebesgue measure on the interval $(0,1)$. Because of (4.2.6) we have the equality

$$\eta_{N1} f_{N1} + \ldots + \eta_{Nk} f_{Nk} \;=\; 1 \quad [\lambda - a.e.]. \tag{4.2.8}$$

Similar to (1.3.18) we may rewrite each $F_i$ in the form

$$F_i \;=\; H_N + \frac{1}{\sqrt{N}} B_{Ni} \circ H_N, \quad 1 \le i \le k, \tag{4.2.9}$$

where

$$B_{Ni}(t) := \int_0^t b_{Ni} \, d\lambda \quad \forall \, t \in [0,1], \quad b_{Ni} := \sqrt{N}(f_{Ni} - 1). \tag{4.2.10}$$

Because of (4.2.7) and (4.2.8) the vector $B_N = (B_{N1}, \ldots, B_{Nk})$ is an element of the set

$$\mathcal{B}_{kN}^0 := \left\{ \begin{array}{l} (B_1, \ldots, B_k) : \\ \text{Each } B_i \text{ is absolutely continuous on } [0,1], \\ B_i(0) = B_i(1) = 0, \quad B_i' \ge -\sqrt{N} \quad [\lambda - a.e.], \\ \text{and } \eta_{N1} B_1 + \ldots + \eta_{Nk} B_k \;=\; 0 \end{array} \right\}. \tag{4.2.11}$$

As in Section 1.3 the representation (4.2.9) induces a new parametrization of the model which is more suitable for our purposes. Putting $( \, \forall \, i = 1, \ldots, k )$

$$F_{i,B,H}^N := H + \frac{1}{\sqrt{N}} B_i \circ H, \quad \text{if } B = (B_1, \ldots, B_k) \in \mathcal{B}_{kN}^0, \; H \in \mathcal{F}_1^c, \tag{4.2.12}$$

we get the $(B,H)$–parametrization of the model according to

$$(\mathcal{F}_1^c)^k \;=\; \{ \, (F_{1,B,H}^N, \ldots, F_{k,B,H}^N) \, : \, B \in \mathcal{B}_{kN}^0, \; H \in F_1^c \, \}. \tag{4.2.13}$$

Using the $(B,H)$–parameter we have the following convenient representations of the null hypothesis of randomness,

$$\mathcal{H}_0^r \;=\; \{ \, (F_{1,B,H}^N, \ldots, F_{k,B,H}^N) \, : \, B \in \mathcal{B}_{kN}^0, \; H \in \mathcal{F}_1^c, \; B = 0 \, \}, \tag{4.2.14}$$

the omnibus alternative,

$$\mathcal{A}_k \;=\; \{ \, (F_{1,B,H}^N, \ldots, F_{k,B,H}^N) \, : \, B \in \mathcal{B}_{kN}^0, \; H \in \mathcal{F}_1^c, \; B \ne 0 \, \}, \tag{4.2.15}$$

and the alternative of trend,

$$\mathcal{A}_k^0 = \{ \, (F_{1,B,H}^N, \ldots, F_{k,B,H}^N) : B \in \mathcal{B}_{kN}^0, \; H \in \mathcal{F}_1^c, \\ B_1 \le \cdots \le B_k, \; B \ne 0 \}. \tag{4.2.16}$$

Each $b = (b_1, ..., b_k)$ with $b_i \in L_2^0(0,1)$ $\forall i$ gives rise to a linear rank statistic

$$S_N(b) = \frac{1}{\sqrt{N}} \sum_{i=1}^{k} \sum_{j=1}^{n_i} b_{Ni}(R_{ij}) \qquad (4.2.17)$$

with the scores

$$b_{Ni}(j) = N \int_{(j-1)/N}^{j/N} b_i \, d\lambda, \quad 1 \leq j \leq N, \quad i = 1, ..., k, \qquad (4.2.18)$$

and with the rank $R_{ij}$ of $X_{ij}$ in the pooled sample, i.e.

$$R_{ij} = \sum_{\kappa=1}^{k} \sum_{\tau=1}^{n_\kappa} 1(X_{\kappa\tau} \leq X_{ij}), \quad 1 \leq j \leq n_i, \quad 1 \leq i \leq k. \qquad (4.2.19)$$

Now let's assume

$$\eta_{Ni} = \frac{n_i(N)}{N} \xrightarrow{N \to \infty} \eta_i \in (0,1) \quad \forall \, 1 \leq i \leq k \qquad (4.2.20)$$

and let $(F_{1,B_N,H_N}^N, ..., F_{k,B_N,H_N}^N)$ be a sequence of $(B_N, H_N)$ –alternatives of the form (4.2.12) with arbitrary $H_N \in \mathcal{F}_1^c$ and almost fixed $B_N \in \mathcal{B}_{kN}^0$ , i.e. for each $1 \leq i \leq k$ we assume the existence of some $b_i \in L_2^0(0,1)$ such that

$$b_{Ni} := B_{Ni}' \xrightarrow{N \to \infty} b_i \quad \text{in} \quad L_2^0(0,1). \qquad (4.2.21)$$

Then the upper test (of asymptotic level $\alpha$ ) based on the linear rank statistic $S_N(b_{N1}, ..., b_{Nk})$ is asymptotically most powerful in the class of all asymptotic level $\alpha$ tests for testing the null hypothesis $\mathcal{H}_0^r$ versus the $(B_N, H_N)$ – sequence of alternatives, if $\sum_{i=1}^{k} \|b_i\|^2 > 0$ . The proof is an easy extension of the proof of Theorem 3.0.1.


## A) Approximating the optimal score function by kernel estimators

Similar to Section 3.1 we'll define kernel estimators (based on the ranks only) of the optimal score functions $b_N = (b_{N1}, ..., b_{Nk})$ . The motivation is an extension of the two sample motivation: For each $1 \leq i \leq k$ the (unobservable) random variables $H_N(X_{i1}), ..., H_N(X_{in_i})$ are i.i.d. with the absolutely continuous distribution function $F_i \circ H_N^{-1} = I + B_{Ni}/\sqrt{N}$ . Thus, if $\tilde{F}_{Ni}$ denotes the empirical distribution function of $H_N(X_{i1}), ..., H_N(X_{in_i})$ , then $\sqrt{N}(\tilde{F}_{Ni} - I)$ is a natural estimator of $B_{Ni}$ . Obviously this is a hypothetical estimator since the $H_N(X_{ij})$ 's are unobservable. However, according to

Section 7.6 (Appendix) we have the inequality

$$\|\breve{F}_{Ni} - F_i \circ H_N^{-1}\|_\infty$$

$$\leq 3\|\tilde{F}_{Ni} - F_i \circ H_N^{-1}\|_\infty + 2 \sum_{\kappa=1}^{k} \frac{n_\kappa}{n_i} \|\tilde{F}_{N\kappa} - F_\kappa \circ H_N^{-1}\|_\infty, \tag{4.2.22}$$

where $\breve{F}_{Ni}$ denotes the empirical distribution function of the **normed ranks** $R_{i1}/N, ..., R_{in_i}/N$ . Because of assumption (4.2.20) inequality (4.2.22) means that we may substitute the unobservable estimator $\sqrt{N}(\tilde{F}_{Ni} - I)$ of the underlying $B_{Ni}$ by the observable rank process $\sqrt{N}(\breve{F}_{Ni} - I)$ . In fact it's more convenient to use the piecewise linearized version $\hat{F}_{Ni}$ of $\breve{F}_{Ni}$ , i.e.

$$\hat{F}_{Ni}(t) = \frac{1}{n_i} \sum_{j=1}^{n_i} d_N(R_{ij}, t), \quad 0 \leq t \leq 1, \tag{4.2.23}$$

where $d_N(r, t)$ is defined as ( $\forall \, r = 1, ..., N, \quad \forall \, 0 \leq t \leq 1$ )

$$d_N(r, t) = 1\big(r \leq [Nt]\big) + \big(Nt - [Nt]\big)\, 1\big(r = [Nt] + 1\big), \tag{4.2.24}$$

and where $[x]$ is the integer part of $x$ . The corresponding ( absolutely continuous ) estimator of the underlying $B_{Ni}$ is the rank process

$$\hat{B}_{Ni}(t) = \sqrt{N}(\hat{F}_{Ni}(t) - t), \quad 0 \leq t \leq 1. \tag{4.2.25}$$

Obviously $\hat{B}_{N1}, ..., \hat{B}_{Nk}$ are $C[0, 1]$ –valued processes and fulfil the side conditions

$$\hat{B}_{Ni}(0) = \hat{B}_{Ni}(1) = 0, \quad \hat{b}_{Ni} \geq -\sqrt{N}, \quad \sum_{i=1}^{k} \eta_{Ni}\hat{B}_{Ni} = 0, \tag{4.2.26}$$

where $\hat{b}_{Ni} := \hat{B}'_{Ni}$ denotes the right continuous version of the derivative of the (absolutely continuous) $\hat{B}_{Ni}$ , i.e. for $r = 1, ..., N$ we have

$$\hat{b}_{Ni}(t) = N\left(\hat{B}_{Ni}(\frac{r}{N}) - \hat{B}_{Ni}(\frac{r-1}{N})\right) \quad \text{if} \quad \frac{r-1}{N} \leq t < \frac{r}{N}. \tag{4.2.27}$$

Completely similar to the two sample case we use $\hat{b}_{Ni}$ as a primitive estimator of the underlying $b_{Ni}$ . This primitive estimator will be modified and smoothed in order to get better and more suitable estimators. Especially in the *case of trend alternatives* (4.2.16) we adjust the processes $\hat{B}_{N1}, ..., \hat{B}_{Nk}$ to the trend condition $\hat{B}_{N1} \leq \hat{B}_{N2} \leq ... \leq \hat{B}_{Nk}$ . This is done by defining the $(C[0, 1])^k$ – valued process $\tilde{B}_N^0 = (\tilde{B}_{N1}^0, ..., \tilde{B}_{Nk}^0)$ according to ( $\forall \, \kappa = 1, ..., k$ )

$$\tilde{B}_{N\kappa}(t) := \max_{1 \leq i \leq \kappa} \hat{B}_{Ni}(t), \quad 0 \leq t \leq 1. \tag{4.2.28}$$

Then the components of $\tilde{B}_N^0$ have absolutely continuous and piecewise linear paths on $[0,1]$, and the following side conditions are fulfilled,

$$\tilde{B}_{Ni}^0(0) = \tilde{B}_{Ni}(1) = 0 \quad \forall\, i, \qquad \tilde{B}_{N1}^0 \leq \tilde{B}_{N2}^0 \leq \cdots \leq \tilde{B}_{Nk}^0. \qquad (4.2.29)$$

Obviously the centering condition $\eta_{N1}B_{N1} + \cdots + \eta_{Nk}B_{Nk} = 0$ isn't automatically fulfilled for $\tilde{B}_N^0$ . Therefore we recenter the components of $\tilde{B}_N^0$ with the weighted average

$$\bar{B}_N^0 := \sum_{i=1}^{k} \eta_{Ni}\tilde{B}_{Ni}^0, \qquad (4.2.30)$$

i.e., we define the trend adjusted $(C[0,1])^k$ –valued rank process $\hat{B}_N^0 = (\hat{B}_{N1}^0, ..., \hat{B}_{Nk}^0)$ according to

$$\hat{B}_{Ni}^0 := \tilde{B}_{Ni}^0 - \bar{B}_N^0, \qquad i = 1, ..., k. \qquad (4.2.31)$$

Again this process has components with absolutely continuous and piecewise linear paths on $[0,1]$, and this process fulfils the side conditions

$$\hat{B}_{Ni}^0(0) = \hat{B}_{Ni}^0(1) = 0 \,\forall\, i, \quad \hat{B}_{N1}^0 \leq \cdots \leq \hat{B}_{Nk}^0, \quad \sum_{i=1}^{k} \eta_{Ni}\hat{B}_{Ni}^0 = 0. \quad (4.2.32)$$

Therefore, in case of trend alternatives (4.2.16), we use $\hat{b}_N^0 = (\hat{b}_{N1}^0, ..., \hat{b}_{Nk}^0)$ as the primitive estimator of $b_N = (b_{N1}, ..., b_{Nk})$, where $\hat{b}_{Ni}^0$ is defined as the (rightcontinuous) derivative of $\hat{B}_{Ni}^0$ .

**Remark 1:** *The trend modification (4.2.31) is a direct generalization of the one-sided projection (3.1.7) for the two sample case.*

In order to verify this remark assume $k = 2$, $n_1 = m$, $n_2 = n$, and (in order to avoid misunderstandings) let $\hat{D}_N$ denote the linearized two sample rank process (3.1.5) and let $\hat{D}_N^0 = \hat{D}_N 1(\hat{D}_N < 0)$ denote the corresponding one-sided projection (3.1.7). Using the terminology of the present section for $k = 2$, $n_1 = m$, and $n_2 = n$ and using $(m\hat{F}_{N1} + n\hat{F}_{N2})/N = I$ , an easy evaluation yields on one hand the equalities

$$\hat{D}_N = \sqrt{\frac{m}{n}}\hat{B}_{N1} = -\sqrt{\frac{n}{m}}\hat{B}_{N2}, \qquad \hat{B}_{N2} = -\frac{m}{n}\hat{B}_{N1}, \qquad (4.2.33)$$

and on the other hand the representations

$$\hat{B}_{N1}^0 = \hat{B}_{N1} 1(\hat{B}_{N1} < 0), \quad \hat{B}_{N2}^0 = -\frac{m}{n}\hat{B}_{N1}^0 = \hat{B}_{N2} 1(\hat{B}_{N2} > 0). \qquad (4.2.34)$$

Combining (4.2.33) and (4.2.34) proves

$$\hat{D}_N^0 = \hat{D}_N 1(\hat{D}_N < 0) = \sqrt{\frac{m}{n}}\,\hat{B}_{N1}^0 = -\sqrt{\frac{n}{m}}\,\hat{B}_{N2}^0, \qquad (4.2.35)$$

which proves the remark. For later reference let's consider the representation of $\tilde{B}_{N1}^0$ and $\tilde{B}_{N2}^0$ in terms of $\hat{D}_N$ and $\hat{D}_N^0$ . Obviously (4.2.33) and (4.2.28) imply

$$\tilde{B}_{N1}^0 = \sqrt{\frac{n}{m}}\hat{D}_N = \sqrt{\frac{n}{m}}\hat{D}_N 1(\hat{D}_N \geq 0) + \sqrt{\frac{n}{m}}\hat{D}_N^0,$$

$$\tilde{B}_{N2}^0 = (\hat{B}_{N1} \vee \hat{B}_{N2}) = \sqrt{\frac{n}{m}}\hat{D}_N 1(\hat{D}_N \geq 0) - \sqrt{\frac{m}{n}}\hat{D}_N^0, \qquad (4.2.36)$$

$$\bar{B}_N^0 = \frac{m}{N}\tilde{B}_{N1}^0 + \frac{n}{N}\tilde{B}_{N2}^0 = \sqrt{\frac{n}{m}}\hat{D}_N 1(\hat{D}_N \geq 0).$$

In order to clarify the meaning of the above definitions let's discuss an example.

### 4.2.1 Example

Assume $k = 3$ and $n_1 = 3$, $n_2 = 2$, $n_3 = 4$, *i.e.* $N = 9$ . In addition let's assume that the following data have been observed,

$$(x_{11}, x_{12}, x_{13}) = (9.2,\ 4.7,\ 5.8),$$
$$(x_{21}, x_{22}) = (3.1,\ 8.8)\ ,$$
$$(x_{31}, x_{32}, x_{33}, x_{34}) = (8.2,\ 3.3,\ 3.0,\ 2.1).$$

Then the corresponding values of the ranks are

$$(r_{11}, r_{12}, r_{13}) = (9,\ 5,\ 6),$$
$$(r_{21}, r_{22}) = (3,\ 8),$$
$$(r_{31}, r_{32}, r_{33}, r_{34}) = (7,\ 4,\ 2,\ 1).$$

The resulting values of $12\,\hat{F}_{N\kappa}(i/N)$, $i = 0, 1, ..., N$ , are

| $12\,\hat{F}_{N1}(i/N)$ : | 0 | 0 | 0 | 0 | 0 | 4 | 8 | 8 | 8 | 12 , |
|---|---|---|---|---|---|---|---|---|---|---|
| $12\,\hat{F}_{N2}(i/N)$ : | 0 | 0 | 0 | 6 | 6 | 6 | 6 | 6 | 12 | 12 , |
| $12\,\hat{F}_{N3}(i/N)$ : | 0 | 3 | 6 | 6 | 9 | 9 | 9 | 12 | 12 | 12 . |

Therefore the values of $12\,\hat{B}_{N\kappa}(i/N)$, $i = 0, 1, ..., N$ , are

| $12\,\hat{B}_{N1}(i/N)$ : | 0 | -4 | -8 | -12 | -16 | -8 | 0 | -4 | -8 | 0 , |
|---|---|---|---|---|---|---|---|---|---|---|
| $12\,\hat{B}_{N2}(i/N)$ : | 0 | -4 | -8 | 6 | 2 | -2 | -6 | -10 | 4 | 0 , |
| $12\,\hat{B}_{N3}(i/N)$ : | 0 | 5 | 10 | 6 | 11 | 7 | 3 | 8 | 4 | 0 . |

The values of $12\,\tilde{B}^0_{N\kappa}(i/N) = \max_{1\leq j\leq\kappa} 12\,\hat{B}_{Nj}(i/N)$, $i = 0,1,...,N$ , are

$12\,\tilde{B}^0_{N1}(i/N):$    0   -4   -8   -12   -16   -8   0   -4   -8   0 ,

$12\,\tilde{B}^0_{N2}(i/N):$    0   -4   -8   6   2   -2   0   -4   4   0 ,

$12\,\tilde{B}^0_{N3}(i/N):$    0   5   10   6   11   7   3   8   4   0 .

Finally the substraction of $12\,\bar{B}^0_N = 12\,(3\tilde{B}^0_{N1} + 2\tilde{B}^0_{N2} + 4\tilde{B}_{N3})/9$ yields the values of $12\,\hat{B}^0_{N\kappa}(i/N)$, $i = 0,1,...,N$ ,

$12\,\hat{B}^0_{N1}(i/N):$    0   -4   -8   -12   -16   -8   -4/3   -16/3   -8   0 ,

$12\,\hat{B}^0_{N2}(i/N):$    0   -4   -8   6   2   -2   -4/3   -16/3   4   0 ,

$12\,\hat{B}^0_{N3}(i/N):$    0   5   10   6   11   7   5/3   20/3   4   0 .


For the asymptotics of the various rank processes we need the following generalization of Theorem 3.1.8.


### 4.2.2 Theorem

*Assume (4.2.20) and $b_i \in L^0_2(0,1)$ for $i = 1,...,k$ . Let $(B_N, H_N) \in \mathcal{B}^0_{kN} \times \mathcal{F}^c_1$, $N \geq k$ , be any sequence which fulfils condition (4.2.21). Then, under the sequence of $(B_N, H_N)$ –alternatives of the form (4.2.12), we have the limiting law*

$$(\hat{B}_{N1},..., \hat{B}_{Nk}) \xrightarrow{\mathcal{L}} (W_1 + B_1,...,W_k + B_k) \tag{4.2.37}$$

*in the space $(C[0,1])^k$ with the usual sup-norm topology, where $(W_1,...,W_k)$ is a $k$ –dimensional centered Gaussian process with paths in $(C[0,1])^k$ and covariance structure*

$$EW_\kappa(s)W_\tau(t) = (s\wedge t - st)(\frac{\delta_{\kappa\tau}}{\sqrt{\eta_\kappa\eta_\tau}} - 1), \qquad \begin{array}{l} 0 \leq s,t \leq 1, \\ 1 \leq \kappa,\tau \leq k, \end{array} \tag{4.2.38}$$

*and where $B_i$ is the integral of $b_i$ ,*

$$B_i(t) = \int_0^t b_i\,d\lambda, \quad 0 \leq t \leq 1, \quad i = 1,...,k. \tag{4.2.39}$$


The *proof* of Theorem 4.2.2 is given in Section 7.7 (Appendix).


Using the derivative $\hat{b}_{Ni}$ of $\hat{B}_{Ni} = \sqrt{N}(\hat{F}_{Ni} - I)$ we can rewrite the optimal linear rank statistic $S_N(b) = S_N(b_1,...,b_k)$ in the form, cf. (3.1.10),

$$S_N(b) = \frac{1}{\sqrt{N}}\sum_{i=1}^k N\int_0^1 b_i(t)\sum_{j=1}^{n_i} 1\Big(\frac{R_{ij}-1}{N} \leq t < \frac{R_{ij}}{N}\Big)\,dt$$

$$= \sqrt{N} \sum_{i=1}^{k} \int_0^1 b_i(t) \frac{n_i}{N} \hat{F}'_{Ni}(t) \, dt$$

$$= \sum_{i=1}^{k} \eta_{Ni} \int_0^1 b_i(t) \sqrt{N} (\hat{F}'_{Ni}(t) - 1) \, dt$$

$$= \sum_{i=1}^{k} \eta_{Ni} < b_i, \hat{b}_{Ni} > . \tag{4.2.40}$$

In the *omnibus model* we use the convolution $\mathcal{K}_a \hat{b}_{Ni}$ , i.e.

$$\mathcal{K}_a \hat{b}_{Ni}(t) = \int_0^1 K_a(s,t) \, \hat{b}_{Ni}(s) \, ds, \quad 0 \le t \le 1, \tag{4.2.41}$$

with convolution kernel (3.1.19) as a suitable rank estimator of $b_i$ . Plugging these estimators into the optimal linear rank statistic (4.2.40) we arrive at the new *omnibus statistic*

$$S_N(\mathcal{K}_a \hat{b}_N) := S_N(\mathcal{K}_a \hat{b}_{N1}, ..., \mathcal{K}_a \hat{b}_{Nk}) = \sum_{i=1}^{k} \eta_{Ni} < \mathcal{K}_a \hat{b}_{Ni}, \hat{b}_{Ni} >, \tag{4.2.42}$$

and because of (3.1.40) and $< \hat{b}_{Ni}, 1 >= 0$ we have the following representation which is useful for proving asymptotic results,

$$\begin{aligned}
S_N(\mathcal{K}_a \hat{b}_N) &= \sum_{\varrho=1}^{\infty} \lambda_\varrho(a) \sum_{i=1}^{k} \eta_{Ni} < \psi_\varrho, \hat{b}_{Ni} >^2 \\
&= \sum_{\varrho=1}^{\infty} \lambda_\varrho(a) \sum_{i=1}^{k} \eta_{Ni} < \psi'_\varrho, \hat{B}_{Ni} >^2 .
\end{aligned} \tag{4.2.43}$$

In case of the *trend model* we use the derivatives $\hat{b}^0_{Ni}$ of the trend adjusted processes $\hat{B}^0_{Ni}$ instead of the unrestricted $\hat{b}_{Ni}$ 's as the primitive estimators, i.e. we plug the convolution $\mathcal{K}_a \hat{b}^0_{Ni}$ into the optimal linear rank statistic (4.2.40), which yields the new *trend statistic*

$$S_N(\mathcal{K}_a \hat{b}^0_N) := S_N(\mathcal{K}_a \hat{b}^0_{N1}, ..., \mathcal{K}_a \hat{b}^0_{Nk}) = \sum_{i=1}^{k} \eta_{Ni} < \mathcal{K}_a \hat{b}^0_{Ni}, \hat{b}_{Ni} > . \tag{4.2.44}$$

For actual computations it's more convenient to avoid the evaluation of the recentered process $\hat{B}^0_N$ defined in (4.2.31), and to use the process $\tilde{B}^0_N$ instead, cf. (4.2.28). The resulting trend statistic is

$$S_N(\mathcal{K}_a \tilde{b}^0_N) := S_N(\mathcal{K}_a \tilde{b}^0_{N1}, ..., \mathcal{K}_a \tilde{b}^0_{Nk}) = \sum_{i=1}^{k} \eta_{Ni} < \mathcal{K}_a \tilde{b}^0_{Ni}, \hat{b}_{Ni} >, \tag{4.2.45}$$

where $\tilde{b}_{Ni}^0$ is defined as the (rightcontinuous) derivative of $\tilde{B}_{Ni}^0$ . In fact formula (4.2.26) implies $\eta_{N1}\hat{b}_{N1} + \cdots + \eta_{Nk}\hat{b}_{Nk} = 0$ and therefore $S_N(\mathcal{K}_a\hat{b}_N^0) = S_N(\mathcal{K}_a\tilde{b}_N^0)$ .

Similar to (4.2.43) we get the following alternative representation of (4.2.44) and (4.2.45), which will be used for the evaluation of the limiting law,

$$S_N(\mathcal{K}_a\hat{b}_N^0) = S_N(\mathcal{K}_a\tilde{b}_N^0)$$

$$= \sum_{\varrho=1}^{\infty} \lambda_\varrho(a) \sum_{i=1}^{k} \eta_{Ni} < \psi_\varrho, \tilde{b}_{Ni}^0 > < \psi_\kappa, \tilde{b}_{Ni} >$$

$$= \sum_{\varrho=1}^{\infty} \lambda_\varrho(a) \sum_{i=1}^{k} \eta_{Ni} < \psi'_\varrho, \tilde{B}_{Ni}^0 > < \psi'_\varrho, \tilde{B}_{Ni} > . \quad (4.2.46)$$

In practice it's more convenient to use a smooth kernel $K$ , e.g. the Parzen-2 kernel, in order to approximate $K_a(s,t)$ by the squarewise constant convolution kernel

$$K_a^N(s,t)$$

$$:= \sum_{i=1}^{N}\sum_{j=1}^{N} k_{Na}(i,j) \, 1\left(\frac{i-1}{N} < s \le \frac{i}{N}\right) 1\left(\frac{j-1}{N} < t \le \frac{j}{N}\right), \quad (4.2.47)$$

where $k_{Na}(i,j) := K_a\big((i-\tfrac{1}{2})/N,(j-\tfrac{1}{2})/N\big)$ . Substituting the operator $\mathcal{K}_a$ in formula (4.2.42) by the corresponding operator $\mathcal{K}_a^N$ yields the *(approximate) omnibus rank statistic*

$$\dot{S}_N(a,K) := \sum_{\kappa=1}^{k} \eta_{N\kappa} < \mathcal{K}_a^N \hat{b}_{N\kappa}, \hat{b}_{N\kappa} > =$$

$$\sum_{\kappa=1}^{k} \eta_{N\kappa} \int_0^1 \int_0^1 K_a^N(s,t)\hat{b}_{N\kappa}(s)\hat{b}_{N\kappa}(t)\, ds\, dt \; = \quad (4.2.48)$$

$$\sum_{i=1}^{N}\sum_{j=1}^{N} k_{Na}(i,j) \sum_{\kappa=1}^{k} \eta_{N\kappa} \left(\hat{B}_{N\kappa}(\tfrac{i}{N}) - \hat{B}_{N\kappa}(\tfrac{i-1}{N})\right)\left(\hat{B}_{N\kappa}(\tfrac{j}{N}) - \hat{B}_{N\kappa}(\tfrac{j-1}{N})\right).$$

The same substitution in (4.2.44) yields the *(approximate) trend rank statistic*

$$S_N^0(a,K) := \sum_{\kappa=1}^{k} \eta_{N\kappa} < \mathcal{K}_a^N \hat{b}_{N\kappa}^0, \hat{b}_{N\kappa} > =$$

$$\sum_{\kappa=1}^{k} \eta_{N\kappa} < \mathcal{K}_a^N \tilde{b}_{N\kappa}^0, \hat{b}_{N\kappa} > - < \mathcal{K}_a^N \bar{b}_N^0, \sum_{\kappa=1}^{k} \eta_{N\kappa} \hat{b}_{N\kappa} > = \qquad (4.2.49)$$

$$\sum_{i=1}^{N} \sum_{j=1}^{N} k_{Na}(i,j) \sum_{\kappa=1}^{k} \eta_{N\kappa} \left( \tilde{B}_{N\kappa}^0 (\frac{i}{N}) - \tilde{B}_{N\kappa}^0 (\frac{i-1}{N}) \right) \left( \hat{B}_{N\kappa}(\frac{j}{N}) - \hat{B}_{N\kappa}(\frac{j-1}{N}) \right).$$

**Remark 2:** *In the two-sample case* $(k = 2, n_1 = m, n_2 = n)$ *let* $\hat{D}_N$ *and* $\hat{D}_N^0$ *denote the two-sample rank processes defined in (3.1.5) and (3.1.7), respectively. Then formula (4.2.33) proves the coincidence of (4.2.48) and (3.1.25) and also the coincidence of (4.2.42) and (3.1.23). Similarly formula (4.2.35) proves the coincidence of (4.2.49) and (3.1.30) and also the coincidence of (4.2.44) and (3.1.26). Thus the statistics of this section are direct generalizations of the respective two-sample statistics.*

The asymptotic distribution of the various test statistics is based on Theorem 4.2.2 and on expansions of the form (4.2.43) and (4.2.46).

### 4.2.3 Theorem

*Assume (4.2.20) and* $b_i \in L_2^0(0,1)$ *for* $i = 1, ..., k$. *Let* $(B_N, H_N) \in \mathcal{B}_{kN}^0 \times \mathcal{F}_1^c$, $N \geq k$, *be any sequence which fulfils condition (4.2.21). Define the corresponding sequence of local asymptotic alternatives*

$$F_{B_N, H_N}^N := (F_{1, B_N, H_N}^N, ..., F_{k, B_N, H_N}^N) \qquad (4.2.50)$$

*according to (4.2.12). Finally, let* $0 < a \leq 1$ *be a given bandwidth for the convolution kernel (3.1.19) with underlying kernel* $K$ *according to (3.1.16)-(3.1.18) and (3.1.31), and let* $\mathcal{K}_a$ *denote the corresponding convolution operator defined in (3.1.33).*

*a) If the eigenvalues* $\lambda_\varrho = \lambda_\varrho(a)$ *of* $\mathcal{K}_a$ *fulfil the condition*

$$\sum_{\varrho=1}^{\infty} |\lambda_\varrho(a)| < \infty, \qquad (4.2.51)$$

*then the following limiting law* $(N \to \infty)$ *holds true,*

$$\mathcal{L}[S_N(\mathcal{K}_a \hat{b}_N) \mid F_{B_N, H_N}^N] \xrightarrow{\mathcal{L}} \mathcal{L}[\sum_{\varrho=1}^{\infty} \lambda_\varrho(a) Y_\varrho^2], \qquad (4.2.52)$$

*where* $Y_\varrho^2$, $\varrho = 1, 2, ...$, *are independent real-valued random variables with (noncentral)* $\chi_{k-1}^2(\delta_\varrho^2)$ *-distributions and respective noncentrality parameters*

$$\delta_\varrho^2 = \sum_{i=1}^{k} \eta_i < \psi_\varrho, b_i >^2, \qquad \varrho \geq 1. \qquad (4.2.53)$$

*b) If the eigenvalues $\lambda_\varrho = \lambda_\varrho(a)$ of $\mathcal{K}_a$ fulfil the stronger condition*

$$\sum_{\varrho=1}^{\infty} \varrho \, |\lambda_\varrho(a)| < \infty, \tag{4.2.54}$$

*then the following limiting law $(N \to \infty)$ holds true,*

$$\mathcal{L}[S_N(\mathcal{K}_a \tilde{b}_N^0) \mid F_{B_N, H_N}^N]$$

$$\xrightarrow{\mathcal{L}} \mathcal{L}[\sum_{\varrho=1}^{\infty} \lambda_\varrho(a) \sum_{\kappa=1}^{k} \eta_\kappa < \psi_\varrho', W_\kappa^0(B) > < \psi_\varrho', W_\kappa + B_\kappa >, \tag{4.2.55}$$

*where*

$$W_\kappa^0(B) = \max_{1 \le i \le \kappa} (W_i + B_i), \quad \kappa = 1, ..., k, \tag{4.2.56}$$

*and the $W_\kappa$'s and $B_\kappa$'s as in Theorem 4.2.2.*

*Proof:* The structure of the proof is the same as in the proof of Theorem 3.1.9.
a) For each $r = 1, 2, ...$ and $f = (f_1, ..., f_k) \in (C[0, 1])^k$ define

$$\mathcal{G}_r(f) = \sum_{\kappa=1}^{r} \lambda_\varrho \sum_{i=1}^{k} < \psi_\varrho', f_i >^2 . \tag{4.2.57}$$

Then each $\mathcal{G}_r$ is a continuous function on $(C[0, 1])^k$ endowed with the usual sup-norm. Therefore Theorem 4.2.2 implies ( $\forall r = 1, 2, ...$) the limiting law

$$\mathcal{G}_r(\sqrt{\eta_{N1}} \hat{B}_{N1}, ..., \sqrt{\eta_{Nk}} \hat{B}_{Nk}) \xrightarrow{\mathcal{L}} \sum_{\varrho=1}^{r} \lambda_\varrho Y_\varrho^2 =: S_r, \tag{4.2.58}$$

where $Y_\varrho^2$ is the real-valued random variable defined by

$$Y_\varrho^2 := \sum_{i=1}^{k} \eta_i < \psi_\varrho', W_i + B_i >^2 . \tag{4.2.59}$$

In a first step let's consider the right hand side of (4.2.58). Similar to (3.1.65) the covariance structure (4.2.38) of $W = (W_1, ..., W_k)$ implies the sequence of random vectors

$$(< \psi_\varrho', \sqrt{\eta_1} W_1 >, ..., < \psi_\varrho', \sqrt{\eta_k} W_k >), \qquad \varrho = 1, 2, ..., \tag{4.2.60}$$

to be jointly normal with zero mean values and covariances according to

$$E( < \psi_\sigma', \sqrt{\eta_i}\, W_i > < \psi_\tau', \sqrt{\eta_j}\, W_j >)$$

$$= \int_0^1 \int_0^1 \psi_\sigma'(s)\, \psi_\tau'(t)\, (s \wedge t - st)\, (\delta_{ij} - \sqrt{\eta_i \eta_j})\, ds\, dt$$

$$= (\delta_{ij} - \sqrt{\eta_i \eta_j})\, (\int_0^1 \psi_\sigma \psi_\tau\, d\lambda - \int_0^1 \psi_\sigma\, d\lambda \int_0^1 \psi_\tau\, d\lambda)$$

$$= (\delta_{ij} - \sqrt{\eta_i \eta_j})\delta_{\sigma\tau}. \tag{4.2.61}$$

Therefore $\eta_1 + \cdots + \eta_k = 1$ and Theorem I.4.1 of Hájek and Šidák (1967) imply $Y_\varrho^2$ to have a noncentral $\chi_{k-1}^2$ –distribution with noncentrality parameter

$$\sum_{i=1}^k < \psi_\varrho', \sqrt{\eta_i} B_i >^2 - (\sum_{i=1}^k < \psi_\varrho', \sqrt{\eta_i} B_i > \sqrt{\eta_i})^2$$

$$= \sum_{i=1}^k \eta_i < \psi_\varrho, b_i >^2 - < \psi_\varrho', \sum_{i=1}^k \eta_i B_i >^2$$

$$= \delta_\varrho^2 \quad \text{as defined in (4.2.53), since} \quad \sum_{i=1}^k \eta_i B_i = 0.$$

Especially this implies

$$EY_\varrho^2 \leq k + 2\delta_\varrho^2 \leq k + 2\sum_{i=1}^k \eta_i \|b_i\|^2 \qquad \forall\, \varrho \geq 1. \tag{4.2.62}$$

Thus assumption (4.2.51) forces the generalized random variable

$$S_\infty := \sum_{\varrho=1}^\infty \lambda_\varrho Y_\varrho^2 \tag{4.2.63}$$

to have finite expectation, and

$$S_r = \sum_{\varrho=1}^r \lambda_\varrho Y_\varrho^2 \overset{r\to\infty}{\longrightarrow} S_\infty \quad [a.s.]. \tag{4.2.64}$$

Finally the independence of the random variables $Y_\varrho^2$, $\varrho = 1, 2, \dots$, is obvious from (4.2.60) and (4.2.61).

Completely similar to the proof of Theorem 3.1.9 the proof of (4.2.52) is concluded, if we prove, cf. (3.1.80),

$$\lim_{r\to\infty} \limsup_{N\to\infty} E_0 |S_N(\mathcal{K}_a \hat{b}_N) - \mathcal{G}_r(\sqrt{\eta_{N1}}\hat{B}_{N1}, \dots, \sqrt{\eta_{Nk}}\hat{B}_{Nk})| = 0, \tag{4.2.65}$$

where $E_0$ means the expectation under the null hypothesis $B_{N1} = \cdots = B_{Nk} = 0$ .

*Proof of (4.2.65):*  From (4.2.43) and (4.2.57) we get the inequality

$$E_0|S_N(\mathcal{K}_a\hat{b}_N) - \mathcal{G}_r(\sqrt{\eta_{N1}}\hat{B}_{N1}, ..., \sqrt{\eta_{Nk}}\hat{B}_{Nk})|$$

$$\leq \sum_{\varrho=r+1}^{\infty} |\lambda_\varrho| \sum_{i=1}^{k} \eta_{Ni} \, E_0 < \psi_\varrho, \hat{b}_{Ni} >^2 . \tag{4.2.66}$$

Thus, because of assumption (4.2.51), the proof of (4.2.65) is concluded, if we prove

$$\sum_{i=1}^{k} \eta_{Ni} \, E_0 < \psi_\varrho, \hat{b}_{Ni} >^2 \leq k \qquad \forall \, \varrho \geq 1. \tag{4.2.67}$$

For the final proof of (4.2.67) let's rewrite $< \psi_\varrho, \hat{b}_{Ni} >$ according to (4.2.40), i.e.

$$< \psi_\varrho, \hat{b}_{Ni} > = \sqrt{N}(\frac{N}{n_i}) \sum_{j=1}^{n_i} \int_0^1 \psi_\varrho(t) \, 1(R_{ij} - 1 \leq Nt < R_{ij}) \, dt,$$

and let's use the abbreviations $f(q) := \int_0^1 \psi_\varrho(t) \, 1(q - 1 \leq Nt < q) \, dt$ , $q = 1, ..., N$ , $m := n_i$ , and $R_j := R_{ij}$ . Then we get

$$E_0 < \psi_\varrho, \hat{b}_{Ni} >^2 = N(\frac{N}{m})^2 E_0( \sum_{j=1}^{m} f(R_j) )^2$$

$$= N(\frac{N}{m})^2 \Big( \frac{m}{N} \sum_{i=1}^{N} f^2(i) + \frac{m(m-1)}{N(N-1)} \sum_{i=1}^{N} \sum_{j=1}^{N} f(i) \, f(j) \, 1(i \neq j) \Big)$$

$$\leq N(\frac{N}{m}) \Big( \sum_{i=1}^{N} f^2(i) + \frac{m-1}{N-1} ( \sum_{i=1}^{N} f(i) )^2 \Big)$$

$$= N(\frac{N}{m}) \Big( \sum_{i=1}^{N} ( \int_{(i-1)/N}^{i/N} \psi_\varrho \, d\lambda )^2 + \frac{m-1}{N-1} ( \int_0^1 \psi_\varrho \, d\lambda )^2 \Big)$$

$$\leq N(\frac{N}{m}) \Big( \frac{1}{N} \int_0^1 \psi_\varrho^2 \, d\lambda + 0 \Big) = \frac{N}{m} = \frac{1}{\eta_{Ni}}, \tag{4.2.68}$$

and therefore (4.2.67).

b) For each $r = 1, 2, ...$ we define a *continuous* function $\mathcal{G}_r^0$ on $(C[0,1])^k$

according to $[\,\forall\, f = (f_1, ..., f_k) \in (C[0,1])^k\,]$

$$\mathcal{G}_r^0(f) = \sum_{\varrho=1}^{r} \lambda_\varrho \sum_{i=1}^{k} \eta_i < \psi_\varrho', m_i(f) > < \psi_\varrho', f_i >,$$

$$\qquad (4.2.69)$$

$$m_i(f) = \max_{1 \le j \le i} f_j.$$

Then, on one hand Theorem 4.2.2 implies (for each $r = 1, 2, ...$) the limiting law (as $N \to \infty$),

$$\mathcal{G}_r^0(\hat{B}_N) \xrightarrow{\mathcal{L}} \sum_{\varrho=1}^{r} \lambda_\varrho \sum_{i=1}^{k} \eta_i \; < \psi_\varrho', W_i^0(B) > < \psi_\varrho', W_i + B_i > \qquad (4.2.70)$$

since

$$W_i^0(B) = \max_{1 \le j \le i} (W_j + B_j) = m_i(W_1 + B_1, ..., W_k + B_k)$$

and

$$\hat{B}_N := (\hat{B}_{N1}, ..., \hat{B}_{Nk}) \xrightarrow{\mathcal{L}} (W_1 + B_1, ..., W_k + B_k).$$

Now let's discuss the right hand side of (4.2.55): Defining the real- valued random variables

$$Y_\varrho^0 := \sum_{i=1}^{k} \eta_i < \psi_\varrho', W_i^0(B) > < \psi_\varrho', W_i + B_i >, \qquad (4.2.71)$$

we get the following chain of inequalities, cf. (4.2.61),

$$E|Y_\varrho^0| \le \sum_{i=1}^{k} \eta_i \sqrt{E < \psi_\varrho', W_i^0(B) >^2} \sqrt{E < \psi_\varrho', W_i + B_i >^2}$$

$$\le \sum_{i=1}^{k} \eta_i \sqrt{\|\psi_\varrho'\|^2 E\|W_i^0(B)\|^2} \sqrt{2E < \psi_\varrho', W_i >^2 + 2 < \psi_\varrho, b_i >^2}$$

$$\le \sum_{i=1}^{k} \eta_i \sqrt{(\pi\varrho)^2 E\|W_i^0(B)\|^2} \sqrt{(2/\eta_i) + 2\|\psi_\varrho\|^2\|b_i\|^2}$$

$$\le \varrho\pi \sum_{i=1}^{k} \sqrt{E\|W_i^0(B)\|^2} \sqrt{2(\eta_i + \eta_i^2\|b_i\|^2)}$$

$$= \varrho\, C(B) < \infty. \qquad (4.2.72)$$

Therefore assumption (4.2.54) implies

$$S_\infty^0 := \sum_{\varrho=1}^{\infty} \lambda_\varrho Y_\varrho^0 \qquad (4.2.73)$$

to be a well-defined [a.s.] real-valued random variable with finite expectation, and

$$S_r^0 := \sum_{\varrho=1}^{r} \lambda_\varrho \, Y_\varrho^0 \; \overset{r \to \infty}{\Longrightarrow} \; S_\infty^0 \quad [a.s]. \tag{4.2.74}$$

Especially, the right hand side of (4.2.55) is well-defined, and similarly to part a) the proof of (4.2.55) is concluded if we prove

$$\limsup_{r \to \infty} \; \limsup_{N \to \infty} \; E_0 |S_N(\mathcal{K}_a \tilde{b}_N^0) - \mathcal{G}_r^0(\hat{B}_N)| = 0, \tag{4.2.75}$$

where $E_0$ means the expectation under the null hypothesis.

As a first step of the proof of (4.2.75) we define the real-valued random variables

$$\mathcal{G}_{Nr}^0(\hat{B}_N) := \sum_{\varrho=1}^{r} \lambda_\varrho \sum_{i=1}^{k} \eta_{Ni} < \psi_\varrho', \tilde{B}_{Ni}^0 > < \psi_\varrho', \hat{B}_{Ni} > \tag{4.2.76}$$

and compare $\mathcal{G}_{Nr}^0(\hat{B}_N)$ with $S_N(\mathcal{K}_a \tilde{b}_N^0)$. Obviously the representation in formula (4.2.46) implies

$$E_0 |S_N(\mathcal{K}_a \tilde{b}_N^0) - \mathcal{G}_{Nr}^0(\hat{B}_N)|$$

$$\leq \sum_{\varrho=r+1}^{\infty} |\lambda_\varrho| \sum_{i=1}^{k} \eta_{Ni} \sqrt{E_0 < \psi_\varrho', \tilde{B}_{Ni}^0 >^2} \; \sqrt{E_0 < \psi_\varrho, \hat{b}_{Ni} >^2}$$

$$\leq \sum_{\varrho=r+1}^{\infty} |\lambda_\varrho| \Big( \sum_{i=1}^{k} \eta_{Ni} E_0 < \psi_\varrho', \tilde{B}_{Ni}^0 >^2 \Big)^{1/2} \Big( \sum_{j=1}^{k} \eta_{Nj} E_0 < \psi_\varrho, \hat{b}_{Nj} >^2 \Big)^{1/2}$$

$$\leq \sum_{\varrho=r+1}^{\infty} |\lambda_\varrho| \Big( E_0 \sum_{i=1}^{k} \eta_{Ni} \| \psi_\varrho' \|^2 \, \| \tilde{B}_{Ni}^0 \|^2 \Big)^{1/2} \sqrt{k}, \qquad cf.(4.2.67)$$

$$\leq \sqrt{k} \sum_{\varrho=r+1}^{\infty} |\lambda_\varrho| \, \varrho \, \pi \Big( E_0 \int_0^1 \sum_{i=1}^{k} \eta_{Ni} \, (\tilde{B}_{Ni}^0)^2 \, d\lambda \Big)^{1/2}$$

$$\leq \pi \sqrt{k} \Big( \sum_{j=1}^{k} E_0 \| \hat{B}_{Nj} \|^2 \Big)^{1/2} \sum_{\varrho=r+1}^{\infty} \varrho \, |\lambda_\varrho|, \qquad cf.(4.2.28).$$

This proves

$$\sup_{N \geq k} E_0 |S_N(\mathcal{K}_a \tilde{b}_N^0) - \mathcal{G}_{Nr}^0(\hat{B}_N)| \leq c_0 \sum_{\varrho=r+1}^{\infty} \varrho \, |\lambda_\varrho| \overset{r \to \infty}{\longrightarrow} 0, \tag{4.2.77}$$

if we manage to prove the inequality

$$\sum_{i=1}^{k} E_0 \|\hat{B}_{Ni}\|^2 \leq \frac{1}{4} \sum_{i=1}^{k} \frac{N}{n_i} \qquad \left[ \xrightarrow{N \to \infty} \frac{1}{4} \sum_{i=1}^{k} \frac{1}{\eta_i} \right]. \qquad (4.2.78)$$

Because of

$$\frac{1}{N} \sum_{i=1}^{k} \sum_{j=1}^{n_i} d_N(R_{ij}, t) = \sum_{i=1}^{k} \eta_{Ni}\, \hat{F}_{Ni}(t) = t \qquad \forall\, 0 \leq t \leq 1$$

we may view each $\sqrt{n_i/(N - n_i)}\ \hat{B}_{Ni}$ under the null hypothesis as a two sample rank process with $m = n_i$ and $n = N - n_i$ , cf. (4.2.33).. Therefore formula (3.1.84) implies ( $\forall\, 0 \leq t \leq 1$) the inequality

$$E_0 \hat{B}_{Ni}^2(t) \leq \frac{1}{4} \left(\frac{N}{N-1}\right)\left(\frac{N - n_i}{n_i}\right) \leq \frac{1}{4} \left(\frac{N}{n_i}\right), \qquad (4.2.79)$$

which obviously proves (4.2.78) and thus (4.2.77).

In a second step we compare $\mathcal{G}_{Nr}^0(\hat{B}_N)$ with $\mathcal{G}_r^0(\hat{B}_N)$ . The definitions given in (4.2.76) and (4.2.69) imply

$$E_0 |\mathcal{G}_{Nr}^0(\hat{B}_N) - \mathcal{G}_r^0(\hat{B}_N)|$$

$$\leq \sum_{\varrho=1}^{r} |\lambda_\varrho| \sum_{i=1}^{k} |\eta_{Ni} - \eta_i|\, E_0| < \psi_\varrho', \tilde{B}_{Ni}^0 > < \psi_\varrho, \hat{b}_{Ni} > |$$

$$\leq \max_{1 \leq i \leq k} |1 - \frac{\eta_i}{\eta_{Ni}}| \sum_{\varrho=1}^{\infty} |\lambda_\varrho| \sum_{i=1}^{k} \eta_{Ni}\, E_0| < \psi_\varrho', \tilde{B}_{Ni}^0 > < \psi_\varrho, \hat{b}_{Ni} > |$$

$$\leq \max_{1 \leq i \leq k} |1 - \frac{\eta_i}{\eta_{Ni}}|\, c_0 \sum_{\varrho=1}^{\infty} \varrho\, |\lambda_\varrho|, \qquad (4.2.80)$$

where the proof of the last inequality is completely similar to the proof of the inequality (4.2.77).

Combining formulae (4.2.77) and (4.2.80), and using the assumptions (4.2.20) and (4.2.54) finally proves (4.2.75). $\square$

**Remark 3:**  *The assertions of Theorem 4.2.3 remain true, if the statistics $S_N(\mathcal{K}_a \hat{b}_N)$ and $S_N(\mathcal{K}_a \tilde{b}_N^0)$ are replaced by the respective approximate statistics $S_N(a, K)$ and $S_N^0(a, K)$ . This may be proved by extending the proof of Corollary 3.1.10 to the k-sample case.*

## B) Approximating the optimal score function by projections

In Section 1.2 we have defined the concept of generalized shift functions. This concept has been used in Section 3.2 in order to construct a finite dimensional cone which is suitable for the approximation of various types of generalized shifts. Here we'll similarly define *generalized regression models* and use this concept for the construction of suitable finite dimensional cones which approximate the one sided alternative of trend (4.2.5).

Let $\varrho_{Ni}$, $i = 1, ..., k$, be given *regression constants* such that the following conditions are fulfilled,

$$\varrho_{N1} \geq \varrho_{N2} \geq \cdots \geq \varrho_{Nk}, \quad \varrho_{N1} \neq \varrho_{Nk}, \quad \sum_{i=1}^{k} \eta_{Ni}\, \varrho_{Ni} = 0, \tag{4.2.81}$$

$$\varrho_{Ni} \overset{N\to\infty}{\longrightarrow} \varrho_i \quad \forall\, i = 1, ..., k, \qquad \varrho_1 \neq \varrho_k.$$

Given the 1-dimensional distribution function $F_0$ with absolutely continuous density $f_0$ and finite Fisher-information $I(f_0)$, we define a sequence of local asymptotic alternatives $(F_{N1,D}, ..., F_{Nk,D})$, $N \geq 1$, according to ( $\forall\, i = 1, ..., k$)

$$F_{Ni,D}(x) = F_0\Big( x - \frac{1}{\sqrt{N}}\, \varrho_{Ni}\, D(x) \Big), \quad x \in \mathbb{R}, \tag{4.2.82}$$

where $D : \mathbb{R} \to [0, \infty)$ is a bounded *shift function* with bounded derivative $d : \mathbb{R} \to \mathbb{R}$, cf. (1.2.2).

Under these assumptions and notations it may be proved that the upper test based on the linear rank statistic

$$S_N(b_1, ..., b_k) = \sum_{i=1}^{k} \eta_{Ni} < b_i, \hat{b}_{Ni} >, \tag{4.2.83}$$

cf. (4.2.17) and (4.2.40), with the score functions $b_i$ according to

$$b_i(t) = \varrho_{Ni}\, \varphi(t, f_0, D), \quad 0 < t < 1, \quad 1 \leq i \leq k, \tag{4.2.84}$$

cf. (1.2.4), is *asymptotically optimal* for testing the null hypothesis of randomness (4.2.3) versus the single sequence of alternatives $(F_{N1,D}, ..., F_{Nk,D})$ defined in (4.2.82).

Obviously the above $b = (b_1, ..., b_k)$ is an element of the Hilbert space

$$\mathcal{H} = \{\, (h_1, ..., h_k) \in (L_2^0(0,1))^k : \sum_{i=1}^{k} \eta_{Ni} h_i = 0 \,\} \tag{4.2.85}$$

with the inner product

$$< g, h >:=< (g_1, ..., g_k), (h_1, ..., h_k) > \; := \; \sum_{i=1}^{k} \eta_{Ni} < g_i, h_i > . \qquad (4.2.86)$$

Using this inner product and putting $\hat{b}_N = (\hat{b}_{N1}, ..., \hat{b}_{Nk})$ we get $\hat{b}_N \in \mathcal{H}$, cf. (4.2.26), and (4.2.83) implies the following representation of the optimal linear rank statistic $S_N(b)$,

$$S_N(b) = < b, \hat{b}_N > . \qquad (4.2.87)$$

Therefore the situation is completely similar to Section 3.2 and we may approximate the unknown $b$ by a projection of $\hat{b}_N$ onto a suitable cone $V \subset \mathcal{H}$. Since formula (1.2.12) implies

$$\int_0^t \varphi(\cdot, f_0, D) \, d\lambda \; \leq \; 0 \qquad \forall \, 0 \leq t \leq 1 \qquad (4.2.88)$$

we get the inequalities, cf. (4.2.81),

$$B_1 \leq B_2 \leq \cdots \leq B_k, \qquad (4.2.89)$$

where $B_i$ is the integral of $b_i$, $i = 1, ..., k$,

$$B_i(t) = \int_0^t b_i \, d\lambda = \varrho_{Ni} \int_0^t \varphi(\cdot, f_0, D) \, d\lambda, \quad 0 \leq t \leq 1. \qquad (4.2.90)$$

Therefore $B = (B_1, ..., B_k)$ corresponds to the general alternative of trend (4.2.16), if the additional condition

$$\varrho_{Ni} \, \varphi(\cdot, f_0, D) \; \geq \; -\sqrt{N}, \quad i = 1, ..., k \qquad (4.2.91)$$

is fulfilled. Choosing $r$ generalized shift functions $D_1, ..., D_r$ such that the corresponding

$$b^{(j)} = (b_1^{(j)}, ..., b_k^{(j)}) = (\varrho_{N1}, ..., \varrho_{Nk})\varphi(\cdot, f_0, D_j), \quad j = 1, ..., r, \qquad (4.2.92)$$

are linearly independent, the $r$–dimensional subcone

$$V := [b^{(1)}, ..., b^{(r)}]^+ = \{ \sum_{j=1}^{r} \vartheta_j b^{(j)} : \vartheta_j \geq 0 \;\; \forall j \} \qquad (4.2.93)$$

of $\mathcal{H}$ seems to be a simple and suitable (r-dimensional) part of the general alternative of trend (4.2.16). According to Section 3.2 the resulting rank statistic for testing the null hypothesis of randomness (4.2.3) versus the one sided alternative of trend (4.2.5) is, cf. (3.2.10),

$$S_N(\Pi_V \hat{b}_N) = < \Pi_V \hat{b}_N, \hat{b}_N > \; = \; \|\Pi_V \hat{b}_N\|^2. \qquad (4.2.94)$$

Because of  ( $\forall\, i,j = 1,...,r$ )

$$< b^{(i)}, b^{(j)} > \; = \sum_{\kappa=1}^{k} \eta_{N\kappa}\, \varrho_{N\kappa}^2 < \varphi(\cdot, f_0, D_i), \varphi(\cdot, f_0, D_j) > \qquad (4.2.95)$$

we may define the positive definite matrix

$$\Gamma = \Big( < \varphi(\cdot, f_0, D_i),\ \varphi(\cdot, f_0, D_j) > \Big)_{i,j=1,...,r}, \qquad (4.2.96)$$

the positive constant

$$\sigma_N^2 = \sum_{\kappa=1}^{k} \eta_{N\kappa}\, \varrho_{N\kappa}^2 , \qquad (4.2.97)$$

and the  $r$ –variate linear rank statistic

$$\vec{S}_N = \frac{1}{\sigma_N}\Big(< b^{(1)}, \hat{b}_N >, ..., < b^{(r)}, \hat{b}_N >\Big)^T$$

$$= \Big(\frac{1}{\sigma_N} \sum_{\kappa=1}^{k} \eta_{N\kappa}\, \varrho_{N\kappa} < \varphi(\cdot, f_0, D_j), \hat{b}_{N\kappa} >, \quad j = 1, ..., r\Big)^T, \qquad (4.2.98)$$

in order to rewrite the statistic (4.2.94) in the form, cf. (3.2.52),

$$S_N(\Pi_V \hat{b}_N) = \|\Pi_V \hat{b}_N\|^2 = \sup_{\vartheta \geq 0} (2\vartheta^T \vec{S}_N - \vartheta^T \Gamma \vartheta). \qquad (4.2.99)$$

For the evaluation of the limiting law of  $S_N(\Pi_V \hat{b}_N)$  let's assume that each $\varphi(\cdot, f_0, D_j)$  is absolutely continuous with derivative  $\varphi'_j$ . Then partial integration yields

$$\vec{S}_N = -\Big(< \varphi'_j, \frac{1}{\sigma_N} \sum_{\kappa=1}^{k} \eta_{N\kappa}\, \varrho_{N\kappa}\, \hat{B}_{N\kappa} >, \quad j = 1, ..., r\Big)^T, \qquad (4.2.100)$$

and Theorem 4.2.2 proves the following limiting law.

### 4.2.4 Corollary

*Under the assumptions of Theorem 4.2.2 we have the limiting law*

$$\vec{S}_N \xrightarrow{\;\mathcal{L}\;} -\Big(< \varphi'_j, \frac{1}{\sigma} \sum_{\kappa=1}^{k} \eta_\kappa\, \varrho_\kappa\, (W_\kappa + B_\kappa) >, \quad j = 1, ..., r\Big)^T, \qquad (4.2.101)$$

*where*  $\sigma^2 = \eta_1 \varrho_1^2 + \cdots + \eta_k \varrho_k^2$ .

Under the null hypothesis we have $B_1 = \cdots = B_k = 0$ and therefore

$$\vec{S}_N \xrightarrow{\mathcal{L}} (S_1, ..., S_r) = \vec{S}, \tag{4.2.102}$$

where

$$S_j = -\frac{1}{\sigma} \sum_{\kappa=1}^{k} \eta_\kappa \, \varrho_\kappa < \varphi_j', W_\kappa >, \quad j = 1, ..., r. \tag{4.2.103}$$

Finally, according to (4.2.61) we get ( $\forall \, i, j = 1, ..., r$ )

$$E(S_i S_j) = \frac{1}{\sigma^2} \sum_{\kappa=1}^{k} \sum_{\tau=1}^{k} \sqrt{\eta_\kappa} \, \varrho_\kappa \, \sqrt{\eta_\tau} \, \varrho_\tau \, (\delta_{\kappa\tau} - \sqrt{\eta_\kappa \eta_\tau}) \int_0^1 \varphi_i \, \varphi_j \, d\lambda$$

$$=< \varphi(\cdot, f_0, D_i), \varphi(\cdot, f_0, D_j) >,$$

which means

$$\mathcal{L}(\vec{S}) = \mathcal{N}(0, \Gamma). \tag{4.2.104}$$

Especially the results of Corollary 3.2.6 and Theorem 3.2.7 are applicable. If we put $r = 3$ and if we take $D_1$, $D_2$, and $D_3$ according to (3.2.86) and $F_0$ according to (3.2.88), then formula (3.2.98) implies the following explicit limiting law under the null hypothesis:

For each $t \geq 0$ we have

$$\lim_{N \to \infty} P_0\{ S_N^* > t \} = \tag{4.2.105}$$

$$0.46765 \, P\{ \chi_1^2 > t \} + 0.23987 \, P\{ \chi_2^2 > t \} + 0.03235 \, P\{ \chi_3^2 > t \},$$

where

$$S_N^* = \sup_{\vartheta \geq 0} \, (2\vartheta^T \vec{S}_N - \vartheta^T \Gamma \vartheta), \tag{4.2.106}$$

and

$$\vec{S}_N = \left( \frac{1}{\sigma_N} \sum_{\kappa=1}^{k} \eta_{N\kappa} \, \varrho_{N\kappa} < b_j, \hat{b}_{N\kappa} >, \quad j = 1, 2, 3 \right)^T \tag{4.2.107}$$

with $b_1$, $b_2$, and $b_3$ given in formula (3.2.89).

Given the observed value $\vec{s} = (s_1, s_2, s_3)^T$ of $\vec{S}_N$ the explicit evaluation of the value $s^*$ of the test statistic $S_N^*$ is given in formula (3.2.73) according to

$$s^* = S_N^*(\vec{s}) \tag{4.2.108}$$

$$= \max\{ \, (\vec{s})_J^T (\Gamma_{J \times J})^{-1} (\vec{s})_J \, : \, (\Gamma_{J \times J})^{-1} (\vec{s})_J \geq 0, \, \emptyset \neq J \subset \{1, 2, 3\} \, \}.$$

The corresponding (approximate) $p$-value is, cf. (4.2.105),

$$p(s^*) = \tag{4.2.109}$$

$$0.46765\, P\{\chi_1^2 > s^*\} + 0.23987\, P\{\chi_2^2 > s^*\} + 0.03235\, P\{\chi_3^2 > s^*\}.$$

Thus the only open point is the actual choice of the regression constants $\varrho_{Ni}$, $i = 1, ..., k$ according to condition (4.2.81). Putting

$$\tilde{\varrho}_{Ni} := -\sum_{j=1}^{i} \eta_{Nj}, \quad i = 1, ..., k, \quad \text{and} \quad \bar{\varrho}_N := \sum_{i=1}^{k} \eta_{Ni}\, \tilde{\varrho}_{Ni}, \tag{4.2.110}$$

we define the $\varrho_{Ni}$ 's according to

$$\varrho_{Ni} := \tilde{\varrho}_{Ni} - \bar{\varrho}_N, \qquad i = 1, ..., k. \tag{4.2.111}$$

Obviously this definition fulfils the condition (4.2.81). In addition definition (4.2.111) is a direct generalization of the two sample case, since in case of $k - 2$, $n_1 = m$, and $n_2 = n$ we get by elementary evaluation

$$\frac{\varrho_{N1}}{\sigma_N} = \sqrt{\frac{n}{m}} \quad \text{and} \quad \frac{\varrho_{N2}}{\sigma_N} = -\sqrt{\frac{m}{n}}. \tag{4.2.112}$$

**C) Dispensing with continuity: Treatment of ties**

For the remainder of this section we'll assume that the underlying distribution functions $F_1, ..., F_k$ may have discontinuities, cf. Section 3.3 for the two sample case. Thus the *general k–sample model* has the obvious parametrization

$$(F_1, ..., F_k) \in \mathcal{F}_1 \times \cdots \times \mathcal{F}_1 = \mathcal{F}_1^k. \tag{4.2.113}$$

Under the general model (4.2.113) the null hypothesis of randomness, the omnibus alternative, and the one sided alternative of trend are still given by the respective conditions (4.2.3), (4.2.4), and (4.2.5). Using the results of Section 3.3 we easily get the following suitable reparametrization of the various testing problems:

The partition of the parameter space $\mathcal{F}_1^k$ corresponding to (3.3.19) or (4.1.54) has the form

$$\mathcal{F}_1^k = \sum_{\mathcal{J} \in \mathcal{J}_{0,1}^*} \mathcal{M}_{kN}(\mathcal{J}), \tag{4.2.114}$$

where the $\mathcal{J}$ 's part $\mathcal{M}_{kN}(\mathcal{J})$ of $\mathcal{F}_1^k$ is defined by

$$\mathcal{M}_{kN}(\mathcal{J}) = \{\, (F_{1,B,H}^N, ..., F_{k,B,H}^N) : \; B \in \mathcal{B}_{kN}^0(\mathcal{J}),\; H \in \mathcal{F}_1(\mathcal{J}) \,\}, \tag{4.2.115}$$

$$F_{i,B,H}^N = H + \frac{1}{\sqrt{N}} B_i \circ H, \quad i = 1, ..., k, \qquad (4.2.116)$$

$$\mathcal{B}_{kN}^0(\mathcal{J}) = \left\{ \begin{array}{c} (B_1, ..., B_k) \in \mathcal{B}_{kN}^0 : \ B_i' \text{ is constant } [\lambda - a.e.] \\ \text{on each } (s,t] \in \mathcal{J}, \ i = 1, ..., k \end{array} \right\}, \qquad (4.2.117)$$

and $\mathcal{F}_1(\mathcal{J})$ as defined in (3.3.8). Thus, completely similar to Proposition 3.3.3, the $\mathcal{J}$'s parts $\mathcal{H}_{0N}^r(\mathcal{J})$, $\mathcal{A}_{kN}(\mathcal{J})$, and $\mathcal{A}_{kN}^0(\mathcal{J})$ of the null hypothesis of randomness $\mathcal{H}_0^r$, the omnibus alternative $\mathcal{A}_k$, and the alternative of trend $\mathcal{A}_k^0$, respectively, have the following representations,

$$\mathcal{H}_{0N}^r(\mathcal{J}) = \{ (F_{1,B,H}^N, ..., F_{k,B,H}^N) \in \mathcal{M}_{kN}(\mathcal{J}) : \ B = 0 \}, \qquad (4.2.118)$$

$$\mathcal{A}_{kN}(\mathcal{J}) = \{ (F_{1,B,H}^N, ..., F_{k,B,H}^N) \in \mathcal{M}_{kN}(\mathcal{J}) : \ B \neq 0 \}, \qquad (4.2.119)$$

$$\mathcal{A}_{kN}^0(\mathcal{J}) = \left\{ \begin{array}{c} (F_{1,B,H}^N, ..., F_{k,B,H}^N) \in \mathcal{M}_{kN}(\mathcal{J}) : \ B \neq 0, \\ B_1 \leq \cdots \leq B_k \end{array} \right\}. \qquad (4.2.120)$$

Given $\mathcal{J} \in \mathcal{J}_{0,1}^*$ the testing problems $\mathcal{H}_{0N}^r(\mathcal{J})$ versus $\mathcal{A}_{kN}^0$ and $\mathcal{H}_{0N}^r(\mathcal{J})$ versus $\mathcal{A}_{kN}(\mathcal{J})$ are quite similar to the respective two sample testing problems $H_{0N}(\mathcal{J})$ versus $H_{1N}(\mathcal{J})$ and $H_{0N}(\mathcal{J})$ versus $\tilde{H}_{1N}(\mathcal{J})$. Especially Proposition 3.3.5 and Proposition 3.3.6 immediately extend to the $k$–sample case. In order to get an extension of the results of Theorem 3.3.7 and Theorem 3.3.8 we need obvious generalizations of the definition (3.3.38) of randomized ranks and of the definition (3.3.43) of averaged scores:

Assume $U_{11}, ..., U_{1n_1}, ..., U_{k1}, ..., U_{kn_k}$ to be i.i.d. random variables with uniform distribution on the unit interval $(0, 1)$ such that the (randomization) vector $U = (U_{11}, ..., U_{kn_k})$ and the observation vector $X = (X_{11}, \cdots, X_{kn_k})$ are stochastically independent. Then we define *randomized ranks* $R_{ij}^* = R_{ij}^*(R, U)$ according to

$$R_{ij}^* = \sum_{\kappa=1}^{k} \sum_{\tau=1}^{n_\kappa} 1(R_{\kappa\tau} + U_{\kappa\tau} \leq R_{ij} + U_{ij}), \quad 1 \leq j \leq n_i, \ 1 \leq i \leq k, \qquad (4.2.121)$$

and similar to (3.3.72) we define

$$R_{ij}^0 = \sum_{\kappa=1}^{k} \sum_{\tau=1}^{n_\kappa} 1(R_{\kappa\tau} < R_{ij}), \quad 1 \leq j \leq n_i, \ 1 \leq i \leq k. \qquad (4.2.122)$$

Obviously the number of observations which are equal to $X_{ij}$ ( *tied with the observation $X_{ij}$* ) is given by $R_{ij} - R_{ij}^0$. Therefore we define the *averaged scores* $b_{Ni}^r(j)$ corresponding to the given scores (4.2.18) for each $i = 1, ..., k$

according to

$$b_{Ni}^{\tau}(j) = \frac{1}{R_{\kappa\tau} - R_{\kappa\tau}^0} \sum_{q=1}^{N} b_{Ni}(q)\, \mathbf{1}(R_{\kappa\tau}^0 < q < R_{\kappa\tau}),\qquad (4.2.123)$$

$$\text{if } R_{\kappa\tau}^0 < j < R_{\kappa\tau}.$$

Finally we define the *k–sample averaged scores linear rank statistic* as

$$S_N^{\tau}(b) = \frac{1}{\sqrt{N}} \sum_{i=1}^{k} \sum_{j=1}^{n_i} b_{Ni}^{\tau}(R_{ij}),\qquad (4.2.124)$$

and the *k–sample randomized linear rank statistic* as

$$S_N^{*}(b) = \frac{1}{\sqrt{N}} \sum_{i=1}^{k} \sum_{j=1}^{n_i} b_{Ni}(R_{ij}^{*}).\qquad (4.2.125)$$

### 4.2.5 Theorem

*Assume* $b = (b_1, ..., b_k) \in (L_2^0(0,1))^k$ *and (4.2.20).*

*a) Under the general null hypothesis* $\mathcal{H}_0^r = \{\, (F_1, ..., F_k) \in \mathcal{F}_1^k : F_1 = \cdots = F_k \,\}$ *the randomized linear rank statistic (4.2.125) has the limiting law*

$$\mathcal{L}[\, S_N^{*}(b) \mid \mathcal{H}_0^r \,] \xrightarrow{\ \mathcal{L}\ } \mathcal{N}\Big( 0, \sum_{i=1}^{k} \eta_i\, \|b_i\|^2 \Big).\qquad (4.2.126)$$

*b) Under the* $\mathcal{J}$ *'s part* $\mathcal{H}_{0N}^r(\mathcal{J})$ *of the null hypothesis* $\mathcal{H}_0^r$ *the averaged scores linear rank statistic (4.2.124) has the limiting law*

$$\mathcal{L}[\, S_N^{\tau}(b) \mid \mathcal{H}_{0N}^r(\mathcal{J}) \,] \xrightarrow{\ \mathcal{L}\ } \mathcal{N}\Big( 0, \sum_{i=1}^{k} \eta_i\, \|L_{\mathcal{J}} b_i\|^2 \Big),\qquad (4.2.127)$$

*where* $L_{\mathcal{J}} b_i$ *is defined according to formula (3.3.49). Additionally we have the following convergence in* $\mathcal{H}_{0N}^r(\mathcal{J})$ *–probability,*

$$\frac{N}{N-1} \sum_{i=1}^{k} \eta_{Ni} \frac{1}{N} \sum_{j=1}^{N} \Big( b_{Ni}^{\tau}(j) \Big)^2 \longrightarrow \sum_{i=1}^{k} \eta_i\, \|L_{\mathcal{J}} b_i\|^2 .\qquad (4.2.128)$$

The *proof* is a straightforward extension of the proof of Theorem 3.3.7.

In order to extend the optimality result of Theorem 3.3.8, too, let's fix the parameter $\mathcal{J} \in \mathcal{J}_{0,1}^{*}$ . Given the $\mathcal{J}$ let's take a sequence of alternatives

$$(F_{1,B_N,H_N}^{N}, ..., F_{k,B_N,H_N}^{N}) \in \mathcal{M}_{kN}(\mathcal{J}),\qquad N \geq k,\qquad (4.2.129)$$

with arbitrary $H_N \in \mathcal{F}_1(\mathcal{J})$ and almost fixed $B_N = (B_{N1}, ..., B_{Nk}) \in \mathcal{B}^0_{kN}(\mathcal{J})$, i.e. for each $i = 1, ..., k$ we assume the existence of some $b_i \in L^0_2(0, 1)$ such that (4.2.21) is fulfilled. From (4.2.117) and (3.3.49) it's clear that the limiting $b = (b_1, ..., b_k)$ has the following additional properties $[\lambda - a.e.]$,

$$\sum_{i=1}^{k} \eta_i \, b_i = 0 \quad \text{and} \quad L_{\mathcal{J}} b_i = b_i, \quad i = 1, ..., k. \tag{4.2.130}$$

### 4.2.6 Theorem

*Assume (4.2.20) and $\mathcal{J} \in \mathcal{J}^*_{0,1}$ . For the underlying distribution functions assume the $(B_N, H_N)$ –parametrization (4.2.129), where the sequence of local parameters $(B_N, H_N) \in \mathcal{B}^0_{kN}(\mathcal{J}) \times \mathcal{F}_1(\mathcal{J})$ fulfils condition (4.2.21).*

*a) For any $h = (h_1, ..., h_k) \in (L^0_2(0, 1))^k$ we have the limiting laws*

$$\mathcal{L}[\, S^*_N(h) \mid (B_N, H_N)\,] \xrightarrow{\mathcal{L}} \mathcal{N}\Big(\sum_{i=1}^{k} \eta_i < h_{i\mathcal{J}}, b_i >, \ \sum_{i=1}^{k} \eta_i \, \|h_i\|^2\Big), \tag{4.2.131}$$

$$\mathcal{L}[\, S^\tau_N(h) \mid (B_N, H_N)\,] \xrightarrow{\mathcal{L}} \mathcal{N}\Big(\sum_{i=1}^{k} \eta_i < h_{i\mathcal{J}}, b_i >, \ \sum_{i=1}^{k} \eta_i \, \|h_{i\mathcal{J}}\|^2\Big). \tag{4.2.132}$$

*b) If $\|b_1\|^2 + \cdots + \|b_k\|^2 > 0$ and if $h = (h_1, ..., h_k) \in (L^0_2(0, 1))^k$ such that $h_{i\mathcal{J}} = b_i$, $i = 1, ..., k$, then the averaged scores linear rank test*

$$\psi^\tau_N(h) = 1\Big(\, S^\tau_N(h) \geq u_\alpha \, \sqrt{\frac{N}{N-1} \sum_{i=1}^{k} \eta_{Ni} \, \frac{1}{N} \sum_{j=1}^{N} \big(h^\tau_{Ni}(j)\big)^2} \ \Big) \tag{4.2.133}$$

*is asymptotically optimal at level $\alpha \in (0, 1)$ for testing $\mathcal{H}^\tau_{0N}(\mathcal{J})$ versus the $(B_N, H_N)$ –sequence of alternatives.*

The *proof* is a straightforward extension of the proof of Theorem 3.3.8.

In practice $B_N = (B_{N1}, ..., B_{Nk}) \in \mathcal{B}^0_{kN}(\mathcal{J})$ is unknown. Therefore we substitute the score functions $b_N = (b_{N1}, ..., b_{Nk}) = (B'_{N1}, ..., B'_{Nk})$ of the optimal averaged scores rank statistic $S^\tau_N(b_N)$ by a suitable (rank) estimator $\hat{b}_N$ of $b_N$, which approximately fulfils the side conditions.

Similar to Section 3.3 and in analogy to the continuous case we start with the piecewise linearized empirical distribution function

$$\hat{F}^*_{Ni}(t) = \frac{1}{n_i} \sum_{j=1}^{n_i} d_N(R^*_{ij}, t), \quad 0 \leq t \leq 1, \tag{4.2.134}$$

of the normed randomized ranks $R^*_{i1}/N, ..., R^*_{in_i}/N$, $i = 1, ..., k$, where the function $d_N(r, t)$ is defined in (4.2.24). In the next step we define ( $\forall\, i = 1, ..., k$ ) the *averaged rank process* $\hat{B}_{Ni}$ according to the following conditional expectation [given the original ranks $R = (R_{11}, ..., R_{kn_k})$ ],

$$\hat{B}_{Ni}(t) = E\big[\, \sqrt{N}\, (\hat{F}^*_{Ni}(t) - t) \mid R \,\big], \quad 0 \le t \le 1. \tag{4.2.135}$$

Obviously each $\hat{B}_{Ni}$ is a $C[0, 1]$–valued process with absolutely continuous and piecewise linear paths, which additionally fulfil the side conditions (4.2.26). Completely similar to the continuous case we use the right continuous version $\hat{b}_{Ni}$ of the derivative of $\hat{B}_{Ni}$ as the primitive estimator of the underlying $b_{Ni}$ , if the omnibus alternative is considered. If the trend alternative is considered, then the trend adjusted processes $\tilde{B}^0_{Ni}$, $i = 1, ..., k$ , corresponding to definition (4.2.28) and their derivatives $\tilde{b}^0_{Ni}$, $i = 1, ..., k$, are used instead of the $\hat{b}_{Ni}$ 's.

Thus, in the *general omnibus model* we use the convolution $\mathcal{K}^N_a \hat{b}_{Ni}$ as the suitable rank estimator of $b_{Ni}$ . The resulting *omnibus rank statistic* $S_N(a, K)$ is given in formula (4.2.48).

In the *general trend model* we use the convolution $\mathcal{K}^N_a \tilde{b}^0_{Ni}$ as the suitable rank estimator of $b_{Ni}$ . The resulting *trend rank statistic* $S^0_N(a, K)$ is given in formula (4.2.49).

For the practical evaluation of $\tilde{B}^0_{N\kappa}(q/N) = \max\{\hat{B}_{Ni}(q/N) : \ 1 \le i \le \kappa\}$ and $\hat{B}_{Ni}(q/N)$ we utilize formula (3.3.85), i.e. for each $i = 1, ..., k$ and each $q = 1, ..., N$ we get the following chain of equalities,

$$\hat{B}_{Ni}\Big(\frac{q}{N}\Big) = E\Big[\, \sqrt{N}\, \Big(\frac{1}{n_i} \sum_{j=1}^{n_i} 1(R^*_{ij} \le q) - \frac{q}{N}\Big) \mid R \,\Big]$$

$$= \frac{\sqrt{N}}{n_i} \Big( \sum_{r=1}^{q} \sum_{j=1}^{n_i} \frac{1(R^0_{ij} < r \le R_{ij})}{R_{ij} - R^0_{ij}} - \frac{n_i\, q}{N} \Big)$$

$$= \frac{\sqrt{N}}{n_i} \sum_{r=1}^{q} \Big( \frac{L_{ir}}{L_{0r}} - \frac{n_i}{N} \Big), \tag{4.2.136}$$

where ( $\forall\, i = 1, ..., k, \quad \forall\, r = 1, ..., N$ )

$$L_{ir} = \sum_{j=1}^{n_i} 1(X_{ij} = X_N^{(r)}) = \sum_{j=1}^{n_i} 1(R^0_{ij} < r \le R_{ij}), \tag{4.2.137}$$

$$L_{0r} = \sum_{i=1}^{k} \sum_{j=1}^{n_i} 1(X_{ij} = X_N^{(r)}) = \sum_{i=1}^{k} L_{ir}, \tag{4.2.138}$$

and $X_N^{(1)} \leq X_N^{(2)} \leq \cdots \leq X_N^{(N)}$ denotes the *ordered pooled sample.*

In accordance with the two sample case we call $L_{ir}$ the *length of the tie of* $X_N^{(r)}$ *in the i-th sample,* $i = 1, ..., k,$ and $L_{0r}$ the *length of the tie of* $X_N^{(r)}$ *in the pooled sample.*

The asymptotics of the statistics $S_N(a, K)$, $S_N(\mathcal{K}_a \hat{b}_N)$, $S_N^0(a, K)$, and $S_N(\mathcal{K}_a \tilde{b}_N^0) = S_N(\mathcal{K}_a \hat{b}_N^0)$ in the general model is based on the following generalization of Theorem 4.2.2. The details are omitted.

### 4.2.7 Theorem

*Assume $\mathcal{J} \in \mathcal{J}_{0,1}^*$, $b = (b_1, ..., b_k) \in (L_2^0(0, 1))^k$, and condition (4.2.20). Let $(B_N, H_N) \in \mathcal{B}_{kN}^0(\mathcal{J}) \times \mathcal{F}_1(\mathcal{J})$, $N \geq k$, be any sequence which fulfils condition (4.2.21). Then, under the sequence of $(B_N, H_N)$ –alternatives of the form (4.2.129), we have the following limiting law [in $(C[0, 1])^k$ ],*

$$(\hat{B}_{N1}, ..., \hat{B}_{Nk}) \xrightarrow{\mathcal{L}} (T_{\mathcal{J}} W_1 + B_1, ..., T_{\mathcal{J}} W_k + B_k), \qquad (4.2.139)$$

*where $(W_1, ..., W_k)$ is defined in Theorem 4.2.2 and where the function $T_{\mathcal{J}} : C[0, 1] \to C[0, 1]$ is defined in formula (3.3.106).*

The *proof* is a straightforward extension of the proofs of Theorem 3.3.13 and Theorem 4.2.2.

## 4.3 Several samples on the circle

In the present section we'll develop a general class of rank tests for comparing circular distributions. The framework and the notation will be the same as in the $k$–sample case of Section 4.2, i.e. we consider $k \geq 2$ independent samples of independent real random variables $X_{i1}, ..., X_{in_i}$, $i = 1, ..., k$, $N :=$ $n_1 + ... + n_k$, taking their values on the unit circle which will be represented by the interval $[0, 2\pi)$. Especially this means that the underlying continuous distribution functions $F_i$ of $X_{ij}$ have the property

$$F_i(0) = 0, \quad F_i(2\pi) = 1 \quad \forall \, i = 1, ..., k. \tag{4.3.1}$$

Since $[0, 2\pi)$ corresponds to the unit circle and since the point of the circle corresponding to zero has no preference to any other point of the circle we'll restrict the discussion to the problem of testing the *null hypothesis of randomness*

$$\mathcal{H}_0^r : \quad F_1 = F_2 = \cdots = F_k \tag{4.3.2}$$

versus the *omnibus alternative*

$$\mathcal{A}_k : \quad F_i \neq F_j \quad \text{for some } i \neq j, \quad 1 \leq i, j \leq k. \tag{4.3.3}$$

Defining $H_N = \eta_{N1} F_1 + ... + \eta_{Nk} F_k$, $f_{Ni} = d(F_i \circ H_N^{-1})/d\lambda$, and $b_{Ni} = \sqrt{N}(f_{Ni} - 1)$ as in Section 4.2 we shall prove the likelihood ratio

$$\frac{dF_i}{dH_N}(X_{ij}) = 1 + \frac{1}{\sqrt{N}} \, b_{Ni} \circ H_N(X_{ij}) \tag{4.3.4}$$

to be *rotational invariant* in the following sense:

If $T_\vartheta : [0, 2\pi) \to [0, 2\pi)$ corresponds to the rotation of the unit circle by the angle $\vartheta \in (0, 2\pi)$ and if the distribution function of the rotated random variables

$$X_{ij}^\vartheta := T_\vartheta(X_{ij}) = \begin{cases} X_{ij} + \vartheta, & \text{if } 0 \leq X_{ij} < 2\pi - \vartheta, \\ X_{ij} + \vartheta - 2\pi, & \text{if } 2\pi - \vartheta \leq X_{ij} < 2\pi, \end{cases} \tag{4.3.5}$$

is denoted by $F_i^\vartheta$, then

$$\frac{dF_i^\vartheta}{dH_N^\vartheta}(X_{ij}^\vartheta) = \frac{dF_i}{dH_N}(X_{ij}) \quad [a.s] \quad \forall \, 1 \leq j \leq n_i \quad \forall \, 1 \leq i \leq k, \tag{4.3.6}$$

where $H_N^\vartheta = \eta_{N1} F_1^\vartheta + \cdots + \eta_{Nk} F_k^\vartheta$ is the corresponding mixture. The invariance property (4.3.6) implies that the Neyman–Pearson test for testing the simple hypothesis $\mathcal{L}[\,(X_{11},...,X_{kn_k})\,|\,(H_N,...,H_N)\,]$ versus the simple alternative $\mathcal{L}[\,(X_{11},...,X_{kn_k})\,|\,(F_1,...,F_k)\,]$ with the test statistic

$$S_N(b_{N1},...,b_{Nk},H_N) := \sum_{i=1}^{k}\sum_{j=1}^{n_i} \log[\,1 + \frac{1}{\sqrt{N}}\,b_{Ni} \circ H_N(X_{ij})\,] \qquad (4.3.7)$$

is invariant under the rotation of all observations and all corresponding distributions by the same angle. Therefore we shall estimate the underlying scores

$$b_{Ni}(r) = N \int_{(r-1)/N}^{r/N} b_{Ni}\,d\lambda = N\left(B_{Ni}(\frac{r}{N}) - B_{Ni}(\frac{r-1}{N})\right) \qquad (4.3.8)$$

of the corresponding asymptotically optimal linear rank statistic

$$S_N(b_{N1},...,b_{Nk}) = \frac{1}{\sqrt{N}}\sum_{i=1}^{k}\sum_{j=1}^{n_i} b_{Ni}(R_{ij}) \qquad (4.3.9)$$

by rank estimators $\hat{g}_{Ni}[r]$ which have the additional invariance property

$$\hat{g}_{Ni}^{\vartheta}[R_{ij}^{\vartheta}] = \hat{g}_{Ni}[R_{ij}], \qquad \forall\, 0 < \vartheta < 2\pi, \quad 1 \leq j \leq n_i, \quad 1 \leq i \leq k, \quad (4.3.10)$$

where the upper index $\vartheta$ denotes the respective quantities computed for the rotated random variables $T_\vartheta(X_{ij})$ instead of $X_{ij}$ .

*Proof of assertion (4.3.6)* :

Obviously the inverse $T_\vartheta^{-1} : [0,2\pi) \to [0,2\pi)$ of $T_\vartheta$ exists and is given by

$$T_\vartheta^{-1}(x) = \begin{cases} x + 2\pi - \vartheta, & \text{if} \quad 0 \leq x < \vartheta, \\ x - \vartheta, & \text{if} \quad \vartheta \leq x < 2\pi. \end{cases} \qquad (4.3.11)$$

Therefore it suffices to prove

$$\frac{dF_i^{\vartheta}}{dH_N^{\vartheta}} = \frac{dF_i}{dH_N} \circ T_\vartheta^{-1} \quad [a.s] \qquad \forall\, 1 \leq i \leq k. \qquad (4.3.12)$$

For $1 \leq i \leq k$ and $0 \leq x \leq 2\pi$ we have

$$F_i^{\vartheta}(x) = P\{X_{ij}^{\vartheta} \leq x\}$$

$$= P\{X_{ij}^{\vartheta} \leq x,\, 0 \leq X_{ij} < 2\pi - \vartheta\} + P\{X_{ij}^{\vartheta} \leq x,\, 2\pi - \vartheta \leq X_{ij} < 2\pi\}$$

$$= P\{X_{ij} \leq x - \vartheta\} + P\{2\pi - \vartheta \leq X_{ij} \leq x + 2\pi - \vartheta\}$$

$$= \begin{cases} F_i(x + 2\pi - \vartheta) - F_i(2\pi - \vartheta), & \text{if} \quad x \leq \vartheta, \\ F_i(x - \vartheta) + 1 - F_i(2\pi - \vartheta), & \text{if} \quad x \geq \vartheta, \end{cases} \qquad (4.3.13)$$

and

$$H_N^\vartheta(x) = \begin{cases} H_N(x + 2\pi - \vartheta) - H_N(2\pi - \vartheta), & \text{if} \quad x \leq \vartheta, \\ H_N(x - \vartheta) + 1 - H_N(2\pi - \vartheta), & \text{if} \quad x \geq \vartheta. \end{cases} \qquad (4.3.14)$$

From (4.3.11), (4.3.14), and (4.3.13) we get on one hand for $0 \leq y \leq \vartheta$

$$\int 1(0 < x \leq y) \frac{dF_i}{dH_N} \circ T_\vartheta^{-1}(x) \, dH_N^\vartheta(x)$$

$$= \int 1(0 < x \leq y) \frac{dF_i}{dH_N}(x + 2\pi - \vartheta) \, dH_N(x + 2\pi - \vartheta)$$

$$= \int 1(2\pi - \vartheta < t \leq y + 2\pi - \vartheta) \frac{dF_i}{dH_N}(t) \, dH_N(t)$$

$$= F_i(y + 2\pi - \vartheta) - F_i(2\pi - \vartheta) \; = \; F_i^\vartheta(y),$$

and therefore on the other hand for $\vartheta < y \leq 2\pi$

$$\int 1(0 < x \leq y) \frac{dF_i}{dH_N} \circ T_\vartheta^{-1}(x) \, dH_N^\vartheta(x)$$

$$= 1 - F_i(2\pi - \vartheta) + \int 1(\vartheta < x \leq y) \frac{dF_i}{dH_N}(x - \vartheta) \, dH_N(x - \vartheta)$$

$$= 1 - F_i(2\pi - \vartheta) + F_i(y - \vartheta) \; = \; F_i^\vartheta(y).$$

Hence assertion (4.3.12) is proved.

For the construction of kernel estimators of the underlying vector of score functions $b_{Ni}$ we procede similarly to Section 4.2 with the only difference that the convolution kernel (3.1.19) must be redefined in order to achieve the invariance property (4.3.10). Therefore, instead of the convolution kernel (3.1.19), in this section we use the *periodic kernel* $K_a : [0,1]^2 \to \mathbb{R}$ defined by ( $\forall s, t \in [0,1]$ )

$$K_a(s,t) = \frac{1}{a}\Big(K(\frac{t-s-1}{a}) + K(\frac{t-s}{a}) + K(\frac{t-s+1}{a})\Big), \qquad (4.3.15)$$

where $K : \mathbb{R} \to \mathbb{R}$ has the properties (3.1.16) to (3.1.18) and (3.1.31), and where $0 < a \leq 1$ is a given bandwidth. Again formula (3.1.33) defines the corresponding linear operator $\mathcal{K}_a : L_2(0,1) \to L_2(0,1)$.

If $g^* : \mathbb{R} \to \mathbb{R}$ is the periodic extension of $g \in L_2(0,1)$ onto $\mathbb{R}$ with period 1 then the definition (4.3.15), $0 < a \leq 1$, and $K(x) = 0$ if $|x| \leq 1$, imply the equality ( $\forall 0 \leq t \leq 1$ )

$$\int_0^1 K_a(s,t) \, g(s) \, d\lambda(s) = \int_{-\infty}^\infty \frac{1}{a} K(\frac{t-x}{a}) \, g^*(x) \, dx. \qquad (4.3.16)$$

Additionally the assumptions (3.1.16) to (3.1.18) produce the following properties of $K_a$ which will be used in order to prove the invariance (4.3.10),

$$K_a(s,t) = K_a(t,s), \qquad K_a(s+u,t+u) = K_a(s,t),$$
$$K_a(s+u-1,t+u) = K_a(s,t), \quad \text{if} \ \ 0 \leq t \leq s \leq 1, \tag{4.3.17}$$

where the last equality holds true since $0 < a \leq 1$ and $0 \leq t \leq s \leq 1$ imply

$$K(\frac{t+1-s+1}{a}) = 0 = K(\frac{t-s-1}{a}).$$

In complete analogy to Section 4.2 the underlying $b_N = (b_{N1}, ..., b_{Nk})$ will be estimated by the rank estimator $\mathcal{K}_a \hat{b}_N = (\mathcal{K}_a \hat{b}_{N1}, ..., \mathcal{K}_a \hat{b}_{Nk})$ , where the primitive estimator $\hat{b}_N = (\hat{b}_{N1}, ..., \hat{b}_{Nk})$ of $b_N$ is defined in (4.2.27). Plugging the estimator $\mathcal{K}_a \hat{b}_N$ into the optimal linear rank statistic (4.3.9) yields the test statistic

$$S_N(\mathcal{K}_a \hat{b}_N) = \sum_{i=1}^{k} \eta_{Ni} < \mathcal{K}_a \hat{b}_{Ni}, \hat{b}_{Ni} >, \tag{4.3.18}$$

cf. (4.2.40) and (4.2.42). In practical applications the kernel $K$ is assumed to be smooth, e.g. the Parzen-2 kernel. Then it's more convenient to approximate $K_a$ by the squarewise constant convolution kernel $K_a^N$ defined by

$$K_a^N(s,t) = K_a(\frac{i-1/2}{N}, \frac{j-1/2}{N}) =: k_{Na}(i,j),$$
$$\text{if} \ \ \frac{i-1}{N} < s \leq \frac{i}{N} \quad \text{and} \quad \frac{j-1}{N} < t \leq \frac{j}{N}. \tag{4.3.19}$$

Substitution of $\mathcal{K}_a$ in formula (4.3.18) by the corresponding operator $\mathcal{K}_a^N$ leads to the (approximate) rank statistic

$$S_N(a, K) = \sum_{\kappa=1}^{k} \eta_{N\kappa} \int_0^1 \int_0^1 K_a^N(s,t) \hat{b}_{N\kappa}(s) \hat{b}_{N\kappa}(t) \, ds \, dt \tag{4.3.20}$$

$$= \sum_{i=1}^{N} \sum_{j=1}^{N} k_{Na}(i,j) \sum_{\kappa=1}^{k} \eta_{N\kappa} \left( \hat{B}_{N\kappa}(\frac{i}{N}) - \hat{B}_{N\kappa}(\frac{i-1}{N}) \right)$$

$$\left( \hat{B}_{N\kappa}(\frac{j}{N}) - \hat{B}_{N\kappa}(\frac{j-1}{N}) \right).$$

Obviously the only difference between the general $k$-sample omnibus rank statistic (4.2.48) and the present rank statistic (4.3.20) is the different evaluation of the respective weights $k_{Na}(i,j)$ via the different definitions of the respective convolution kernels $K_a$ , cf. (3.1.19) and (4.3.15).

Now let's prove the invariance property (4.3.10):

In a first step we get from (4.2.27) and (4.2.25) the following explicit evaluation of $\hat{b}_{Ni}$ ,

$$\hat{b}_{Ni}(t) = N^{3/2} \sum_{r=1}^{N} 1\left(\frac{r-1}{N} \leq t < \frac{r}{N}\right) \left(\frac{1}{n_i} \sum_{j=1}^{n_i} 1(R_{ij} = r) - \frac{1}{N}\right). \quad (4.3.21)$$

Therefore the respective estimators $\hat{g}_{Ni}[r]$ and $\breve{g}_{Ni}[r]$ of the score $b_{Ni}(r)$ have the form

$$\hat{g}_{Ni}[r] := N \int_{(r-1)/N}^{r/N} \mathcal{K}_a \hat{b}_{Ni}(t) \, dt = N \int_{(r-1)/N}^{r/N} \left(\int_0^1 K_a(s,t) \, \hat{b}_{Ni}(s) \, ds\right) dt$$

$$= N \sum_{\varrho=1}^{N} \int_{(\varrho-1)/N}^{\varrho/N} \hat{b}_{Ni}(s) \left(\int_{(r-1)/N}^{r/N} K_a(s,t) \, dt\right) ds \quad (4.3.22)$$

$$= N^{5/2} \sum_{\varrho=1}^{N} \left(\frac{1}{n_i} \sum_{j=1}^{n_i} 1(R_{ij} = \varrho) - \frac{1}{N}\right) \int_{(\varrho-1)/N}^{\varrho/N} \int_{(r-1)/N}^{r/N} K_a(s,t) \, ds \, dt$$

$$= N^{5/2} \sum_{\varrho=1}^{N} \left(\frac{1}{n_i} \sum_{j=1}^{n_i} 1(R_{ij} = \varrho) - \frac{1}{N}\right) \int_{-1/N}^{0} \int_{-1/N}^{0} K_a(x + \frac{\varrho}{N}, y + \frac{r}{N}) \, dx \, dy$$

and similarly

$$\breve{g}_{Ni}[r] := N \int_{(r-1)/N}^{r/N} \mathcal{K}_a^N \hat{b}_{Ni}(t) \, dt$$

$$= N^{1/2} \sum_{\varrho=1}^{N} \left(\frac{1}{n_i} \sum_{j=1}^{n_i} 1(R_{ij} = \varrho) - \frac{1}{N}\right) K_a(\frac{-1}{2N} + \frac{\varrho}{N}, \frac{-1}{2N} + \frac{r}{N}). \quad (4.3.23)$$

In the next step we define the rotation $Z_q : \{1,...,N\} \rightarrow \{1,...,N\}$ for any given $q \in \{0,1,...,N\}$ according to

$$Z_q(r) = \begin{cases} r+q, & \text{if } r \leq N-q, \\ r+q-N, & \text{if } r > N-q. \end{cases} \quad (4.3.24)$$

Especially we get $Z_0(j) = Z_N(j) = j \;\; \forall\, j = 1,...,N$ . Depending on the observations $X_{\kappa\tau}$ and on $\vartheta \in (0, 2\pi)$ we define

$$q := \sum_{\kappa=1}^{k} \sum_{\tau=1}^{n_\kappa} 1(X_{\kappa\tau} \geq 2\pi - \vartheta) \;\; \in \{0,1,...,N\} \quad (4.3.25)$$

and use the following properties of $T_\vartheta$ ,

$$[\, T_\vartheta(y) \le T_\vartheta(x) \iff y \le x \text{ or } y \ge 2\pi - \vartheta \,], \quad \text{if } 0 \le x < 2\pi - \vartheta, \qquad (4.3.26)$$

$$[\, T_\vartheta(y) \le T_\vartheta(x) \iff 2\pi - \vartheta \le y \le x \,], \quad \text{if } 2\pi - \vartheta \le x < 2\pi, \qquad (4.3.27)$$

in order to derive the following representation of the transformed ranks $R_{ij}^\vartheta$ ,

$$
\begin{aligned}
R_{ij}^\vartheta &= \sum_{\kappa=1}^{k} \sum_{\tau=1}^{n_\kappa} 1\big(T_\vartheta(X_{\kappa\tau}) \le T_\vartheta(X_{ij})\big) \\[2mm]
&= \begin{cases} R_{ij} + q, & \text{if } X_{ij} < 2\pi - \vartheta, \\ R_{ij} + q - N, & \text{if } X_{ij} \ge 2\pi - \vartheta, \end{cases} \\[2mm]
&= \begin{cases} R_{ij} + q, & \text{if } R_{ij} + q \le N, \\ R_{ij} + q - N, & \text{if } R_{ij} + q > N, \end{cases} \\[2mm]
&= Z_q(R_{ij}). \qquad\qquad (4.3.28)
\end{aligned}
$$

Obviously $Z_q$ has an inverse $Z_q^{-1} : \{1, ..., N\} \to \{1, ..., N\}$ which is given by

$$
Z_q^{-1}(r) = \begin{cases} r - q + N, & \text{if } r \le q, \\ r - q, & \text{if } r > q. \end{cases} \qquad (4.3.29)
$$

Therefore we have the following general type of equality,

$$
\sum_{\varrho=1}^{N} \Big(1(Z_q(r) = \varrho) + c\Big)\, h(\varrho) = \sum_{\tau=1}^{N} \Big(1(r = \tau) + c\Big)\, h(Z_q(\tau)). \qquad (4.3.30)
$$

Considering the rotated random variables $T_\vartheta(X_{ij})$ and the corresponding ranks $R_{ij}^\vartheta$ and assuming $R_{ij} = r$ we get from (4.3.28), (4.3.22), and (4.3.30)

$$
N^{-5/2}\, \hat{g}_{Ni}^\vartheta[R_{ij}^\vartheta] = N^{-5/2}\, \hat{g}_{Ni}^\vartheta[Z_q(r)] \qquad (4.3.31)
$$

$$
= \sum_{\varrho=1}^{N} \Big(\frac{1}{n_i} \sum_{j=1}^{n_i} 1(R_{ij}^\vartheta = \varrho) - \frac{1}{N}\Big) \int_{-1/N}^{0} \int_{-1/N}^{0} K_a\Big(x + \frac{\varrho}{N},\ y + \frac{Z_q(r)}{N}\Big)\, dx\, dy
$$

$$
= \sum_{\tau=1}^{N} \Big(\frac{1}{n_i} \sum_{j=1}^{n_i} (R_{ij} = \tau) - \frac{1}{N}\Big) \int_{\frac{-1}{N}}^{0} \int_{\frac{-1}{N}}^{0} K_a\Big(x + \frac{Z_q(\tau)}{N},\ y + \frac{Z_q(r)}{N}\Big)\, dx\, dy.
$$

Completely similar we get from (4.3.23)

$$
N^{-1/2}\, \breve{g}_{Ni}^\vartheta[R_{ij}^\vartheta] = N^{-1/2}\, \breve{g}_{Ni}^\vartheta[Z_q(r)]
$$

$$
= \sum_{\tau=1}^{N} \Big(\frac{1}{n_i} \sum_{j=1}^{n_i} 1(R_{ij} = \tau) - \frac{1}{N}\Big)\, K_a\Big(\frac{-1}{2N} + \frac{Z_q(\tau)}{N},\ \frac{-1}{2N} + \frac{Z_q(r)}{N}\Big). \qquad (4.3.32)
$$

Finally, for $-1/N < x < 0$, $-1/N < y < 0$, and $r, \tau \in \{1, ..., N\}$ we get from (4.3.17) and (4.3.24) the equality

$$K_a\left(x + \frac{Z_q(\tau)}{N}, y + \frac{Z_q(r)}{N}\right) = K_a\left(x + \frac{\tau}{N}, y + \frac{r}{N}\right). \tag{4.3.33}$$

Combining (4.3.33), (4.3.31), and (4.3.22) concludes the proof of

$$\hat{g}^{\vartheta}_{Ni}[R^{\vartheta}_{ij}] = \hat{g}_{Ni}[R_{ij}], \tag{4.3.34}$$

and combining (4.3.33), (4.3.32), and (4.3.23) concludes the proof of

$$\breve{g}^{\vartheta}_{Ni}[R^{\vartheta}_{ij}] = \breve{g}_{Ni}[R_{ij}]. \tag{4.3.35}$$

Because of

$$S_N(\mathcal{K}_a \hat{b}_N) = \frac{1}{\sqrt{N}} \sum_{i=1}^{k} \sum_{j=1}^{n_i} \hat{g}_{Ni}[R_{ij}] \tag{4.3.36}$$

and

$$S_N(a, K) = S_N(\mathcal{K}_a^N \hat{b}_N) = \frac{1}{\sqrt{N}} \sum_{i=1}^{k} \sum_{j=1}^{n_i} \breve{g}_{Ni}[R_{ij}] \tag{4.3.37}$$

the rotational invariance of the statistics $S_N(\mathcal{K}_a \hat{b}_N)$ and $S_N(a, K)$ has been proved, too.

Our next goal is a limiting theorem which is analogous to Theorem 4.2.3. Since we use the periodic kernel (4.3.15) instead of the symmetrized kernel (3.1.19) we have to compute the spectral representation of the corresponding operator $\mathcal{K}_a$ , cf. Lemma 3.1.1.

### 4.3.1 Lemma

*For each $\varrho = 1, 2, ...$ we define $\psi_{1\varrho}, \psi_{2\varrho} \in L_2^0(0, 1)$ by*

$$\psi_{1\varrho}(t) = \sqrt{2}\sin(2\pi\varrho t), \quad \psi_{2\varrho}(t) = \sqrt{2}\cos(2\pi\varrho t), \quad 0 < t < 1, \tag{4.3.38}$$

*and for each $0 < a \leq 1$ we put*

$$\lambda_\varrho(a) = \int_{-1}^{1} K(t)\cos(2\pi\varrho a t)\, dt. \tag{4.3.39}$$

*Then the following identity holds true for all $b \in L_2(0, 1)$ ,*

$$\mathcal{K}_a b = <b, 1> + \sum_{\varrho=1}^{\infty} \lambda_\varrho(a)(<\psi_{1\varrho}, b> \psi_{1\varrho} + <\psi_{2\varrho}, b> \psi_{2\varrho}), \tag{4.3.40}$$

where the convergence of the infinite series in (4.3.40) holds with respect to $L_2(0,1)$ .

*Proof* : From (4.3.16) we get $\mathcal{K}_a 1 = \int_{-\infty}^{\infty} K(x)\, dx = 1$ . Since the system $\{1, \psi_{1\varrho}, \psi_{2\varrho} : \varrho = 1, 2, ...\}$ is a complete orthonormal basis in $L_2(0,1)$ it suffices to prove

$$\mathcal{K}_a \psi_{1\varrho} = \lambda_\varrho(a)\, \psi_{1\varrho} \quad \text{and} \quad \mathcal{K}_a \psi_{2\varrho} = \lambda_\varrho(a)\, \psi_{2\varrho}. \quad \forall\, \varrho \geq 1 \tag{4.3.41}$$

Since $\psi_{1\varrho}^*(x) = \sqrt{2}\sin(2\pi\varrho x)$ and $\psi_{2\varrho}^*(x) = \sqrt{2}\cos(2\pi\varrho x)$ are the periodic extensions of $\psi_{1\varrho}$ and $\psi_{2\varrho}$ onto $\mathbb{R}$ we get from (4.3.16) and $K(x) = K(-x)$ the equalities ( $\forall\, 0 \leq t \leq 1$)

$$\mathcal{K}_a \psi_{1\varrho}(t) = \int_{-\infty}^{\infty} K(x)\, \sqrt{2}\, \sin\big(2\pi\varrho(ax + t)\big)\, dx, \tag{4.3.42}$$

$$\mathcal{K}_a \psi_{2\varrho}(t) = \int_{-\infty}^{\infty} K(x)\, \sqrt{2}\, \cos\big(2\pi\varrho(ax + t)\big)\, dx. \tag{4.3.43}$$

Therefore the trigonometric formulae

$$\begin{aligned}
\sin(\alpha + \beta) &= \sin(\alpha)\cos(\beta) + \cos(\alpha)\sin(\beta), \\
\cos(\alpha + \beta) &= \cos(\alpha)\cos(\beta) - \sin(\alpha)\sin(\beta),
\end{aligned} \tag{4.3.44}$$

and $\int K(x)\sin(2\pi\varrho ax)\, dx = 0$ imply

$$\mathcal{K}_a \psi_{1\varrho}(t) = \sqrt{2}\sin(2\pi\varrho t)\int_{-\infty}^{\infty} K(x)\, \cos(2\pi\varrho ax)\, dx = \psi_{1\varrho}(t)\, \lambda_\varrho(a),$$

$$\mathcal{K}_a \psi_{2\varrho}(t) = \sqrt{2}\cos(2\pi\varrho t)\int_{-\infty}^{\infty} K(x)\, \cos(2\pi\varrho ax)\, dx = \psi_{2\varrho}(t)\, \lambda_\varrho(a),$$

which concludes the proof of (4.3.31). $\square$

The formulation of the limiting law of $S_N(\mathcal{K}_a \hat{b}_N)$ and $S_N(a, K)$ uses the $(B, H)$ – parametrization (4.2.13) of the general $k$ –sample problem. However, because of the additional assumption (4.3.2) we have to substitute the set $\mathcal{F}_1^c$ of all continuous distribution functions by the smaller set

$$\mathcal{F}_{circ}^c := \{\, H \in \mathcal{F}_1^c : H(0) = 0,\ H(2\pi) = 1\, \}. \tag{4.3.45}$$

Using this definition the null hypothesis (4.3.2) and the omnibus alternative (4.3.3) have the form, cf. (4.2.14) and (4.2.15),

$$\mathcal{H}_0^r = \{\, (F_{1,B,H}^N, ..., F_{k,B,H}^N) :\ B \in \mathcal{B}_{kN}^0,\ H \in \mathcal{F}_{circ}^c,\ B = 0\, \} \tag{4.3.46}$$

$$\mathcal{A}_k = \{ (F^N_{1,B,H}, ..., F^N_{k,B,H}) : \; B \in \mathcal{B}^0_{kN}, \; H \in \mathcal{F}^c_{circ}, \; B \neq 0 \}, \qquad (4.3.47)$$

where $F^H_{i,B,H}$ and $\mathcal{B}^0_{kN}$ are defined in (4.2.12) and (4.2.11).

### 4.3.2 Theorem

Assume $\eta_{N\kappa} = n_\kappa/N \rightarrow \eta_\kappa \in (0,1)$ as $N \rightarrow \infty$ and $b_\kappa \in L^0_2(0,1)$ for $\kappa = 1,...,k$. Let $(B_N, H_N) \in \mathcal{B}^0_{kN} \times \mathcal{F}^c_{circ}$, $N \geq k$, be any sequence which fulfils condition (4.2.21), i.e. for $1 \leq \kappa \leq k$ we assume $b_{N\kappa} = B'_{N\kappa} \rightarrow b_\kappa$ in $L_2(0,1)$ as $N \rightarrow \infty$. Define the corresponding sequence of local asymptotic alternatives

$$F^N_{B_N,H_N} = (F^N_{1,B_N,H_N}, ..., F^N_{k,B_N H_N}) \in (\mathcal{F}^c_{circ})^k, \qquad N \geq k. \qquad (4.3.48)$$

Finally, let $0 < a \leq 1$ be a given bandwidth for the periodic kernel (4.3.15) with underlying kernel $K$ according to (3.1.16)–(3.1.18) and (3.1.31), and let $\mathcal{K}_a$ denote the corresponding convolution operator defined as in formula (3.1.33).

If the eigenvalues $\lambda_\varrho = \lambda_\varrho(a)$ of $\mathcal{K}_a$ given in (4.3.39) fulfil the condition

$$\sum_{\varrho=1}^{\infty} |\lambda_\varrho(a)| < \infty, \qquad (4.3.49)$$

then the following limiting law $(N \rightarrow \infty)$ holds true,

$$\mathcal{L}[\, S_N(\mathcal{K}_a \hat{b}_N) \mid F^N_{B_N,H_N} \,] \; \xrightarrow{\;\mathcal{L}\;} \; \mathcal{L}[\, \sum_{\varrho=1}^{\infty} \lambda_\varrho(a) \, Y^2_\varrho \,], \qquad (4.3.50)$$

where $Y^2_\varrho$, $\varrho = 1,2,...$, are independent real-valued random variables with (noncentral) $\chi^2_{2(k-1)}(\delta^2_\varrho)$ - distributions and respective noncentrality parameters

$$\delta^2_\varrho = \sum_{\kappa=1}^{k} \eta_\kappa \left( < \psi_{1\varrho}, b_\kappa >^2 + < \psi_{2\varrho}, b_\kappa >^2 \right), \qquad \varrho \geq 1. \qquad (4.3.51)$$

*Proof*: From (4.3.18), (4.3.40), and $< \hat{b}_{N\kappa}, 1 >= 0$ we have the representation

$$S_N(\mathcal{K}_a \hat{b}_N) = \sum_{\varrho=1}^{\infty} \lambda_\varrho(a) \sum_{\kappa=1}^{k} \eta_{N\kappa} \left( < \psi_{1\varrho}, \hat{b}_{N\kappa} >^2 + < \psi_{2\varrho}, \hat{b}_{N\kappa} >^2 \right)$$

$$= \sum_{\varrho=1}^{\infty} \lambda_\varrho(a) \sum_{\kappa=1}^{k} \eta_{N\kappa} \left( < \psi'_{1\varrho}, \hat{B}_{N\kappa} >^2 + < \psi'_{2\varrho}, \hat{B}_{N\kappa} >^2 \right). \qquad (4.3.52)$$

By application of Theorem 4.2.2 we get completely similar to the proof of Theorem 4.2.3 the limiting law

$$S_N(\mathcal{K}_a \hat{b}_N)$$

$$\xrightarrow{\mathcal{L}} \sum_{\varrho=1}^{\infty} \lambda_\varrho \sum_{\kappa=1}^{k} \eta_\kappa \left( <\psi'_{1\varrho}, W_\kappa + B_\kappa >^2 + <\psi'_{2\varrho}, W_\kappa + B_\kappa >^2 \right). \qquad (4.3.53)$$

According to the proof of Theorem 4.2.3, especially (4.2.61), the two - dimensional random variables

$$(Y_{1\varrho}^2, Y_{2\varrho}^2) := \sum_{\kappa=1}^{k} \eta_\kappa \left( <\psi'_{1\varrho}, W_\kappa + B_\kappa >^2, <\psi'_{2\varrho}, W_\kappa + B_\kappa >^2 \right), \qquad (4.3.54)$$

$\varrho = 1, 2, \ldots$ , and also the random variables $Y_{1\varrho}^2, Y_{2\varrho}^2$ are stochastically independent, and the distribution of $Y_{i\varrho}^2$ is a noncentral $\chi_{k-1}^2$ -distribution with noncentrality parameter

$$\delta_{i\varrho}^2 = \sum_{\kappa=1}^{k} \eta_\kappa <\psi_{i\varrho}, b_\kappa >^2 . \qquad (4.3.55)$$

Thus $Y_\varrho^2 := Y_{1\varrho}^2 + Y_{2\varrho}^2$ , $\varrho = 1, 2, \ldots$ , are stochastically independent random variables with (noncentral) $\chi_{2(k-1)}^2(\delta_{1\varrho}^2 + \delta_{2\varrho}^2)$ -distributions, and (4.3.53) concludes the proof of (4.3.50). $\square$

*Remark 1: As in the general $k$ –sample case the assertions of Theorem 4.3.2 remain true if the statistic $S_N(\mathcal{K}_a \hat{b}_N)$ is replaced by the approximate statistic $S_N(a, K)$ .*

*Remark 2: The treatment of ties in the $k$ –sample model on the circle is completely similar to the treatment of ties in the general $k$ –sample model of Section 4.2. If we substitute $\mathcal{F}_1$ by*

$$\mathcal{F}_{circ} = \{ H \in \mathcal{F}_1 : H(0) = 0, H(2\pi) = 1 \} \qquad (4.3.56)$$

*and $\mathcal{F}_1(\mathcal{J})$ by*

$$\mathcal{F}_{circ}(\mathcal{J}) = \{ H \in \mathcal{F}_{circ} : \mathcal{I}(H) = \mathcal{J} \} \qquad (4.3.57)$$

*then all assertions of Theorem 4.2.5, Theorem 4.2.6, and Theorem 4.2.7 remain true and may be utilized in order to derive the asymptotics of the statistics $S_N(\mathcal{K}_a \hat{b}_N)$ and $S_N(a, K)$ of the present section.*

## 4.4  Two samples under type II censoring

The present section is an extension of the usual two-sample case discussed in Chapter 3 to the situation of censored data. Therefore we'll use the notations and the results of Chapter 3.

Especially $X_1, ..., X_m$ and $X_{m+1}, ..., X_{m+n}$ will denote the two samples with the respective underlying (continuous)distribution functions $F \in \mathcal{F}_1^c$ and $G \in \mathcal{F}_1^c$, and all components of the pooled sample $X = (X_1, ..., X_N)$, $N = m+n$ are assumed to be stochastically independent.

As before we want to test the *null hypothesis of randomness*

$$\mathcal{H}_0^r = \{ (F, G) \in \mathcal{F}_1^c \times \mathcal{F}_1^c : F = G \} \tag{4.4.1}$$

versus the *one-sided alternative*

$$\mathcal{A}_2^0 = \{ (F, G) \in \mathcal{F}_1^c \times \mathcal{F}_1^c : F \leq G, F \neq G \}, \tag{4.4.2}$$

which means that the first sample is stochastically larger than the second sample, or we want to test the null hypothesis $\mathcal{H}_0^r$ versus the *omnibus alternative*

$$\mathcal{A}_2 = \{ (F, G) \in \mathcal{F}_1^c \times \mathcal{F}_1^c : F \neq G \}. \tag{4.4.3}$$

But in contrast to Chapter 3 the random variables $X_1, ..., X_N$ are *unobservable*, i.e. we have the following additional assumption.


**4.4.1 Assumption**  (type II censoring)

*Only the $r$ smallest values of the pooled sample $X_1, ..., X_N$ can actually be observed, where $1 \leq r < N$ is a given constant.*


If $X_N^{(1)} < X_N^{(2)} < \cdots < X_N^{(N)}$ denotes the ordered pooled sample, then Assumption 4.4.1 means that the values of $X_N^{(1)}, ..., X_N^{(r)}$ are observable and that for each $i \in \{1, ..., N\}$ the random variable $X_i$ is censored if and only if $X_i > X_N^{(r)}$ holds true. Restricting the discussion to *rank procedures* we have the situation that exactly the ranks of the uncensored random variables are observable whereas the ranks of the censored random variables are known to be larger than $r$, i.e.

$$R_i = \sum_{j=1}^{N} 1(X_j \leq X_i) = \sum_{j=1}^{r} 1(X_N^{(j)} \leq X_i) \leq r, \quad \text{if } X_i \text{ is uncensored,}$$

$$r < R_i \leq N, \quad \text{if } X_i \text{ is censored.} \tag{4.4.4}$$

Given any score function $b \in L_2^0(0,1)$ we define the corresponding *r-censored scores* $b_N^r(i)$ , $i = 1, ..., N$ , according to

$$
b_N^r(i) = \begin{cases} N \int_{(i-1)/N}^{i/N} b \, d\lambda, & \text{if } 1 \leq i \leq r, \\[2mm] \frac{N}{N-r} \int_{r/N}^{1} b \, d\lambda, & \text{if } r < i \leq N, \end{cases}
\tag{4.4.5}
$$

and the corresponding *r-censored scores linear rank statistic* $S_N^r(b)$ according to

$$
S_N^r(b) = \sum_{i=1}^{N} c_{Ni} \, b_N^r(R_i),
\tag{4.4.6}
$$

where the two-sample coefficients $c_{Ni}$ are given in formula (1.1.10). Because of $b_N^r(i) = b_N(i) \ \forall \ 1 \leq i \leq r$, $\sum c_{Ni} = 0$, and $\int_0^1 b \, d\lambda = 0$ the rank statistic $S_N^r(b)$ can be written in the form

$$
S_N^r(b) = \sum_{i=1}^{N} 1(R_i \leq r) \, c_{Ni} \left( b_N(R_i) + \frac{N}{N-r} \int_0^{r/N} b \, d\lambda \right).
\tag{4.4.7}
$$

Thus, the r - censored scores linear rank statistic $S_N^r(b)$ can be evaluated on the basis of the *ranks of the uncensored observations* together with the restriction of the score function $b$ onto the interval $(0, r/N)$ .

In order to formulate and prove a suitable local asymptotic optimality of $S_N^r(b)$ for the above testing problems let's assume

$$
\eta_N := \frac{m}{N} \xrightarrow{n \to \infty} \eta \quad \text{and} \quad p_N := \frac{r}{N} \xrightarrow{N \to \infty} p,
\tag{4.4.8}
$$

for some $0 < \eta < 1$ and $0 < p < 1$ . Defining $b_N^r \in L_2^0(0,1)$ and $b_p \in L_2^0(0,1)$ according to ( $\forall \ 0 < t < 1$)

$$
\tilde{b}_N^r(t) = \sum_{i=1}^{N} b_N^r(i) \, 1(\frac{i-1}{N} \leq t < \frac{i}{N})
\tag{4.4.9}
$$

and

$$
\begin{aligned}
b_p(t) &= b(t)1(0 < t \leq p) + \left( \frac{1}{1-p} \int_p^1 b \, d\lambda \right) 1(p < t < 1) \\[3mm]
&= b(t)1(0 < t \leq p) - \left( \frac{1}{1-p} \int_0^p b \, d\lambda \right) 1(p < t < 1)
\end{aligned}
\tag{4.4.10}
$$

the assumption (4.4.8) implies

$$
\|\tilde{b}_N^r - b_p\| \xrightarrow{N \to \infty} 0
\tag{4.4.11}
$$

and

$$\|\tilde{b}_N^r\|^2 \overset{N\to\infty}{\longrightarrow} \|b_p\|^2 = \int_0^p b^2\, d\lambda + \frac{1}{1-p}\Big(\int_0^p b\, d\lambda\Big)^2. \tag{4.4.12}$$

Therefore Theorem 3.3.7 yields the limiting law

$$\mathcal{L}[\, S_N^r(b) \mid \mathcal{H}_0^r \,] \overset{\mathcal{L}}{\longrightarrow} \mathcal{N}(0, \|b_p\|^2) \tag{4.4.13}$$

and for any given $\alpha \in (0,1)$ the *r-censored scores linear rank test*

$$\psi_N^r(b) = 1(\, S_N^r(b) \geq u_\alpha \|b_p\| \,) \tag{4.4.14}$$

is of asymptotic level $\alpha$ for testing the null hypothesis $\mathcal{H}_0^r$, if the condition $\int_0^p b^2 d\lambda > 0$ holds true.

Using the terminology of Section 3.3 we put $\mathcal{J}_p := \{\, (p,1] \,\} \in \mathcal{J}_{0,1}^*$ and define the sequence $[\,(F_N, G_N),\ N \geq 2\,]$ of local alternatives corresponding to $b \in L_2^0(0,1)$ and $p \in (0,1)$ according to

$$(F_N, G_N) = (\, H_N + c_{N1} B_N \circ H_N,\ H_N + c_{NN} B_N \circ H_N \,), \tag{4.4.15}$$

where

$$(B_N, H_N) \in \mathcal{B}_N^0(\mathcal{J}_p) \times \mathcal{F}_1(\mathcal{J}_p) \tag{4.4.16}$$

and

$$b_N = B_N' \overset{N\to\infty}{\longrightarrow} b_{\mathcal{J}_p} = L_{\mathcal{J}_p} b \quad \text{in}\ \ L_2(0,1), \tag{4.4.17}$$

cf. definition (3.3.49). Obviously the special form of $\mathcal{J}_p$ implies the equality, cf. (3.3.49) and (4.4.10),

$$b_{\mathcal{J}_p} = b_p. \tag{4.4.18}$$

### 4.4.2 Theorem

*Assume (4.4.8) and (4.4.15) to (4.4.17) for given $b \in L_2^0(0,1)$ .*

*a) For any score function $h \in L_2^0(0,1)$ we have the limiting law*

$$\mathcal{L}[\, S_N^r(h) \mid (F_N, G_N) \,] \overset{\mathcal{L}}{\longrightarrow} \mathcal{N}(< h_p, b_p >, \|h_p\|^2), \tag{4.4.19}$$

*where $b_p$ and $h_p$ are defined according to (4.4.10).*

*b) If $\int_0^p b^2 d\lambda > 0$ then the $r$ -censored scores linear rank test $\psi_N^r(b)$ defined in (4.4.14) is asymptotically optimal at level $\alpha$ for testing $\mathcal{H}_0^r$ versus the sequence of local alternatives $[\,(F_N, G_N),\ N \geq 2\,]$ .*

*Proof*: From (4.4.12) it's obvious that the assumption $\int_0^p b^2 d\lambda > 0$ implies $\|b_p\|^2 > 0$ . Thus, because of (4.4.13), (4.4.18), Theorem 3.3.8, and the contiguity of the sequence of $(F_N, G_N)$ -alternatives with respect to the sequence of

$(H_N, H_N)$ -hypothesis-points, the proof is complete if we prove the assertion (as $N \to \infty$)

$$S_N^r(h) - S_N^r(h_p) \longrightarrow 0 \qquad \text{in} \quad (H_N, H_N) - \text{probability} , \qquad (4.4.20)$$

where $S_N^r(h_p)$ denotes the *averaged scores linear rank statistic* corresponding to the score function $h_p$ , cf. (3.3.40) and (3.3.44). Because of (3.3.56) and the equality $(h_p)_{\mathcal{J}_p} = h_p$ the proof of assertion (4.4.20) is concluded if we prove (as $N \to \infty$ )

$$S_N^r(h) - S_N^*(h_p) \longrightarrow 0 \qquad \text{in} \quad (H_N, H_N) - \text{probability} , \qquad (4.4.21)$$

where $S_N^*(h_p)$ is the corresponding *randomized linear rank statistic* according to (3.3.39), (3.3.38), and (3.3.37). Because of $H_N \in \mathcal{F}_1(\mathcal{J}_p)$ and Proposition 3.3.5 it suffices to prove (4.4.21) under the special assumption ( $\forall N \geq 2$)

$$H_N(x) = F_p(x) = x1(0 \leq x < p) + 1(x \geq p), \qquad x \in \mathbb{R}. \qquad (4.4.22)$$

If $E_p(\cdots)$ denotes the expectation with respect to the $(F_p, F_p)$ -hypothesis then $\sum c_{Ni} = 0$ and $\sum c_{Ni}^2 = 1$ imply

$$E_p\big( S_N^r(h) - S_N^*(h_p)\big)^2 = E_p\Big(\sum_{i=1}^{N} c_{Ni}\big(h_N^r(R_i) - h_{pN}(R_i^*)\big)\Big)^2$$

$$= E_p\big(h_N^r(R_1) - h_{pN}(R_1^*)\big)^2 + E_p\big(h_N^r(R_1) - h_{pN}(R_1^*)\big)\big(h_N^r(R_2) - h_{pN}(R_2^*)\big)$$

$$\leq 2\, E_p\big(h_N^r(R_1) - h_{pN}(R_1^*)\big)^2$$

$$= 2\, A_{N1} + 2\, A_{N2}, \qquad (4.4.23)$$

where
$$A_{N1} := E_p\big(h_N^r(R_1) - h_{pN}(R_1^*)\big)^2 1(X_1 < p) \qquad (4.4.24)$$
and
$$A_{N2} := E_p\big(h_N^r(R_1) - h_{pN}(R_1^*)\big)^2 1(X_1 = p). \qquad (4.4.25)$$

For the final proof of $A_{N1} \to 0$ and $A_{N2} \to 0$ we get in a first step from (4.4.9) and (4.4.11)

$$A_{N1} = E_p\big(h_N^r(R_1^*) - h_{pN}(R_1^*)\big)^2 1(X_1 < p)$$

$$\leq E_p\big(h_N^r(R_1^*) - h_{pN}(R_1^*)\big)^2$$

$$= \frac{1}{N} \sum_{i=1}^{N} \big(h_N^r(i) - h_{pN}(i)\big)^2 \overset{N \to \infty}{\longrightarrow} 0. \qquad (4.4.26)$$

Since $X_1 = p$ implies $R_1 = N$ , we get in a second step the equalities

$$A_{N2} = E_p\big(h_N^r(N) - h_{pN}(R_1^*)\big)^2 1\big(X_N^{(R_1^*)} = p\big)$$

$$= \frac{1}{N} \sum_{i=1}^{N} \big(h_N^r(N) - h_{pN}(i)\big)^2 P_p\{X_N^{(i)} = p\},$$

$$\tag{4.4.27}$$

where the last equality holds true, since the randomized ranks and the order statistic are stochastically independent under the general null hypothesis $(F, F) \in \mathcal{F}_1 \times \mathcal{F}_1$ ,cf. Lemma 3.3.9.

In a third step the properties

$$h_{pN}(i) = N \int_{(i-1)/N}^{i/N} h_p \, d\lambda = \frac{1}{1-p} \int_p^1 h \, d\lambda =: d \quad \forall \, i > [Np] + 1, \tag{4.4.28}$$

$$d_N := h_N^r(N) = \frac{1}{1-p_N} \int_{p_N}^1 h \, d\lambda \overset{N \to \infty}{\longrightarrow} d, \tag{4.4.29}$$

$$\int_0^1 \big(h_{pN}([Nx] + 1) - h_p(x)\big)^2 \, dx \overset{N \to \infty}{\longrightarrow} 0, \tag{4.4.30}$$

and (4.4.27) imply

$$A_{N2} = \int_0^1 \big(d_N - h_{pN}([Nx] + 1)\big)^2 P_p\{X_N^{([Nx]+1)} = p\} \, dx$$

$$= \int_0^p \big(d - h_p(x)\big)^2 P_p\{X_N^{([Nx]+1)} = p\} \, dx + o(1). \tag{4.4.31}$$

Finally, for any $x \in (0, p)$  we have

$$P_p\{X_N^{([Nx]+1)} = p\} = \sum_{i=0}^{[Nx]} \binom{N}{i} p^i (1-p)^{N-i}$$

$$\leq \frac{1}{(Np - [Nx])^2} \, \mathrm{Var}[\,\mathrm{Binomial(N,p)}\,]$$

$$\leq \frac{1}{N} \frac{p(1-p)}{(p-x)^2} \overset{N \to \infty}{\longrightarrow} 0. \tag{4.4.32}$$

Therefore (4.4.31) and the dominated convergence theorem imply $A_{N2} \to 0$ as $N \to \infty$ , which concludes the proof of Theorem 4.4.2. $\square$

Notice, the evaluation of $S_N^r(h)$  is based only on the ranks of the $r$  smallest observations, but there is no explicit assumption of censoring in the formulation of Theorem 4.4.2. In analogy to Theorem 3.0.1 [cf. the proof in Section 7.1 (Appendix)] the asymptotic optimality of the rank test $\psi_N^r(b)$ means that the sequence ( $\psi_N^r(b)$, $N \geq 2$ ) is asymptotically equivalent to the sequence ( $\psi_{N\alpha}^*$, $N \geq 2$ ) of the (optimal) Neyman-Pearson tests $\psi_{N\alpha}^*$ at the given level $\alpha$ for testing the null hypothesis $(H_N, H_N)$ versus the alternative $(F_N, G_N)$, and the Neyman-Pearson test is *based on all of the observations* $X_1, ..., X_N$. So where is the censoring? Obviously the assumption $H_N \in \mathcal{F}_1(\mathcal{J}_p)$ means (as $N \to \infty$ )

$$\frac{1}{N} \sum_{i=1}^{N} 1\big(H_N(X_i) = p\big) \longrightarrow 1 - p \quad \text{in } (H_N, H_N) - \text{probability}, \quad (4.4.33)$$

i.e. under the null hypothesis $(H_N, H_N)$ as well as under the alternative $(F_N, G_N)$ approximately $100(1 - p)\%$ of the observations are "censored" at $H_N^{-1}(p)$ .

If we assume continuous underlying distribution functions $H_N \in \mathcal{F}_1^c$ , and $B_N \in \mathcal{B}_N^0$ such that $\|B_N' - b\| \longrightarrow 0$ for some $b \in L_2^0(0,1)$ , then (4.4.8) implies (as $N \to \infty$ )

$$H_N(X_N^{(r)}) \longrightarrow p \qquad \text{in } (H_N, H_N) - \text{probability}, \qquad (4.4.34)$$

i.e. under the null hypothesis $(H_N, H_N)$ as well as under the alternative $(F_N, G_N)$ the censored observations

$$X_1 1(X_1 \leq X_N^{(r)}), \; \ldots \, , X_N 1(X_N \leq X_N^{(r)}) \qquad (4.4.35)$$

do not contain information about

$$H_N(x) 1\big(H_N(x) > p\big) \qquad \text{and} \qquad b(t) 1(t > p). \qquad (4.4.36)$$

In this sense Theorem 4.4.2 is the best optimality result which can be achieved for the $r$ -censored scores linear rank test $\psi_N^r(b)$ , cf. Mehrotra and Johnson (1976) for the special case of shift alternatives.

As in the usual two-sample case we'll substitute the unknown scores $b_N^r(i)$ of the optimal linear rank statistic $S_N^r(b)$ by suitable rank estimators. Since only the $r$ smallest values of the pooled sample are observable, these estimators must be based on the ranks of the uncensored observations. These ranks may be represented in the form

$$R_i^r := \begin{cases} R_i \leq r, & \text{if } X_i \text{ is uncensored}, \\ N, & \text{if } X_i \text{ is censored}. \end{cases} \qquad (4.4.37)$$

If $\hat{B}_{N2}$ denotes the piecewise linear rank process (3.1.5) defined on the basis of the (unobservable) complete set of ranks $R_1, ..., R_N$ , then obviously the modified rank process

$$\hat{B}^r_{N2}(t) = \begin{cases} \hat{B}_{N2}(t), & \text{if } 0 \leq t \leq p_N, \\ \frac{1-t}{1-p_N}\hat{B}_{N2}(p_N), & \text{if } p_N \leq t \leq 1, \end{cases} \qquad (4.4.38)$$

is the corresponding natural estimator of the (unknown) underlying $B_N \in \mathcal{B}^0_N(\mathcal{J}_p)$ , since $B_N$ has the property

$$B_N(t) = \begin{cases} B_N(t), & \text{if } 0 \leq t \leq p, \\ \frac{1-t}{1-p}B_N(p), & \text{if } p \leq t \leq 1. \end{cases} \qquad (4.4.39)$$

Fortunately the process $\hat{B}^r_{N2}$ is observable since we have the representation

$$\hat{B}^r_{N2}(\frac{j}{N}) = \sum_{i=1}^N c_{Ni}1(R_i \leq j) = \sum_{i=1}^N c_{Ni}1(R^r_i \leq j), \quad j = 0, 1, ..., r,$$

$$\hat{B}^r_{N2} \quad \text{linear on } [\frac{j-1}{N}, \frac{j}{N}] \text{ for any } 1 \leq j \leq r,$$

$$\hat{B}^r_{N2}(t) = \frac{1-t}{1-p_N} \, \hat{B}^r_{N2}(\frac{r}{N}) \qquad \forall \, t \in [\frac{r}{N}, 1]. \qquad (4.4.40)$$

Additionally the following limiting law of $\hat{B}^r_{N2}$ holds true in $C[0,1]$.

### 4.4.3 Theorem

*Assume $b \in L^0_2(0,1)$ and (4.4.8). Take any sequence $[\,(F_N, G_N),\, N \geq 2\,]$ of corresponding local alternatives, i.e. take $(F_N, G_N)$ according to*

$$F_N = H_N + c_{N1}B_N \circ H_N, \; G_N = H_N + c_{NN}B_N \circ H_N, \; B_N \in \mathcal{B}^0_N,$$

$$H_N \in \mathcal{F}^c_1, \qquad b_N = B'_N \xrightarrow{N \to \infty} b \; \text{ in } \; L_2(0,1). \qquad (4.4.41)$$

*Then we have the limiting law*

$$\mathcal{L}[\,\hat{B}^r_{N2} \mid (F_N, G_N)\,] \xrightarrow{\mathcal{L}} \mathcal{L}[W_{0p} + B_p] \; \text{ in } C[0,1], \qquad (4.4.42)$$

*where ( $\forall \, 0 \leq t \leq 1$ )*

$$W_{0p}(t) = W_0(t)1(0 \leq t < p) + \frac{1-t}{1-p} \, W_0(p)1(p \leq t \leq 1) \qquad (4.4.43)$$

*and $W_0$ the Brownian-bridge on $C[0,1]$, and where ( $\forall \, 0 \leq t \leq 1$ )*

$$B_p(t) = B(t)1(0 \leq t < p) + \frac{1-t}{1-p} \, B(p)1(p \leq t \leq 1) \qquad (4.4.44)$$

and $B : [0,1] \to \mathbb{R}$ according to

$$B(t) = \int_0^t b \, d\lambda, \qquad 0 \le t \le 1. \tag{4.4.45}$$

*Proof:* Define $T_p : C[0,1] \to C[0,1]$ according to ( $\forall f \in C[0,1]$ )

$$T_p f(t) = \begin{cases} f(t), & \text{if } 0 \le t < p, \\ \frac{1-t}{1-p} f(p) + \frac{t-p}{1-p} f(1), & \text{if } p \le t \le 1. \end{cases} \tag{4.4.46}$$

Then $T_p$ is linear and continuous on $C[0,1]$, cf. proof of Theorem 3.3.13. Since $W_0(1) = 0$ and $B(1) = 0$ imply $T_p(W_0 + B) = T_p W_0 + T_p B = W_{0p} + B_p$ and since Theorem 3.1.8 implies

$$\mathcal{L}[\, T_p \hat{B}_{N2} \mid (F_N, G_N)\,] \xrightarrow{\mathcal{L}} \mathcal{L}[\, T_p(W_0 + B)\,] \qquad \text{in } C[0,1], \tag{4.4.47}$$

the proof of (4.4.42) is concluded if we prove the assertion ( as $N \to \infty$ )

$$\|\hat{B}_{N2}^r - T_p \hat{B}_{N2}\|_\infty \longrightarrow 0 \qquad \text{in } (F_N, G_N) - \text{probability}. \tag{4.4.48}$$

Because of $\hat{B}_{N2}(1) = 0$ we get from (4.4.38) and (4.4.46)

$$\hat{B}_{N2}^r(t) - T_p \hat{B}_{N2}(t) = \begin{cases} 0, & \text{if } 0 \le t < p \wedge p_N, \\[2mm] \hat{B}_{N2}(t) - \frac{1-t}{1-p} \hat{B}_{N2}(p), & \text{if } p \le t \le p_N, \\[2mm] \frac{1-t}{1-p_N} \hat{B}_{N2}(p_N) - \hat{B}_{N2}(t), & \text{if } p_N \le t \le p, \\[2mm] \frac{1-t}{1-p_N} \hat{B}_{N2}(p_N) - \frac{1-t}{1-p} \hat{B}_{N2}, & \text{if } p \vee p_N < t \le 1, \end{cases} \tag{4.4.49}$$

and therefore by easy calculation,

$$\|\hat{B}_{N2}^r - T_p \hat{B}_{N2}\|_\infty$$
$$\le \sup_{|s-t| \le |p - p_N|} |\hat{B}_{N2}(s) - \hat{B}_{N2}(t)| + \frac{|p - p_N|}{1-p} \|\hat{B}_{N2}\|_\infty. \tag{4.4.50}$$

According to Theorem 3.1.8 we have $\hat{B}_{N2} \xrightarrow{\mathcal{L}} W_0 + B$ in $C[0,1]$, which implies $\|\hat{B}_{N2}\|_\infty \xrightarrow{\mathcal{L}} \|W_0 + B\|_\infty$ and also the tightness of the sequence ( $\mathcal{L}[\hat{B}_{N2} \mid (F_N, G)]$, $N \ge 2$ ). Therefore inequality (4.4.50), $|p - p_N| \to 0$, and Theorem 8.2 of Billingsley (1968) conclude the proof of (4.4.48). $\square$

Now the definitions of the kernel estimators of $b_p$ are completely similar to the definitions in Chapter 3. In *case of the omnibus alternative* (4.4.3) the rank estimator of the underlying $b_p$ is defined as the convolution $\mathcal{K}_a \hat{b}^r_{N2}$ where $\hat{b}^r_{N2}$ is the right continuous version of the derivative of $\hat{B}^r_{N2}$ , i.e.

$$\hat{b}^r_{N2}(t) = \begin{cases} \hat{b}_{N2}(t), & \text{if } 0 \le t < p_N, \\ -\hat{B}_{N2}(p_N)/(1 - p_N), & \text{if } p_N \le t < 1, \end{cases} \tag{4.4.51}$$

and because of [ cf. (4.4.7), (4.4.51), and (4.4.10) ]

$$S^r_N(b) \; = \; < b, \hat{b}^r_{N2} > \; = \; < b_{p_N}, \hat{b}^r_{N2} > \; = \; S^r_N(b_{p_N}) \tag{4.4.52}$$

the corresponding *omnibus rank statistic* is given by

$$S^r_N(\mathcal{K}_a \hat{b}^r_{N2}) \; = \; < \mathcal{K}_a \hat{b}^r_{N2}, \hat{b}^r_{N2} > \tag{4.4.53}$$

$$= \sum_{i=1}^{N} \sum_{j=1}^{N} \left( \hat{B}^r_{N2}(\frac{i}{N}) - \hat{B}^r_{N2}(\frac{i-1}{N}) \right) \left( \hat{B}^r_{N2}(\frac{j}{N}) - \hat{B}^r_{N2}(\frac{j-1}{N}) \right) \bar{k}_{Na}(i, j),$$

where the weights $\bar{k}_{Na}(i, j)$ are defined in (3.1.24).

In practical applications we use a smooth kernel $K$ and substitute the weights $\bar{k}_{Na}(i, j)$ by the simpler approximations $k_{Na}(i, j) = K_a\big((i-\frac{1}{2})/N, (j-\frac{1}{2})/N\big)$ , i.e. we use the omnibus rank statistic

$$S^r_N(a, K) \tag{4.4.54}$$

$$= \sum_{i=1}^{N} \sum_{j=1}^{N} \left( \hat{B}^r_{N2}(\frac{i}{N}) - \hat{B}^r_{N2}(\frac{i-1}{N}) \right) \left( \hat{B}^r_{N2}(\frac{j}{N}) - \hat{B}^r_{N2}(\frac{j-1}{N}) \right) k_{Na}(i, j).$$

In *case of the one-sided alternative* (4.4.2) we adapt the process $\hat{B}^r_{N2}$ to the side conditon $B_p \le 0$ , i.e. we define

$$\hat{B}^{r0}_{N2} \; := \; \hat{B}^r_{N2} 1(\hat{B}^r_{N2} < 0). \tag{4.4.55}$$

Since $\hat{B}^r_{N2}(t)$, $0 \le t \le 1$ , is absolutely continuous and piecewise linear we may take the derivative

$$\hat{b}^{r0}_{N2} \; = \; \hat{b}^r_{N2} 1(\hat{B}^r_{N2} < 0) \tag{4.4.56}$$

as the primitive estimator of $b_p$ . The corresponding kernel estimator is $\mathcal{K}_a \hat{b}^{r0}_{N2}$ and the corresponding *one-sided rank statistic* is given by

$$S^r_N(\mathcal{K}_a \hat{b}^{r0}_{N2}) \; = \; < \mathcal{K}_a \hat{b}^{r0}_{N2}, \hat{b}^r_{N2} >$$

$$= \int_0^1 \int_0^1 \hat{b}^r_{N2}(s)\, \hat{b}^r_{N2}(t)\, K_a(s,t)\, 1\big(\hat{B}^r_{N2}(s) < 0\big)\, ds\, dt \qquad (4.4.57)$$

$$= \sum_{i=1}^{N} \sum_{j=1}^{N} \Big(\hat{B}^r_{N2}(\tfrac{i}{N}) - \hat{B}^r_{N2}(\tfrac{i-1}{N})\Big)\Big(\hat{B}^r_{N2}(\tfrac{j}{N}) - \hat{B}^r_{N2}(\tfrac{j-1}{N})\Big)\hat{k}^r_{Na}(i,j),$$

where the (random) one-sided weights $\hat{k}^r_{Na}(i,j)$ are defined by

$$\hat{k}^r_{Na}(i,j) = N^2 \int_{(i-1)/N}^{i/N} \int_{(j-1)/N}^{j/N} K_a(s,t)\, 1\big(\hat{B}^r_{N2}(s) < 0\big)\, ds\, dt. \qquad (4.4.58)$$

Thus, also in the one-sided censored case the rank statistic $S^r_N(\mathcal{K}_a \hat{b}^{r0}_{N2})$ is derived from the (uncensored) one-sided rank statistic (3.1.26), (3.1.27) by substituting $\hat{B}_{N2}$ by $\hat{B}^r_{N2}$ .

In practical applications we use a smooth kernel $K$ , e.g. the Parzen-2 kernel, and substitute the weights $\hat{k}^r_{Na}(i,j)$ by the approximations

$$k_{Na}(i,j)N \int_{(i-1)/N}^{i/N} 1\big(\hat{B}^r_{N2}(s) < 0\big)\, ds. \qquad (4.4.59)$$

Since on each interval $((i-1)/N, i/N)$ the derivative $\hat{b}^r_{N2}$ equals the constant $N\big(\hat{B}^r_{N2}(i/N) - \hat{B}^r_{N2}((i-1)/N)\big)$ we get from (4.4.56) the resulting *(approximate) one-sided rank statistic*

$$S^{r0}_N(a, K) \qquad (4.4.60)$$

$$= \sum_{i=1}^{N} \sum_{j=1}^{N} \Big(\hat{B}^{r0}_{N2}(\tfrac{i}{N}) - \hat{B}^{r0}_{N2}(\tfrac{i-1}{N})\Big)\Big(\hat{B}^r_{N2}(\tfrac{j}{N}) - \hat{B}^r_{N2}(\tfrac{j-1}{N})\Big)k_{Na}(i,j)$$

for testing the null hypothesis $\mathcal{H}^r_0 : F = G$ versus the one-sided alternative $\mathcal{A}^0_2 : F \leq G,\ F \neq G$ under Assumption 4.4.1, cf. formula (3.1.30) for the uncensored situation.

Substituting $\hat{B}_{N2}$ by $\hat{B}^r_{N2}$ and substituting Theorem 3.1.8 by Theorem 4.4.3 the proof of Theorem 3.1.9 yields the following limiting laws for the omnibus statistic (4.4.53) and for the one-sided statistic (4.4.57).

### 4.4.4 Theorem

*Assume $b \in L^0_2(0,1)$ and (4.4.8). Take any sequence $[\ (F_N, G_N),\ N \geq 2\ ]$ of corresponding local alternatives (4.4.41). Let $0 < a \leq 1$ be a given bandwidth for the convolution kernel (3.1.19) with underlying kernel $K$ according* *to (3.1.16)–(3.1.18) and (3.1.31), and let $\mathcal{K}_a$ denote the corresponding convolution operator defined in (3.1.33).*

*a) If the eigenvalues (3.1.38) of $\mathcal{K}_a$ fulfil the condition*

$$\sum_{\varrho=1}^{\infty} |\lambda_{\varrho}(a)| < \infty, \tag{4.4.61}$$

*then the following limiting law $(N \to \infty)$ holds true,*

$$\mathcal{L}[\, S_N^r(\mathcal{K}_a \hat{b}_{N2}^r) \,|\, (F_N, G_N)\,] \xrightarrow{\mathcal{L}} \mathcal{L}[\sum_{\varrho=1}^{\infty} \lambda_{\varrho}(a) < \psi_{\varrho}', W_{0p} + B_p >^2 \,], \tag{4.4.62}$$

*with $W_{0p}$ and $B_p$ as in Theorem 4.4.3.*

*b) If the eigenvalues (3.1.38) of $\mathcal{K}_a$ fulfil the stronger condition*

$$\sum_{\varrho=1}^{\infty} \varrho\, |\lambda_{\varrho}(a)| < \infty, \tag{4.4.63}$$

*then the following limiting law $(N \to \infty)$ holds true,*

$$\mathcal{L}[\, S_N^r(\mathcal{K}_a \hat{b}_{N2}^{r0}) \,|\, (F_N, G_N)\,] \tag{4.4.64}$$

$$\xrightarrow{\mathcal{L}} \mathcal{L}[\sum_{\varrho=1}^{\infty} \lambda_{\varrho}(a) < \psi_{\varrho}', W_{0p} + B_p > < \psi_{\varrho}', (W_{0p} + B_p)^0 > \,],$$

*where*

$$(W_{0p} + B_p)^0 = (W_{0p} + B_p)1(W_{0p} + B_p < 0). \tag{4.4.65}$$

*Proof:* Because of $\hat{B}_{N2}^r(0) = \hat{B}_{N2}^r(1) = 0$ we get from (4.4.53) and (3.1.39)

$$S_N^r(\mathcal{K}_a \hat{b}_{N2}^r) = \sum_{\varrho=1}^{\infty} \lambda_{\varrho}(a) < \psi_{\varrho}, \hat{b}_{N2}^r >^2$$

$$= \sum_{\varrho=1}^{\infty} \lambda_{\varrho}(a) < \psi_{\varrho}', \hat{B}_{N2}^r >^2. \tag{4.4.66}$$

Similarly, from $\hat{B}_{N2}^{r0}(0) = \hat{B}_{N2}^{r0}(1) = 0$ , (4.4.57), and (3.1.39) we get

$$S_N^r(\mathcal{K}_a \hat{b}_{N2}^{r0}) = \sum_{\varrho=1}^{\infty} \lambda_{\varrho}(a) < \psi_{\varrho}, \hat{b}_{N2}^r > < \psi_{\varrho}, \hat{b}_{N2}^{r0} > \tag{4.4.67}$$

$$= \sum_{\varrho=1}^{\infty} \lambda_{\varrho}(a) < \psi_{\varrho}', \hat{B}_{N2}^r > < \psi_{\varrho}', \hat{B}_{N2}^r 1(\hat{B}_{N2}^r < 0) >.$$

Using Theorem 4.4.3 instead of Theorem 3.1.8 the proof is immediate from the proof of Theorem 3.1.9. $\square$

### 4.4.5 Remark

*If we substitute $S_N^r(\mathcal{K}_a\hat{b}_{N2}^r)$ in (4.4.62) by $S_N^r(a, K)$ and if we substitute $S_N^r(\mathcal{K}_a\hat{b}_{N2}^{r0})$ in (4.4.64) by $S_N^{r0}(a, K)$ then the assertions of Theorem 4.4.4 remain true. The proof is an adaptation of the proof of Corollary 3.1.10 to the present situation.*

### 4.4.6 Treatment of ties

The above theorems have been proved under the assumption that the uncensored observations have no ties, i.e. the observable ranks form a permutation of $(1, ..., r)$ . Asymptotically $(N \to \infty,\ m/N \to \eta,\ r/N \to p)$ this corresponds to the assumption of underlying *continuous* distribution functions at least on the interval $\left(-\infty, H^{-1}(p)\right)$ , where $H = \eta F + (1 - \eta)G$ .

Additionally, in the continuous case the censoring of $X_N^{(r+1)}, ..., X_N^{(N)}$ is shown to be equivalent to $X_N^{(1)} < \cdots < X_N^{(r)}$ and

$$X_N^{(r+1)} = \cdots = X_N^{(N)} > X_N^{(r)}, \tag{4.4.68}$$

cf. Theorem 4.4.3 and Theorem 3.3.13, and the assumption (4.4.68) is shown to be asymptotically equivalent to

$$(B_N, H_N) \in \mathcal{B}_N^0(\mathcal{J}_p) \times \mathcal{F}_1(\mathcal{J}_p). \tag{4.4.69}$$

Therefore we may consider the treatment of ties under type II censoring as a special case of Section 3.3, i.e. the ties in the uncensored observations are treated as before and the (unobservable) censored observations are treated according to the additional tie (4.4.68). Using the assumption (4.4.68) the corresponding (tied and censored) ranks $R_i^r$ may be represented in the form

$$R_i^r = \sum_{j=1}^{N} 1(X_j \leq X_i) = \sum_{j=1}^{r} 1(X_N^{(j)} \leq X_i), \quad \text{if } X_i \text{ is uncensored,}$$

$$R_i^r = \sum_{j=1}^{N} 1(X_j \leq X_i) = N, \quad \text{if } X_i \text{ is censored.} \tag{4.4.70}$$

According to (3.3.41), (3.3.42), and Proposition 3.3.12 the corresponding *(r-censored) averaged two-sample rank process* $\hat{B}_{N2}^r$ is given by

$$\hat{B}_{N2}^r(\frac{T_i}{N}) = \sqrt{\frac{mn}{N}} \left(\frac{1}{m}\sum_{j=1}^{m} 1(R_j^r \leq T_i) - \frac{1}{n}\sum_{j=m+1}^{N} 1(R_j \leq T_i)\right), \quad 0 \leq i \leq d,$$

$$\hat{B}^r_{N2} \quad \text{linear on} \quad [\frac{T_{i-1}}{N}, \frac{T_i}{N}] \qquad \forall\, i = 1, ..., d, \tag{4.4.71}$$

where

$$T_0 = 0, \quad T_i = T_{i-1} + \tau_i, \ \ i = 1, ..., d-1, \quad T_d = N, \tag{4.4.72}$$

and where $d-1$ and $\tau_1, ..., \tau_{d-1}$ are the (random) number of distinct values and the lengths of ties in the observable ordered pooled sample $X_N^{(1)}, ..., X_N^{(r)}$, respectively.

The asymptotics of the resulting *averaged scores rank statistics with estimated scores*, cf. (3.3.88), (3.3.89), (3.3.92), and (3.3.97), is based on the following adaptation of Theorem 3.3.13. Details are omitted.

### 4.4.7 Theorem

*Assume (4.4.8) and $\mathcal{J} \in \mathcal{J}^*_{0,1}$ such that*

$$\forall\, J \in \mathcal{J} \quad \text{either} \quad J \cap (p, 1] = \emptyset \quad \text{or} \quad J \cap (p, 1] = J. \tag{4.4.73}$$

*Take any sequence $[\,(F_N, G_N),\ N \le 2\,]$ of local alternatives according to*

$$(F_N,\ G_N) = (H_N + c_{N1} B_N \circ H_N,\ H_N + c_{NN} B_N \circ H_N),\ \ H_N \in \mathcal{F}_1(\mathcal{J}),$$

$$B_N \in \mathcal{B}^0_N(\mathcal{J}), \quad \|B'_N - b\| \stackrel{N \to \infty}{\longrightarrow} 0, \quad b \in L^0_2(0, 1). \tag{4.4.74}$$

*Then we have the limiting law (in $C[0, 1]$ )*

$$\mathcal{L}[\,\hat{B}^r_{N2} \mid (F_N, G_N)\,] \stackrel{\mathcal{L}}{\longrightarrow} \mathcal{L}[\,T_{\mathcal{J}} W_{0p} + B_p\,], \tag{4.4.75}$$

*where $W_{0p}$ and $B_p$ are defined in Theorem 4.4.3 and where the mapping $T_{\mathcal{J}} : C[0, 1] \to C[0, 1]$ is defined in (3.3.106).*

# Chapter 5

# The hypothesis of symmetry

As in Chapter 3 the basic problem is the comparison of a new treatment with a standard, but in contrast to Chapter 3 we don't have two independent samples. Usually we have paired observations $(Y_i, Z_i)$, $i = 1, ..., n$, where the $Y$-components are the measurements under the standard treatment whereas the $Z$-components are the measurements under the new treatment. It's assumed that the design of the experiment allows the pairs $(Y_1, Z_1), ..., (Y_n, Z_n)$ to be independent random variables with values in $\mathbb{R}^2$. It's not realistic to assume the stochastic independence of the components $Y_i$ and $Z_i$, since the $Y_i$-measurement and the $Z_i$-measurement are usually taken at matched pairs or even at the same experimental unit, e.g. left-right treatments or pre-post treatments, but we assume the differences $Z_1 - Y_1, ..., Z_n - Y_n$ to be i.i.d. random variables.

In this setting the *equality of the two treatments* may be modelled by the assumption

$$\mathcal{L}(Y_i, Z_i) = \mathcal{L}(Z_i, Y_i), \qquad 1 \leq i \leq n, \tag{5.0.1}$$

which especially implies the distribution of $X_i := Z_i - Y_i$ to be *symmetric about zero*. If the new treatment is better (produces larger measurements) than the standard treatment, the distribution of $X_i$ should be stochastically larger than the distribution of $-X_i$. Therefore the comparison is based upon the i.i.d. real random variables $X_1, ..., X_n$.

## 5.1   Linear rank tests

According to the above discussion we assume $X_1, ..., X_n$ to be i.i.d. real random variables with underlying *continuous* distribution function $F$ on $\mathbb{R}$, and the null hypothesis is assumed to be the *hypothesis of symmetry*

$$\mathcal{H}_0^s = \{ \, F \in \mathcal{F}_1^c : \ F = F_- \, \}, \tag{5.1.1}$$

where $F_-$ denotes the distribution function of $-X_i$, i.e.

$$F_-(x) = P\{-X_i \leq x\} = 1 - P\{X_i < -x\} = 1 - F(-x), \quad x \in \mathbb{R}, \tag{5.1.2}$$

since $F$ is assumed to be continuous.

Let $R^+ = (R_1^+, ..., R_n^+)$ denote the vector of ranks of the absolute values $|X_1|, ..., |X_n|$, i.e.

$$R_i^+ = \sum_{j=1}^{n} 1(|X_j| \leq |X_i|), \qquad 1 \leq i \leq n, \tag{5.1.3}$$

and let $\text{sign}(X_i)$ denote the sign of $X_i$ according to the definition

$$\text{sign}(x) = \begin{cases} -1, & \text{if } x < 0, \\ \phantom{-}0, & \text{if } x = 0, \\ \phantom{-}1, & \text{if } x > 0, \end{cases} \tag{5.1.4}$$

then it's well-known that under the null hypothesis $\mathcal{H}_0^s$ the random variables $R^+$, $\text{sign}(X_1), ..., \text{sign}(X_n)$ are stochastically independent, $R^+$ is uniformly distributed on the $n!$ permutations of $(1, ..., n)$, and $(\forall\, i = 1, ..., n)$

$$P[\,\text{sign}(X_i) = -1 \mid \mathcal{H}_0^s\,] \ = \ P[\,\text{sign}(X_i) = 1 \mid \mathcal{H}_0^s\,] \ = \ \frac{1}{2}. \tag{5.1.5}$$

Consequently, each test based on $\text{sign}(X_1), ..., \text{sign}(X_n)$ and $R^+$ is distribution – free under the null hypothesis of symmetry.

Assuming $F_0 \in \mathcal{H}_0^s$ and defining the corresponding *shift alternative* $\mathcal{H}_1(F_0)$ according to

$$\mathcal{H}_1(F_0) = \{ \, F_{0,\vartheta} \in \mathcal{F}_1^c : \ \vartheta > 0, \ F_{0,\vartheta}(x) = F_0(x - \vartheta) \ \ \forall\, x \in \mathbb{R} \, \} \tag{5.1.6}$$

it's shown in Hájek and Šidák(1967) that the upper test of asymptotic level $\alpha \in (0,1)$ based on the *signed linear rank statistic*

$$S_N^+(f_0) = \frac{1}{\sqrt{n}} \sum_{i=1}^{n} a_n(R_i^+) \, \text{sign}(X_i) \tag{5.1.7}$$

is asymptotically uniformly most powerful in the class of all asymptotic level $\alpha$ tests for testing $F_0$ versus any local sequence ( $F_{0,\varrho/\sqrt{n}}$, $n \geq 1$ ) from $\mathcal{H}_1(F_0)$ , if the scores fulfil the condition

$$\int_0^1 \left( a_n(1 + [nx]) - \varphi(\tfrac{1}{2} + \tfrac{1}{2}x, f_0) \right)^2 d\lambda(x) \overset{n \to \infty}{\longrightarrow} 0, \qquad (5.1.8)$$

where $\varphi(\cdot, f_0)$ is defined in (1.1.8).

If we could really trust in shift models, then it might be a good idea to estimate the underlying optimal scores in case of unknown $F_0$ . But completely similar to Section 1.2 we may define *generalized shift alternatives* $\mathcal{H}_1(F_0, D)$ according to

$$\mathcal{H}_1(F_0, D)$$
$$= \left\{ F_{0,D,\vartheta} \in \mathcal{F}_1^c : \vartheta > 0, \ F_{0,D,\vartheta}(x) = F_0(x - \vartheta D(x)) \ \forall \, x \in \mathbb{R} \right\}, \qquad (5.1.9)$$

where $D : \mathbb{R} \to [0, \infty)$ is a bounded shift function with bounded derivate $d : \mathbb{R} \to \mathbb{R}$ . In this case the asymptotically optimal score function has the form

$$\varphi(u, f_0, D^*) = \varphi(u, f_0) D^*(F_0^{-1}(u)) - d^*(F_0^{-1}(u)), \quad 0 < u < 1, \qquad (5.1.10)$$

where

$$D^*(x) = \frac{D(x) + D(-x)}{2}, \qquad d^*(x) = \frac{d(x) - d(-x)}{2}, \qquad x \in \mathbb{R}, \qquad (5.1.11)$$

cf. formula (1.2.4). This means: if the scores $a_n(i)$ fulfil the condition

$$\int_0^1 \left( a_n(1 + [nx]) - \varphi(\tfrac{1}{2} + \tfrac{1}{2}x, f_0, D^*) \right)^2 d\lambda(x) \overset{n \to \infty}{\longrightarrow} 0 \qquad (5.1.12)$$

then the upper level $\alpha$ test based on the corresponding signed linear rank statistic (5.1.7) can be shown to be asymptotically equivalent to the sequence of (optimal) Neyman-Pearson tests at level $\alpha$ for testing the simple null hypothesis $H_{0,D,\varrho/\sqrt{n}} \in \mathcal{H}_0^s$ versus the (local) alternative $F_{0,D,\varrho/\sqrt{n}} \in \mathcal{H}_1(F_0, D)$ , where

$$H_{0,D,\varrho/\sqrt{n}}(x) = \frac{1}{2} F_0\big(x - \frac{\varrho}{\sqrt{n}} D(x)\big) + \frac{1}{2} F_0\big(x + \frac{\varrho}{\sqrt{n}} D(-x)\big) \qquad (5.1.13)$$

and $\varrho > 0$ .

As in Section 1.2 the shape of the optimal score function $\varphi(\cdot, f_0, D^*)$ is substantially influenced by the density $f_0$ *and* the shift function $D$ . Since there is no dependence between $f_0$ and $D$, the estimation of the classical shift score function $\varphi(\cdot, f_0)$ may lead to nonsense if the strict shift model isn't true. Therefore we'll consider a general nonparametric model, identify the corresponding optimal score function, and construct suitable rank estimators of the unknown underlying score function, cf. the two-sample case in Section 3.1. The generalized shift alternatives (5.1.9) are utilized in Section 5.3 in order to construct representative score functions for the projection estimators, cf. Section 3.2.

In the sequel let's assume the more general *alternative of positive unsymmetry*

$$\mathcal{A}_1^0 = \{\, F \in \mathcal{F}_1^c : \ F \leq F_-, \ F \neq F_- \,\}, \tag{5.1.14}$$

which means that under $F \in \mathcal{A}_1^0$ the distribution of $X_i$ is stochastically larger than the distribution of $-X_i$. Obviously $F \leq F_-$ is equivalent to the inequality $F \leq H$, where

$$H := \frac{1}{2}F + \frac{1}{2}F_- \ \in \mathcal{H}_0^s. \tag{5.1.15}$$

In analogy to Section 1.3 we'll reparametrize $\mathcal{F}_1^c$ in order to identify the optimal score functions for testing $\mathcal{H}_0^s$ versus $\mathcal{A}_1^0$.

For arbitrary $F \in \mathcal{F}_1^c$ define $H$ as in (5.1.15) and $B_n : [0,1] \to \mathbb{R}$ according to

$$B_n = \sqrt{n}\,(F \circ H^{-1} - I), \tag{5.1.16}$$

where $I$ denotes the identity on $[0,1]$, and where $(F \circ H^{-1})(0) = 0$, and $(F \circ H^{-1})(1) = 1$. Putting $G = F_-$, $m = n$, and $N = 2n$, the results of Section 1.3 imply $B_n$ to be absolutely continuous on $[0,1]$ with derivative

$$-\sqrt{n} \leq B_n' \leq \sqrt{n} \qquad [\lambda - a.e.] \tag{5.1.17}$$

and

$$F = H + \frac{1}{\sqrt{n}}\,B_n \circ H, \qquad F_- = H - \frac{1}{\sqrt{n}}\,B_n \circ H. \tag{5.1.18}$$

Because of (5.1.2) and $H \in \mathcal{H}_0^s$ we get in addition ($\forall\, x \in \mathbb{R}$)

$$\frac{2}{\sqrt{n}}B_n(H(x)) = F(x) - F_-(x) = F(-x) - F_-(-x)$$

$$= \frac{2}{\sqrt{n}}\,B_n(H(-x)) = \frac{2}{\sqrt{n}}\,B_n(1 - H(x))$$

which implies

$$B_n(t) = B_n(1-t) \qquad \forall\, 0 \leq t \leq 1. \tag{5.1.19}$$

Thus, defining

$$\mathcal{B}_n^s := \left\{ B : \begin{array}{l} B \text{ is an absolutely continuous function on } [0,1], \\ B(0) = B(1) = 0, \ B(t) = B(1-t) \ \forall \ 0 \le t \le 1, \\ -\sqrt{n} \le B' \le \sqrt{n} \quad [\lambda - a.e.] \end{array} \right\} \qquad (5.1.20)$$

we obtain the following parametrization of $\mathcal{F}_1^c$ ,

$$\mathcal{F}_1^c = \{ \ H + \frac{1}{\sqrt{n}} \ B \circ H : \ (B,H) \in \mathcal{B}_n^s \times \mathcal{H}_0^s \ \}, \qquad (5.1.21)$$

where ( $\forall \ x \in \mathbb{R}$)

$$(H + \frac{1}{\sqrt{n}} \ B \circ H)_-(x) = 1 - H(-x) - \frac{1}{\sqrt{n}} \ B(H(-x))$$

$$= H(x) - \frac{1}{\sqrt{n}} \ B(1 - H(x)) = (H - \frac{1}{\sqrt{n}} \ B \circ H)(x) \qquad (5.1.22)$$

and therefore

$$\frac{1}{2}\Big(H + \frac{1}{\sqrt{n}} \ B \circ H\Big) + \frac{1}{2}\Big(H + \frac{1}{\sqrt{n}} \ B \circ H\Big)_- = H. \qquad (5.1.23)$$

Obviously $F = F_-$ is equivalent to $B = 0$ and $F \le F_-$ is equivalent to $B \le 0$ . Therefore the null hypothesis $\mathcal{H}_0^s$ and the alternative $\mathcal{A}_1^0$ may be written in the form

$$\mathcal{H}_0^s = \{ \ H + \frac{1}{\sqrt{n}} \ B \circ H : \ (B,H) \in \mathcal{B}_n^s \times \mathcal{H}_0^s, \ B = 0 \ \}, \qquad (5.1.24)$$

$$\mathcal{A}_1^0 = \{ \ H + \frac{1}{\sqrt{n}} \ B \circ H : \ (B,H) \in \mathcal{B}_n^s \times \mathcal{H}_0^s, \ B \le 0, \ B \ne 0 \ \}, \qquad (5.1.25)$$

which reveals $H \in \mathcal{H}_0^s$ to be a nuisance parameter while $B \in \mathcal{B}_n^s$ is the parameter under test.

For the sake of completeness we include the discussion of the *omnibus alternative*

$$\mathcal{A}_1 = \{ \ F \in \mathcal{F}_1^c : \ F \ne F_- \ \}$$

$$= \{ \ H + \frac{1}{\sqrt{n}} \ B \circ H : \ (B,H) \in \mathcal{B}_n^s \times \mathcal{H}_0^s, \ B \ne 0 \ \}. \qquad (5.1.26)$$

Given any $b \in L_2^0(0,1)$ with the properties $b(t) = -b(1-t) \ \forall \ 0 < t < 1$ and $\|b\| > 0$ we define the corresponding *signed linear rank statistic* $S_n(b)$ according to

$$S_n(b) = \frac{1}{\sqrt{n}} \sum_{i=1}^{n} b_n(R_i^+) \ \mathrm{sign}(X_i), \qquad (5.1.27)$$

where

$$b_n(i) := n \int_{(i-1)/n}^{i/n} b(\frac{1}{2} + \frac{t}{2})\, dt, \qquad 1 \le i \le n, \qquad (5.1.28)$$

and the corresponding *signed linear rank test* $\psi_n(b)$ according to

$$\psi_n(b) = 1(\, S_n(b) \ge u_\alpha \|b\| \,), \qquad (5.1.29)$$

where $0 < \alpha < 1$ and $u_\alpha = \Phi^{-1}(1 - \alpha)$ is the upper $\alpha$-quantile of the standard normal distribution.

Taking any sequence $(F_n, n \ge 1)$ of local alternatives corresponding to $b$ and $0 < \varrho \le 1$, i.e.

$$F_n = H_n + \frac{\varrho}{\sqrt{n}} B_n \circ H_n, \quad B_n \in \mathcal{B}_n^s, \quad H_n \in \mathcal{H}_0^s, \quad \|B_n' - b\| \overset{n \to \infty}{\longrightarrow} 0, \quad (5.1.30)$$

then it's proved in Behnen (1972) that the signed linear rank test $\psi_n(b)$ is asymptotically equivalent to the (optimal) Neyman- Pearson test for testing $H_n$ versus $F_n$ at asymptotic level $\alpha$. Especially it's proved

$$\lim_{n \to \infty} E[\, \psi_n(b) \mid \mathcal{H}_0^s \,] = \alpha, \qquad (5.1.31)$$

and

$$\lim_{n \to \infty} E[\, \psi_n(b) \mid F_n \,] = 1 - \Phi(u_\alpha - \varrho\|b\|). \qquad (5.1.32)$$

In practice the alternative may be any $F \in \mathcal{A}_1$, i.e. the corresponding optimal score function $B_n' = \sqrt{n}\,(F \circ H^{-1} - I)'$ is unknown. Therefore we'll construct a suitable estimator of $B_n'$ and plug it into the optimal signed linear rank statistic $S_n(B_n')$.

If $k : \mathbb{R} \to \mathbb{R}$ is a bijective and strictly increasing function with $k(x) = -k(-x)$ $\forall\, x \in \mathbb{R}$, then the optimal score function corresponding to the transformed observations $k(X_i)$, $1 \le i \le n$, is the same as for the original observations $X_i$, $1 \le i \le n$. Since the statistic $(R^+, \operatorname{sign}(X_1), ..., \operatorname{sign}(X_n))$ is maximal invariant under the group of such transformations $k$, we'll try to estimate $B_n'$ on the basis of the signed ranks $(R^+, \operatorname{sign}(X_1), ..., \operatorname{sign}(X_n))$.

## 5.2 Kernel estimators of the score function

According formulae to (5.1.18), (1.3.10), and (1.3.11) the transformed random variables $H(X_1)$, ..., $H(X_n)$ are i.i.d. with the absolutely continuous distribution function $F \circ H^{-1} = I + n^{-1/2}B_n$ and the transformed random variables $H(-X_1), ..., H(-X_n)$ are i.i.d. with the absolutely continuous distribution function $F_- \circ H^{-1} = I - n^{-1/2}B_n$ . Using these (unobservable) 'observations' the standard estimator of the integrated score function

$$\tilde{B}_n(t) := 2\, B_n(\frac{1}{2} + \frac{t}{2}), \qquad 0 \le t \le 1, \tag{5.2.1}$$

would be a smoothed version of the 'empirical' process $(0 \le t \le 1)$

$$\sqrt{n}\, \Big(\frac{1}{n}\sum_{i=1}^{n} 1\big(H(X_i) \le \frac{1}{2} + \frac{t}{2}\big) - \frac{1}{n}\sum_{i=1}^{n} 1\big(H(-X_i) \le \frac{1}{2} + \frac{t}{2}\big)\Big). \tag{5.2.2}$$

Unfortunately the $H(X_i)$ 's and the $H(-X_i)$ 's are unobservable in real applications since $F$ and $H = (F + F_-)/2$ are unknown parameters. But fortunately we can rewrite (5.2.2) according to

$$1\big(H(X_i) \le 1/2 + t/2\big) - 1\big(H(-X_i) \le 1/2 + t/2\big)$$

$$= 1\big(2H(-X_i) - 1 > t\big) - 1\big(2H(X_i) - 1 > t\big)$$

$$= 1\big(2H(|X_i|) - 1 > t\big)1(X_i < 0) - 1\big(2H(|H_i|) - 1 > t\big)1(X_i > 0)$$

$$= -1\big(2H(|X_i|) - 1 > t\big)\, \operatorname{sign}(X_i),$$

and since $2H(y) - 1$ , $y \ge 0$ , is the distribution function of the $|X_j|$ 's we may substitute $2H(|X_i|) - 1$ by the empirical distribution function of the random variables $|X_1|, ..., |X_n|$ at the point $|X_i|$ , which is nothing but $R_i^+/n$ . Therefore the resulting *signed rank process*

$$\hat{B}_{n1}(t) := -\frac{1}{\sqrt{n}}\, \sum_{i=1}^{n} 1\Big(\frac{R_i^+}{n} > t\Big)\, \operatorname{sign}(X_i), \qquad 0 \le t \le 1, \tag{5.2.3}$$

should be a suitable estimator of $\tilde{B}_n$ .

For technical reasons it's more convenient to start with the piecewise linearized version $\hat{B}_{n2}$ of $\hat{B}_{n1}$ given by the following linear interpolation of $\hat{B}_{n1}$ ,

$$\hat{B}_{n2}(\frac{i}{n}) := \hat{B}_{n1}(\frac{i}{n}) = -\frac{1}{\sqrt{n}}\, \sum_{j=1}^{n} 1(R_j^+ > i)\, \operatorname{sign}(X_j), \quad 0 \le i \le n,$$

$$\hat{B}_{n2} \quad \text{linear on } \big[\frac{i-1}{n}, \frac{i}{n}\big] \text{ for any } 1 \le i \le n. \tag{5.2.4}$$

The following limiting law under local alternatives of the type (5.1.30) shows that the signed rank process $\hat{B}_{n2}$ is indeed a suitable estimator of the integrated limiting score function.


### 5.2.1 Theorem

*Assume $b \in L_2^0(0,1)$ with the property $b(t) = -b(1-t) \ \forall \ 0 < t < 1$, and assume $0 \leq \varrho \leq 1$. Define a corresponding sequence $(F_n, n \geq 1)$ of local alternatives $F_n$ according to formula (5.1.30), i.e. $F_n = H_n + (\varrho/\sqrt{n})B_n \circ H_n$ Let $W$ denote the standard Brownian-motion on $C[0,1]$, and define $W^*$ according to*

$$W^*(t) = W(1-t), \qquad 0 \leq t \leq 1. \tag{5.2.5}$$

*Then we have the following limiting law $(n \to \infty)$ in $C[0,1]$,*

$$\mathcal{L}[\hat{B}_{n2} \mid F_n] \xrightarrow{\ \mathcal{L}\ } \mathcal{L}[W^* + \varrho\tilde{B}], \tag{5.2.6}$$

*where $\tilde{B}(t) = 2B(1/2 + t/2)$, $B(t) = \int_0^t b\, d\lambda \ \ \forall\, 0 \leq t \leq 1$.*


The *proof* of Theorem 5.2.1 is given in Appendix 7.8.


Since the process $\hat{B}_{n2}$ is an absolutely continuous estimator of $\tilde{B}$ and since the scores $b_n(i)$ are defined by the derivative $\tilde{b}(t) = b(1/2 + t/2)$ of $\tilde{B}(t) = 2B(1/2 + t/2)$, cf. formula (5.1.28), we may use the right continuous derivative $\hat{b}_{n2}$ of $\hat{B}_{n2}$ as a primitive estimator of $\tilde{b}$, i.e. for $i = 1, ..., n$ we have

$$\hat{b}_{n2}(t) = n\left(\hat{B}_{n2}(\frac{i}{n}) - \hat{B}_{n2}(\frac{i-1}{n})\right) \qquad \text{if} \quad \frac{i-1}{n} \leq t < \frac{i}{n},$$

$$= \sqrt{n} \sum_{j=1}^{n} 1(R_j^+ = [nt] + 1)\, \text{sign}(X_j). \tag{5.2.7}$$

This primitive estimator will be modified and smoothed in order to get better estimators of the underlying scores $b_n(i)$. In case of the *one-sided* alternative (5.1.25) of positive unsymmetry $\mathcal{A}_1^0$ the function $\tilde{B}$ has the property $\tilde{B} \leq 0$. In this case we start with the "projection"

$$\hat{B}_{n2}^0 := \hat{B}_{n2}\, 1(\hat{B}_{n2} < 0) \tag{5.2.8}$$

of $\hat{B}_{n2}$ onto the cone of nonpositive functions on $[0,1]$ and use it's special derivative

$$\hat{b}_{n2}^0 = \hat{b}_{n2}\, 1(\hat{B}_{n2} < 0) \tag{5.2.9}$$

as the corresponding primitive estimator of $\tilde{b}$.

Using the derivative (5.2.7) of $\hat{B}_{n2}$ the optimal signed linear rank statistic $S_n(b)$ defined in (5.1.27) and (5.1.28) has the following simple form,

$$S_n(b) = \frac{1}{\sqrt{n}} \sum_{i=1}^{n} \text{sign}(X_i) \, n \int_0^1 \tilde{b}(t) \, 1(R_i^+ - 1 \le nt < R_i^+) \, dt$$

$$= \int_0^1 \tilde{b}(t) \, \sqrt{n} \, \sum_{i=1}^{n} 1(R_i = [nt] + 1) \, \text{sign}(X_i) \, dt$$

$$= \; <\tilde{b}, \hat{b}_{n2}> \; . \tag{5.2.10}$$

Since $\tilde{b}$ is an unknown parameter we'll construct suitable kernel estimators of $\tilde{b}$ and plug these estimators into the representation (5.2.10). The resulting nonlinear rank statistics should have good power properties for testing $\mathcal{H}_0^s$ versus the respective one-sided and two-sided alternatives.

Since $\hat{B}_{n2}$ and $\hat{B}_{n2}^0$ are the respective estimators of $\tilde{B}$ corresponding to the omnibus testing problem and to the one-sided testing problem, we may proceed completely similar to Section 3.1:

Let $K : \mathbb{R} \to \mathbb{R}$ be a given kernel with the properties (3.1.16) to (3.1.18) and (3.1.31). For the fixed bandwidth $0 < a \le 1$ define the convolution kernel $K_a : [0,1]^2 \to \mathbb{R}$ according to (3.1.19) and the corresponding linear operator $\mathcal{K}_a : L_2(0,1) \to L_2(0,1)$ according to (3.1.33). Then

$$\mathcal{K}_a \hat{b}_{n2}(t) = \int_0^1 K_a(s,t) \, \hat{b}_{n2}(s) \, ds, \qquad 0 \le t \le 1, \tag{5.2.11}$$

will be used as the estimator of $\tilde{b}$ in case of testing the null hypothesis $\mathcal{H}_0^s$ versus the omnibus alternative $\mathcal{A}_1$ , and

$$\mathcal{K}_a \hat{b}_{n2}^0(t) = \int_0^1 K_a(s,t) \, \hat{b}_{n2}^0(s) \, ds, \qquad 0 \le t \le 1, \tag{5.2.12}$$

will be used as the estimator of $\tilde{b}$ in case of testing $\mathcal{H}_0^s$ versus the one-sided alternative $\mathcal{A}_1^0$ .

Substitution of $\tilde{b}$ by $\mathcal{K}_a \hat{b}_{n2}$ in the optimal signed linear rank statistic (5.2.10) yields the (nonlinear) signed rank statistic $\bar{S}_n(a, K)$ for testing the null hypothesis of symmetry $\mathcal{H}_0^s$ versus the omnibus alternative $\mathcal{A}_1$ , i.e.

$$\bar{S}_n(a, K) \; = \; <\mathcal{K}_a \hat{b}_{n2}, \hat{b}_{n2}>$$

$$= \int_0^1 \left( \int_0^1 K_a(s,t) \, \hat{b}_{n2}(s) \, ds \right) \hat{b}_{N2}(t) \, dt \tag{5.2.13}$$

$$= \sum_{i=1}^{n}\sum_{j=1}^{n}\left(\hat{B}_{n2}(\tfrac{i}{n}) - \hat{B}_{n2}(\tfrac{i-1}{n})\right)\left(\hat{B}_{n2}(\tfrac{j}{n}) - \hat{B}_{n2}(\tfrac{j-1}{n})\right)\bar{k}_{na}(i,j),$$

where the last equality is a consequence of (5.2.7) and where the weights $\bar{k}_{na}(i,j)$ are defined in (3.1.24). As in Section 3.1 we'll use instead of (5.2.13) the approximation

$$S_n(a,K) \tag{5.2.14}$$

$$= \sum_{i=1}^{n}\sum_{j=1}^{n}\left(\hat{B}_{n2}(\tfrac{i}{n}) - \hat{B}_{n2}(\tfrac{i-1}{n})\right)\left(\hat{B}_{n2}(\tfrac{j}{n}) - \hat{B}_{n2}(\tfrac{j-1}{n})\right)k_{na}(i,j),$$

if the kernel $K$ is smooth, e.g. the Parzen-2 kernel. The weights $k_{na}(i,j)$ are given by

$$k_{na}(i,j) = K_a\left(\frac{i-1/2}{n}, \frac{j-1/2}{n}\right), \qquad 1 \le i,j \le n. \tag{5.2.15}$$

Similarly the substitution of $\tilde{b}$ in the optimal statistic (5.2.10) by the one-sided estimator $\mathcal{K}_a\hat{b}_{n2}^0$ yields the (nonlinear) signed rank statistic $\hat{S}_n^0(a,K)$ for testing the null hypothesis of symmetry $\mathcal{H}_0^s$ versus the one-sided alternative of positive unsymmetry $\mathcal{A}_1^0$, i.e.

$$\hat{S}_n^0(a,K) = \; <\mathcal{K}_a\hat{b}_{n2}^0, \hat{b}_{n2}>$$

$$= \int_0^1\left(\int_0^1 K_a(s,t)\,\hat{b}_{n2}(s)\,1\bigl(\hat{B}_{n2}(s) < 0\bigr)\,ds\right)\hat{b}_{n2}(t)\,dt \tag{5.2.16}$$

$$= \sum_{i=1}^{n}\sum_{j=1}^{n}\left(\hat{B}_{n2}(\tfrac{i}{n}) - \hat{B}_{n2}(\tfrac{i-1}{n})\right)\left(\hat{B}_{n2}(\tfrac{j}{n}) - \hat{B}_{n2}(\tfrac{j-1}{n})\right)\hat{k}_{na}(i,j),$$

where the random weights $\hat{k}_{na}(i,j)$ are defined as in Section 3.1, i.e.

$$\hat{k}_{na}(i,j) = n^2 \int_{(i-1)/n}^{i/n}\int_{(j-1)/n}^{j/n} K_a(s,t)\,1\bigl(\hat{B}_{n2}(s) < 0\bigr)\,ds\,dt. \tag{5.2.17}$$

Again, if the kernel $K$ is smooth, we'll approximate the weights $\hat{k}_{na}(i,j)$ by

$$k_{na}(i,j) = n\int_{(i-1)/n}^{i/n} 1\bigl(\hat{B}_{n2}(s) < 0\bigr)\,ds. \tag{5.2.18}$$

Because of (5.2.7) and (5.2.9) the substitution of the weights $\hat{k}_{na}(i,j)$ in (5.2.16) by the approximations (5.2.18) yields the signed rank statistic

$$S_n^0(a,k) \tag{5.2.19}$$

$$= \sum_{i=1}^{n}\sum_{j=1}^{n}\left(\hat{B}_{n2}^0(\tfrac{i}{n}) - \hat{B}_{n2}^0(\tfrac{i-1}{n})\right)\left(\hat{B}_{n2}(\tfrac{j}{n}) - \hat{B}_{n2}(\tfrac{j-1}{n})\right)k_{na}(i,j)$$

for testing $\mathcal{H}_0^s$ versus the one-sided alternative $\mathcal{A}_1^0$ .

### 5.2.2 Theorem

*Assume $0 \leq \varrho \leq 1$ and $b \in L_2^0(0,1)$ such that $b(t) = -b(1-t) \; \forall \; 0 < t < 1$ . Define the sequence of local alternatives $(F_n, n \geq 1)$ according to*

$$F_n = H_n + (\varrho/\sqrt{n})B_n \circ H_n, \quad B_n \in \mathcal{B}_n^s, \quad H_n \in \mathcal{H}_0^s, \quad \|B_n' - b\| \to 0.$$

*Let $0 < a \leq 1$ be a given bandwidth for the convolution kernel (3.1.19) with underlying kernel $K$ according to (3.1.16)-(3.1.18) and (3.1.31), and let $\mathcal{K}_a$ denote the corresponding convolution operator defined in (3.1.33) with representation (3.1.39).*

*a) If the eigenvalues $\lambda_\kappa(a)$ corresponding to the eigenfunctions $\psi_\kappa$ fulfil the condition (3.1.60), then the following limiting law $(n \to \infty)$ holds true,*

$$\mathcal{L}[\bar{S}_n(a, K) \mid F_n] \xrightarrow{\mathcal{L}} \mathcal{L}[\sum_{\kappa=0}^{\infty} \lambda_\kappa(a) \, (Z_\kappa + \varrho < \psi_\kappa, \tilde{b} >)^2 \,], \qquad (5.2.20)$$

*where $Z_0, Z_1, \dots$ are i.i.d. random variables with standard normal distribution, $\psi_0 \equiv 1$ , $\lambda_0(a) = 1$ , and $\tilde{b}(t) = b(1/2 + t/2) \quad \forall \; 0 < t < 1$ .*

*b) If the eigenvalues $\lambda_\kappa(a)$ fulfil the stronger condition (3.1.63), then the following limiting law $(n \to \infty)$ holds true,*

$$\mathcal{L}[\hat{S}_n^0(a, K) \mid F_n] \xrightarrow{\mathcal{L}} \mathcal{L}[\sum_{\kappa=0}^{\infty} \lambda_\kappa(a)Y_\kappa(\varrho\tilde{B}) \,], \qquad (5.2.21)$$

*where*

$$Y_0(\varrho\tilde{B}) := \Big((W^*(0) + \varrho\tilde{B}(0)) \wedge 0\Big)\Big(W^*(0) + \varrho\tilde{B}(0)\Big) \qquad (5.2.22)$$

*and ( $\forall \, \kappa \geq 1$ )*

$$Y_\kappa(\varrho\tilde{B}) := < \psi_\kappa', T_0\big((W^* + \varrho\tilde{B}) \wedge 0\big) > < \psi_\kappa', T_0(W^* + \varrho\tilde{B}) >, \qquad (5.2.23)$$

$$(T_0 f)(s) = f(s) - (1-s)f(0) \qquad \forall \; 0 \leq s \leq 1 \quad \forall \, f \in C[0,1], \qquad (5.2.24)$$

*$W^*$ and $\tilde{B}$ as defined in Theorem 5.2.1.*

*Proof:* a) Lemma 3.1.1 and formula (5.2.13) imply

$$\bar{S}_n(a, K) = < 1, \hat{b}_{n2} >^2 + \sum_{\kappa=1}^{\infty} \lambda_\kappa(a) < \psi_\kappa, \hat{b}_{n2} >^2 . \qquad (5.2.25)$$

Using $\hat{B}_{n2}(1) = 0$ and $\int_0^1 \psi_\kappa(s)\, ds = 0$ partial integration easily proves

$$
\begin{aligned}
< \psi_\kappa, \hat{b}_{N2} > &= -\int_0^1 \psi'_\kappa(s)\left(\hat{B}_{n2}(s) - (1-s)\hat{B}_{n2}(0)\right) ds \\[2mm]
&= -\int_0^1 \psi'_\kappa(t)\left(\hat{B}_{n2}(1-t) - t\hat{B}_{n2}(0)\right) dt,
\end{aligned}
\tag{5.2.26}
$$

where the last equality follows from $\psi'_\kappa(t) = \psi'_\kappa(1-t)$ $\forall\, 0 < t < 1$. Additionally we have

$$
< 1, \hat{b}_{n2} > = -\hat{B}_{n2}(0).
\tag{5.2.27}
$$

Therefore, using Theorem 5.2.1 instead of Theorem 3.1.8 and defining the process $W_0(t) = W(t) - tW(1)$ $\forall\, 0 \le t \le 1$ we get completely similar to the proof of Theorem 3.1.9 the limiting law

$$
\mathcal{L}[\bar{S}_n(a, K) \mid F_n] \xrightarrow{\mathcal{L}} \mathcal{L}[X(a, \varrho, b)],
\tag{5.2.28}
$$

where

$$
X(a, \varrho, b) = \left(W(1) + \varrho\tilde{B}(0)\right)^2
\tag{5.2.29}
$$

$$
+ \sum_{\kappa=1}^{\infty} \lambda_\kappa(a)\left(< \psi'_\kappa, W_0 > + \varrho\int_0^1 \psi'_\kappa(t)\left(\tilde{B}(1-t) - t\tilde{B}(0)\right) dt\right)^2
$$

and where the inequality (3.1.80) has to be substituted by the corresponding inequality ( $E_0$ = expectation under the null hypothesis $\mathcal{H}_0^s$ )

$$
\sum_{\kappa=k+1}^{\infty} |\lambda_\kappa|\, E_0 < \psi_\kappa, \hat{b}_{n2} >^2
$$

$$
\begin{aligned}
&= \sum_{\kappa=k+1}^{\infty} |\lambda_\kappa|\, E_0\Big(\sum_{j=1}^{n} \operatorname{sign}(X_j)\int_{(R_j-1)/n}^{R_j/n} \psi_\kappa(t)\, dt\Big)^2 \\[3mm]
&= \sum_{\kappa=k+1}^{\infty} |\lambda_\kappa|\, n\sum_{j=1}^{n} E_0\Big(\int_{(R_j-1)/n}^{R_j/n} \psi_\kappa(t)\, dt\Big)^2 \\[3mm]
&\le \sum_{\kappa=k+1}^{\infty} |\lambda_\kappa|\, E_0\sum_{j=1}^{n} \int_{(R_j-1)/n}^{R_j/n} \psi_\kappa^2(t)\, dt \\[3mm]
&= \sum_{\kappa=k+1}^{\infty} |\lambda_\kappa|\, \|\psi_\kappa\|^2 = \sum_{\kappa=k+1}^{\infty} |\lambda_\kappa|.
\end{aligned}
\tag{5.2.30}
$$

Defining $Z_0 := -W(1)$ and $Z_\kappa := - < \psi'_\kappa, W_0 >$ $\forall\, \kappa \ge 1$, we get from Theorem 3.1.9 and $EZ_0 Z_\kappa = 0$ $\forall\, \kappa \ge 1$ that $Z_0, Z_1, \ldots$ are i.i.d. random variables with standard normal distribution. Additionally partial integration, $\int_0^1 \psi_\kappa(t)\, dt = 0$ , and $\psi_\kappa(t) = -\psi_\kappa(1-t)$ $\forall\, 0 \le t \le 1$ imply

$$\int_0^1 \psi_\kappa'(t)\Big(\tilde{B}(1-t) - t\tilde{B}(0)\Big)\, dt \; = \; - <\psi_\kappa, \tilde{b}> \qquad \forall\, \kappa \ge 1 \qquad (5.2.31)$$

and

$$\tilde{B}(0) \; = \; -\int_0^1 \tilde{b}(t)\, dt \; = \; - <1, \tilde{b}> . \qquad (5.2.32)$$

Therefore (5.2.28) to (5.2.32) conclude the proof of part (a).

b) Using $\psi_0 \equiv 1$ and $\lambda_0(a) = 1$ , formula (5.2.16) and Lemma 3.1.1 imply

$$\hat{S}_n^0(a, K) \; = \; \sum_{\kappa=0}^{\infty} \lambda_\kappa(a) \; <\psi_\kappa, \hat{b}_{n2}^0> \; <\psi_\kappa, \hat{b}_{n2}> . \qquad (5.2.33)$$

Completely similar to part (a) we have ( $\forall\, \kappa \ge 1$ )

$$<\psi_\kappa, \hat{b}_{n2}^0> \; = \; -\int_0^1 \psi_\kappa'(s)\, \Big(\hat{B}_{n2}^0(s) - (1-s)\hat{B}_{n2}^0(0)\Big)\, ds \qquad (5.2.34)$$

and

$$<\psi_0, \hat{b}_{n2}^0> \; = \; -\hat{B}_{n2}^0(0). \qquad (5.2.35)$$

Hence (5.2.24) implies the representation

$$\hat{S}_n^0(a, K) = \hat{B}_{n2}^0(0)\hat{B}_{n2}(0) + \sum_{\kappa=1}^{\infty} \lambda_\kappa <\psi_\kappa', T_0\hat{B}_{n2}^0> \; <\psi_\kappa', T_0\hat{B}_{n2}> . \qquad (5.2.36)$$

Obviously the mapping $T_0 : C[0,1] \to C[0,1]$ defined in (5.2.24) is linear and continuous. Therefore, again using Theorem 5.2.1 instead of Theorem 3.1.8, we get completely similar to the proof of Theorem 3.1.9 the limiting law (5.2.21), where the inequality (3.1.82) has to be substituted by the corresponding inequality

$$\sum_{\kappa=k+1}^{\infty} |\lambda_\kappa|\, E_0 \, | <\psi_\kappa, \hat{b}_{n2}^0> \; <\psi_\kappa, \hat{b}_{n2}> |$$

$$\le \sum_{\kappa=k+1}^{\infty} |\lambda_\kappa| \, \sqrt{E_0 <\psi_\kappa, \hat{b}_{n2}^0>^2} \, \sqrt{E_0 <\psi_\kappa, \hat{b}_{n2}>^2}$$

$$\le \sum_{\kappa=k+1}^{\infty} |\lambda_\kappa| \, \sqrt{E_0 <\psi_\kappa', T_0\hat{B}_{n2}^0>^2}$$

$$\leq \sum_{\kappa=k+1}^{\infty} |\lambda_\kappa| \sqrt{\|\psi_\kappa'\|^2 \, E_0 \|T_0 \hat{B}_{n2}^0\|^2}$$

$$\leq \pi \sqrt{2 \, E_0 \|\hat{B}_{n2}\|^2 + 2 \, E_0 \hat{B}_{n2}^2(0)} \, \sum_{\kappa=k+1}^{\infty} \kappa \, |\lambda_\kappa|$$

$$\leq 2\pi \sum_{\kappa=k+1}^{\infty} \kappa \, |\lambda_\kappa| \, . \tag{5.2.37}$$

Here the last inequality holds true, since formula (5.2.4) implies

$$E_0 \hat{B}_{n2}^2(t) \leq 1 \qquad \forall \, 0 \leq t \leq 1. \tag{5.2.38}$$

This concludes the proof of part (b). $\square$

**Remark:** *The assertions of Theorem 5.2.2 remain true, if the rank statistics* $\bar{S}_n(a, K)$ *and* $\hat{S}_n^0(a, K)$ *are replaced by the respective approximate statistics* $S_n(a, K)$ *and* $S_n^0(a, K)$. *This may be proved by adjusting the proof of Theorem 5.2.2 to the empirical (jump) processes*

$$\tilde{B}_{n1}(t) = -\frac{1}{\sqrt{n}} \sum_{j=1}^{n} 1\left(\frac{R_j^+ - 1/2}{n} > t\right) \operatorname{sign}(X_j), \quad 0 \leq t \leq 1, \tag{5.2.39}$$

*and*

$$\tilde{B}_{n1}^0 = \tilde{B}_{n1} 1(\tilde{B}_{n1} < 0), \tag{5.2.40}$$

*cf. the proof of Corollary 3.1.10.*

## 5.3   Projection estimators of the score function

In the two-sample problem we've seen in Section 2.1 that the projections onto very large cones, cf. Examples 3.2.2 to 3.2.4, produce rank tests which have rather low power for large classes of alternatives. Since many relevant types of alternatives can be approximated by a given finite set of suitable score functions of the form

$$\tilde{b}(t) = \varphi(\frac{1}{2} + \frac{t}{2}, f_0, D^*), \qquad 0 < t < 1, \tag{5.3.1}$$

where $\varphi(\cdot, f_0, D^*)$ and $D^*$ are defined in (5.1.10) and (5.1.11), we'll restrict the discussion to the case of finite dimensional cones.

According to Section 5.1 the signed linear rank statistic $S_n(b) = \, < \tilde{b}, \hat{b}_{n2} >$ , cf. (5.2.10), is asymptotically optimal for testing the null hypothesis $\mathcal{H}_0^s$ versus the one-sided alternative of positive unsymmetry $\mathcal{A}_1^0$ if the defining function $\tilde{b}(t) = b(1/2 + t/2)$ has the additional property

$$\int_t^1 \tilde{b}(x)\, dx \geq 0 \qquad \forall\, 0 \leq t \leq 1. \tag{5.3.2}$$

Therefore, in the one-sided situation, we'll estimate the unknown $\tilde{b}$ by the projection $\pi_V \hat{b}_{n2}$ of the primitive estimator $\hat{b}_{n2}$ onto the $r$ –dimensional $L_2(0,1)$ –cone

$$V := [\tilde{b}_1, ..., \tilde{b}_r]^+ = \left\{ \sum_{\varrho=1}^{r} \vartheta_\varrho \, \tilde{b}_\varrho : \vartheta_\varrho \geq 0 \ \ \forall\, \varrho \right\}, \tag{5.3.3}$$

where the functions $\tilde{b}_1, ..., \tilde{b}_r$ are linearly independent in $L_2(0,1)$ and fulfil the condition (5.3.2). According to formula (3.2.10) the resulting signed rank statistic $S_n(V)$ has the form

$$S_n(V) = \, < \pi_V \hat{b}_{n2}, \hat{b}_{n2} > = \sup_{g \in V} \left( 2 \, < g, \hat{b}_{n2} > - \, \|g\|^2 \right)$$

$$= \sup_{\vartheta \geq 0} \left( 2\vartheta^T \vec{S}_n - \vartheta^T \Gamma \vartheta \right), \tag{5.3.4}$$

where $\vartheta = (\vartheta_1, ..., \vartheta_r)^T \in \mathbb{R}^r$ ,

$$\vec{S}_n = \left( < \tilde{b}_1, \hat{b}_{n2} >, ..., < \tilde{b}_r, \hat{b}_{n2} > \right)^T, \tag{5.3.5}$$

and

$$\Gamma = \left( < \tilde{b}_\kappa, \tilde{b}_\varrho > \right)_{\kappa, \varrho = 1, ..., r}. \tag{5.3.6}$$

Using the representation (5.2.10) of $< \tilde{b}_\varrho, \hat{b}_{n2} >$ as a signed linear rank statistic the Cramér-Wold device and Behnen (1972) imply

$$\mathcal{L}[\vec{S}_n \mid F_n] \xrightarrow{\mathcal{L}} \mathcal{N}\left( (< \tilde{b}_1, \tilde{g} >, ..., < \tilde{b}_r, \tilde{g} >)^T, \ \Gamma \right) \tag{5.3.7}$$

if $F_n = H_n + n^{-1/2} G_n \circ H_n$ , $G_n \in \mathcal{B}_n^s$ , $H_n \in \mathcal{H}_0^s$ , $\|G_n' - g\| \to 0$ .

Therefore, on one hand we can use the results of Section 3.2.C in order to evaluate the statistic $S_n(V)$ on the basis of $\vec{S}_n$ and $\Gamma$ , cf. formula (3.2.73), on the other hand we may apply the results of Section 3.2.C in order to derive the limiting law of $S_n(V)$ under local alternatives, cf. Corollary 3.2.6 and Theorem 3.2.7.

Now let's construct a suitable set of score functions of the type (5.3.1). As in Section 3.2 we'll consider the shift functions (3.2.86) in the case of underlying logistic distribution function (3.2.88). Obviously the lower shift function $D_1 = 1 - F_0$ and the upper shift function $D_3 = F_0$ yield the same symmetrized shift function (5.1.11) according to ( $\forall\, x \in \mathbb{R}$)

$$D_1^*(x) \;=\; D_3^*(x) \;=\; \frac{F_0(x) + F_0(-x)}{2} \;=\; \frac{1}{2}\,, \tag{5.3.8}$$

i.e. lower shift and upper shift coincide with exact shift. Additionally we have $D_2^* = D_2$ . Therefore, in case of testing the null hypothesis of symmetry $\mathcal{H}_0^s$ versus the one-sided alternative $\mathcal{A}_1^0$ , we'll only use the two representative score functions $b_1 = \varphi(\cdot\,, f_0, 1/2)$ and $b_2 = \varphi(\cdot\,, f_0, D_2)$ , i.e.

$$\tilde{b}_1(t) \;=\; b_1(\frac{1}{2} + \frac{t}{2}) \;=\; t, \qquad 0 \le t \le 1, \tag{5.3.9}$$

and

$$\tilde{b}_2(t) \;=\; b_2(\frac{1}{2} + \frac{t}{2}) \;=\; 2t(1 - t^2), \qquad 0 \le t \le 1. \tag{5.3.10}$$

The evaluation of the covariance matrix $\Gamma = ( < \tilde{b}_\kappa, \tilde{b}_\varrho > )$ and its inverse $\Gamma^{-1}$ yields

$$\Gamma = \frac{1}{105} \begin{pmatrix} 35 & 28 \\ 28 & 32 \end{pmatrix}, \qquad \Gamma^{-1} = \frac{5}{16} \begin{pmatrix} 32 & -28 \\ -28 & 35 \end{pmatrix} . \tag{5.3.11}$$

Using the abbreviations

$$S_1 \;=\; < \tilde{b}_1, \hat{b}_{n2} > \;=\; \frac{1}{\sqrt{n}} \sum_{i=1}^{n} \frac{R_i^+ - 1/2}{n} \, \text{sign}(X_i) \tag{5.3.12}$$

and

$$S_2 = \; < \tilde{b}_2, \hat{b}_{n2} > \; = \; \frac{1}{\sqrt{n}} \sum_{i=1}^{n} b_{2n}(R_i^+) \, \text{sign}(X_i), \qquad (5.3.13)$$

where

$$b_{2n}(i) = n \int_{(i-1)/n}^{i/n} 2t^2(1-t^2)\, dt \approx 2\Big(\frac{i-1/2}{n}\Big)^2 \Big(1 - \big(\frac{i-1/2}{n}\big)^2\Big), \quad (5.3.14)$$

formula (3.2.73) implies the following explicit representation of the signed rank statistic (5.3.4),

$$\begin{aligned}
S_n^* \; &:= \; S_n([\tilde{b}_1, \tilde{b}_2]^+) \\[4pt]
&= \max\Big[\, 3S_1^2 1(S_1 \geq 0),\; \frac{105}{32} S_2^2 1(S_2 \geq 0), \qquad\qquad (5.3.15) \\[6pt]
&\qquad \frac{5}{16}\,(32 S_1^2 + 35 S_2^2 - 56 S_1 S_2)\, 1(7S_2 \leq 8S_1 \leq 10 S_2)\,\Big].
\end{aligned}$$

Finally Corollary 3.2.6, Theorem 3.2.7, and formula (3.2.95) imply the following limiting law under $\mathcal{H}_0^s$,

$$\lim_{n\to\infty} P_0\{\, S_n^* = 0 \,\} \; = \; \frac{1}{2\pi}\, \arccos\Big(\frac{-28}{\sqrt{32}\,\sqrt{35}}\Big) \; = \; 0.40775 \qquad (5.3.16)$$

and ( $\forall\, x \geq 0$ )

$$\lim_{n\to\infty} P_0\{\, S_n^* > x \,\} \; = \; \frac{1}{2} P\{\, \chi_1^2 > x \,\} + 0.09225\, P\{\, \chi_2^2 > x \,\}. \qquad (5.3.17)$$

## 5.4   Treatment of ties

In this section we'll assume that the underlying distribution function $F$ may have discontinuities, i.e. we assume $\mathcal{F}_1$ as the underlying parameter space, cf. (3.3.1). We want to test the general null hypothesis of symmetry

$$\mathcal{H}_0^s = \{\, F \in \mathcal{F}_1 :\ F = F_- \,\} \tag{5.4.1}$$

versus the general one-sided alternative of positive unsymmetry

$$\mathcal{A}_1^0 = \{\, F \in \mathcal{F}_1 :\ F \leq F_-;\ F \neq F_- \,\} \tag{5.4.2}$$

or $\mathcal{H}_0^s$ versus the general omnibus alternative

$$\mathcal{A}_1 = \{\, F \in \mathcal{F}_1 :\ F \neq F_- \,\}. \tag{5.4.3}$$

Here the distribution function $F_-$ of the random variable $-X_i$ has the form

$$F_-(x) = 1 - \lim_{k \to \infty} F\left(-x - \frac{1}{k}\right) = 1 - F\big((-x)_-\big), \quad x \in \mathbb{R}. \tag{5.4.4}$$

In order to identify and estimate the score function of the general model we'll construct a reparametrization of the model which contains the (continuous) parametrization (5.1.21) as a special case.

Throughout this section the results and the terminology of Section 3.3 will be used. Additionally we define

$$\mathcal{J}_{0,1}^s = \{\, \mathcal{J} \in \mathcal{J}_{0,1}^* :\ (1 - y, 1 - x] \in \mathcal{J}\ \ \text{if}\ \ (x,y] \in \mathcal{J} \,\}$$

$$= \{\, \mathcal{I}(H):\ H \in \mathcal{H}_0^s \,\} \tag{5.4.5}$$

and

$$\mathcal{H}_0^s(\mathcal{J}) \ := \ \mathcal{H}_0^s \cap \mathcal{F}_1(\mathcal{J}). \tag{5.4.6}$$

Then we have the following disjoint partition of the null hypothesis $\mathcal{H}_0^s$ ,

$$\mathcal{H}_0^s = \sum_{\mathcal{J} \in \mathcal{J}_{0,1}^s} \mathcal{H}_0^s(\mathcal{J}). \tag{5.4.7}$$

Putting $G = F_-$ , $m = n$ , and $N = 2n$ the results of Section 3.3 imply the disjoint partition

$$\mathcal{F}_1 = \sum_{\mathcal{J} \in \mathcal{J}_{0,1}^s} \mathcal{F}_1^s(\mathcal{J}), \tag{5.4.8}$$

where

$$\mathcal{F}_1^s(\mathcal{J}) \ := \ \{\, H + \frac{1}{\sqrt{n}} B \circ H :\ (B, H) \in \mathcal{B}_n^s(\mathcal{J}) \times \mathcal{H}_0^s(\mathcal{J}) \,\} \tag{5.4.9}$$

and

$$\mathcal{B}_n^s(\mathcal{J}) := \big\{ \, B \in \mathcal{B}_n^s \, : \, B' \text{ is } \lambda\text{-a.e. constant on each } (x,y] \in \mathcal{J} \, \big\}. \qquad (5.4.10)$$

Additionally the proof of Proposition 3.3.3 yields the equivalences

$$
\begin{aligned}
F = F_- &\iff B = 0 \,, \\
F \le F_- &\iff B \le 0 \,, \\
F \ge F_- &\iff B \ge 0 \,.
\end{aligned}
\qquad (5.4.11)
$$

Therefore we get the corresponding $(B, H)$–parametrization of the alternatives according to

$$\mathcal{A}_1^0 = \sum_{\mathcal{J} \in \mathcal{J}_{0,1}^s} \mathcal{A}_1^0(\mathcal{J}),$$

$$\mathcal{A}_1^0(\mathcal{J}) = \{ \, H + \frac{1}{\sqrt{n}} B \circ H \in \mathcal{F}_1^s(\mathcal{J}) : \ B \le 0, \ B \neq 0 \, \}, \qquad (5.4.12)$$

and

$$\mathcal{A}_1 = \sum_{\mathcal{J} \in \mathcal{J}_{0,1}^s} \mathcal{A}_1(\mathcal{J}),$$

$$\mathcal{A}_1(\mathcal{J}) = \{ \, H + \frac{1}{\sqrt{n}} B \circ H \in \mathcal{F}_1^s(\mathcal{J}) : \ B \neq 0 \, \}. \qquad (5.4.13)$$

Again we'll show that for arbitrary $\mathcal{J} \in \mathcal{J}_{0,1}^s$ there is a complete analogy between the *continuous* testing problem of Section 5.1 and the problem of testing $\mathcal{H}_0^s(\mathcal{J})$ versus $\mathcal{A}_1^0(\mathcal{J})$ or $\mathcal{A}_1(\mathcal{J})$. Note, the continuous model coincides with the special case $\mathcal{J} = \emptyset$.

Defining $R^+ = (R_1^+, ..., R_n^+)$ and $\mathrm{sign}(X) = \big(\mathrm{sign}(X_1), ..., \mathrm{sign}(X_n)\big)$ according to (5.1.3) and (5.1.4) the next proposition will prove the distribution of $\big(R^+, \mathrm{sign}(X)\big)$ to be independent of the underlying $H \in \mathcal{H}_0^s(\mathcal{J})$. Under $B = 0$ this result is an extension of the well-known fact that the statistic $\big(R^+, \mathrm{sign}(X)\big)$ is distribution-free under the continuous null hypothesis of symmetry.

### 5.4.1 Proposition

*For any given* $\mathcal{J} \in \mathcal{J}_{0,1}^s$ , $B \in \mathcal{B}_n^s(\mathcal{J})$ , *and* $H_1, H_2 \in \mathcal{H}_0^s(\mathcal{J})$ *we have*

$$\mathcal{L}[ \, (R^+, \mathrm{sign}(X)) \mid (B, H_1) \, ] = \mathcal{L}[ \, (R^+, \mathrm{sign}(X)) \mid (B, H_2) \, ]. \qquad (5.4.14)$$

*Proof:* Assume $\mathcal{J} \in \mathcal{J}_{0,1}^s$ , $B \in \mathcal{B}_n^s(\mathcal{J})$ and $H \in \mathcal{H}_0^s(\mathcal{J})$ .

Since the assumption $F = H + n^{-1/2} B \circ H$ implies $F_- = H - n^{-1/2} B \circ H$ , the distribution function $\tilde{F}$ of $|X_i|$ under $(B, H)$ is given by

$$\tilde{F}(x) = F(x) + F_-(x) - 1 = 2H(x) - 1 \qquad \forall \, x \geq 0. \qquad (5.4.15)$$

This implies $[(B, H) - a.e.]$ the equality

$$R_i^+ = \sum_{j=1}^n 1\Big( |X_j| \leq |X_i| \Big) = \sum_{j=1}^n 1\Big( h(|X_j|) \leq h(|X_i|) \Big), \qquad (5.4.16)$$

where the non-decreasing function $h : \mathbb{R} \to \mathbb{R}$ is defined by

$$h(x) = H(x) - H(-x), \qquad x \in \mathbb{R}. \qquad (5.4.17)$$

Since $h(x) = -h(-x)$ implies $h(|X_j|) = h(X_j)\,\text{sign}(X_j)$ , and since $H \in \mathcal{H}_0^s$ implies

$$\text{sign}(X_j) = \text{sign}\big(h(X_j)\big) \qquad [(B, H) - a.e.], \qquad (5.4.18)$$

the proof is concluded if we prove the assertion

$$\mathcal{L}[\, h(X_j) \mid (B, H)\,] = Q_j \qquad \forall \, H \in \mathcal{H}_0^s(\mathcal{J}), \qquad (5.4.19)$$

where $Q_j$ is a suitable probability measure on $\mathbb{R}$ which does not depend on $H \in \mathcal{H}_0^s(\mathcal{J})$ .

Since $F = H + n^{-1/2} B \circ H$ is the distribution function of $X_j$ the set $D(H) :=$ $\{\, H^{-1}(t) : \emptyset \neq (s, t] \in \mathcal{J}\,\}$ contains all discontinuity points of $F$ , and for each $\emptyset \neq (s, t] \in \mathcal{J}$ we have

$$P_{(B,H)}\{\, X_j = H^{-1}(t)\,\} = (t - s) + n^{-1/2}\big(B(t) - B(s)\big) \qquad (5.4.20)$$

and

$$h\big(H^{-1}(t)\big) = t + s - 1. \qquad (5.4.21)$$

Now let's consider the set $\{\, X_j \in \mathbb{R} \backslash D(H)\,\}$ . On this set we have on one hand $h(X_j) = 2H(X_j) - 1$ , on the other hand the restriction of $H(X_j)$ to this set has an absolutely continuous distribution with Lebesgue-density $1 + n^{-1/2} B'$ on $[0, 1] \backslash \bigcup \{J \in \mathcal{J}\}$ , cf. proof of Proposition 3.3.5. Combining the results we've proved the assertion (5.4.19). $\square$

Similar to Proposition 3.3.6 the following invariance properties of the signed ranks may be proved. In order to formulate the result we call a mapping $k : \mathbb{R} \to \mathbb{R}$ a *symmetric transformation of the measurement scale*, if $k$ is strictly increasing and continuous such that $k(x) = -k(-x) \;\; \forall \, x \in \mathbb{R}$ and $k(\mathbb{R}) = \mathbb{R}$ .

### 5.4.2 Proposition

a) For each $\mathcal{J} \in \mathcal{J}_{0,1}^s$ the hypotheses $\mathcal{H}_0^s(\mathcal{J})$, $\mathcal{A}_1^0(\mathcal{J})$, and $\mathcal{A}_1(\mathcal{J})$ are *invariant under the group of symmetric transformations of the measurement scale.*

b) *The vector of signed ranks* $(R^+, \text{sign}(X))$ *is maximal invariant with respect to the group of symmetric transformations of the measurement scale.*

Fixing $\mathcal{J} \in \mathcal{J}_{0,1}^s$ and considering local asymptotic alternatives corresponding to the testing problem $\mathcal{H}_0^s(\mathcal{J})$ versus $\mathcal{A}_1^0(\mathcal{J})$ or $\mathcal{A}_1(\mathcal{J})$ we'll get an optimality result which is completely similar to the continuous result (5.1.31), (5.1.32). For the formulation of the asymptotic optimality result we assume an arbitrary score function $b \in L_2^0(0,1)$ such that $b(t) = -b(1-t)$ $\forall\, 0 < t < 1$ and $\|b\| > 0$. The corresponding scores $b_n(i)$, $i = 1, ..., n$, are defined in (5.1.28).

Let $|X|^{(1)} \le |X|^{(2)} \le \cdots \le |X|^{(n)}$ denote the ordered absolute values of the sample $X_1, ..., X_n$. In a first step we define the *length* $\tau_0$ *of the zero-tie* according to

$$\tau_0 \;:=\; T_0 \;:=\; \sum_{j=1}^{n} 1(\,|X_j| = 0\,). \qquad (5.4.22)$$

If $\tau_{k-1}$ and $T_{k-1}$ have been defined and if $T_{k-1} < n$, then we define the *length* $\tau_k$ *of the k-th tie* according to

$$\tau_k \;:=\; \sum_{j=1}^{n} 1\Big(\,|X_j| = |X|^{(T_{k-1}+1)}\,\Big), \qquad T_k := \tau_k + T_{k-1}. \qquad (5.4.23)$$

Finally let $d$ denote the (random) number of non-zero ties, which implies $0 \le d \le n$ and $T_d = n$. Additionally we get $0 \le \tau_0 \le n$, and $d \ge 1$ if $\tau_0 < n$, whereas $\tau_0 = n$ means $X_1 = ... = X_n = 0$. Since $\tau_0 + ... + \tau_d = n$, the number of zeros $\tau_0$ is determined by the tie-lengths vector $\tau = (\tau_1, ..., \tau_d)$ of the non-zero values among $|X|^{(1)} \le \cdots \le |X|^{(n)}$. Because of formulae (5.4.22) and (5.4.23) the vectors $\tau$ and $T = (T_0, ..., T_d)$ are related by a one-to-one correspondence.

Using $\tau_0, ..., \tau_d$ and $T_0, ..., T_d$ we define the *averaged scores* $b_n^\tau(i)$ corresponding to the given scores $b_n(i)$, $i = 1, ..., n$, according to

$$b_n^\tau(i) \;=\; 0, \qquad \text{if } i \le T_0, \qquad (5.4.24)$$

$$b_n^\tau(i) \;=\; \frac{1}{\tau_k} \sum_{j=1}^{n} b_n(j)\, 1(T_{k-1} < j \le T_k),$$

$$\text{if } T_{k-1} < i \le T_k \text{ for some } k \ge 1, \qquad (5.4.25)$$

and the *averaged scores signed linear rank statistic* $S_n^\tau(b)$ is defined as

$$S_n^\tau(b) \; = \; \frac{1}{\sqrt{n}} \sum_{i=1}^{n} b_n^\tau(R_i^+)\,\mathrm{sign}(X_i). \qquad (5.4.26)$$

Similar to Section 3.3 we'll utilize a suitable randomization procedure in order to prove the asymptotic properties of $S_n^\tau(b)$ .

Assume $U_1, ..., U_n$ to be i.i.d. random variables with uniform distribution on the unit interval, assume $V_1, ..., V_n$ to be i.i.d. random variables with distribution given by $P\{V_i = -1\} = P\{V_i = 1\} = 1/2$ , and assume the random vectors $(U_1, ..., U_n)$ , $(V_1, ..., V_n)$ , $(X_1, ..., X_n)$ to be stochastically independent. We use the $U$ 's in order to define the *randomized ranks* $Q_i^* = Q_i^*(R^+, U)$ of the absolute values of the observations according to

$$Q_i^* \; = \sum_{j=1}^{n} 1(\, R_j^+ + U_j \leq R_i^+ + U_i \,), \qquad 1 \leq i \leq n, \qquad (5.4.27)$$

and we use the $V$ 's in order to define the *randomized signs*

$$\Delta_i^* \; = \mathrm{sign}(X_i) + 1(\, X_i = 0 \,)\, V_i, \qquad 1 \leq i \leq n, \qquad (5.4.28)$$

of the observations $X_1, ..., X_n$ . Obviously the combination of randomized ranks and averaged scores yields the identity

$$S_n^\tau(b) \; = \; \frac{1}{\sqrt{n}} \sum_{i=1}^{n} b_n^\tau(Q_i^*)\, \Delta_i^*, \qquad (5.4.29)$$

and the following lemma will show that the randomized ranks and the randomized signs may be viewed as the usual ranks of the absolute values and the usual signs, respectively, of suitable i.i.d. random variables $X_1^*, ..., X_n^*$ with absolutely continuous distribution function.


### 5.4.3 Lemma

Assume $\mathcal{J} \in \mathcal{J}_{0,1}^s$ and $(B, H) \in \mathcal{B}_n^s(\mathcal{J}) \times \mathcal{H}_0^s(\mathcal{J})$ . Let $X_1, ..., X_n$ be i.i.d. random variables with distribution function $F = H + n^{-1/2} B \circ H$ . Using the above notations and assumptions we define the randomized random variables $X_1^*, ..., X_n^*$ according to

$$X_i^* \; = \; T(X_i, U_i, V_i), \qquad i = 1, ..., n, \qquad (5.4.30)$$

where $T : \mathbb{R} \times [0, 1] \times \{-1, 1\} \rightarrow \mathbb{R}$ is defined by

$$T(x, u, v) = \Big( u\big(2H(|x|) - 1\big) + (1 - u)\big(2H(|x|_-) - 1\big) \Big)\, \mathrm{sign}(x)$$
$$+ u\big(2H(0) - 1\big)v\,1(x = 0). \qquad (5.4.31)$$

*Then the following assertions hold true:*

*a)* $X_1^*, ..., X_n^*$ *are i.i.d. random variables with values in* $[-1, 1]$ .

*b) For each* $1 \leq i \leq n$ *we have with probability 1*

$$\sum_{j=1}^{n} 1( \, |X_j^*| \leq |X_i^*| \,) \; = \; \sum_{j=1}^{n} 1( \, R_j^+ + U_j \leq R_i^+ + U_i \,) \; = \; Q_i^* \qquad (5.4.32)$$

*and*

$$\mathrm{sign}(X_i^*) \; = \; \mathrm{sign}(X_i) + 1(X_i = 0) \, V_i \; = \; \Delta_i^*. \qquad (5.4.33)$$

*c)* $\mathcal{L}(X_i^*)$ *is absolutely continuous with distribution function*

$$P\{X_i^* \leq x\} \; = \; \frac{1}{2} + \frac{x}{2} + \frac{1}{\sqrt{n}} \, B\Big(\frac{1}{2} + \frac{x}{2}\Big) \qquad \forall \, x \in [-1, 1], \qquad (5.4.34)$$

*which especially implies*

$$\mathcal{L}( \, |X_i^*| \,) \; = \; \mathcal{R}(0, 1) \qquad \forall \, B \in \mathcal{B}_n^s(\mathcal{J}). \qquad (5.4.35)$$

*d) Under the hypothesis* $B = 0$ *the random vectors* $|X^*| := (|X_1^*|, ..., |X_n^*|)$
*and* $\mathrm{sign}(X^*) := (\mathrm{sign}(X_1^*), ..., \mathrm{sign}(X_n^*))$ *are stochastically independent.*

*e) Under the hypothesis* $B = 0$ *the random vectors* $\Delta^* = (\Delta_1^*, ..., \Delta_n^*)$ ,
$Q^* = (Q_1^*, ..., Q_n^*)$ , *and* $\tau = (\tau_1, ..., \tau_d)$ *are stochastically independent.*

*Proof:* Assertion a) is obvious from the definitions (5.4.30) and (5.4.31). For
the proof of assertion b) we notice that with probability 1 the assumption
$|X_j| < |X_i|$ implies $|X_j^*| < |X_i^*|$ . Therefore (5.4.33) is immediate. For the
proof of (5.4.32) we get with probability 1 the following chain of equalities,

$$\sum_{j=1}^{n} 1\Big( \, |X_j^*| \leq |X_i^*| \,\Big)$$

$$= \sum_{j=1}^{n} 1\Big( \, |X_i| < |X_i| \,\Big) + \sum_{j=1}^{n} 1\Big( \, |X_j^*| \leq |X_i^*|, \; |X_j| = |X_i| \,\Big)$$

$$= \sum_{j=1}^{n} 1\Big( \, R_j^+ < R_i^+ \,\Big)$$

$$+ \sum_{j=1}^{n} \Big( 1( \, |X_j| = |X_i| = 0, \; U_j \leq U_i \,) + 1( \, |X_j| = |X_i| > 0, \; U_j \leq U_i \,) \Big)$$

$$= \sum_{j=1}^{n} 1\left( R_j^+ + U_j < R_i^+ + U_i \right) + \sum_{j=1}^{n} 1\left( R_j^+ = R_i^+,\ U_j \le U_i \right)$$

$$= \sum_{j=1}^{n} 1\left( R_j^+ + U_j \le R_i^+ + U_i \right)\ =\ Q_i^*. \tag{5.4.36}$$

For the proof of c) we notice that $H = H_-$ implies

$$T(x, u, v) = \begin{cases} (1 - u)(2H(x) - 1) + u(2H(x_-) - 1), & \text{if } x < 0, \\ u(2H(0) - 1)v\ =\ -u(2H(0_-) - 1)v, & \text{if } x = 0, \\ u(2H(x) - 1) + (1 - u)(2H(x_-) - 1), & \text{if } x > 0. \end{cases} \tag{5.4.37}$$

Thus, for arbitrary $u_1, u_2 \in [0, 1]$, $v_1, v_2 \in \{-1, 1\}$, and $x_1, x_2 \in \mathbb{R}$ we have the implication

$$x_1 < x_2 \quad \Longrightarrow \quad T(x_1, u_1, v_1) \le T(x_2, u_2, v_2). \tag{5.4.38}$$

In order to prove (5.4.34) we consider the $[0, 1]$ –valued random variable $Z :=(X_i^* + 1)/2$ and prove

$$P\{\ Z \le t\ \}\ =\ t + n^{-1/2}B(t) \qquad \forall\, t \in [0, 1]. \tag{5.4.39}$$

Obviously (5.4.39) proves assertion c) since (5.4.35) is an immediate consequence of (5.4.34) and $B(t) = B(1 - t)$ $\forall\, 0 \le t \le 1$. For the proof of (5.4.39) we'll consider different cases:

1) Assume $(1 - y, y] \in \mathcal{J}$ for some $1/2 < y \le 1$ and $t \in (1 - y, y]$. In this case we get from (5.4.37) and (5.4.38)

$$P\{Z \le t\} = P\{X_i < 0\} + P\{\ X_i = 0,\ U_i(y - 1/2)V_i + 1/2 \le t\ \}$$

$$= F(0_-) + (F(0) - F(0_-))P\{\ U_i V_i \le (2t - 1)/(2y - 1)\ \}$$

$$= (1 - y) + \frac{1}{\sqrt{n}}B(1 - y) + (2y - 1)\frac{t + y - 1}{2y - 1}, \quad \text{since } B(y) = B(1 - y),$$

$$= t + n^{-1/2}B(t),$$

since $B$ is linear on $[1 - y, y]$ and $B(1 - y) = B(y)$.

2) Assume $(x, y] \in \mathcal{J}$ such that $y < 1/2$ and $t \in (x, y]$. In this case we get from (5.4.37) and (5.4.38)

$$P\{Z \le t\} = P\{X_i < H^{-1}(y)\} + P\{\ X_i = H^{-1}(y),\ (1 - U_i)y + U_i x \le t\ \}$$

$$= x + n^{-1/2}B(x) + \left( y - x + n^{-1/2}(B(y) - B(x)) \right)\frac{t - x}{y - x}$$

$$= t + n^{-1/2}\left( B(x) + \frac{t - x}{y - x}(B(y) - B(x)) \right)\ =\ t + n^{-1/2}B(t),$$

since $B$ is linear on $[x, y]$ .

3) Assume $(x, y] \in \mathcal{J}$ such that $x > 1/2$ and $t \in (x, y]$ . Here we get completely similar to case 2) the equalities

$$P\{Z \leq t\} = P\{X_i < H^{-1}(y)\} + P\{\ X_i = H^{-1}(y),\ U_i y + (1 - U_i)x \leq t\ \}$$
$$= t + n^{-1/2} B(t).$$

4) In the final step we assume $t \in [0, 1] \setminus \bigcup \{(x, y] \in \mathcal{J}\}$ . In this case (5.4.37) and (5.4.38) imply

$$P\{Z \leq t\} = P\{X_i \leq H^{-1}(t)\} = F\left(H^{-1}(t)\right) = t + n^{-1/2} B(t).$$

This concludes the proof of assertion (5.4.39).

Under the null hypothesis $B = 0$ , i.e. $F = H \in \mathcal{H}_0^s(\mathcal{J})$ , we get from (5.4.33)

$$
\begin{aligned}
P\{\ \mathrm{sign}(X_i^*) = -1\ \} &= P\{X_i < 0\} + \frac{1}{2}\, P\{X_i = 0\} = \frac{1}{2}, \\
P\{\ \mathrm{sign}(X_i^*) = +1\ \} &= P\{X_i > 0\} + \frac{1}{2}\, P\{X_i = 0\} = \frac{1}{2}
\end{aligned}
\tag{5.4.40}
$$

Since $X_1^*, ..., X_n^*$ are independent random variables the proof of assertion d) is concluded if we prove the independence of $|X_i^*|$ and $\mathrm{sign}(X_i^*)$ under the hypothesis $B = 0$ . For each $0 \leq t \leq 1$ we get from (5.4.34) and (5.4.35)

$$P\{\ |X_i^*| \leq t,\ X_i^* < 0\ \} = \frac{t}{2} = \frac{1}{2}\, P\{\ |X_i^*| \leq t\ \},$$

$$P\{\ |X_i^*| \leq t,\ X_i^* > 0\ \} = \frac{t}{2} = \frac{1}{2}\, P\{\ |X_i^*| \leq t\ \}.$$

Therefore (5.4.40) proves the independence of $|X_i^*|$ and $\mathrm{sign}(X_i^*)$ under the hypothesis $B = 0$ .

According to formula (5.4.32) we may view the vector $Q^*$ as the vector of ranks of $|X_1^*|, ..., |X_n^*|$ , which are i.i.d. random variables with continuous distribution function. Thus the independence of $Q^*$ and $(|X^*|^{(1)}, ..., |X^*|^{(n)})$ is well-known, cf. Theorem II.1.2.a of Hájek and Šidák (1967). Since formula (5.4.33) and assertion d) imply the independence of $\Delta^*$ and $|X^*|$ under the hypothesis $B = 0$ the proof of assertion e) is concluded if the ordered absolute values $|X|^{(1)}, ..., |X|^{(n)}$ can be evaluated with probability 1 from the ordered absolute values $|X^*|^{(1)}, ..., |X^*|^{(n)}$ . This is quite easy since (5.4.37) implies with probability 1

$$X_i = H^{-1}\left(\frac{1}{2}X_i^* + \frac{1}{2}\right), \qquad i = 1, ..., n. \tag{5.4.41}$$

Therefore the proof of Lemma 5.4.3 is complete. $\square$

The next theorem will prove the asymptotic normality of the averaged scores signed linear rank statistic $S_n^\tau(b)$ under the null hypothesis. Since the variance of the limiting law will depend on the underlying $J \in \mathcal{J}_{0,1}^s$ we'll prove in addition that $S_n^\tau(b)$ may be standardized by its conditional variance given $\tau$, in order to have an asymptotically distribution-free test under $\mathcal{H}_0^s$ .

### 5.4.4 Theorem

*Assume* $b \in L_2^0(0,1)$ *such that* $b(t) = -b(1-t)$ $\forall\, 0 < t < 1$ .

*a) Under the general null hypothesis* $\mathcal{H}_0^s = \{F \in \mathcal{F}_1 : F = F_-\}$ *the randomized signed linear rank statistic*

$$S_n^*(b) \;=\; \frac{1}{\sqrt{n}} \sum_{i=1}^{n} b_n(Q_i^*)\Delta_i^* \tag{5.4.42}$$

*with* $b_n(i)$ *from (5.1.28) has the limiting law*

$$\mathcal{L}\big[\, S_n^*(b) \mid \mathcal{H}_0^s \,\big] \;\xrightarrow{\;\mathcal{L}\;}\; \mathcal{N}(0, \|b\|^2). \tag{5.4.43}$$

*b) Under the* $J$ *–part* $\mathcal{H}_{0n}^s(J) = \{F : F \in \mathcal{F}_1^s(J)\}$ *of the general null hypothesis* $\mathcal{H}_0^s$ *the averaged scores signed linear rank statistic* $S_n^\tau(b)$ *defined in (5.4.26) has the limiting law*

$$\mathcal{L}\big[\, S_n^\tau(b) \mid \mathcal{H}_{0n}^s(J) \,\big] \;\xrightarrow{\;\mathcal{L}\;}\; \mathcal{N}(0, \|L_J b\|^2), \tag{5.4.44}$$

*where* $L_J : L_2(0,1) \to L_2(0,1)$ *is defined in (3.3.49).*

*Obviously the assumption* $b \in L_2^0(0,1)$ *implies* $L_J b \in L_2^0(0,1)$ . *Additionally we have the following convergence in* $\mathcal{H}_{0n}^s(J)$ *–probability,*

$$\frac{1}{n} \sum_{i=1}^{n} \big(b_n^\tau(i)\big)^2 \;\longrightarrow\; \|L_J b\|^2 \tag{5.4.45}$$

*with* $b_n^\tau(i)$ *from (5.4.24) and (5.4.25).*

*Proof of a)* : According to Lemma 5.4.3 we may construct i.i.d. random variables $X_1^*, ..., X_n^*$ with continuous distribution functions such that $Q_i^*$ is equal to the rank of $|X_i^*|$ in the absolute $X^*$ –values and $\Delta_i^* = \text{sign}(X_i^*)$ . Therefore the limiting law (5.4.43) is the well-known limiting law of signed linear rank statistics under the continuous null hypothesis. $\square$

Before proving part b) let's give some remarks:

Obviously the vector $\tau = (\tau_1, ..., \tau_d)$ of the tie-lengths of the non-zero values among $|X|^{(1)} \le ... \le |X|^{(n)}$ defines an element $\mathcal{J}_n^\tau$ of $\mathcal{J}_{0,1}^s$ according to

$$\mathcal{J}_n^\tau := \left\{ \left( \frac{1}{2} - \frac{1}{2}\frac{T_j}{n}, \frac{1}{2} - \frac{1}{2}\frac{T_{j-1}}{n} \right], \left( \frac{1}{2} + \frac{1}{2}\frac{T_{j-1}}{n}, \frac{1}{2} + \frac{1}{2}\frac{T_j}{n} \right], 1 \le j \le n \right\}. \quad (5.4.46)$$

Using $\mathcal{J}_n^\tau$ and the definition (3.3.49) of $L_{\mathcal{J}_n^\tau}$ we get from (5.4.29) and (5.1.28) the equalities

$$S_n^\tau(b) = S_n^*(L_{\mathcal{J}_n^\tau} b) \quad (5.4.47)$$

and

$$\|L_{\mathcal{J}_n^\tau} b\|^2 = \frac{1}{n} \sum_{i=1}^{n} \left( b_n^\tau(i) \right)^2. \quad (5.4.48)$$

Here and below we define for any distribution function $F \in \mathcal{F}_1$ with $F(0) \ge 0$ a distribution function $F^* \in \mathcal{H}_0^s$ by

$$F^*(t) = \begin{cases} \frac{1}{2} + \frac{1}{2} F(t), & \text{if } t \ge 0, \\ \frac{1}{2} - \frac{1}{2} F\left((-t)_-\right), & \text{if } t < 0. \end{cases} \quad (5.4.49)$$

If $\hat{H}_n$ is the empirical distribution function of the absolute values $|X_1|, ..., |X_n|$ and if $\mathcal{I}(\hat{H}_n^*) = \left\{ \left( \hat{H}_n^*(x_-), \hat{H}_n^*(x) \right] : x \in \mathbb{R} \right\}$ is the family of jump-intervals of $\hat{H}_n^*$, then we have in addition

$$\mathcal{I}(\hat{H}_n^*) = \mathcal{J}_n^\tau, \quad (5.4.50)$$

i.e. the family $\mathcal{I}(\hat{H}_n^*) \in \mathcal{J}_{0,1}^s$ only depends on $\tau$.

*Proof of part b) :*

Assume $F = F_- \in \mathcal{H}_0^s(\mathcal{J})$, i.e. $\mathcal{J} = \mathcal{I}(F)$, and let $\hat{H}_n$ denote the empirical distribution function of the absolute observations $|X_1|, ..., |X_n|$ having distribution function H, say. Then $\|\hat{H}_n - H\|_\infty \to 0$ in probability as well as $\|\hat{H}_n^* - F\|_\infty \to 0$ in probability, since $H^* = F$. Now Lemma 7.5.5 implies

$$\|L_{\mathcal{J}_n^\tau} b - L_{\mathcal{J}} b\| \overset{n \to \infty}{\longrightarrow} 0 \quad \text{in } F\text{-probability.} \quad (5.4.51)$$

Since the right-hand side of (5.4.48) is distribution-free under $\mathcal{H}_0^s(\mathcal{J})$, c.f. Proposition 5.4.1, the proof of (5.4.45) follows from (5.4.48) and (5.4.51).

For the proof of (5.4.44) we prove

$$S_n^\tau(b) - S_n^*(L_{\mathcal{J}} b) \overset{n \to \infty}{\longrightarrow} 0 \quad \text{in } \mathcal{H}_{0n}^s(\mathcal{J})\text{-probability} \quad (5.4.52)$$

and apply part a) with $L_{\mathcal{J}} b$ instead of $b$.

For the final proof of (5.4.52) we apply (5.4.48) and the independence of $Q^*$, $\Delta^*$, and $\tau$ under $\mathcal{H}_{0n}^s$ in order to get

$$E_0\left[\left(S_n^\tau(b)-S_n^*(L_{\mathcal{J}}b)\right)^2 \mid \tau\right] = E_0\left[\left(S_n^*(L_{\mathcal{J}_n^*}b-L_{\mathcal{J}}b)\right)^2 \mid \tau\right] \leq \|L_{\mathcal{J}_n^*}b-L_{\mathcal{J}}b\|^2.$$

Therefore (5.4.51) implies (5.4.52). $\square$

For any given level $0 < \alpha < 1$ Theorem 5.4.4 implies the *randomized signed linear rank test*

$$\psi_n^*(b) = 1\left(S_n^*(b) \geq u_\alpha\,\|b\|\right) \tag{5.4.53}$$

and the *averaged scores signed linear rank test*

$$\psi_n^\tau(b) = 1\left(S_n^\tau(b) \geq u_\alpha\sqrt{\frac{1}{n}\sum_{i=1}^n\left(b_n^\tau(i)\right)^2}\right) \tag{5.4.54}$$

to be asymptotically of level $\alpha$ for testing the null hypothesis $\mathcal{H}_{0n}^s(\mathcal{J})$, if the condition $\|L_{\mathcal{J}}b\| > 0$ holds true.

**Remark:** *In Theorem 5.4.4 and in the definition of the asymptotic tests (5.4.53) and (5.4.54) the score function $b \in L_2^0(0,1)$ may be substituted by any $b_n \in L_2^0(0,1)$ such that $\|b_n - b\| \to 0$ as $n \to \infty$.*

### 5.4.5 Theorem

*Assume $\mathcal{J} \in \mathcal{J}_{0,1}^s$ and let $(H_n,\ n \geq 1)$ be any sequence of hypothesis points $H_n \in \mathcal{H}_0^s(\mathcal{J})$. Define a corresponding sequence of local asymptotic alternatives $(F_n,\ n \geq 1)$ according to*

$$F_n = H_n + n^{-1/2}B_n \circ H_n, \qquad B_n \in \mathcal{B}_n^s(\mathcal{J}),$$

$$\|B_n' - b\| \overset{n\to\infty}{\longrightarrow} 0 \quad \text{for some } b \in L_2^0(0,1). \tag{5.4.55}$$

*a) For any score function $h \in L_2^0(0,1)$ such that $h(t) = -h(1-t)$ for $\lambda$-almost all $t \in (0,1)$, we have the limiting law (as $n \to \infty$)*

$$\mathcal{L}[\,S_n^\tau(h) \mid F_n\,] \overset{\mathcal{L}}{\longrightarrow} \mathcal{N}(\,<L_{\mathcal{J}}h,b>,\ \|L_{\mathcal{J}}h\|^2\,), \tag{5.4.56}$$

*where $L_{\mathcal{J}}h \in L_2^0(0,1)$ is defined by formula (3.3.49).*

*b) If $\|b\| > 0$ and if $h \in L_2^0(0,1)$ such that $L_{\mathcal{J}}h = b$ $[\lambda - a.e.]$, then the averaged scores signed linear rank test $\psi_n^\tau(h)$ defined by formula (5.4.54) is asymptotically optimal at level $\alpha$ for testing the null hypothesis $\mathcal{H}_0^s(\mathcal{J})$ versus $(F_n,\ n \geq 1)$.*

*Proof:* Because of Proposition 5.4.1,

$$\tau_0 \;=\; n - \sum_{j=1}^{n} \, |\mathrm{sign}(X_j)| \,,$$

and

$$\{R_1^\dagger, ..., R_n^\dagger\} \;=\; \begin{cases} \{T_0, ..., T_d\}, & \text{if } \tau_0 > 0, \\[2mm] \{T_1, ..., T_d\}, & \text{if } \tau_0 = 0, \end{cases}$$

we may assume without loss of generality $H_n = H \in \mathcal{H}_0^s(\mathcal{J}) \;\; \forall\, n \geq 1$ . On the basis of $H$ define the randomized random variables $X_1^*, ..., X_n^*$ according to Lemma 5.4.3. If the null hypothesis $B_n = 0$ is assumed, then $X_1^*, ..., X_n^*$ are i.i.d. random variables with absolutely continuous distribution function $H^*(x) = (x+1)/2 \;\; \forall\, x \in [-1, 1]$ . Under the alternative $F_n = H + n^{-1/2} B_n \circ H$ the random variables $X_1^*, ..., X_n^*$ are i.i.d. with absolutely continuous distribution function $F_n^* = H^* + n^{-1/2} B_n \circ H^*$ . Especially we get the result that the sequence ( $\mathcal{L}[(X_1^*, ..., X_n^*) \mid F_n], \; n \geq 1$ ) is contiguous to the sequence ( $\mathcal{L}[(X_1^*, ..., X_n^*) \mid H], \; n \geq 1$ ) , cf. Section 7.1.

a) Using the abbreviation $h_{\mathcal{J}} = L_{\mathcal{J}} h$ we define

$$S_n^*(h_{\mathcal{J}}) \;:=\; \frac{1}{\sqrt{n}} \sum_{i=1}^{n} h_{\mathcal{J}n}(Q_i^*) \, \mathrm{sign}(X_i^*) \tag{5.4.57}$$

and

$$T_n^*(h_{\mathcal{J}}) \;:=\; \frac{1}{\sqrt{n}} \sum_{i=1}^{n} h_{\mathcal{J}}\left(\frac{1}{2} + \frac{1}{2}|X_i^*|\right) \, \mathrm{sign}(X_i^*)$$

$$\tag{5.4.58}$$

$$\overset{a.s.}{=} \; \frac{1}{\sqrt{n}} \sum_{i=1}^{n} h_{\mathcal{J}}\left(\frac{1}{2} + \frac{1}{2}X_i^*\right).$$

Then, on one hand formula (5.4.52) implies

$$S_n^\tau(h) - S_n^*(h_{\mathcal{J}}) \;\overset{n \to \infty}{\longrightarrow}\; 0 \quad \text{in } H - \text{probability} \,, \tag{5.4.59}$$

on the other hand (5.4.35) and Section V.1.7 of Hájek and Šidák (1967) imply

$$S_n^*(h_{\mathcal{J}}) - T_n^*(h_{\mathcal{J}}) \;\overset{n \to \infty}{\longrightarrow}\; 0 \quad \text{in } H - \text{probability} \,. \tag{5.4.60}$$

Combining (5.4.59) and (5.4.60), and using contiguity we get

$$S_n^\tau(h) - T_n^*(h_{\mathcal{J}}) \;\overset{n \to \infty}{\longrightarrow}\; 0 \quad \text{in } F_n - \text{probability} \,. \tag{5.4.61}$$

Since $1/2 + X_1^*/2, ..., 1/2 + X_n^*/2$ are i.i.d. random variables with Lebesgue-density $g_n^*$ according to

$$g_n^*(t) = 1 + n^{-1/2} B_n'(t), \qquad 0 < t < 1, \tag{5.4.62}$$

and since $B_n \in \mathcal{B}_n^s(\mathcal{J})$ implies $[\lambda - a.e.]$

$$B_n'(t) = -B_n'(1-t) \qquad \text{and} \qquad -\sqrt{n} \le B_n'(t) \le \sqrt{n}, \qquad (5.4.63)$$

the central limit theorem and $\|B_n' - b\| \to 0$ prove

$$\mathcal{L}[\, T_n^*(h_{\mathcal{J}}) \mid F_n \,] \xrightarrow{\mathcal{L}} \mathcal{N}(\, <h_{\mathcal{J}}, b>, \, \|h_{\mathcal{J}}\|^2 \,). \qquad (5.4.64)$$

Obviously (5.4.64) and (5.4.61) yield (5.4.56).

b) Because of $B_n \in \mathcal{B}_n^s(\mathcal{J})$ and $\|B_n' - b\| \to 0$ we have for $\lambda$-almost all $t \in (0,1)$

$$b(t) = -b(1-t) \qquad \text{and} \qquad b_{\mathcal{J}}(t) = b(t). \qquad (5.4.65)$$

Completely similar to the proof of Lemma 3.3.9.c we get from (5.4.62) and $H = (F_n + F_{n-})/2$

$$\frac{dF_n}{dH} \circ H^{-1} = 1 + n^{-1/2} B_n' \qquad [\lambda - a.e.]. \qquad (5.4.66)$$

Thus formula (5.4.41) implies with probability 1

$$\frac{dF_n}{dH}(X_i) = 1 + n^{-1/2} B_n'\left(\frac{1}{2} + \frac{1}{2} X_i^*\right), \qquad i = 1, ..., n, \qquad (5.4.67)$$

which implies that the (optimal) log-likelihood-ratio statistic $L_n$ for testing the simple hypothesis $H$ versus the simple alternative $F_n$ has the form

$$\begin{aligned}
L_n &= \sum_{i=1}^{n} \log\left(1 + n^{-1/2} B_n'\left(\frac{1}{2} + \frac{1}{2} X_i^*\right)\right) \\
&= \frac{1}{\sqrt{n}} \sum_{i=1}^{n} b\left(\frac{1}{2} + \frac{1}{2} X_i^*\right) - \frac{1}{2}\|b\|^2 + \Delta_n, \qquad (5.4.68)
\end{aligned}$$

where $\Delta_n \to 0$ in $H$-probability, and because of contiguity,

$$\Delta_n \xrightarrow{n \to \infty} 0 \qquad \text{in} \quad F_n - \text{probability} . \qquad (5.4.69)$$

Therefore (5.4.65) and (5.4.61) imply

$$L_n + \frac{1}{2}\|b\|^2 - S_n^{\tau}(h) \xrightarrow{n \to \infty} 0 \qquad \text{in} \quad F_n - \text{probability} \qquad (5.4.70)$$

and in $H$-probability, too. Because of $\|b\| = \|b_{\mathcal{J}}\|^2 > 0$ the convergence (5.4.70) and Theorem 5.4.4 imply that $\psi_n^{\tau}(h)$ is asymptotically equivalent to the (optimal) Neyman-Pearson test at level $\alpha$ for testing the simple hypothesis $H$ versus the simple alternative $F_n$ . $\square$

### 5.4.6 Corollary

*Under the assumptions of Theorem 5.4.5.a we have the limiting law* $(n \to \infty)$

$$\mathcal{L}[\, S_n^*(h) \mid F_n \,] \xrightarrow{\mathcal{L}} \mathcal{N}(\, < L_{\mathcal{J}} h, b >,\ \|h\|^2\, ), \tag{5.4.71}$$

*where* $S_n^*(h)$ *is the randomized signed linear rank statistic*

$$S_n^*(h) = \frac{1}{\sqrt{n}} \sum_{i=1}^{n} h_n(Q_i^*)\, \Delta_i^* \tag{5.4.72}$$

*with* $Q_i^*$ *and* $\Delta_i^*$ *defined in (5.4.27) and (5.4.28), and* $h_n(i)$, $i = 1, ..., n$, *defined by formula (5.1.28).*

*Proof:* From $L_{\mathcal{J}} b = b$ $[\lambda - a.e.]$ we get $< h, b > \, = \, < L_{\mathcal{J}} h, b >$ . Defining $T_n^*(h)$ as in (5.4.58) the proof of Theorem 5.4.5.a yields $S_n^*(h) - T_n^*(h) \to 0$ in $F_n$ –probability and $\mathcal{L}[\, T_n^*(h) \mid F_n \,] \xrightarrow{\mathcal{L}} \mathcal{N}(\, < h, b >,\ \|h\|^2\, ) \cdot \square$

Under the assumptions of Theorem 5.4.5 the averaged scores signed linear rank statistic $S_n^{\tau}(b)$ is asymptotically optimal for testing $\mathcal{H}_0^s(\mathcal{J})$ versus $F_n = H + n^{-1/2} B_n \circ H$ . Since Lemma 5.4.3 implies that the corresponding randomized random variables $X_1^*, ..., X_n^*$ are i.i.d. with absolutely continuous distribution function $H_0 + n^{-1/2} B_n \circ H_0$ , Section 5.2 yields that the *randomized signed rank process* $\hat{B}_{n2}^*$ according to

$$\hat{B}_{n2}^*\Big(\frac{i}{n}\Big) = -\frac{1}{\sqrt{n}} \sum_{j=1}^{n} 1(Q_j^* > i)\, \Delta_j^*, \qquad 0 \le i \le n, \tag{5.4.73}$$

$$\hat{B}_{n2}^* \ \text{ linear on } \ [\frac{i-1}{n}, \frac{i}{n}] \ \text{ for any } \ 1 \le i \le n,$$

will be a suitable estimator of the underlying $\tilde{B}_n(t) = 2 B_n(1/2 + t/2)$ , and the right-continuous derivative

$$\hat{b}_{n2}^*(t) = \sum_{i=1}^{n} n\Big(\hat{B}_{n2}^*(\frac{i}{n}) - \hat{B}_{n2}^*(\frac{i-1}{n})\Big)\, 1\Big(\frac{i-1}{n} \le t < \frac{i}{n}\Big) \tag{5.4.74}$$

$$= \sqrt{n} \sum_{i=1}^{n} 1(\, Q_i^* = [nt] + 1\, )\, \Delta_i^*$$

may serve as a primitive estimator of $\tilde{b}(t) = b(1/2 + t/2)$ . Additionally the properties

$$E[\, \Delta_i^* \mid \text{sign}(X_i)\, ] = \text{sign}(X_i), \tag{5.4.75}$$

$$P[\, Q_i^* = j \mid R^+ \,] \;=\; \frac{1}{T_k}\, 1(\, T_{k-1} < j \le T_k\,), \quad \text{if } R_i^+ = T_k, \qquad (5.4.76)$$

and formula (5.2.10) imply the representation

$$\begin{aligned}
S_n^\tau(b) &= E[\, S_n^*(b) \mid R^+, \mathrm{sign}(X)\,] = E[\, <\tilde b, \hat b_{n2}^* > \mid R^+, \mathrm{sign}(X)\,] \\
&= <\tilde b,\ E[\hat b_{n2}^* \mid R^+, \mathrm{sign}(X)] > \; = \; <\tilde b, \hat b_{n2}>,
\end{aligned} \qquad (5.4.77)$$

where

$$\hat b_{n2}(t) \;=\; \sum_{i=1}^{n} n\, \left( \hat B_{n2}(\frac{i}{n}) - \hat B_{n2}(\frac{i-1}{n}) \right) 1(\, \frac{i-1}{n} \le t < \frac{i}{n}\,) \qquad (5.4.78)$$

is the right-continuous derivative of the *averaged signed rank process* $\hat B_{n2}$ which is defined by

$$\hat B_{n2}(t) \;=\; E[\, \hat B_{n2}^*(t) \mid R^+, \mathrm{sign}(X)\,], \qquad 0 \le t \le 1. \qquad (5.4.79)$$

With the same arguments as in Section 3.3 we'll estimate the unknown $\tilde b$ in case of the omnibus alternative $B_n \ne 0$ by the kernel estimator $\mathcal{K}_a \hat b_{n2}$ , and in case of the one-sided alternative $B_n \le 0$, $B_n \ne 0$ by the kernel estimator $\mathcal{K}_a \hat b_{n2}^0$ , where

$$\hat b_{n2}^0 \;=\; \hat b_{n2} 1(\hat B_{n2} < 0) \qquad (5.4.80)$$

is a special version of the derivative of the "projection" $\hat B_{n2}^0$ of $\hat B_{n2}$ onto the cone of nonpositive functions on $[0,1]$ , i.e.

$$\hat B_{n2}^0 = \hat B_{n2} 1(\hat B_{n2} < 0). \qquad (5.4.81)$$

As in the two-sample case the processes (5.4.79) and (5.4.81) coincide with the respective processes (5.2.4) and (5.2.8) if no ties are present.

Because of (5.4.78) and (5.2.13) the resulting *omnibus statistic*

$$\bar S_n(a, K) \;:=\; <\mathcal{K}_a \hat b_{n2}, \hat b_{n2}> \qquad (5.4.82)$$

can be represented in the form (5.2.13). Completely similar to (5.2.14) we may use the approximation

$$S_n(a, K) \;:= \qquad\qquad\qquad\qquad\qquad\qquad\qquad\qquad (5.4.83)$$

$$\sum_{i=1}^{n}\sum_{j=1}^{n} \left( \hat B_{n2}(\frac{i}{n}) - \hat B_{n2}(\frac{i-1}{n}) \right) \left( \hat B_{n2}(\frac{j}{n}) - \hat B_{n2}(\frac{j-1}{n}) \right) k_{na}(i,j),$$

if the kernel $K$ is smooth, e.g. the Parzen-2 kernel.

Also the resulting *one-sided statistic*

$$\hat{S}_n^0(a, K) \ := \ < \mathcal{K}_a \hat{b}_{n2}^0, \hat{b}_{n2} > \tag{5.4.84}$$

can be represented in the form (5.2.16) and, in case of smooth kernels, can be approximated by

$$S_n^0(a, K) \ := \tag{5.4.85}$$

$$\sum_{i=1}^{n} \sum_{j=1}^{n} \left( \hat{B}_{n2}^0(\frac{i}{n}) - \hat{B}_{n2}^0(\frac{i-1}{n}) \right) \left( \hat{B}_{n2}(\frac{j}{n}) - \hat{B}_{n2}(\frac{j-1}{n}) \right) k_{na}(i,j).$$

**Evaluation of the Tests:**

In the case of underlying discontinuous distribution functions the rank vector $R^+ = (R_1^+, ..., R_n^+)$ and the sign vector $\text{sign}(X) = (\text{sign}(X_1), ..., \text{sign}(X_n))$ are not distribution-free under the general null hypothesis $\mathcal{H}_0^s : F = F_-$, c.f. Proposition 5.4.1. But according to Lemma 5.4.3 the vector of randomized ranks $Q^* = (Q_1^*, ..., Q_n^*)$ and the vector of randomized signs $\Delta^* = (\Delta_1^*, ..., \Delta_n^*)$ are distribution-free under $\mathcal{H}_0^s$.

Therefore the conditional distribution of $R^+$ and $\text{sign}(X)$ given the tie-lengths vector $\tau = (\tau_1, ..., \tau_d)$ of the non-zero values among the ordered absolute observations $|X|^{(1)} \leq \cdots \leq |X|^{(n)}$ will be distribution-free according to the subsequent lemma.

**Lemma 5.4.7**

*Let the components of $X = (X_1, ..., X_n)$ be i.i.d. real random variables with arbitrary underlying distribution function $F \in \mathcal{F}_1$.*

*Let the vectors $R^+ = (R_1^+, ..., R_n^+)$, $\text{sign}(X) = (\text{sign}(X_1), ..., \text{sign}(X_n))$, and $\tau = (\tau_1, ..., \tau_d)$ denote the ranks of the absolute values $|X| = (|X_1|, ..., |X_n|)$, the signs of the $X$ –components, and the lengths of the ties of the non-zero values in the ordered $|X|$ –sample, respectively.*

*Then $\mathcal{L}_F[(\text{sign}(X), R^+) \mid \tau]$, the conditional distribution of $(\text{sign}(X), R^+)$ given $\tau$, is the same for any null hypothesis point $F \in \mathcal{H}_0^s$, i.e. the conditional distribution of $(\text{sign}(X), R^+)$ given $\tau$ is distribution-free under the general null hypotheses of symmetry (5.4.1).*

*Proof* : Define the randomized ranks $Q^* = (Q_1^*, ..., Q_n^*)$ and the randomized signs $\Delta^* = (\Delta_1^*, ..., \Delta_n^*)$ according to (5.4.27) and (5.4.28). Then, under the assumption $F \in \mathcal{H}_0^s$, we get from Lemma 5.4.3 $(B = 0)$ :

- $\Delta^*$ , $Q^*$ , $\tau$ are stochastically independent,

- $\Delta^*$ is uniformly distributed on $\{-1, 1\}^n$ ,

- $Q^*$ is uniformly distributed on the permutations of $(1, ..., n)$ .

Therefore the following representations of $\text{sign}(X)$ and $R^+$ in terms of the random variables $\Delta^*$ , $Q^*$ , and $\tau$ concludes the proof:

If $\tau_0 = n$ , then we have ( $\forall\, i = 1, ..., n$ )

$$\text{sign}(X_i) = 0 \qquad \text{and} \qquad R_i^+ = n. \tag{5.4.86}$$

If $\tau_0 < n$ , then we have $d \geq 1$ and ( $\forall\, i = 1, ..., n$ )

$$\text{sign}(X_i) = \Delta_i^* \, 1(Q_i^* > \tau_0),$$

$$R_i^+ = \tau_0 \, 1(Q_i^* \leq \tau_0) + \sum_{j=1}^{d} T_j \, 1\big(T_{j-1} < Q_i^* \leq T_j\big), \tag{5.4.87}$$

where $T_j = \tau_0 + \cdots + \tau_j$ , cf. (5.4.22) and (5.4.23). $\square$

Obviously (5.4.86) and (5.4.87) may be used in order to simulate the conditional distribution of $\big(\text{sign}(X), R^+\big)$ given $\tau$ under the null hypothesis $\mathcal{H}_0^s$ of symmetry:

Since $\tau_0 = n$ implies $\text{sign}(X_i) = 0$ and $R_i^+ = n$ for any $i = 1, ..., n$ , we may restrict the discussion to the non-degenerate case $\tau_0 < n$ .

Let $\tau = (\tau_1, ..., \tau_d)$ with $\tau_0 < n$ $(d \geq 1)$ be given and let $U_1, ..., U_n$ be i.i.d. random variables with uniform distribution on the interval $(-1, 1)$ . Evaluate $\Delta^* = (\Delta_1^*, ..., \Delta_n^*)$ and $Q^* = (Q_1^*, ..., Q_n^*)$ according to

$$\Delta_i^* = \text{sign}(U_i), \qquad Q_i^* = \sum_{j=1}^{n} 1\big(|U_j| \leq |U_i|\big). \tag{5.4.88}$$

On the basis of $\Delta^*$ , $Q^*$ , and the given $\tau$ evaluate $\Delta = (\Delta_1, ..., \Delta_n)$ and $Q = (Q_1, ..., Q_n)$ according to

$$\Delta_i = \begin{cases} 0, & \text{if } Q_i^* \leq \tau_0, \\ \Delta_i^*, & \text{if } Q_i^* > \tau_0, \end{cases} \tag{5.4.89}$$

$$Q_i = \begin{cases} \tau_0, & \text{if } Q_i^* \leq \tau_0, \\ T_j, & \text{if } T_{j-1} < Q_i^* \leq T_j \quad \text{for some } j \in \{1, ..., d\}, \end{cases} \tag{5.4.90}$$

where  $T_j = \tau_0 + \cdots + \tau_j$ .

Then we have

$$\mathcal{L}\big[(\Delta, Q)\big] \;=\; \mathcal{L}_{\mathcal{H}_0^*}\big[(\operatorname{sign}(X), R^+) \mid \tau\big]. \tag{5.4.91}$$

For the practical evaluation of the signed rank process  $\hat{B}_{n2}$  the following representation seems to be most convenient:

Because of (5.4.73) and (5.4.79)  $\hat{B}_{n2}$  is linear on  $[(i-1)/n, i/n]$  for any  $i = 1, ..., n$ , and

$$\Delta \hat{B}_{n2}(\frac{i}{n}) \;:=\; \hat{B}_{n2}(\frac{i}{n}) - \hat{B}_{n2}(\frac{i-1}{n})$$
$$= \; \frac{1}{\sqrt{n}} \sum_{j=1}^{n} P\big[Q_j^* = i \mid R^+, \operatorname{sign}(X)\big] \operatorname{sign}(X_j). \tag{5.4.92}$$

In case of  $i \le \tau_0$  obviously the assumption  $\operatorname{sign}(X_j) \ne 0$  implies the equality  $P[Q_j^* = i \mid R^+, \operatorname{sign}(X)] = 0$ , i.e. we have

$$\Delta \hat{B}_{n2}(\frac{i}{n}) = 0 \quad \text{if} \quad 1 \le i \le \tau_0. \tag{5.4.93}$$

Now let's assume  $d \ge 1$  and  $T_{q-1} < i \le T_q$  for some  $1 \le q \le d$ . In this case we get from (5.4.76)

$$\Delta \hat{B}_{n2}(\frac{i}{n}) \;=\; \frac{1}{\sqrt{n}} \sum_{j=1}^{n} \frac{1}{\tau_q} \, 1(R_j^+ = T_q) \, \operatorname{sign}(X_j)$$
$$= \; \frac{1}{\sqrt{n}} \frac{\tau_q^+ - \tau_q^-}{\tau_q} \quad (\ \text{if} \quad T_{q-1} < i \le T_q\ ), \tag{5.4.94}$$

where

$$\tau_q^+ \;=\; \sum_{j=1}^{n} 1\big(X_j > 0,\, R_j^+ = T_q\big),$$
$$\tau_q^- \;=\; \sum_{j=1}^{n} 1\big(X_j < 0,\, R_j^+ = T_q\big). \tag{5.4.95}$$

Obviously  $\tau_q^+$  and  $\tau_q^-$  are the respective numbers of *positive*  and *negative*  observations which contribute to the  $q$ –th tie of the ordered absolute values of  $X$ . Because of  $\hat{B}_{n2}(1) = 0$  or

$$\hat{B}_{n2}(0) \;=\; -\frac{1}{\sqrt{n}} \sum_{j=1}^{n} \operatorname{sign}(X_j) \;=\; -\frac{1}{\sqrt{n}} \sum_{q=1}^{d} \big(\tau_q^+ - \tau_q^-\big) \tag{5.4.96}$$

the evaluation of $\hat{B}_{n2}$ is complete.

As a consequence of Lemma 5.4.7 we'll use the conditional tests (given $\tau$ )
based on the nonlinear signed rank statistics $S_n(a, K)$ (in the omnibus case)
and $S_n^0(a, K)$ (in the one-sided case), i.e.

$$\psi_{n\alpha} = 1\Big( S_n(a, K) > q_\alpha(\tau) \Big),$$

$$\psi_{n\alpha}^0 = 1\Big( S_n^0(a, K) > q_\alpha^0(\tau) \Big). \tag{5.4.97}$$

Formulae (5.4.88) to (5.4.91) show an explicit way for constructing the conditional critical values $q_\alpha(\tau)$ and $q_\alpha^0(\tau)$ or the conditional p-values of the
respective test statistics $S_n(a, K)$ and $S_n^0(a, K)$. But the evaluation may become very time-consuming for larger sample sizes. Starting with an observed $\tau$
and using a $\mathcal{R}(-1, 1)$ random generator we utilize the representation (5.4.91)
in order to simulate the conditional p-values of $S_n^0(a, K)$ for the Parzen-2
kernel (3.1.49) and the bandwidth a=0.40 . An illustration of the practical
evaluation of the process $\hat{B}_{n2}$ via (5.4.94) to (5.4.96) is given in Numerical
Example 2.6.1.


**The Asymptotic Distribution:**

The asymptotics of the signed rank statistics (5.4.82)-(5.4.85) is based on the
following limiting law of the averaged signed rank process $\hat{B}_{n2}$ .


**5.4.8 Theorem**

*Assume $\mathcal{J} \in \mathcal{J}_{0,1}^s$ and let $(H_n, n \geq 1)$ be any sequence of hypothesis points
$H_n \in \mathcal{H}_0^s(\mathcal{J})$ . Define a corresponding sequence of local alternatives $(F_n, n \geq
1)$ according to formula (5.4.55). Then the following limiting law ( $n \to \infty$ )
holds true in $C[0, 1]$ ,*

$$\mathcal{L}[\ \hat{B}_{n2} \mid F_n\ ] \xrightarrow{\mathcal{L}} \mathcal{L}[\ S_{\mathcal{J}} W^* + \tilde{B}\ ] \tag{5.4.98}$$

*with $W^*$ and $\tilde{B}$ as in Theorem 5.2.1 and $S_{\mathcal{J}} : C[0, 1] \to C[0, 1]$ according
to ( $\forall f \in C[0, 1]$ )*

$$(S_{\mathcal{J}} f)(t) = \begin{cases} f(2y - 1), & \text{if } 1/2 + t/2 \in (1 - y, y] \in \mathcal{J}, \\ \frac{2y-1-t}{2y-2x} f(2x - 1) + \frac{1+t-2x}{2y-2x} f(2y - 1), \\ \qquad \text{if } 1/2 + t/2 \in (x, y] \in \mathcal{J} \text{ and } x > 1/2, \\ f(t), & \text{elsewhere.} \end{cases} \tag{5.4.99}$$

Under the special null hypothesis $F = F_- \in \mathcal{H}_0^s(\mathcal{J})$ we have the (conditional) limiting law

$$\mathcal{L}_F[\,\hat{B}_{n2} \mid \tau\,] \xrightarrow{\;\mathcal{L}\;} \mathcal{L}[\,S_{\mathcal{J}} W^*\,] \qquad [a.s.] \qquad\qquad (5.4.100)$$

in $\left(C[0,1],\, \mathcal{B}(C[0,1])\right)$ .

*Proof:* Define the bijective, linear mapping

$$\sigma : C[0,1] \to C^*[0,1] := \left\{ f \in C[0,1] : f(t) = f(1-t) \quad \forall\, t \in [0,1] \right\}$$

by

$$(\sigma f)\left(\frac{1}{2} + \frac{1}{2}t\right) \;=\; (\sigma f)\left(\frac{1}{2} - \frac{1}{2}t\right) \;=\; f(t) \qquad \forall\, t \in [0,1].$$

Then, for any $\mathcal{J} \in \mathcal{J}_{0,1}^s$ , we have the representation $S_{\mathcal{J}} = \sigma^{-1} \circ T_{\mathcal{J}} \circ \sigma$ with the mapping $T_{\mathcal{J}}$ defined in (3.3.106). According to formulae (5.4.94) to (5.4.96) the signed rank process $\hat{B}_{n2}$ fulfils the side condition $\hat{B}_{n2}(0) = \hat{B}_{n2}(T_0/n)$ and is linear on $[T_{i-1}/n, T_i/n]$ with $\hat{B}_{n2}(T_i/n) = \hat{B}_{n2}^*(T_i/n)$ for $i = 1, ..., d$ , c.f. (5.4.73) and (5.4.87). In other words,

$$\hat{B}_{n2} = S_{\mathcal{J}_n^\tau} \hat{B}_{n2}^* \qquad\qquad (5.4.101)$$

with $\mathcal{J}_n^\tau = \mathcal{I}(\hat{H}_n^*)$ , c.f. (5.4.46) and (5.4.50). Because of $\|\sigma^{-1} g\| = \|g\|$ $\forall\, g \in C^*[0,1]$ , $\omega(\sigma f, \varepsilon) = \omega(f, 2\varepsilon)$ $\forall\, f \in C[0,1]$ , $\mathcal{I}(H_n) = \mathcal{J}$ , formula (5.4.101), and Lemma 7.5.4 we have the inequality

$$P\big\{\, \|\hat{B}_{n2} - S_{\mathcal{J}} \hat{B}_{n2}^*\|_\infty \geq \eta \,\big\}$$
$$\qquad\qquad (5.4.102)$$
$$\leq\; P\big\{\, \omega(\hat{B}_{n2}^*, 2\varepsilon) \geq \eta/23 \,\big\} \;+\; P\big\{\, \|\hat{H}_n^* - H_n\|_\infty \geq \varepsilon/2 \,\big\} ,$$

for any $\eta > 0$ and any $\varepsilon \in (0,1)$ .
Therefore the continuity of $f \to \omega(f, \varepsilon)$ , the continuity of $W^* + \tilde{B}$ , and $\|\hat{H}_n^* - H_n\|_\infty \to 0$ in probability, prove $\|\hat{B}_{n2} - S_{\mathcal{J}} \hat{B}_{n2}^*\|_\infty \to 0$ in probability. Hence (5.4.98) follows from the continuity of $S_{\mathcal{J}}$ and from $\hat{B}_{n2}^* \xrightarrow{\;\mathcal{L}\;} W^* + \tilde{B}$ , where the latter result is a consequence of Theorem 5.2.1 and Lemma 5.4.3.

Under $F = F_- \in \mathcal{H}_0^s(\mathcal{J})$ we have $\|\hat{H}_n^* - F\|_\infty \to 0$ $[a.s.]$ . For any fixed sequence $(\hat{H}_n^*, n \geq 1)$ with $\|\hat{H}_n^* - F\|_\infty \to 0$ let $(\mathcal{J}_n := \mathcal{I}(\hat{H}_n^*), n \geq 1)$ and $(\hat{\tau}_n, n \geq 1)$ be the corresponding sequences of (non-random) jump-interval systems and tie-lengths vectors. Then formula (5.4.101), $\mathcal{I}(F) = \mathcal{J}$ , Lemma 7.5.4, and $\hat{B}_{n2}^* \xrightarrow{\;\mathcal{L}\;} W^*$ imply

$$\mathcal{L}_F[\,\hat{B}_{n2} \mid \tau = \hat{\tau}_n\,] \;=\; \mathcal{L}_0[\,S_{\mathcal{J}_n} \hat{B}_{n2}^*\,] \;\xrightarrow{\;\mathcal{L}\;}\; \mathcal{L}[\,S_{\mathcal{J}} W^*\,]. \qquad \square$$

**Approximating the optimal score function by projections onto finite dimensional cones in the presence of ties**

Our aim is to extend the (continuous) results of Section 5.3 to the case when ties are present. In accordance with the continuous case let $b_1, ..., b_r \in L_2^0(0,1)$ be special score functions such that $\tilde{b}_i(t) := b(1/2 + t/2)$, $0 < t < 1$, $1 \leq i \leq r$, fulfil the side condition (5.3.2) of the one-sided model. Define

$$\vec{S}_n = (< \tilde{b}_1, \hat{b}_{n2} >, ..., < \tilde{b}_r, \hat{b}_{n2} >)^T \tag{5.4.103}$$

as in (5.3.5), but with the general $\hat{b}_{n2}$ defined in (5.4.78). If no ties are present, $\hat{b}_{n2}$ coincides with its original definition (5.2.7). Using (5.4.77), (5.4.47), (5.4.74), and (5.2.10) we get

$$< \tilde{b}_i, \hat{b}_{n2} > = S_n^\tau(b_i) = S_n^*(L_{\mathcal{J}_n^\tau} b_i) = < (L_{\mathcal{J}_n^\tau} b_i)^\sim, \hat{b}_{n2}^* > \tag{5.4.104}$$

with $\mathcal{J}_n^\tau$ from (5.4.46) and $\tau = (\tau_1, ..., \tau_d)$ the tie-lengths' vector of the non-zero values among $|X|^{(1)} \leq ... \leq |X|^{(n)}$ .

Now we are in the position to prove the following convergence result.

### 5.4.9 Theorem

*Assume $\mathcal{J} \in \mathcal{J}_{0,1}^s$ and let $(H_n, n \geq 1)$ be any sequence of hypothesis points $H_n \in \mathcal{H}_0^s(\mathcal{J})$ . Define a corresponding sequence of local alternatives $(F_n,\ n \geq 1)$ according to formula (5.4.55).*

*a) Under the above assumptions we have the limiting law*

$$\mathcal{L}[\vec{S}_n \mid F_n] \xrightarrow{\mathcal{L}} \mathcal{N}(\vec{\mu}_{\mathcal{J}}, \Gamma_{\mathcal{J}}) \tag{5.4.105}$$

*with mean vector $\vec{\mu}_{\mathcal{J}} = (< L_{\mathcal{J}} b_1, b >, ..., < L_{\mathcal{J}} b_r, b >)^T$ and covariance matrix $\Gamma_{\mathcal{J}} = (< L_{\mathcal{J}} b_i, L_{\mathcal{J}} b_j >)_{i,j=1,...,r}$ .*

*b) Under the null hypothesis $F = F_- \in \mathcal{H}_0^s(\mathcal{J})$ we have the conditional limiting law*

$$\mathcal{L}_F[\vec{S}_n \mid \tau] \xrightarrow{\mathcal{L}} \mathcal{N}(0, \Gamma_{\mathcal{J}}) \quad [a.s]. \tag{5.4.106}$$

*c) Let $\hat{H}_n$ denote the empirical distribution function of the absolute values $|X_1|, ..., |X_n|$ , let $\hat{H}_n^*$ be its symmetrization according to (5.4.49) implying $\mathcal{I}(\hat{H}_n^*) = \mathcal{J}_n^\tau$ , c.f. (5.4.50). Then, under the null hypothesis $F = F_- \in \mathcal{H}_0^s(\mathcal{J})$ , we have the convergence*

$$\Gamma_{\mathcal{J}_n^\tau} := \left( < L_{\mathcal{J}_n^\tau} b_i, L_{\mathcal{J}_n^\tau} b_j > \right)_{i,j=1,...,r} \xrightarrow{a.s.} \Gamma_{\mathcal{J}}. \tag{5.4.107}$$

*Proof:* According to (5.4.52) the vectors $\vec{S}_n$ and $(S_n^*(L_{\mathcal{J}}b_1), \cdots, S_n^*(L_{\mathcal{J}}b_r))$ have the same limiting distribution under $\mathcal{H}_0^s(\mathcal{J})$ and by contiguity under $F_n$, too. Now Corollary 5.4.6 and the Cramér-Wold device imply part a).

Under $F = F_- \in \mathcal{H}_0^s(\mathcal{J})$ we have $\|\hat{H}_n^* - F\|_\infty \to 0$ [a.s.]. For any fixed sequence $\hat{H}_n^*$ with $\|\hat{H}_n^* - F\|_\infty \to 0$ let $\{\hat{\tau}_n\}$ be the corresponding *(non-random)* $\tau$–vector and $\hat{\mathcal{J}}_n := \mathcal{I}(\hat{H}_n^*)$. Then, because of $\mathcal{I}(F) = \mathcal{J}$, $<\tilde{b}_i, \hat{b}_{n2}> = S_n^*(L_{\hat{\mathcal{J}}_n}b_i)$, $1 \leq i \leq r$, Lemma 7.5.5, and the independence of $\tau$ and $(\Delta^*, Q^*)$ we get

$$\mathcal{L}_F[\vec{S}_n \mid \tau = \hat{\tau}_n] = \mathcal{L}_0[(S_n^*(L_{\hat{\mathcal{J}}_n}b_i),\ i=1,...,r)^T] \xrightarrow{\mathcal{L}} \mathcal{N}(0, \Gamma_{\mathcal{J}}), \quad (5.4.108)$$

where the convergence in law follows from Corollary 5.4.6 and the Cramér-Wold device just as in part a). Since Lemma 7.5.5 implies $L_{\hat{\mathcal{J}}_n}b_i \to L_{\mathcal{J}}b_i$ as $n \to \infty$, $1 \leq i \leq r$, part b) is proved.

Part c) is an immediate consequence of Lemma 7.5.5. $\square$

In order to emphasize the $\Gamma$–dependence of the function $f_0$ defined in formula (3.2.53), we explicitly write

$$f(x, \Gamma) = \sup_{\vartheta \geq 0}\left(2\vartheta^T x - \vartheta^T \Gamma \vartheta\right), \qquad x = (x_1, ..., x_r)^T \in \mathbb{R}^r. \quad (5.4.109)$$

Then, according to (3.2.52), we'll use

$$S_n^\tau(b_1, ..., b_r) := f(\vec{S}_n, \Gamma_{\mathcal{J}_n^\tau}) \quad (5.4.110)$$

as the *projection rank statistic* if ties are present.

Notice, even in the continuous case (no ties) the original statistic $f_0(\vec{S}_n) = f(\vec{S}_n, \Gamma)$ defined in (3.2.52) differs from the above statistic $f(\vec{S}_n, \Gamma_{\mathcal{J}_n^\tau})$, since the limiting $\Gamma$ is substituted by its approximation $\Gamma_{\mathcal{J}_n^\tau}$. In the continuous model the matrix $\Gamma_{\mathcal{J}_n^\tau}$ is non-random, since $d = n$ and $\tau_1 = ... = \tau_n = 1$ in this case. In the general model the matrix $\Gamma_{\mathcal{J}_n^\tau}$ will depend on the random vector of tie-lengths $\tau$.

According to Lemma 7.5.7 the function $f(x, \Gamma)$ is jointly continuous in $x$ and $\Gamma$. Therefore, under the null hypothesis $F = F_- \in \mathcal{H}_0^s(\mathcal{J})$ and if the matrix $\Gamma_{\mathcal{J}}$ is non-singular, Theorem 5.4.9 implies the limiting law

$$\mathcal{L}_F\left[f(\vec{S}_n, \Gamma_{\mathcal{J}_n^\tau})\right] \xrightarrow{\mathcal{L}} \mathcal{L}\left[f(X, \Gamma_{\mathcal{J}})\right] \quad (5.4.111)$$

and also the conditional limiting law

$$\mathcal{L}_F\left[f(\vec{S}_n, \Gamma_{\mathcal{J}_n^\tau}) \mid \tau\right] \xrightarrow{\mathcal{L}} \mathcal{L}\left[f(X, \Gamma_{\mathcal{J}})\right] \quad [a.s.], \quad (5.4.112)$$

where $X = (X_1, ..., X_r)^T$ is a random vector with $r$-variate normal distribution $\mathcal{N}(0, \Gamma_{\mathcal{J}})$.

The $p$-value of the limiting distribution may be computed (approximately) by using formula (3.2.78) with $\Gamma$ replaced by $\Gamma_{\mathcal{J}_n^\tau}$. The more ties are present the more questionable the above approximation usually is. In contrast, the computation of the $p$-value of the conditional distribution $\mathcal{L}_F[f(\vec{S}_n, \Gamma_{\mathcal{J}_n^\tau}) \mid \tau]$ given the observed $\tau$ is always possible by Monte-Carlo simulation.

# Chapter 6

# The hypothesis of independence

In this chapter we assume i.i.d. $\mathbb{R}^2$ -valued random variables $X_1, ..., X_n$ with unknown (continuous) 2-dimensional distribution function $F$ . Throughout the chapter we'll use the following *conventions and notations* :

If $X_i$ is a $\mathbb{R}^2$ -valued random variable, then the components of $X_i$ will be denoted by $Y_i$ and $Z_i$ , i.e. $X_i = (Y_i, Z_i)$ .

If $F$ is a 2-dimensional distribution function, then the respective first and second marginal distribution functions will be denoted by $G$ and $H$ , i.e.

$$G(y) = F(y, \infty) \quad \forall\, y \in \mathbb{R}, \qquad H(z) = F(\infty, z) \quad \forall\, z \in \mathbb{R}. \qquad (6.0.1)$$

If $G$ and $H$ are 1-dimensional distribution functions, then the distribution function of the corresponding product measure is denoted by $G \times H$ , i.e.

$$(G \times H)(y, z) = G(y)H(z) \quad \forall\, (y, z) \in \mathbb{R}^2. \qquad (6.0.2)$$

Finally, let $\mathcal{F}_2$ denote the set of all distribution functions $F$ on $\mathbb{R}^2$ such that the corresponding probability measure $F$ is dominated by the product $G \times H$ of the marginal measures $G$ and $H$ ,

$$\mathcal{F}_2 = \{\, F :\ F \text{ is a distribution function on } \mathbb{R}^2 \text{ and } F \ll G \times H \,\}, \qquad (6.0.3)$$

and let $\mathcal{F}_2^c$ denote the set of continuous distribution functions in $\mathcal{F}_2$ ,

$$\mathcal{F}_2^c = \{\, F \in \mathcal{F}_2 :\ G \text{ and } H \text{ are continuous } \}. \qquad (6.0.4)$$

In a first step we want to test the continuous *null hypothesis of independence*

$$\mathcal{H}_0^i = \{ \, F \in \mathcal{F}_2^c : \ F = G \times H \, \} \tag{6.0.5}$$

versus the continuous *omnibus alternative*

$$\mathcal{A}_d = \{ \, F \in \mathcal{F}_2^c : \ F \neq G \times H \, \} \tag{6.0.6}$$

or versus the continuous one-sided *alternative of positive quadrant dependence*

$$\mathcal{A}_d^0 = \{ \, F \in \mathcal{F}_2^c : \ F \geq G \times H, \ F \neq G \times H \, \}. \tag{6.0.7}$$

Similar to the two-sample case of Chapter 3 we shall see that linear rank tests are asymptotically optimal for testing $\mathcal{H}_0^i$ versus a special local asymptotic alternative, if the type of the alternative corresponds to the score function of the linear rank statistic.

Given any score function $h \in L_2\big((0,1) \times (0,1)\big)$ such that

$$\int_0^1 h(\cdot, x)\, dx = 0 \qquad \text{and} \qquad \int_0^1 h(x, \cdot)\, dx = 0$$

we define the corresponding linear rank statistic $S_n(h)$ according to

$$S_n(h) = \frac{1}{\sqrt{n}} \sum_{i=1}^n h_n(R_{1i}, R_{2i}), \tag{6.0.8}$$

where

$$h_n(i, j) = n^2 \int_{(i-1)/n}^{i/n} \int_{(j-1)/n}^{j/n} h(s, t)\, d\lambda_2(s, t), \tag{6.0.9}$$

and where $(R_{11}, ..., R_{1n})$ and $(R_{21}, ..., R_{2n})$ are the ranks of $(Y_1, ..., Y_n)$ and $(Z_1, ..., Z_n)$, respectively,

$$R_i = (R_{1i}, R_{2i}) = \Big( \sum_{j=1}^n 1(Y_j \leq Y_i), \ \sum_{j=1}^n 1(Z_j \leq Z_i) \Big). \tag{6.0.10}$$

Then Theorem 2.1 of Behnen (1972) implies the limiting law

$$\mathcal{L}[\, S_n(h) \mid \mathcal{H}_0^i \,] \xrightarrow{\ \mathcal{L}\ } \mathcal{N}(\, 0, \ \|h\|^2 \,). \tag{6.0.11}$$

Therefore the *linear rank test*

$$\psi_n(h) = 1(\, S_n(h) \geq u_\alpha \|h\| \,) \tag{6.0.12}$$

is an asymptotic level $\alpha$ test for testing the null hypothesis $\mathcal{H}_0^i$ , if $\|h\|^2 = \int_0^1 \int_0^1 h^2(s, t)\, d\lambda_2(s, t) > 0$ .

The asymptotic optimality result for linear rank tests is given in the next section after having given a suitable reparametrization of the model and of the hypotheses.

# 6.1 Linear rank tests

If $F$ is any distribution function on $\mathbb{R}^2$ with marginal distribution functions $G$ and $H$, then the well-known inequality ($\forall (y,z),(s,t) \in \mathbb{R}^2$)

$$|F(y,z) - F(s,t)| \leq |G(y) - G(s)| + |H(z) - H(t)| \qquad (6.1.1)$$

and the identities $G = G \circ G^{-1} \circ G$, $H = H \circ H^{-1} \circ H$ imply

$$F = F \circ (G^{-1}, H^{-1}) \circ (G, H). \qquad (6.1.2)$$

If $X = (Y, Z)$ is a $\mathbb{R}^2$–valued random variable with distribution function $F$, then obviously the transformed random variables $G(Y)$ and $H(Z)$ have a $\mathcal{R}(0,1)$-distribution if the marginal distribution functions $G$ and $H$ are continuous. In this case the joint distribution of $(G(Y), H(Z))$ has the distribution function $F \circ (G^{-1}, H^{-1})$ on the unit square $[0,1]^2$.

### 6.1.1 Proposition

*Assume $F$ to be a distribution function on $\mathbb{R}^2$ with continuous marginals $G$ and $H$ such that $F$ is dominated by the product measure $G \times H$, i.e. $F \in \mathcal{F}_2^c$. Then the following assertions hold true:*

*a) If $X = (Y, Z)$ has the distribution function $F$, then the $[0,1]^2$–valued transformed random variable $(G(Y), H(Z))$ has a $\lambda_2$–density $f^*$ on $[0,1]^2$ according to*

$$f^* := \frac{dF}{d(G \times H)} \circ (G^{-1}, H^{-1}), \qquad (6.1.3)$$

*and the marginals of $f^*$ correspond to the uniform distribution on $(0,1)$, i.e.*

$$\int_0^1 f^*(\cdot, x)\, dx = \int_0^1 f^*(x, \cdot)\, dx = 1 \quad [\lambda - a.e.]. \qquad (6.1.4)$$

*b) The distribution function $F$ has the representation*

$$F = F^* \circ (G, H), \qquad (6.1.5)$$

*where $F^* : [0,1]^2 \to [0,1]$ is the distribution function corresponding to the $\lambda_2$–density $f^*$,*

$$F^*(s,t) = \int_0^s \int_0^t f^*(y,z)\, d\lambda_2(y,z) \quad \forall\, s,t \in [0,1]. \qquad (6.1.6)$$

*Proof:* Since $G \times H$ is the distribution function of $\mathcal{L}[(G^{-1}, H^{-1}) \mid \lambda_2]$ we get from definition (6.1.3) for any $(y, z) \in \mathbb{R}^2$ the following chain of equalities,

$$
F^*(G(y), H(z)) = \int_0^{G(y)} \int_0^{H(z)} f^* \, d\lambda_2
$$

$$
= \int_{-\infty}^{y} \int_{-\infty}^{z} \frac{dF}{d(G \times H)} \, d(G \times H) = F(y, z),
$$

$$(6.1.7)$$

which proves (6.1.5). Using (6.1.2) we get in addition

$$
F^* \circ (G, H) = F \circ (G^{-1}, H^{-1}) \circ (G, H). \tag{6.1.8}
$$

Therefore the continuity of the 1-dimensional distribution functions $G$ and $H$ implies $F^* = F \circ (G^{-1}, H^{-1})$. This concludes the proof, since the distribution function of $(G(Y), H(Z))$ is $F \circ (G^{-1}, H^{-1})$. $\square$

According to Proposition 6.1.1 we may write each $F \in \mathcal{F}_2^c$ in the form

$$
F = G \times H + n^{-1/2} B \circ (G, H), \tag{6.1.9}
$$

where

$$
B(s, t) = \int_0^s \int_0^t b(u, v) \, d\lambda_2(u, v), \quad (s, t) \in [0, 1]^2, \tag{6.1.10}
$$

$$
b \geq -\sqrt{n} \qquad [\lambda_2 - a.e.], \tag{6.1.11}
$$

$$
\int_0^1 b(\cdot, x) \, dx = \int_0^1 b(x, \cdot) \, dx = 0 \qquad [\lambda - a.e.], \tag{6.1.12}
$$

and $(G(Y), H(Z))$ has the $\lambda_2$-density $1 + n^{-1/2}b$ if $X = (Y, Z)$ has the distribution function $F$.

Conversely, for all distribution functions $G, H \in \mathcal{F}_1^c$ and any measurable function $b : (0, 1)^2 \to \mathbb{R}$ with the properties (6.1.11) and (6.1.12) the definition (6.1.10) implies $G \times H + n^{-1/2} B \circ (G, H)$ to be a 2-dimensional distribution function with respective continuous marginal distribution functions $G$ and $H$, and the corresponding probability measure is dominated by the product measure $G \times H$.

Thus we've proved the following reparametrization of $\mathcal{F}_2^c$,

$$
\mathcal{F}_2^c = \{ G \times H + \frac{1}{\sqrt{n}} B \circ (G, H) : B \in \mathcal{B}_n^i, \ G, H \in \mathcal{F}_1^c \}, \tag{6.1.13}
$$

where $\mathcal{F}_1^c$ is the set of all continuous distribution functions on $\mathbb{R}$ and

$$\mathcal{B}_n^i = \left\{ B : \begin{array}{l} B \text{ has the representation (6.1.10)} \\ \text{for some measurable } b : (0,1)^2 \to \mathbb{R} \\ \text{with the properties (6.1.11) and (6.1.12)} \end{array} \right\}. \qquad (6.1.14)$$

Obviously each $B \in \mathcal{B}_n^i$ is absolutely continuous on $[0,1]^2$ and has the properties

$$B(\cdot,0) = B(0,\cdot) = B(\cdot,1) = B(1,\cdot) = 0. \qquad (6.1.15)$$

Since $G$ and $H$ are the continuous marginal distribution functions of the distribution function $F = G \times H + n^{-1/2} B \circ (G,H)$ if $B \in \mathcal{B}_n^i$ and $G, H \in \mathcal{F}_1^c$, the new parametrization (6.1.13) induces the following analogous reparametrization of the hypotheses (6.0.5) to (6.0.7),

$$\mathcal{H}_0^i = \{G \times H + \frac{1}{\sqrt{n}} B \circ (G,H) : B \in \mathcal{B}_n^i, \ G, H \in \mathcal{F}_1^c, \ B = 0\}, \qquad (6.1.16)$$

$$\mathcal{A}_d = \{G \times H + \frac{1}{\sqrt{n}} B \circ (G,H) : B \in \mathcal{B}_n^i, \ G, H \in \mathcal{F}_1^c, \ B \neq 0\}, \qquad (6.1.17)$$

$$\mathcal{A}_d^0 = \{G \times H + \frac{1}{\sqrt{n}} B \circ (G,H) \in \mathcal{A}_d : \ B \geq 0\}. \qquad (6.1.18)$$

Again $B \in \mathcal{B}_n^i$ is the parameter under test, whereas $(G,H) \in \mathcal{F}_i^c \times \mathcal{F}_1^c$, is a nuisance parameter. Moreover, the hypotheses $\mathcal{H}_0^i$, $\mathcal{A}_d$, and $\mathcal{A}_d^0$ are invariant under the group of transformations of the measurement scale of each of the two components, i.e. each potential observation $X_i = (Y_i, Z_i)$ is transformed into $T(X_i) = \big(T_1(Y_i), T_2(Z_i)\big)$, where $T_1$ and $T_2$ are strictly increasing and continuous mappings from $\mathbb{R}$ to $\mathbb{R}$ such that $T_1(\mathbb{R}) = T_2(\mathbb{R}) = \mathbb{R}$, and the vector of ranks $(R_1, ..., R_n)$ defined in formula (6.0.10) is *maximal invariant* with respect to the above group of transformations. Therefore we may expect good rank procedures for testing the hypothesis of independence $\mathcal{H}_0^i$ versus the alternatives $\mathcal{A}_d$ or $\mathcal{A}_d^0$. Especially the distribution of the ranks does not depend on the nuisance parameter $(G,H) \in \mathcal{F}_1^c \times \mathcal{F}_1^c$:

For any $B \in \mathcal{B}_n^i$, and any $(G_1, H_1), (G_2, H_2) \in \mathcal{F}_1^c \times \mathcal{F}_1^c$ we have

$$\mathcal{L}[\,(R_1, ..., R_n) \mid (B, G_1, H_1)\,] = \mathcal{L}[\,(R_1, ..., R_n) \mid (B, G_2, H_2)\,]. \qquad (6.1.19)$$

The *proof* is an easy consequence of Proposition 6.1.1.

The next theorem will prove the asymptotic optimality of the linear rank test $\psi_n(b)$ for testing $\mathcal{H}_0^i$ versus local asymptotic alternatives of type $b$.

### 6.1.2 Theorem

*Assume $b \in L_2\big((0,1)^2\big)$ such that $\int_0^1 b(\cdot, x)\, dx = 0$ and $\int_0^1 b(x, \cdot)\, dx = 0$. Let $(G_n \times H_n \in \mathcal{H}_0^i,\ n \geq 1)$ be any sequence of null hypothesis points.*

Define a corresponding sequence of local asymptotic alternatives $(F_n,\ n \geq 1)$ according to

$$F_n = G_n \times H_n + n^{-1/2} B_n \circ (G_n, H_n), \quad B_n \in \mathcal{B}_n^i, \quad \|b_n - b\| \overset{n \to \infty}{\longrightarrow} 0. \quad (6.1.20)$$

a) For any score function $h \in L_2((0,1) \times (0,1))$ such that $\int_0^1 h(\cdot, x)\, dx = 0$ and $\int_0^1 h(x, \cdot)\, dx = 0$ we have the limiting law ( $n \to \infty$ )

$$\mathcal{L}[\, S_n(h) \mid F_n \,] \overset{\mathcal{L}}{\longrightarrow} \mathcal{N}(\, < h, b >,\ \|h\|^2\, ), \qquad (6.1.21)$$

where $S_n(h)$ is the linear rank statistic defined in (6.0.8) and (6.0.9).

b) If $h = b$ and $\|b\| > 0$ , then the linear rank test $\psi_n(h)$ defined in (6.0.12) is asymptotically optimal at level $\alpha$ for testing the null hypothesis $\mathcal{H}_0^i$ versus the sequence of alternatives $(F_n,\ n \geq 1)$ .

The *proof* is an immediate consequence of Theorem 2.1 of Behnen (1972).

Usually the underlying optimal nonparametric score functions $b_n$ and $b$ are unknown in reality. Therefore we'll substitute the $b$ in $S_n(b)$ by suitable rank estimators of $b_n$ .

## 6.2 Kernel estimators of the score function

If $X_i = (Y_i, Z_i)$, $i = 1, ..., n$, are i.i.d $\mathrm{IR}^2$ –valued random variables with distribution function $F = G \times H + n^{-1/2} B \circ (G, H) \in \mathcal{F}_2^c$, then Proposition 6.1.1 implies that the pairs $(G(Y_i), H(Z_i))$, $i = 1, ..., n$, are i.i.d. $[0, 1]^2$ – valued random variables with $\lambda_2$ –density $f^* = 1 + n^{-1/2} b$, and the integrated score function $B$ has the representation

$$
\begin{aligned}
B(s, t) &= \sqrt{n}\left(F^*(s, t) - st\right) \\
&= \sqrt{n}\left(F^*(s, t) - sF^*(1, t) - tF^*(s, 1) + st\right).
\end{aligned}
\tag{6.2.1}
$$

Using the "observations" $(G(Y_i), H(Z_i))$, $i = 1, ..., n$, the standard estimator of $B$ would be a smoothed version of the corresponding "empirical" process

$$
\frac{1}{\sqrt{n}} \sum_{i=1}^{n} \left(1(G(Y_i) \leq s) - s\right)\left(1(H(Z_i) \leq t) - t\right), \qquad s, t \in [0, 1].
\tag{6.2.2}
$$

Since $G$ and $H$ are unknown parameters of the model, the random variables $G(Y_i)$ and $H(Z_i)$ are unobservable, but obviously the observable pair of the normed ranks

$$
\begin{aligned}
\left(\frac{R_{1i}}{n}, \frac{R_{2i}}{n}\right) &= \left(\frac{1}{n} \sum_{j=1}^{n} 1(Y_j \leq Y_i), \frac{1}{n} \sum_{j=1}^{n} 1(Z_j \leq Z_i)\right) \\
&\overset{a.e.}{=} \left(\frac{1}{n} \sum_{j=1}^{n} 1(G(Y_j) \leq G(Y_i)), \frac{1}{n} \sum_{j=1}^{n} 1(H(Z_j) \leq H(Z_i))\right)
\end{aligned}
\tag{6.2.3}
$$

is a uniformly good estimator of $(G(Y_i), H(Z_i))$.

Therefore the rank process

$$
\hat{B}_{n1}(s, t) := \frac{1}{\sqrt{n}} \sum_{i=1}^{n} \left(1(R_{1i} \leq [ns]) - \frac{[ns]}{n}\right)\left(1(R_{2i} \leq [nt]) - \frac{[nt]}{n}\right)
\tag{6.2.4}
$$

will be our basic estimator of the integrated score function $B$, if the omnibus testing problem $\mathcal{H}_0^i$ versus $\mathcal{A}_d$ is considered. Since $B$ is absolutely continuous, it's more suitable to start with the following square-wise interpolated version $\hat{B}_{n2}$ of the process $\hat{B}_{n1}$,

$$
\begin{aligned}
\hat{B}_{n2}(s, t) &= (1 - ns + [ns])(1 - nt + [nt])\,\hat{B}_{n1}\left(\frac{[ns]}{n}, \frac{[nt]}{n}\right) \\
&\quad + (ns - [ns])(1 - nt + [nt])\,\hat{B}_{n1}\left(\frac{[ns] + 1}{n}, \frac{[nt]}{n}\right)
\end{aligned}
$$

$$+ (1 - ns + [ns])\ (nt - [nt])\ \hat{B}_{n1}\big(\frac{[ns]}{n}, \frac{[nt]+1}{n}\big)$$

$$+ (ns - [ns])\ (nt - [nt])\ \hat{B}_{n1}\big(\frac{[ns]+1}{n}, \frac{[nt]+1}{n}\big)$$

$$= \frac{1}{\sqrt{n}} \sum_{i=1}^{n} \Big( 1(\ R_{1i} \le [ns]\ ) + (ns - [ns])\ 1(\ R_{1i} = [ns]+1\ ) - s\Big)$$

$$\times \Big( 1(\ R_{2i} \le [nt]\ ) + (nt - [nt])\ 1(\ R_{2i} = [nt]+1\ ) - t\Big). \qquad (6.2.5)$$

Obviously $\hat{B}_{n2}$ is a $C(\ [0,1]^2\ )$ –valued rank process which fulfils the side conditions (6.1.15), i.e.

$$\hat{B}_{n2}(\cdot, 0)\ =\ \hat{B}_{n2}(0, \cdot)\ =\ \hat{B}_{n2}(\cdot, 1)\ =\ \hat{B}_{n2}(1, \cdot)\ =\ 0, \qquad (6.2.6)$$

and which can be written in the form

$$\hat{B}_{n2}(s,t)\ =\ \int_0^s \int_0^t \hat{b}_{n2}(u,v)\ du\ dv \qquad \forall\ s,t \in [0,1], \qquad (6.2.7)$$

with $\lambda_2$ –derivative $\hat{b}_{n2}$ according to

$$\hat{b}_{n2}(s,t) = \frac{1}{\sqrt{n}} \sum_{i=1}^{n} \Big( n\ 1(R_{1i} = [ns]+1) - 1\Big)\Big( n\ 1(R_{2i} = [nt]+1) - 1\Big)$$

$$= n^{3/2}\Big( -\frac{1}{n} + \sum_{i=1}^{n} 1\big( \frac{R_{1i}-1}{n} \le s < \frac{R_{1i}}{n},\ \frac{R_{2i}-1}{n} \le t < \frac{R_{2i}}{n}\big) \Big), \qquad (6.2.8)$$

which means that $\hat{b}_{n2}$ fulfils the side conditions (6.1.11) and (6.1.12).

As in the two-sample model of Section 3.1 the process $\hat{b}_{n2}$ may be viewed as a primitive estimator of $b$ in the omnibus case. Starting from $\hat{b}_{n2}$ we'll substitute the unknown optimal score function of the linear rank statistic $S_n(b)$ by kernel estimators of the type $\mathcal{K}_a \hat{b}_{n2}$, i.e. we define $(\ \forall\ s,t \in [0,1]\ )$

$$\mathcal{K}_a \hat{b}_{n2}(s,t) = \int_0^1 \int_0^1 \hat{b}_{n2}(x,y)\ K_a(x,s)\ K_a(y,t)\ dx\ dy, \qquad (6.2.9)$$

where $K_a : [0,1]^2 \to \mathbb{R}$ is the convolution kernel (3.1.19) corresponding to the fixed bandwidth $0 < a \le 1$ and to the kernel $K : \mathbb{R} \to \mathbb{R}$ with the properties (3.1.16)-(3.1.18) and (3.1.31).

Given any $b \in L_2(\ (0,1)^2\ )$ such that $\int_0^1 \int_0^1 b\ d\lambda_2 = 0$ we get from definition (6.0.8) and formula (6.2.8) the following representation of the linear rank statistic $S_n(b)$ ,

$$S_n(b) = n^{3/2} \sum_{i=1}^{n} \int_{(R_{1i}-1)/n}^{R_{1i}/n} \int_{(R_{2i}-1)/n}^{R_{2i}/n} b(s,t) \, d\lambda_2(s,t)$$

$$= \int_0^1 \int_0^1 b(s,t) \, \hat{b}_{n2}(s,t) \, d\lambda_2(s,t) \; = \; < b, \hat{b}_{n2} > . \qquad (6.2.10)$$

According to formula (6.2.8) the process $\hat{b}_{n2}$ is constant on each square of the form $[(i-1)/n, i/n) \times [(j-1)/n, j/n)$, $i,j = 1,...,n$ . Therefore the substitution of $b$ by $\mathcal{K}_a \hat{b}_{n2}$ yields the (nonlinear) *omnibus rank statistic*

$$S_n(\mathcal{K}_a \hat{b}_{n2}) = \int_0^1 \int_0^1 \int_0^1 \int_0^1 \hat{b}_{n2}(x,y) \, \hat{b}_{n2}(s,t) \, K_a(x,s) \, K_a(y,t) \, dx \, dy \, ds \, dt$$

$$= \sum_{i=1}^{n} \sum_{j=1}^{n} \sum_{q=1}^{n} \sum_{r=1}^{n} \hat{b}_{n2}\Big(\frac{i-1}{n}, \frac{j-1}{n}\Big) \, \hat{b}_{n2}\Big(\frac{q-1}{n}, \frac{r-1}{n}\Big) \, \bar{k}_{na}(i,j,q,r), \quad (6.2.11)$$

where the weights $\bar{k}_{na}(i,j,q,r)$ are defined by

$$\bar{k}_{na}(i,j,q,r) = \qquad\qquad\qquad\qquad\qquad\qquad\qquad\qquad\qquad (6.2.12)$$

$$\int_{(i-1)/n}^{i/n} \int_{(q-1)/n}^{q/n} K_a(x,s) \, dx \, ds \int_{(j-1)/n}^{j/n} \int_{(r-1)/n}^{r/n} K_a(y,t) \, dy \, dt.$$

If the kernel $K$ is smooth, we use instead of the weights (6.2.12) the simpler approximations, cf Section 3.1,

$$k_{na}(i,j,q,r) = \frac{1}{n^2} \, K_a\Big(\frac{i-1/2}{n}, \frac{q-1/2}{n}\Big) \frac{1}{n^2} \, K_a\Big(\frac{j-1/2}{n}, \frac{r-1/2}{n}\Big)$$

$$= \; n^{-4} \, k_{na}(i,q) \, k_{na}(j,r). \qquad\qquad\qquad (6.2.13)$$

In case of the one-sided testing problem $\mathcal{H}_0^i$ versus the alternative of positive quadrant dependence $\mathcal{A}_d^0$ the integrated score function $B$ has the additional property $B \geq 0$ . In this case the corresponding "projection" $\hat{B}_{n1}^0$ of $\hat{B}_{n1}$ will be our basic estimator of $B$ , i.e.

$$\hat{B}_{n1}^0 \; := \; \hat{B}_{n1} \, 1(\hat{B}_{n1} > 0). \qquad\qquad\qquad (6.2.14)$$

As in the omnibus case we'll smooth $\hat{B}_{n1}^0$ by square-wise bilinear interpolation, i.e. we define the smooth estimator $\hat{B}_{n2}^0 \geq 0$ of $B$ according to

$$\hat{B}_{n2}^0(s,t) \; = \; \big(1 - ns + [ns]\big) \big(1 - nt + [nt]\big) \, \hat{B}_{n1}^0\Big(\frac{[ns]}{n}, \frac{[nt]}{n}\Big)$$

$$+ \left(ns - [ns]\right) \left(1 - nt + [nt]\right) \hat{B}^0_{n1}\big(\frac{[ns]+1}{n}, \frac{[nt]}{n}\big)$$

$$+ \left(1 - ns + [ns]\right) \left(nt - [nt]\right) \hat{B}^0_{n1}\big(\frac{[ns]}{n}, \frac{[nt]+1}{n}\big)$$

$$+ \left(ns - [ns]\right) \left(nt - [nt]\right) \hat{B}^0_{n1}\big(\frac{[ns]+1}{n}, \frac{[nt]+1}{n}\big). \qquad (6.2.15)$$

Again $\hat{B}^0_{n2}$ is a $C\big([0,1]^2\big)$ –valued rank process which fulfils the side conditions (6.1.15), i.e.

$$\hat{B}^0_{n2}(\cdot,0) \;=\; \hat{B}^0_{n2}(0,\cdot) \;=\; \hat{B}^0_{n2}(\cdot,1) \;=\; \hat{B}^0_{n2}(1,\cdot) \;=\; 0, \qquad (6.2.16)$$

and which can be written as an integral, too,

$$\hat{B}^0_{n2}(s,t) \;=\; \int_0^s \int_0^t \hat{b}^0_{n2}(u,v)\; du\; dv \qquad \forall\, s,t \in [0,1], \qquad (6.2.17)$$

where the square-wise constant $\lambda_2$ –derivative $\hat{b}^0_{n2}$ of $\hat{B}^0_{n2}$ is given by

$$\hat{b}^0_{n2}(s,t) = n^2 \left(\; \hat{B}^0_{n1}\big(\frac{[ns]+1}{n}, \frac{[nt]+1}{n}\big) - \hat{B}^0_{n1}\big(\frac{[ns]+1}{n}, \frac{[nt]}{n}\big) \right.$$
$$\left. - \hat{B}^0_{n1}\big(\frac{[ns]}{n}, \frac{[nt]+1}{n}\big) + \hat{B}^0_{n1}\big(\frac{[ns]}{n}, \frac{[nt]}{n}\big) \;\right). \qquad (6.2.18)$$

Notice, the simpler "projection" $\tilde{B}^0_{n2} = \hat{B}_{n2}\, 1(\hat{B}_{n2} > 0)$ isn't suitable in the present setting, since in general $\tilde{B}^0_{n2}$ doesn't have an integral representation of the form (6.2.17). But according to the subsequent Lemma 6.2.3 we have $\|\hat{B}^0_{n2} - \tilde{B}^0_{n2}\|_\infty \leq 8/\sqrt{n}$ , where $\|\cdot\|_\infty$ denotes the sup-norm on $C\big([0,1]^2\big)$ .

Similar to the omnibus case the process $\hat{b}^0_{n2}$ may be viewed as a primitive estimator of $b$ in the one-sided model. Starting from $\hat{b}^0_{n2}$ we'll substitute the unknown optimal score function of the linear rank statistic $S_n(b)$ by kernel estimators of the type $\mathcal{K}_a \hat{b}^0_{n2}$ , i.e. we define ( $\forall\, s,t \in [0,1]$ )

$$\mathcal{K}_a \hat{b}^0_{n2}(s,t) = \int_0^1 \int_0^1 \hat{b}^0_{n2}(x,y)\, K_a(x,s)\, K_a(y,t)\; dx\; dy\; . \qquad (6.2.19)$$

The substitution of $b$ by $\mathcal{K}_a \hat{b}^0_{n2}$ in formula (6.2.10) yields the (nonlinear) *one-sided rank statistic*

$$S_n(\mathcal{K}_a \hat{b}^0_{n2}) = \int_0^1 \int_0^1 \int_0^1 \int_0^1 \hat{b}^0_{n2}(x,y)\, \hat{b}_{n2}(s,t)\, K_a(x,s)\, K_a(y,t)\; dx\; dy\; ds\; dt$$

$$= \sum_{i=1}^n \sum_{j=1}^n \sum_{q=1}^n \sum_{r=1}^n \hat{b}^0_{n2}\big(\frac{i-1}{n}, \frac{j-1}{n}\big)\, \hat{b}_{n2}\big(\frac{q-1}{n}, \frac{r-1}{n}\big)\, \bar{k}_{na}(i,j,q,r), \qquad (6.2.20)$$

with the same weights $\bar{k}_{na}(i,j,q,r)$ as before, cf. (6.2.12).

If the kernel $K$ is smooth, we use instead of the weights $\bar{k}_{na}(i,j,q,r)$ the simpler approximations $n^{-4}\,k_{na}(i,q)\,k_{na}(j,r)$ given in formula (6.2.13). This yields the (approximate) *one-sided rank statistic*

$$S_n^0(a,K) \;=$$

$$\sum_{i=1}^{n}\sum_{j=1}^{n}\sum_{q=1}^{n}\sum_{r=1}^{n} \Delta\hat{B}_{n2}^0(i,j)\,\Delta\hat{B}_{n2}(q,r)\,k_{na}(i,q)\,k_{na}(j,r), \tag{6.2.21}$$

where

$$\Delta\hat{B}_{n2}^0(i,j) = \frac{1}{n^2}\,\hat{b}_{n2}^0\Big(\frac{i-1}{n},\frac{j-1}{n}\Big) \tag{6.2.22}$$

$$= \hat{B}_{n1}^0\Big(\frac{i}{n},\frac{j}{n}\Big) - \hat{B}_{n1}^0\Big(\frac{i}{n},\frac{j-1}{n}\Big) - \hat{B}_{n1}^0\Big(\frac{i-1}{n},\frac{j}{n}\Big) + \hat{B}_{n1}^0\Big(\frac{i-1}{n},\frac{j-1}{n}\Big)$$

and

$$\Delta\hat{B}_{n2}(q,r) = \frac{1}{n^2}\,\hat{b}_{n2}\Big(\frac{q-1}{n},\frac{r-1}{n}\Big)$$

$$= \hat{B}_{n1}\Big(\frac{q}{n},\frac{r}{n}\Big) - \hat{B}_{n1}\Big(\frac{q}{n},\frac{r-1}{n}\Big) - \hat{B}_{n1}\Big(\frac{q-1}{n},\frac{r}{n}\Big) + \hat{B}_{n1}\Big(\frac{q-1}{n},\frac{r-1}{n}\Big)$$

$$= \frac{1}{\sqrt{n}}\Big( -\frac{1}{n} + \sum_{i=1}^{n} 1\big( R_{1i} = q,\ R_{2i} = r \big) \Big). \tag{6.2.23}$$

Obviously (6.2.16) and (6.2.18) imply

$$\int_0^1 \hat{b}_{n2}^0(\cdot,x)\,dx \;=\; \int_0^1 \hat{b}_{n2}^0(x,\cdot)\,dx \;=\; 0\,, \tag{6.2.24}$$

and formula (6.2.8) yields

$$\int_0^1 \hat{b}_{n2}(\cdot,x)\,dx \;=\; \int_0^1 \hat{b}_{n2}(x,\cdot)\,dx \;=\; 0\,, \tag{6.2.25}$$

i.e. both estimators $\hat{b}_{n2}^0$ and $\hat{b}_{n2}$ of $b$ fulfil the side condition (6.1.12). Under the additional assumption

$$\sum_{\kappa=1}^{\infty} |\lambda_\kappa(a)| \;<\; \infty, \tag{6.2.26}$$

with $\lambda_\kappa(a)$ defined in formula (3.1.38), we may use Lemma 3.1.1 and (6.2.24), (6.2.25) in order to rewrite the rank statistics $S_n(\mathcal{K}_a \hat{b}^0_{n2})$ and $S_n(\mathcal{K}_a \hat{b}_{n2})$ in the following form,

$$S_n(\mathcal{K}_a \hat{b}^0_{n2}) \;=\; <\mathcal{K}_a \hat{b}^0_{n2}, \hat{b}_{n2}>$$

$$= \sum_{\kappa=1}^{\infty} \sum_{\tau=1}^{\infty} \lambda_\kappa(a)\, \lambda_\tau(a)\; <\hat{b}^0_{n2}, \psi_\kappa \times \psi_\tau> \; <\hat{b}_{n2}, \psi_\kappa \times \psi_\tau>$$

$$= \sum_{\kappa=1}^{\infty} \sum_{\tau=1}^{\infty} \lambda_\kappa(a)\, \lambda_\tau(a)\; <\hat{B}^0_{n2}, \psi'_\kappa \times \psi'_\tau> \; <\hat{B}_{n2}, \psi'_\kappa \times \psi'_\tau>, \quad (6.2.27)$$

where

$$(\psi_\kappa \times \psi_\tau)(s,t) \;=\; \psi_\kappa(s)\, \psi_\tau(t) \;=\; 2\, \cos(\pi\kappa s)\, \cos(\pi\tau t),$$
$$(\psi'_\kappa \times \psi'_\tau)(s,t) \;=\; \psi'_\kappa(s)\, \psi'_\tau(t) \;=\; 2\, \pi\kappa\, \sin(\pi\kappa s)\, \pi\tau\, \sin(\pi\tau t),$$

$$(6.2.28)$$

and similarly

$$S_n(\mathcal{K}_a \hat{b}_{n2}) \;=\; <\mathcal{K}_a \hat{b}_{n2}, \hat{b}_{n2}>$$

$$= \sum_{\kappa=1}^{\infty} \sum_{\tau=1}^{\infty} \lambda_\kappa(a)\, \lambda_\tau(a)\; <\hat{b}_{n2}, \psi_\kappa \times \psi_\tau>^2$$

$$= \sum_{\kappa=1}^{\infty} \sum_{\tau=1}^{\infty} \lambda_\kappa(a)\, \lambda_\tau(a)\; <\hat{B}_{n2}, \psi'_\kappa \times \psi'_\tau>^2 . \quad (6.2.29)$$

The asymptotics of the above rank statistics is derived from the following limiting law of the process $\hat{B}_{n2}$ .

### 6.2.1 Theorem

*Assume $b \in L_2((0,1)^2)$ with the additional properties $\int_0^1 b(\cdot, x)\, dx = 0$ and $\int_0^1 b(x, \cdot)\, dx = 0$. Let $(G_n \times H_n \in \mathcal{H}^i_0,\; n \geq 1)$ be any sequence of null hypothesis points. Define a corresponding sequence $(F_n,\; n \geq 1)$ of local asymptotic alternatives $F_n$ according to formula (6.1.20), i.e.*

$$F_n = G_n \times H_n + n^{-1/2} B_n \circ (G_n, H_n)\,, \quad B_n \in \mathcal{B}^i_n\,, \text{ and } \|b_n - b\| \to 0\,.$$

*Then the following limiting law $(n \to \infty)$ holds true in $C([0,1]^2)$ ,*

$$\mathcal{L}[\, \hat{B}_{n2} \mid F_n\,] \xrightarrow{\;\mathcal{L}\;} \mathcal{L}[\, W_0 + B\,], \quad (6.2.30)$$

*where $W_0$ is a $C([0,1]^2)$ –valued centered Gaussian process with covariance structure*

$$EW_0(s,t)W_0(u,v) = (s \wedge u - su)\,(t \wedge v - tv)$$
$$\forall\, (s,t), (u,v) \in [0,1]^2, \quad (6.2.31)$$

*and where  $B$  is the integral of  $b$ ,*

$$B(s,t) = \int_0^s \int_0^t b(u,v)\, d\lambda_2(u,v) \qquad \forall\,(s,t) \in [0,1]^2. \qquad (6.2.32)$$

*Proof*: Since the existence of  $W_0$  is well-known, it suffies to prove the tightness of the sequence  $(\mathcal{L}[\hat{B}_{n2} \mid F_n],\ n \geq 1)$  and the convergence in distribution of the finite dimensional distributions of  $\hat{B}_{n2}$  under  $F_n$  to the corresponding finite dimensional distributions of  $W_0 + B$ .

Because of equality (6.1.19) we may assume  $G_n = H_n = H_0\ \forall\,n \geq 1$ , where  $H_0$  corresponds to the uniform distribution on the interval  $(0,1)$ . Then, because of  $\|b_n - b\| \to 0$ , we see—completely similar to Section 7.1—that the sequence of alternatives  $(Q_n := \mathcal{L}[(X_1,...,X_n) \mid F_n],\ n \geq 1)$  is contiguous to the null hypothesis sequence  $(P_n := \mathcal{L}[(X_1,...,X_n) \mid H_0 \times H_0],\ n \geq 1)$ . Therefore it's sufficient to prove the tightness of the special sequence  $(\mathcal{L}[\hat{B}_{n2} \mid H_0 \times H_0],\ n \geq 1)$ .

If  $E_0(\cdots)$  denotes the expectation of  $(\cdots)$  under  $H_0 \times H_0$ , then Section 7.9 (Appendix) proves the inequality  ( $\forall\,n \geq 4$ )

$$E_0\left(\hat{B}_{n2}(u,v) - \hat{B}_{n2}(u,t) - \hat{B}_{n2}(s,v) + \hat{B}_{n2}(s,t)\right)^4 \leq k_0\left((u-s)(v-t)\right)^{3/2}$$

$$\forall\ \ 0 \leq s \leq u \leq 1,\ \ 0 \leq t \leq v \leq 1, \qquad (6.2.33)$$

where  $k_0$  is a finite constant which doesn't depend on  $n$  and  $s,t,u,v$ . Since  $\hat{B}_{n2}$  vanishes along the boundary of  $[0,1]^2$ , cf. formula (6.2.6), the inequality (6.2.33) and Theorem 3 of Bickel and Wichura (1971) prove the tightness of  $\hat{B}_{n2}$  under  $\mathcal{H}_0^i$ .

In order to prove the convergence in distribution of the finite dimensional distributions of  $\hat{B}_{n2}$  under  $F_n$  we'll compare  $\hat{B}_{n2}(s,t)$  with

$$T_n(s,t) := \frac{1}{\sqrt{n}} \sum_{i=1}^n \left(1(Y_i \leq s) - s\right)\left(1(Z_i \leq t) - t\right). \qquad (6.2.34)$$

Notice, because of  $H_0 \sim \mathcal{R}(0,1)$  and  $X_i = (Y_i, Z_i)$  the random variables  $Y_1,...,Y_n$  and also the random variables  $Z_1,...,Z_n$  are i.i.d. with  $\mathcal{R}(0,1)$ -distribution under  $F_n$  as well as under  $H_0 \times H_0$ . Using the abbreviation  $a_n(i,t)$  given in formula (7.9.1) we get from (6.2.5) the representation

$$\hat{B}_{n2}(s,t) - T_n(s,t) = S_{n1} + S_{n2} + S_{n3}, \qquad (6.2.35)$$

where

$$S_{n1} = n^{-1/2} \sum_{i=1}^{n} \Big( a_n(R_{1i}, s) + s - 1(Y_i \leq s) \Big) \Big( a_n(R_{2i}, t) + t - 1(Z_i \leq t) \Big),$$

$$S_{n2} = n^{-1/2} \sum_{i=1}^{n} \Big( a_n(R_{1i}, s) + s - 1(Y_i \leq s) \Big) \Big( 1(Z_i \leq t) - t \Big),$$

$$S_{n3} = n^{-1/2} \sum_{i=1}^{n} \Big( 1(Y_i \leq s) - s \Big) \Big( a_n(R_{2i}, t) + t - 1(Z_i \leq t) \Big),$$

and

$$E_0 S_{n3}^2 = \frac{1}{n} \sum_{i=1}^{n} E_0 \Big( 1(Y_i \leq s) - s \Big)^2 E_0 \Big( a_n(R_{2i}, t) + t - 1(Z_i \leq t) \Big)^2$$

$$= s(1 - s) \, E_0 \Big( a_n(R_{21}, t) + t - 1(Z_1 \leq t) \Big)^2$$

$$\stackrel{n \to \infty}{\longrightarrow} 0, \tag{6.2.36}$$

cf. (7.7.14) to (7.7.17). Completely similar we get

$$E_0 S_{n2}^2 = t(1 - t) \, E_0 \Big( a_n(R_{11}, s) + s - 1(Y_1 \leq s) \Big)^2 \stackrel{n \to \infty}{\longrightarrow} 0. \tag{6.2.37}$$

Because of

$$E_0 S_{N1}^2 = E_0 \Big( a_n(R_{11}, s) + s - 1(Y_1 \leq s) \Big)^2 E_0 \Big( a_n(R_{21}, t) + t - 1(Z_1 \leq t) \Big)^2$$

$$+ (n - 1) \, E_0 \Big( a_n(R_{11}, s) + s - 1(Y_1 \leq s) \Big) \Big( a_n(R_{12}, s) + s - 1(Y_2 \leq s) \Big)$$

$$\times E_0 \Big( a_n(R_{21}, t) + t - 1(Z_1 \leq t) \Big) \Big( a_n(R_{22}, t) + t - 1(Z_2 \leq t) \Big)$$

and

$$E_0 \Big( a_n(R_{11}, s) + s - 1(Y_1 \leq s) \Big) \Big( a_n(R_{12}, s) + s - 1(Y_2 \leq s) \Big)$$

$$= \frac{1}{n(n - 1)} \sum_{i \neq j} E_0 \Big( a_n(i, s) + s - 1(Y_n^{(i)} \leq s) \Big) \Big( a_n(j, s) + s - 1(Y_n^{(j)} \leq s) \Big)$$

$$= \frac{1}{n(n - 1)} \, E_0 \Big( \sum_{i=1}^{n} a_n(i, s) - \sum_{i=1}^{n} \big( 1(Y_i \leq s) - s \big) \Big)^2$$

$$- \frac{1}{n(n-1)} E_0 \sum_{i=1}^{n} \Big( a_n(i,s) + s - 1(Y_n^{(i)} \le s) \Big)^2$$

$$= \frac{1}{n(n-1)} E_0 \Big( \sum_{i=1}^{n} \big( 1(Y_i \le s) - s \big) \Big)^2$$

$$+ \frac{1}{n-1} E_0 \Big( a_n(R_{11},s) + s - 1(Y_1 \le s) \Big)^2$$

$$= \frac{1}{n-1} \Big( s(1-s) + E_0 \big( a_n(R_{11},s) + s - 1(Y_1 \le s) \big)^2 \Big)$$

we finally get from (6.2.36) and (6.2.37)

$$E_0 S_{n1}^2 \overset{n \to \infty}{\longrightarrow} 0. \tag{6.2.38}$$

Combining (6.2.35) to (6.2.38) and using contiguity we've proved $\forall\, s,t \in [0,1]$

$$\hat{B}_{n2}(s,t) - T_n(s,t) \overset{n \to \infty}{\longrightarrow} 0 \quad \text{in} \quad F_n - \text{probability}. \tag{6.2.39}$$

Therefore it's sufficient to prove

$$\mathcal{L}\big[ \,(T_n(s_1,t_1), ..., T_n(s_k,t_k)) \mid F_n \,\big]$$
$$\overset{\mathcal{L}}{\longrightarrow} \mathcal{L}\big[\, ((W_0 + B)(s_1,t_1), ..., (W_0 + B)(s_k,t_k)) \,\big], \tag{6.2.40}$$

for any $1 \le k < \infty$ and $(s_1,t_1), ..., (s_k,t_k) \in [0,1]^2$ . Using the Cramér-Wold device the proof of (6.2.40) is quite obvious, since on one hand the assumptions $F_n(s,t) = st + n^{-1/2} B_n(s,t)$ and $\|b_n - b\| \to 0$ imply ( $\forall\, c_1, ..., c_k \in \mathbb{R}$)

$$E[\, \sum_{\kappa=1}^{k} c_\kappa\, T_n(s_\kappa,t_\kappa) \mid F_n \,]$$

$$= \sum_{\kappa=1}^{k} c_\kappa \frac{1}{\sqrt{n}} \sum_{i=1}^{n} \int_0^1 \big(1(y \le s_\kappa) - s_\kappa\big)\big(1(z \le t_\kappa) - t_\kappa\big)\Big(1 + \frac{1}{\sqrt{n}} b_n(y,z)\Big)\, dy\, dz$$

$$= \sum_{\kappa=1}^{k} c_\kappa\, B_n(s_\kappa,t_\kappa) \overset{n \to \infty}{\longrightarrow} \sum_{\kappa=1}^{k} c_\kappa\, B(s_\kappa,t_\kappa) \tag{6.2.41}$$

and

$$\mathrm{Var}\,\big[ \sum_{\kappa=1}^{k} c_\kappa\, T_n(s_\kappa,t_\kappa) \mid F_n \,\big]$$

$$= \mathrm{Var}\,\Big[\, \sum_{\kappa=1}^{k} c_\kappa \,\big(1(Y_1 \le s_\kappa) - s_\kappa\big)\,\big(1(Z_1 \le t_\kappa) - t_\kappa\big) \mid F_n \,\Big]$$

$$= \sum_{\kappa=1}^{k}\sum_{\tau=1}^{k} c_\kappa\, c_\tau\, C_n(s_\kappa, t_\kappa, s_\tau, t_\tau), \tag{6.2.42}$$

where

$$C_n(s,t,u,v)$$

$$= \mathrm{Cov}_{F_n}\Big[\big(1(Y_1 \le s) - s\big)\big(1(Z_1 \le t) - t\big),\, \big(1(Y_1 \le u) - u\big)\big(1(Z_1 \le v) - v\big)\Big]$$

$$= E_{F_n}\Big[\big(1(Y_1 \le s) - s\big)\big(1(Y_1 \le u) - u\big)\big(1(Z_1 \le t) - t\big)\big(1(Z_1 \le v) - v\big)\Big]$$

$$- n^{-1/2}\, B_n(s,t)\, n^{-1/2}\, B_n(u,v)$$

$$= \big(s \wedge u - su\big)\,\big(t \wedge v - tv\big) + O(n^{-1/2}), \tag{6.2.43}$$

i.e. the central limit theorem yields

$$\mathcal{L}\Big[\, \sum_{\kappa=1}^{k} c_\kappa\, T_n(s_\kappa, t_\kappa) \mid F_n \,\Big] \tag{6.2.44}$$

$$\xrightarrow{\ \mathcal{L}\ } \mathcal{N}\Big(\sum_{\kappa=1}^{k} c_\kappa\, B(s_\kappa, t_\kappa),\, \sum_{\kappa=1}^{k}\sum_{\tau=1}^{k} c_\kappa\, c_\tau\, \big(s_\kappa \wedge s_\tau - s_\kappa s_\tau\big)\big(t_\kappa \wedge t_\tau - t_\kappa t_\tau\big)\Big),$$

on the other hand the right hand side of (6.2.44) obviously is the distribution of $\sum_{\kappa=1}^{k} c_\kappa\,\big(W_0(s_\kappa, t_\kappa) + B(s_\kappa, t_\kappa)\big)$. Thus the proof of Theorem 6.2.1 is complete.
$\square$

Now let's combine Theorem 6.2.1 and the representation (6.2.29) in order to prove a limiting law of the omnibus rank statistic $S_n(\mathcal{K}_a \hat{b}_{n2})$ under local asymptotic alternatives of the form (6.1.20).

### 6.2.2 Theorem

*Under the assumptions and notations of Theorem 6.2.1 and under the additional assumption (6.2.26) we have the limiting law $(n \to \infty)$*

$$\mathcal{L}[\, S_n(\mathcal{K}_a \hat{b}_{n2}) \mid F_n \,]$$

$$\xrightarrow{\ \mathcal{L}\ } \mathcal{L}\Big[\, \sum_{\kappa=1}^{\infty}\sum_{\tau=1}^{\infty} \lambda_\kappa(a)\, \lambda_\tau(a)\, \big(Z_{\kappa\tau} + <b, \psi_\kappa \times \psi_\tau>\big)^2 \,\Big], \tag{6.2.45}$$

*where*

$$Z_{\kappa\tau} = \langle W_0, \psi'_\kappa \times \psi'_\tau \rangle, \qquad (\kappa, \tau) \in \mathbb{N}^2, \tag{6.2.46}$$

*are i.i.d. random variables with standard normal distribution.*

*Proof:* Since $W_0$ is a $C([0,1]^2)$ –valued centered Gaussian process with covariance structure (6.2.31) the $Z_{\kappa\tau}$, $\kappa, \tau \geq 1$, are real-valued random variables with joint normal distribution, zero expectation, and covariances according to

$$E(Z_{\alpha\beta} Z_{\kappa\tau})$$

$$= \int_0^1 \int_0^1 \psi'_\alpha(s)\psi'_\kappa(u)(s \wedge u - su) \, ds \, du \int_0^1 \int_0^1 \psi'_\beta(t)\psi'_\tau(v)(t \wedge v - tv) \, dt \, dv$$

$$= \left( \int_0^1 \psi_\alpha \psi_\kappa \, d\lambda - \int_0^1 \psi_\alpha \, d\lambda \int_0^1 \psi_\kappa \, d\lambda \right)\left( \int_0^1 \psi_\beta \psi_\tau \, d\lambda - \int_0^1 \psi_\beta \, d\lambda \int_0^1 \psi_\tau \, d\lambda \right)$$

$$= \delta_{\alpha\kappa} \, \delta_{\beta\tau}, \tag{6.2.47}$$

cf. (6.2.31) and formula (3.1.65). Thus the $Z_{\kappa\tau}$, $\kappa, \tau \geq 1$, are i.i.d. random variables with standard normal distribution.

Because of $\langle b, \psi_\kappa \times \psi_\tau \rangle = \langle B, \psi'_\kappa \times \psi'_\tau \rangle$ the proof of (6.2.45) is completely similar to the proof of Theorem 3.1.9, if we substitute the inequality (3.1.80) by the corresponding inequality ($\forall \, n \geq 2, \, \forall \, k \geq 1$)

$$E_0 \Big| \sum_{\kappa+\tau>k} \lambda_\kappa(a) \, \lambda_\tau(a) \, \langle \hat{b}_{n2}, \psi_\kappa \times \psi_\tau \rangle^2 \Big|$$

$$\leq \sum_{\kappa+\tau>k} |\lambda_\kappa(a)\lambda_\tau(a)| \, E_0 S_n^2(\psi_\kappa \times \psi_\tau)$$

$$\leq 2 \sum_{\kappa+\tau>k} |\lambda_\kappa(a)\lambda_\tau(a)| \xrightarrow{k\to\infty} 0, \tag{6.2.48}$$

where the last inequality holds true, since Section II.3.2 of Hájek and Šidák (1967) implies ($\forall \, g, h \in L_2^0(0,1)$)

$$E_0 S_n^2(g \times h) = \frac{n}{n-1} \, \frac{1}{n} \sum_{i=1}^n \big( g_n(i) \big)^2 \, \frac{1}{n} \sum_{j=1}^n \big( h_n(j) \big)^2, \tag{6.2.49}$$

and since we have in addition

$$\frac{1}{n} \sum_{i=1}^n \big( g_n(i) \big)^2 = \frac{1}{n} \sum_{i=1}^n \left( n \int_{(i-1)/n}^{i/n} g \, d\lambda \right)^2 \leq \int_0^1 g^2 \, d\lambda, \tag{6.2.50}$$

i.e. $E_0 S_n^2(\psi_\kappa \times \psi_\tau) \leq \frac{n}{n-1}\|\psi_\kappa\|^2\|\psi_\tau\|^2 \leq 2 \qquad \forall\, n \geq 2 .$ $\square$

Under the null hypothesis of independence $\mathcal{H}_0^i$ we have the special limiting law

$$\mathcal{L}[\, S_n(\mathcal{K}_a \hat{b}_{n2})\,|\,\mathcal{H}_0^i\,] \;\xrightarrow{\;\mathcal{L}\;}\; \mathcal{L}\Big[\, \sum_{\kappa=1}^{\infty}\sum_{\tau=1}^{\infty} \lambda_\kappa(a)\,\lambda_\tau(a)\,Z_{\kappa\tau}^2 \,\Big], \tag{6.2.51}$$

i.e. the asymptotic critical values of the omnibus $S_n(\mathcal{K}_a\hat{b}_{n2})$ –test may be computed by standard numerical methods. We'll conclude this section by proving

a corresponding limiting law for the one-sided rank statistic $S_n(\mathcal{K}_a\hat{b}_{n2}^0)$ . In a first step we compare the $C(\,[0,1]^2\,)$ –valued rank process $\hat{B}_{n2}^0$ defined in (6.2.15) and (6.2.14) with the auxiliary rank process

$$\tilde{B}_{n2}^0 := \hat{B}_{n2}1(\hat{B}_{n2} > 0) = T(\hat{B}_{n2}), \tag{6.2.52}$$

where

$$T(f) = f1(f > 0) \qquad \forall\, f \in C(\,[0,1]^2\,) \tag{6.2.53}$$

obviously defines a mapping from $C(\,[0,1]^2\,)$ to $C(\,[0,1]^2\,)$ which is continuous with respect to the sup-norm $\|\cdot\|_\infty$ on $C(\,[0,1]^2\,)$ , cf. inequality (6.2.55).

### 6.2.3 Lemma

*For any $n \geq 1$ we have the inequality*

$$\|\hat{B}_{n2}^0 - \tilde{B}_{n2}^0\|_\infty \;\leq\; 8/\sqrt{n}. \tag{6.2.54}$$

*Proof:* Obviously

$$|x1(x > 0) - y1(y > 0)| \;\leq\; |x - y| \qquad \forall\, x,y \in \mathbb{R}. \tag{6.2.55}$$

As a first step we get from definition (6.2.5) the inequality ( $\forall\, (s,t) \in [0,1]^2$ )

$$|\hat{B}_{n2}(s,t) - \hat{B}_{n1}(s,t)|$$

$$\leq \max_{0\leq i<n}\max_{0\leq j\leq n}\big|\hat{B}_{n1}(\tfrac{i+1}{n},\tfrac{j}{n}) - \hat{B}_{n1}(\tfrac{i}{n},\tfrac{j}{n})\big|$$

$$+ \max_{0\leq i\leq n}\max_{0\leq j<n}\big|\hat{B}_{n1}(\tfrac{i}{n},\tfrac{j+1}{n}) - \hat{B}_{n1}(\tfrac{i}{n},\tfrac{j}{n})\big|$$

$$\leq \max_{0\leq i<n}\max_{0\leq j\leq n}\frac{1}{\sqrt{n}}\sum_{q=1}^{n}\left|\left(1(R_{1q}=i+1)-\frac{1}{n}\right)\left(1(R_{2q}\leq j)-\frac{j}{n}\right)\right|$$

$$+\max_{0\leq i\leq n}\max_{0\leq j<n}\frac{1}{\sqrt{n}}\sum_{q=1}^{n}\left|\left(1(R_{1q}\leq i)-\frac{i}{n}\right)\left(1(R_{2q}=j+1)-\frac{1}{n}\right)\right|$$

$$\leq \frac{2}{\sqrt{n}}+\frac{2}{\sqrt{n}}=\frac{4}{\sqrt{n}}. \tag{6.2.56}$$

In a second step we use (6.2.56) and (6.2.55) in order to get

$$\sup_{0\leq s\leq 1}\sup_{0\leq t\leq 1}|\tilde{B}^0_{n2}(s,t)-\hat{B}^0_{n1}(s,t)| \leq 4/\sqrt{n}, \tag{6.2.57}$$

where $\hat{B}^0_{n1}=\hat{B}_{n1}1(\hat{B}_{n1}>0)$, cf. (6.2.14).

In a final step (6.2.15), (6.2.55), and (6.2.56) imply for any $(s,t)\in[0,1]^2$ the inequality

$$|\hat{B}^0_{n2}(s,t)-\hat{B}^0_{n1}(s,t)|$$

$$\leq \max_{0\leq i<n}\max_{0\leq j\leq n}|\hat{B}^0_{n1}(\frac{i+1}{n},\frac{j}{n})-\hat{B}^0_{n1}(\frac{i}{n},\frac{j}{n})|$$

$$+\max_{0\leq i\leq n}\max_{0\leq j<n}|\hat{B}^0_{n1}(\frac{i}{n},\frac{j+1}{n})-\hat{B}^0_{n1}(\frac{i}{n},\frac{j}{n})|$$

$$\leq 4/\sqrt{n}. \tag{6.2.58}$$

Combining (6.2.58), (6.2.57), and the triangle inequality obviously proves the assertion (6.2.54). $\square$

### 6.2.4 Theorem

*In addition to the assumptions and notations of Theorem 6.2.1 let's assume*

$$\sum_{\kappa=1}^{\infty}\kappa\,|\lambda_{\kappa}(a)|<\infty. \tag{6.2.59}$$

*Then the following limiting law* $(n\to\infty)$ *holds true,*

$$\mathcal{L}[\,S_n(\mathcal{K}_a\hat{b}^0_{n2})\mid F_n\,]\xrightarrow{\mathcal{L}}$$

$$\mathcal{L}\Big[\sum_{\kappa=1}^{\infty}\sum_{\tau=1}^{\infty}\lambda_{\kappa}(a)\,\lambda_{\tau}(a)\,<(W_0+B)1(W_0+B>0),\psi'_{\kappa}\times\psi'_{\tau}>$$

$$<W_0+B,\psi'_{\kappa}\times\psi'_{\tau}>\Big] \tag{6.2.60}$$

*Proof:*  Using (6.2.53) and defining

$$S_n^0 := \sum_{\kappa=1}^{\infty} \sum_{\tau=1}^{\infty} \lambda_\kappa(a) \ < T(\hat{B}_{n2}), \psi_\kappa' \times \psi_\tau' > \ < \hat{B}_{n2}, \psi_\kappa' \times \psi_\tau' > \qquad (6.2.61)$$

the proof of assertion (6.2.60) is concluded if we prove

$$S_n(\mathcal{K}_a \hat{b}_{n2}^0) - S_n^0 \ \stackrel{n \to \infty}{\longrightarrow} \ 0 \qquad \text{in} \ \ F_n - \text{probability}, \qquad (6.2.62)$$

and

$$\mathcal{L}[\ S_n^0 \mid F_n \ ] \ \stackrel{\mathcal{L}}{\longrightarrow} \qquad\qquad\qquad\qquad\qquad\qquad (6.2.63)$$

$$\mathcal{L}\Big[ \sum_{\kappa=1}^{\infty} \sum_{\tau=1}^{\infty} \lambda_\kappa(a)\, \lambda_\tau(a) \ < T(W_0 + B), \psi_\kappa' \times \psi_\tau' > \ < W_0 + B, \psi_\kappa' \times \psi_\tau' > \Big].$$

Because of contiguity, cf. the proof of Theorem 6.2.1, the proof of assertion (6.2.62) is complete if we prove

$$E_0 |S_n(\mathcal{K}_a \hat{b}_{n2}^0) - S_n^0| \ \stackrel{n \to \infty}{\longrightarrow} \ 0, \qquad (6.2.64)$$

where $E_0(\cdots)$ denotes the expectation under $\mathcal{H}_0^i$ .

Using the representation (6.2.27) we have the inequality

$$E_0 |S_n(\mathcal{K}_a \hat{b}_{n2}^0) - S_n^0| \qquad\qquad\qquad\qquad\qquad\qquad (6.2.65)$$

$$\leq \sum_{\kappa=1}^{\infty} \sum_{\tau=1}^{\infty} |\lambda_\kappa \lambda_\tau| \ E_0| < \hat{B}_{n2}^0 - \tilde{B}_{n2}^0, \psi_\kappa' \times \psi_\tau' > \ < \hat{B}_{n2}, \psi_\kappa' \times \psi_\tau' >|$$

$$\leq \sum_{\kappa=1}^{\infty} \sum_{\tau=1}^{\infty} |\lambda_\kappa \lambda_\tau| \ \sqrt{E_0 < \hat{B}_{n2}^0 - \tilde{B}_{n2}^0, \psi_\kappa' \times \psi_\tau' >^2} \ \sqrt{E_0 < \hat{B}_{n2}, \psi_\kappa' \times \psi_\tau' >^2} \ .$$

Additionally we get from (6.2.49) and (6.2.50)

$$E_0(< \hat{B}_{n2}, \psi_\kappa' \times \psi_\tau' >)^2 = E_0(< \hat{b}_{n2}, \psi_\kappa \times \psi_\tau >)^2$$

$$= E_0 S_n^2(\psi_\kappa \times \psi_\tau) \ \leq \ 2, \qquad (6.2.66)$$

and from Lemma 6.2.3

$$E_0(< \hat{B}_{n2}^0 - \tilde{B}_{n2}^0, \psi_\kappa' \times \psi_\tau' >)^2 \ \leq \ \|\psi_\kappa' \times \psi_\tau'\|^2 \ E_0 \|\hat{B}_{n2}^0 - \tilde{B}_{n2}^0\|^2$$

$$\leq \ (\pi\kappa)^2 (\pi\tau)^2 (8/\sqrt{n})^2 . \qquad (6.2.67)$$

Combining (6.2.65) to (6.2.67) and assumption (6.2.59) we've proved assertion (6.2.64) and thus (6.2.62).

The final proof of (6.2.63) is completely similar to the corresponding proof of Theorem 3.1.9 since for each $k \geq 1$

$$\sum_{\kappa+\tau \leq k} \lambda_\kappa \, \lambda_\tau \; <T(f), \psi'_\kappa \times \psi'_\tau> \; <f, \psi'_\kappa \times \psi'_\tau>, \quad f \in C([0,1]^2), \quad (6.2.68)$$

defines a continuous mapping from $\left(C([0,1]^2), \|\cdot\|_\infty\right)$ to $\mathbb{R}$ and since the inequality (3.1.82) can be substituted by the following corresponding inequality, cf. (6.2.66) and (6.2.67),

$$E_0 \left| \sum_{\kappa+\tau>k} \lambda_\kappa \, \lambda_\tau \; <T(\hat{B}_{n2}), \psi'_\kappa \times \psi'_\tau> \; <\hat{B}_{n2}, \psi'_\kappa \times \psi'_\tau> \right|$$

$$\leq \sum_{\kappa+\tau>k} |\lambda_\kappa \lambda_\tau| \, \sqrt{E_0(<T(\hat{B}_{n2}), \psi'_\kappa \times \psi'_\tau>)^2} \, \sqrt{E_0(<\hat{B}_{n2}, \psi'_\kappa \times \psi'_\tau>)^2}$$

$$\leq \sum_{\kappa+\tau>k} |\lambda_\kappa \lambda_\tau| \, \sqrt{(\pi\kappa)^2(\pi\tau)^2 \, E_0\|T(\hat{B}_{n2})\|^2} \, \sqrt{2}$$

$$\leq \sqrt{2} \, \pi^2 \, \sqrt{E_0\|\hat{B}_{n2}\|^2} \sum_{\kappa+\tau>k} \kappa \, |\lambda_\kappa(a)| \, \tau \, |\lambda_\tau(a)|$$

$$\leq 2 \, \pi^2 \sum_{\kappa+\tau>k} \kappa \, |\lambda_\kappa(a)| \, \tau \, |\lambda_\tau(a)| \; \overset{k\to\infty}{\longrightarrow} \; 0, \qquad (6.2.69)$$

where the last inequality holds true, since for each $(s,t) \in [0,1]^2$ we get from formula (6.2.5) the inequality

$$E_0 \hat{B}_{n2}^2(s,t) \; \leq \; \max_{0 \leq i \leq n} \max_{0 \leq j \leq n} E_0 \hat{B}_{n1}^2(i/n, j/n), \qquad (6.2.70)$$

and since for each $n \geq 2$ we get from Section II.3.2 of Hájek and Šidák (1967) and from definition (6.2.4) the inequality

$$E_0 \hat{B}_{n1}^2(\frac{i}{n}, \frac{j}{n})$$

$$= E_0 \Big( \frac{1}{\sqrt{n}} \sum_{q=1}^{n} \big(1(R_{1q} \leq i) - \frac{i}{n}\big)\big(1(R_{2q} \leq j) - \frac{j}{n}\big) \Big)^2$$

$$= \frac{n}{n-1} \frac{1}{n} \sum_{q=1}^{n} \Big(1(q \leq i) - \frac{i}{n}\Big)^2 \frac{1}{n} \sum_{r=1}^{n} \Big(1(r \leq j) - \frac{j}{n}\Big)^2$$

$$\leq 2. \qquad (6.2.71)$$

Therefore the proof of Theorem 6.2.4 is complete. $\square$

## 6.3    Projection estimators of the score function

As in the symmetry case of Section 5.3 we'll restrict the discussion to the projection of the primitive estimator $\hat{b}_{n2}$ onto suitable finite dimensional cones.

According to Theorem 6.1.2 the linear rank statistic $S_n(b) = \,<b,\hat{b}_{n2}>$ , cf. formula (6.2.10), is asymptotically optimal for testing the null hypothesis $\mathcal{H}_0^i$ versus a special direction of the *one-sided alternative of positive quadrant dependence* $\mathcal{A}_d^0$ , if the defining score function $b$ has the properties .

$$
\int_0^1 b(\cdot,x)\,dx = \int_0^1 b(x,\cdot)\,dx = 0,
$$

$$
\int_0^s \int_0^t b(u,v)\,d\lambda_2(u,v) \geq 0 \qquad \forall\,(s,t) \in [0,1]^2.
$$

(6.3.1)

Therefore we choose suitable score functions $h_1,...,h_r \in L_2(\,(0,1)^2\,)$ which are linearly independent and which fulfil the conditions (6.3.1), and we estimate the unknown $b$ by the projection $\pi_V \hat{b}_{n2}$ of the primitive estimator $\hat{b}_{n2}$ onto the $r$ –dimensional $L_2(\,(0,1)^2\,)$ –cone $V$ corresponding to (6.3.1), i.e.

$$
V = [h_1,...,h_r]^+ = \{\,\sum_{\varrho=1}^{r} \vartheta_\varrho h_\varrho : \vartheta_\varrho \geq 0 \quad \forall\,\varrho\,\}.
$$

(6.3.2)

According to formula (3.2.10) the resulting rank statistic has the form

$$
S_n(V) := S_n(\pi_V \hat{b}_{n2}) = \,<\pi_V \hat{b}_{n2}, \hat{b}_{n2}>
$$

$$
= \sup_{h \in V}\left(\,2\,<h,\hat{b}_{n2}>\,-\|h\|^2\,\right) \;=\; \sup_{\vartheta \geq 0}\left(\,2\,\vartheta^T \vec{S}_n - \vartheta^T \Gamma \vartheta\,\right),
$$

(6.3.3)

where $\vartheta = (\vartheta_1,...,\vartheta_r)^T \in \mathbb{R}^r$ ,

$$
\vec{S}_n = (<h_1,\hat{b}_{n2}>,...,<h_r,\hat{b}_{n2}>)^T = \left(S_n(h_1),...,S_n(h_r)\right)^T,
$$

(6.3.4)

and

$$
\Gamma = \left(<h_\kappa, h_\varrho>\right)_{\kappa,\varrho=1,...,r.}
$$

(6.3.5)

Using the Cramér-Wold device and Theorem 6.1.2 we get the limiting law $(n \to \infty)$

$$
\mathcal{L}[\,\vec{S}_n \mid F_n\,] \xrightarrow{\;\mathcal{L}\;} \mathcal{N}\left(\,(<h_1,b>,...,<h_r,b>)^T,\,\Gamma\,\right)
$$

(6.3.6)

if $F_n = G_n \times H_n + n^{-1/2} B_n \circ (G_n, H_n)$ , $G_n \times H_n \in \mathcal{H}_0^i$ , $B_n \in \mathcal{B}_n^i$ , and $\|b_n - b\| \to 0$ .

According to Lemma 7.5.7

$$f_0(x) = \sup_{\vartheta \geq 0} \left( \; 2 \, \vartheta^T x - \vartheta^T \Gamma \vartheta \; \right), \qquad x = (x_1, ..., x_r)^T \in \mathbb{R}^r, \qquad (6.3.7)$$

defines a continuous function from $\mathbb{R}^r$ to $\mathbb{R}$ , and $S_n(V) = f_0(\vec{S}_n)$ .
Therefore (6.3.6) implies the limiting law

$$\mathcal{L}[\, S_n(V) \,|\, F_n \,] \xrightarrow{\mathcal{L}} \mathcal{L}[\, f_0(X + <\vec{h}, b>) \,], \qquad (6.3.8)$$

where $X$ is a $\mathbb{R}^r$ –valued random variable with $\mathcal{L}(X) = \mathcal{N}(0, \Gamma)$ , and where

$$<\vec{h}, b> \; = \; (<h_1, b>, ..., <h_r, b>)^T \, , \quad \vec{h} = (h_1, ..., h_r)^T \, .$$

Especially under the null hypothesis $\mathcal{H}_0^i$ we have the limiting law

$$\mathcal{L}[\, S_n(V) \,|\, \mathcal{H}_0^i \,] \xrightarrow{\mathcal{L}} \mathcal{L}[\, f_0(X) \,]. \qquad (6.3.9)$$

For the explicit evaluation of the rank statistic $S_n(V) = f_0(\vec{S}_n)$ on the basis
of the vector $\vec{S}_n$ of simple linear rank statistics $S_n(h_1), ..., S_n(h_r)$ we can use
the results of Section 3.2.C, i.e.

$$S_n(V) = \qquad\qquad\qquad\qquad\qquad\qquad\qquad\qquad\qquad (6.3.10)$$

$$\max\{ \, (\vec{S}_n)_J^T (\Gamma_{J \times J})^{-1} (\vec{S}_n)_J \, : \, (\Gamma_{J \times J})^{-1} (\vec{S}_n)_J \geq 0, \; \emptyset \neq J \subset R \, \}$$

where $R = \{1, ..., r\}$ and $\max \emptyset = 0$ .

In order to include a variety of different types of positive quadrant dependencies
we'll take $r = 4$ and score functions $h_\varrho$ , $\varrho = 1, ..., 4$ , according to

$$\begin{aligned}
h_1(s,t) &= \; b_1(s) \, b_1(t), & h_2(s,t) &= \; b_1(s) \, b_3(t), \\
h_3(s,t) &= \; b_3(s) \, b_1(t), & h_4(s,t) &= \; b_3(s) \, b_3(t),
\end{aligned} \qquad (6.3.11)$$

where $b_1(u)$ , $0 \leq u \leq 1$ , and $b_3(v)$ , $0 \leq v \leq 1$ , are defined in formula
(3.2.89), i.e.

$$b_3(u) \; = \; -b_1(1-u) \; = \; u \, (3u - 2), \qquad 0 \leq u \leq 1. \qquad (6.3.12)$$

Obviously the score functions $h_1, ..., h_4$ fulfil the side conditions (6.3.1). Because of the property $b_1(u) + b_3(u) \equiv 2u - 1$ also the score function $h(s,t) \equiv$
$(s - 1/2)(t - 1/2)$ , which corresponds to Spearman's rho, is contained in the
cone $V = [\, h_1, h_2, h_3, h_4 \,]^+$ . The covariance matrix $\Gamma = (\, <h_i, h_j> \,)$ may
be evaluated from (6.3.11) and (3.2.90) as the Kronecker product

$$\Gamma = \begin{pmatrix} \frac{4}{30} & \frac{1}{30} \\ \frac{1}{30} & \frac{4}{30} \end{pmatrix} \otimes \begin{pmatrix} \frac{4}{30} & \frac{1}{30} \\ \frac{1}{30} & \frac{4}{30} \end{pmatrix} = \frac{1}{900} \begin{pmatrix} 16 & 4 & 4 & 1 \\ 4 & 16 & 1 & 4 \\ 4 & 1 & 16 & 4 \\ 1 & 4 & 4 & 16 \end{pmatrix} \qquad (6.3.13)$$

and the inverse $\Gamma^{-1}$ of $\Gamma$ is given by

$$\Gamma^{-1} = \begin{pmatrix} 8 & -2 \\ -2 & 8 \end{pmatrix} \otimes \begin{pmatrix} 8 & -2 \\ -2 & 8 \end{pmatrix} = 4 \begin{pmatrix} 16 & -4 & -4 & 1 \\ -4 & 16 & 1 & -4 \\ -4 & 1 & 16 & -4 \\ 1 & -4 & -4 & 16 \end{pmatrix}. \qquad (6.3.14)$$

Now let's assume that the observed value of $S_n(V)$ is $s_0 \geq 0$, cf. (6.3.10). Then formula (6.3.9) and Theorem 3.2.7 imply that the corresponding asymptotic $p$–value is given by

$$P\{ f_0(X) > s_0 \} = \sum_{k=1}^{4} w_k \, P\{ \chi_k^2 > s_0 \} \qquad (6.3.15)$$

with weights $w_k$ according to

$$w_k = \sum_{J \subset R, \ |J|=k} P\{ Y_J \geq 0 \} \, P\{ \tilde{Y}_{R \setminus J} \geq 0 \}, \qquad R = \{1, 2, 3, 4\}, \qquad (6.3.16)$$

where $( \forall J \subset R )$

$$\mathcal{L}(Y_J) = \mathcal{N}\big(0, \, (\Gamma_{J \times J})^{-1}\big), \quad \mathcal{L}(\tilde{Y}_J) = \mathcal{N}\big(0, \, ((\Gamma^{-1})_{J \times J})^{-1}\big). \qquad (6.3.17)$$

According to (6.3.14) the *correlation matrix* $\Sigma = (\sigma_{\kappa \varrho})$ corresponding to $\Gamma^{-1}$ fulfils the condition

$$\sigma_{12} = \sigma_{34} = \alpha, \qquad \sigma_{13} = \sigma_{24} = \beta, \qquad \sigma_{14} = \sigma_{23} = \alpha\beta, \qquad (6.3.18)$$

where $\alpha = \beta = -1/4$. Therefore formula (50) of Chapter 35 in Johnson and Kotz (1972) implies

$$P\{\, Y_{\{1,2,3,4\}} \geq 0 \,\} = \frac{1}{16} + \frac{1}{4\pi}\, [\, \arcsin(\alpha) + \arcsin(\beta) + \arcsin(\alpha\beta)\, ]$$

$$+ \frac{1}{4\pi^2}\, [\, \arcsin^2(\alpha) + \arcsin^2(\beta) - \arcsin^2(\alpha\beta)\, ]$$

$$= 0.030397\,. \tag{6.3.19}$$

In case of $J \subset \{1,2,3,4\}$ and $|J| \leq 3$ we can evaluate $P\{\, Y_J \geq 0 \,\}$ and $P\{\, \tilde{Y}_J \geq 0 \,\}$ from formula (3.2.95). Doing the respective routine calculations we get

$$w_1 = 0.327979, \qquad w_2 = 0.358775,$$
$$w_3 = 0.172021, \qquad w_4 = 0.030397. \tag{6.3.20}$$

## 6.4    Treatment of ties

As before we assume i.i.d. $\mathbb{R}^2$ –valued random variables $X_1, ..., X_n$ with unknown $2$ –dimensional distribution function $F$ , but now there may be ties with respect to the first components $Y_1, ..., Y_n$ or ties with respect to the second components $Z_1, ..., Z_n$ , i.e. we allow the marginal distribution functions $G$ and $H$ of $F$ to have discontinuities. In this case we assume $\mathcal{F}_2$ defined in formula (6.0.3) to be the underlying parameter space, i.e. the underlying distribution $F$ of $X_i$ may be any $2$ –dimensional distribution which is dominated by the product measure $G \times H$ of the $1$ –dimensional marginal distributions $G$ and $H$ of $F$ .

Using the notations and the results of Section 3.3 we'll extend the results and the procedures of the previous sections to this general model.

Given any $\mathcal{J}_1, \mathcal{J}_2 \in \mathcal{J}_{0,1}^*$ , cf. definition (3.3.7), we define the $(\mathcal{J}_1, \mathcal{J}_2)$ –part of $\mathcal{F}_2$ as

$$\mathcal{F}_2(\mathcal{J}_1, \mathcal{J}_2) := \left\{ F \in \mathcal{F}_2 : G \in \mathcal{F}_1(\mathcal{J}_1),\ H \in \mathcal{F}_1(\mathcal{J}_2) \right\}, \qquad (6.4.1)$$

where $\mathcal{F}_1(\mathcal{J})$ is defined in (3.3.8) and (3.3.6). Then the partition (3.3.11) of $\mathcal{F}_1$ obviously implies the following disjoint partition of $\mathcal{F}_2$ ,

$$\mathcal{F}_2 = \sum_{\mathcal{J}_1 \in \mathcal{J}_{0,1}^*} \sum_{\mathcal{J}_2 \in \mathcal{J}_{0,1}^*} \mathcal{F}_2(\mathcal{J}_1, \mathcal{J}_2). \qquad (6.4.2)$$

Similar to the two-sample model of Chapter 3 the $(\emptyset, \emptyset)$ –part of $\mathcal{F}_2$ corresponds to the continuous case,

$$\mathcal{F}_2^c = \left\{ F \in \mathcal{F}_2 : G \text{ and } H \text{ continuous } \right\} = \mathcal{F}_2(\emptyset, \emptyset). \qquad (6.4.3)$$

The partition (6.4.2) induces a corresponding disjoint partition of the respective hypotheses in the general model, i.e. the *general null hypothesis of independence* has the form

$$\mathcal{H}_0^i = \{ F \in \mathcal{F}_2 : F = G \times H \} = \sum_{\mathcal{J}_1} \sum_{\mathcal{J}_2} \mathcal{H}_0^i(\mathcal{J}_1, \mathcal{J}_2),$$
$$\mathcal{H}_0^i(\mathcal{J}_1, \mathcal{J}_2) = \{ F \in \mathcal{F}_2(\mathcal{J}_1, \mathcal{J}_2) : F = G \times H \}, \qquad (6.4.4)$$

the *general omnibus alternative of dependence* has the form

$$\mathcal{A}_d = \{ F \in \mathcal{F}_2 : F \neq G \times H \} = \sum_{\mathcal{J}_1} \sum_{\mathcal{J}_2} \mathcal{A}_d(\mathcal{J}_1, \mathcal{J}_2),$$
$$\mathcal{A}_d(\mathcal{J}_1, \mathcal{J}_2) = \{ F \in \mathcal{F}_2(\mathcal{J}_1, \mathcal{J}_2) : F \neq G \times H \}, \qquad (6.4.5)$$

and the *general one-sided alternative of positive quadrant dependence* has the form

$$\mathcal{A}_d^0 = \{\, F \in \mathcal{F}_2 : F \geq G \times H, \; F \neq G \times H \,\} = \sum_{\mathcal{J}_1} \sum_{\mathcal{J}_2} \mathcal{A}_d^0(\mathcal{J}_1, \mathcal{J}_2),$$

$$\mathcal{A}_d^0(\mathcal{J}_1, \mathcal{J}_2) = \{\, F \in \mathcal{F}_2(\mathcal{J}_1, \mathcal{J}_2) : F \geq G \times H, \; F \neq G \times H \,\}. \tag{6.4.6}$$

Given any $\mathcal{J}_1, \mathcal{J}_2 \in \mathcal{J}_{0,1}^*$ the next proposition will prove a reparametrization of the $(\mathcal{J}_1, \mathcal{J}_2)$ –part $\mathcal{F}_2(\mathcal{J}_1, \mathcal{J}_2)$ of the general model which is completely similar to the reparametrization (6.1.13) to (6.1.18) of the continuous model.

### 6.4.1 Proposition

*Given any sample size* $n$ *and any* $\mathcal{J}_1, \mathcal{J}_2 \in \mathcal{J}_{0,1}^*$ *and using the definition*

$$\mathcal{B}_n^i(\mathcal{J}_1, \mathcal{J}_2) = \left\{ \begin{array}{ll} B \in \mathcal{B}_n^i : & \forall\, y \in [0,1] \text{ the function } B(\cdot, y) \text{ is} \\ & \text{linear on each interval } (s,t] \in \mathcal{J}_1, \\ & \forall\, x \in [0,1] \text{ the function } B(x, \cdot) \text{ is} \\ & \text{linear on each interval } (u,v] \in \mathcal{J}_2 \end{array} \right\} \tag{6.4.7}$$

*with* $\mathcal{B}_n^i$ *given in formula (6.1.14), we have the equality*

$$\mathcal{F}_2(\mathcal{J}_1, \mathcal{J}_2) = \{\, F_{B,G,H}^n : B \in \mathcal{B}_n^i(\mathcal{J}_1, \mathcal{J}_2), \, G \in \mathcal{F}_1(\mathcal{J}_1), \, H \in \mathcal{F}_1(\mathcal{J}_2) \,\}, \tag{6.4.8}$$

*where*

$$F_{B,G,H}^n = G \times H + \frac{1}{\sqrt{n}}\, B \circ (G, H). \tag{6.4.9}$$

*Additionally we have the representations*

$$\mathcal{H}_0^i(\mathcal{J}_1, \mathcal{J}_2) = \{\, F_{B,G,H}^n \in \mathcal{F}_2(\mathcal{J}_1, \mathcal{J}_2) : B = 0 \,\}, \tag{6.4.10}$$

$$\mathcal{A}_d(\mathcal{J}_1, \mathcal{J}_2) = \{\, F_{B,G,H}^n \in \mathcal{F}_2(\mathcal{J}_1, \mathcal{J}_2) : B \neq 0 \,\}, \tag{6.4.11}$$

$$\mathcal{A}_d^0(\mathcal{J}_1, \mathcal{J}_2) = \{\, F_{B,G,H}^n \in \mathcal{F}_2(\mathcal{J}_1, \mathcal{J}_2) : B \geq 0, \; B \neq 0 \,\}. \tag{6.4.12}$$

*Proof:* a) Assume $B \in \mathcal{B}_n^i(\mathcal{J}_1, \mathcal{J}_2)$, $G \in \mathcal{F}_1(\mathcal{J}_1)$, and $H \in \mathcal{F}_1(\mathcal{J}_2)$. Define $F^* : [0,1]^2 \to \mathbb{R}$ according to $F^*(s,t) = st + n^{-1/2}B(s,t)$ and put $\tilde{F} = F^* \circ (G, H) = G \times H + n^{-1/2}B \circ (G, H)$. Obviously the assumptions (6.1.10) to (6.1.12) imply $F^*$ to be an absolutely continuous distribution function on $[0,1]^2$ with uniform marginal distributions. Therefore $\tilde{F}$ is a $2$ –dimensional distribution function with the property

$$\tilde{F}(y,z) = G(y)H(z) + \frac{1}{\sqrt{n}} \int_0^{G(y)} \int_0^{H(z)} b \; d\lambda_2 \quad \forall\, (y,z) \in \mathbb{R}^2, \tag{6.4.13}$$

i.e. the marginals $\tilde{G}$ and $\tilde{H}$ of $\tilde{F}$ fulfil the conditions $\tilde{G} = G \in \mathcal{F}_1(\mathcal{J}_1)$ and $\tilde{H} = H \in \mathcal{F}_1(\mathcal{J}_2)$ .

Additionally the measure $\tilde{F}$ is dominated by the product measure $\tilde{G} \times \tilde{H} = G \times H$ , since formula (6.4.13) implies

$$\tilde{F}(A) = \int 1\big( (G^{-1}, H^{-1}) \in A \big) \left(1 + \frac{1}{\sqrt{n}}\, b\right) d\lambda_2 \qquad \forall\, A \in \mathbb{B}^2 \qquad (6.4.14)$$

and since the induced measure of $(G^{-1}, H^{-1})$ under $\lambda_2$ is $G \times H$ , i.e. $(G \times H)(A) = 0$ implies $\tilde{F}(A) = 0$ . Combining the results we've proved $\tilde{F} \in \mathcal{F}_2(\mathcal{J}_1, \mathcal{J}_2)$ .

In addition the respective assumptions $B = 0$ and $B \geq 0$ obviously imply the respective assertions $\tilde{F} = \tilde{G} \times \tilde{H}$ and $\tilde{F} \geq \tilde{G} \times \tilde{H}$ . Finally, the assumption $\tilde{F} = G \times H$ implies $B = 0$ on the set $G(\mathbb{R}) \times H(\mathbb{R})$ and because of continuity also $B = 0$ on the set

$$\overline{G(\mathbb{R})} \times \overline{H(\mathbb{R})} = \left( [0,1] \setminus \bigcup_{(s,t] \in \mathcal{J}_1} (s,t) \right) \times \left( [0,1] \setminus \bigcup_{(u,v] \in \mathcal{J}_2} (u,v) \right). \qquad (6.4.15)$$

Therefore the linearity of $B(\cdot, y)$ on each $(s,t] \in \mathcal{J}_1$ and the linearity of $B(x, \cdot)$ on each $(u,v] \in \mathcal{J}_2$ implies $B = 0$ on $[0,1]^2$ .

Thus the right-hand sides of formulae (6.4.8) and (6.4.10) to (6.4.12) are subsets of the respective left-hand sides.

b) For the proof of the reverse inclusions let's assume $F \in \mathcal{F}_2(\mathcal{J}_1, \mathcal{J}_2)$ , i.e. $F \in \mathcal{F}_2$ , $G \in \mathcal{F}_1(\mathcal{J}_1)$ , and $H \in \mathcal{F}_1(\mathcal{J}_2)$ . Put $f^* : [0,1]^2 \to \mathbb{R}$ according to

$$f^* = \frac{dF}{d(G \times H)} \circ (G^{-1}, H^{-1}), \qquad (6.4.16)$$

where $f^*(s,t) := 0$ if $\big(G^{-1}(s), H^{-1}(t)\big)$ is not defined. Obviously the function $f^*$ is measurable and $f^* \geq 0$ . Defining $F^* : [0,1]^2 \to [0, \infty]$ according to

$$F^*(s,t) = \int_0^s \int_0^t f^*(u,v)\, d\lambda_2(u,v) \qquad \forall\, 0 \leq s,t \leq 1, \qquad (6.4.17)$$

we get for any $(y,z) \in \mathbb{R}^2$ the following chain of equalities,

$$F^*(G(y), H(z)) = \int f^*(u,v)\, 1\big(u \leq G(y), v \leq H(z)\big)\, d\lambda_2(u,v)$$

$$= \int \frac{dF}{d(G \times H)}(G^{-1}(u), H^{-1}(v))\, 1\big(G^{-1}(u) \leq y, H^{-1}(v) \leq z\big)\, d\lambda_2(u,v)$$

$$= \int \frac{dF}{d(G \times H)}(s,t)\, 1(-\infty < s \leq y,\; -\infty < t \leq z)\, d(G \times H)(s,t)$$

$$= F(y,z),$$

i.e. we've proved

$$F^* \circ (G, H) = F. \tag{6.4.18}$$

Especially formulae (6.4.17) and (6.4.18) imply

$$\int f^* \, d\lambda_2 = F^*(1,1) = F(\infty, \infty) = 1. \tag{6.4.19}$$

Thus $f^*$ is a probability density on $[0,1]^2$ with respect to the Lebesgue measure $\lambda_2$ on $[0,1]^2$, and $F^*$ is the corresponding distribution function on the unit square. Because of (6.4.18) the marginal distribution functions $G^*$ and $H^*$ of $F^*$ have the properties

$$G^*\big(G(y)\big) = F^*\big(G(y), 1\big) = F(y, \infty) = G(y) \qquad \forall \, y \in \mathbb{R},$$
$$H^*\big(H(z)\big) = F^*\big(1, H(z)\big) = F(\infty, z) = H(z) \qquad \forall \, z \in \mathbb{R}. \tag{6.4.20}$$

Additionally, the assumptions $G \in \mathcal{F}_1(\mathcal{J}_1)$ and $H \in \mathcal{F}_1(\mathcal{J}_2)$ imply

$$G^{-1} \text{ is constant on } (s,t] \text{ for any } (s,t] \in \mathcal{J}_1,$$

$$H^{-1} \text{ is constant on } (u,v] \text{ for any } (u,v] \in \mathcal{J}_2.$$

Therefore (6.4.16) and (6.4.17) imply the properties

$$\begin{cases} \forall \, y \in [0,1] \text{ the function } F^*(\cdot, y) \text{ is linear} \\ \qquad \text{on each interval } (s,t] \in \mathcal{J}_1, \\[2mm] \forall \, x \in [0,1] \text{ the function } F^*(x, \cdot) \text{ is linear} \\ \qquad \text{on each interval } (u,v] \in \mathcal{J}_2. \end{cases} \tag{6.4.21}$$

Since $G^*$ and $H^*$ are continuous and since $G \in \mathcal{F}_1(\mathcal{J}_1)$, $H \in \mathcal{F}_1(\mathcal{J}_2)$, we get from (6.4.20) and (6.4.21)

$$G^*(t) = H^*(t) = t \qquad \forall \, 0 \leq t \leq 1, \tag{6.4.22}$$

and therefore from (6.4.17)

$$\int_0^1 f^*(\cdot, t) \, d\lambda(t) = \int_0^1 f^*(s, \cdot) \, d\lambda(s) = 1 \qquad [\lambda - a.e.]. \tag{6.4.23}$$

Defining

$$b_n := \sqrt{n} \, (f^* - 1) \tag{6.4.24}$$

and

$$B_n(s,t) := \int_0^s \int_0^t b_n(u,v) \, d\lambda_2(u,v) \qquad \forall \, (s,t) \in [0,1]^2 \tag{6.4.25}$$

we've proved $B_n \in \mathcal{B}_n^i(\mathcal{J}_1, \mathcal{J}_2)$ and

$$F = F^* \circ (G, H) = G \times H + \frac{1}{\sqrt{n}} \, B_n \circ (G, H), \qquad (6.4.26)$$

which means that $F$ is in the right-hand side of (6.4.8).

Therefore the left-hand sides of (6.4.10) to (6.4.12) are subsets of the respective right-hand sides if we prove the following three implications

$$F = G \times H \quad \Longrightarrow \quad B_n = 0, \qquad (6.4.27)$$

$$F \neq G \times H \quad \Longrightarrow \quad B_n \neq 0, \qquad (6.4.28)$$

$$F \geq G \times H \quad \Longrightarrow \quad B_n \geq 0. \qquad (6.4.29)$$

For the proof of (6.4.27) we get from $F = G \times H$ and definition (6.4.16) the property $f^* = 1 \; [\lambda_2 - a.e.]$. This implies $B_n = 0$ because of (6.4.24) and (6.4.25).

For the proof of (6.4.28) assume $F \neq G \times H$. Then (6.4.26) implies $B_n \circ (G, H) \neq 0$ and thus $B_n \neq 0$.

For the final proof of (6.4.29) assume $F \geq G \times H$. Then (6.4.26) proves $B_n \circ (G, H) \geq 0$. By continuity of $B_n$ we get $B_n \geq 0$ on $\overline{G(\mathbb{R})} \times \overline{H(\mathbb{R})}$, and because of $G \in \mathcal{F}_1(\mathcal{J}_1)$ and $H \in \mathcal{F}_1(\mathcal{J}_2)$ we have the equality (6.4.15). Therefore the linearity of $B_n(\cdot, y)$ on each $(s, t] \in \mathcal{J}_1$ and the linearity of $B_n(x, \cdot)$ on each $(u, v] \in \mathcal{J}_2$ imply $B \geq 0$ on the unit square $[0, 1]^2$. $\square$

As a consequence of the proof of Proposition 6.4.1 we see that the parametrization $(B, G, H) \in \mathcal{B}_n^i(\mathcal{J}_1, \mathcal{J}_2) \times \mathcal{F}_1(\mathcal{J}_1) \times \mathcal{F}_1(\mathcal{J}_2)$ is injective:

### 6.4.2 Corollary

Assume $\mathcal{J}_1, \mathcal{J}_2 \in \mathcal{J}_{0,1}^*$, $B, \tilde{B} \in \mathcal{B}_n^i(\mathcal{J}_1, \mathcal{J}_2)$, $G, \tilde{G} \in \mathcal{F}_1(\mathcal{J}_1)$, and $H, \tilde{H} \in \mathcal{F}_1(\mathcal{J}_2)$. Then we have the implication

$$(B, G, H) \neq (\tilde{B}, \tilde{G}, \tilde{H})$$

$$\Longrightarrow \quad G \times H + \frac{1}{\sqrt{n}} \, B \circ (G, H) \neq \tilde{G} \times \tilde{H} + \frac{1}{\sqrt{n}} \, \tilde{B} \circ (\tilde{G}, \tilde{H}). \qquad (6.4.30)$$

*Proof:* Let's assume equality on the right-hand side of formula (6.4.30). Because of (6.1.15) and $H(\infty) = \tilde{H}(\infty) = 1$ we get $G = \tilde{G}$, and similarly $H = \tilde{H}$. Additionally this implies $B \circ (G, H) = \tilde{B} \circ (\tilde{G}, \tilde{H}) = \tilde{B} \circ (G, H)$, and because of the continuity of $B$ and $\tilde{B}$ the equality $B = \tilde{B}$ on the set $\overline{G(\mathbb{R})} \times \overline{H(\mathbb{R})}$. Since $G \in \mathcal{F}_1(\mathcal{J}_1)$ and $H \in \mathcal{F}_1(\mathcal{J}_2)$ imply the equality (6.4.15), the linearity of $(\tilde{B} - B)(\cdot, y)$ on each $(s,t] \in \mathcal{J}_1$ and the linearity of $(\tilde{B} - B)(x, \cdot)$ on each $(u, v] \in \mathcal{J}_2$ yield $B = \tilde{B}$ on $[0,1]^2$. $\square$

As in the two-sample case the $(\mathcal{J}_1, \mathcal{J}_2)$ –part of the testing problem is completely similar to the continuous testing problem. For any $(\mathcal{J}_1, \mathcal{J}_2) \in \mathcal{J}_{0,1}^* \times \mathcal{J}_{0,1}^*$ the hypotheses $\mathcal{H}_0^i(\mathcal{J}_1, \mathcal{J}_2)$, $\mathcal{A}_d^0(\mathcal{J}_1, \mathcal{J}_2)$, and $\mathcal{A}_d(\mathcal{J}_1, \mathcal{J}_2)$ are invariant under the group of transformations of the measurement scale of each of the two components, and the ranks $R_i = (R_{1i}, R_{2i})$, $i = 1, ..., n$, as defined in formula (6.0.10) are *maximal invariant* with respect to this group. Additionally the distribution of $(R_1, ..., R_n)$ does not depend on the nuisance parameter $(G, H) \in \mathcal{F}_1(\mathcal{J}_1) \times \mathcal{F}_1(\mathcal{J}_2)$:

### 6.4.3 Proposition

Assume $\mathcal{J}_1, \mathcal{J}_2 \in \mathcal{J}_{0,1}^*$, $B \in \mathcal{B}_n^i(\mathcal{J}_1, \mathcal{J}_2)$, $G_1, G_2 \in \mathcal{F}_1(\mathcal{J}_1)$, $H_1, H_2 \in \mathcal{F}_1(\mathcal{J}_2)$. Then we have

$$\mathcal{L}\big[\, (R_1, ..., R_n) \mid (B, G_1, H_1)\,\big] = \mathcal{L}\big[\, (R_1, ..., R_n) \mid (B, G_2, H_2)\,\big]. \quad (6.4.31)$$

*Proof:* Since $X_1, ..., X_n$ are independent and since for each $i = 1, ..., n$ we have $[\,(B_n, G, H) - a.e.\,]$

$$(R_{1i}, R_{2i}) = \Big(\, \sum_{j=1}^{n} 1(Y_j \le Y_i),\ \sum_{j=1}^{n} 1(Z_j \le Z_i)\,\Big)$$

$$= \Big(\, \sum_{j=1}^{n} 1(G(Y_j) \le G(Y_i)),\ \sum_{j=1}^{n} 1(H(Z_j) \le H(Z_i))\,\Big)$$

it suffices to prove the assertion

$$\mathcal{L}\big[\, (G(Y_i), H(Z_i)) \mid (B, G, H)\,\big] = Q_i \quad \forall\, (G, H) \in \mathcal{F}_1(\mathcal{J}_1) \times \mathcal{F}_1(\mathcal{J}_2), \quad (6.4.32)$$

where $Q_i$ is some suitable distribution on $[0,1]^2$.

In a first step we get for any $(s, t) \in [0,1]^2$ and $P$ corresponding to $(B, G, H)$

the equalities

$$P\{\, G(Y_i) < s,\ H(Z_i) < t \,\} \;=\; P\{\, Y_i < G^{-1}(s),\ Z_i < H^{-1}(t) \,\}$$

$$= \lim_{0<\varepsilon\to 0} P\{\, Y_i \le G^{-1}(s) - \varepsilon,\ Z_i \le H^{-1}(t) - \varepsilon \,\}$$

$$= \lim_{0<\varepsilon\to 0} \Big[\, G(G^{-1}(s) - \varepsilon)\, H(H^{-1}(t) - \varepsilon)$$

$$+ \frac{1}{\sqrt{n}}\, B\Big(G(G^{-1}(s) - \varepsilon),\ H(H^{-1}(t) - \varepsilon)\Big) \Big].$$

Therefore the proof of (6.4.32) is concluded if we prove for all $\mathcal{J} \in \mathcal{J}_{0,1}^*$ and for all $H \in \mathcal{F}_1(\mathcal{J})$ the equality

$$\lim_{0<\varepsilon\to 0} H(H^{-1}(u) - \varepsilon) = \begin{cases} u, & \text{if } u \in [0,1]\setminus\cup\{(s,t] \in \mathcal{J}\}, \\ s, & \text{if } u \in (s,t] \in \mathcal{J}, \end{cases} \qquad (6.4.33)$$

which especially means that the left-hand side of (6.4.33) is the same function for any $H \in \mathcal{F}_1(\mathcal{J})$.

If $u \in (s,t] \in \mathcal{J}$ then $H(H^{-1}(u)) = t$ and $H(H^{-1}(u) - \varepsilon) \le s\ \forall\, \varepsilon > 0$, i.e. $H(H^{-1}(u)_-) = s$.

If $u \in [0,1]\setminus\cup\{(s,t] \in \mathcal{J}\}$ and $u \in H(\mathbb{R})$ then $H^{-1}(u)$ is a continuity point of $H$, i.e. $H(H^{-1}(u)_-) = H(H^{-1}(u)) = u$.

If $(s,t] \in \mathcal{J}$, $u = s$, and $u \notin H(\mathbb{R})$ then $H^{-1}(u) = H^{-1}(t)$, i.e. we have $H(H^{-1}(u)_-) = s = u$. $\square$

Fixing $\mathcal{J}_1, \mathcal{J}_2 \in \mathcal{J}_{0,1}^*$ and considering local asymptotic alternatives corresponding to

$$F_n = G_n \times H_n + \frac{1}{\sqrt{n}}\, B_n \circ (G_n, H_n), \quad B_n \in \mathcal{B}_n^i(\mathcal{J}_1, \mathcal{J}_2),$$

$$(6.4.34)$$

$$G_n \in \mathcal{F}_1(\mathcal{J}_1), \quad H_n \in \mathcal{F}_1(\mathcal{J}_2), \quad \text{and} \quad \|b_n - b\| \overset{n\to\infty}{\longrightarrow} 0,$$

where $b_n$ is the given $\lambda_2$–derivative of $B_n$, cf. (6.1.14), and where $b \in L_2\big((0,1)^2\big)$ is some limiting direction such that

$$\int_0^1 b(\cdot, x)\, dx \;=\; \int_0^1 b(x, \cdot)\, dx \;= 0 \qquad [\lambda - a.e.], \qquad (6.4.35)$$

we'll prove an optimality result for *averaged scores linear rank tests* which is completely similar to the continuous result of Theorem 6.1.2. On the basis of this result we'll substitute the unknown optimal averaged scores by suitable rank estimators, cf. Section 3.3 for the two-sample case.

In a first step we define $\tau_{11}, ..., \tau_{1d_1}$ as the *lengths of the ties of the first components* $Y_1, ..., Y_n$ of $X_i = (Y_i, Z_i)$ and $\tau_{21}, ..., \tau_{2d_2}$ as the *lengths of the ties of the second components* $Z_1, ..., Z_n$, cf. formula (3.3.42). Using the definition (6.0.10) of the ranks $R_i = (R_{1i}, R_{2i})$, $i = 1, ..., n$, we notice that $(R_{11}, ..., R_{1n})$ completely determines $(\tau_{11}, ...\tau_{1d_1})$, and similarly $(R_{21}, ..., R_{2n})$ completely determines $(\tau_{21}, ..., \tau_{2d_2})$. Specifically, if $T_{\kappa 1} < \cdots < T_{\kappa d_\kappa}(= n)$ are the ordered *distinct* values of $R_{\kappa 1}, ..., R_{\kappa n}$, $\kappa \in \{1, 2\}$, then we have

$$T_{\kappa i} = \sum_{j=1}^{i} \tau_{\kappa j}, \qquad i = 1, ..., d_\kappa, \quad \kappa = 1, 2. \qquad (6.4.36)$$

Now the *averaged scores* $b_n^\tau(i, j)$, $i, j = 1, ..., n$, corresponding to the score function $b$ are defined as

$$b_n^\tau(i, j) = \frac{n^2}{\tau_{1q} \tau_{2r}} \int_{(T_{1q} - \tau_{1q})/n}^{T_{1q}/n} \int_{(T_{2r} - \tau_{2r})/n}^{T_{2r}/n} b \, d\lambda_2, \qquad (6.4.37)$$

$$\text{if} \quad T_{1q} - \tau_{1q} < i \leq T_{1q} \quad \text{and} \quad T_{2r} - \tau_{2r} < j \leq T_{2r}.$$

Using the scores $b_n(i, j)$ as defined in (6.0.9) we obviously get the representation

$$n^2 \int_{(T_{1q}-\tau_{1q})/n}^{T_{1q}/n} \int_{(T_{2r}-\tau_{2r})/n}^{T_{2r}/n} b \, d\lambda_2 = \sum_{i=T_{1q}-\tau_{1q}+1}^{T_{1q}} \sum_{j=T_{2r}-\tau_{2r}+1}^{T_{2r}} b_n(i, j). \qquad (6.4.38)$$

Especially we have, cf. assumption (6.4.35),

$$\sum_{i=1}^{n} b_n^\tau(i, j) = 0 \quad \forall j \qquad \text{and} \qquad \sum_{j=1}^{n} b_n^\tau(i, j) = 0 \quad \forall i, \qquad (6.4.39)$$

and

$$\sigma_n^2(\tau, b) := \frac{1}{n^2} \sum_{i=1}^{n} \sum_{j=1}^{n} \left( b_n^\tau(i, j) \right)^2 \leq \frac{1}{n^2} \sum_{i=1}^{n} \sum_{j=1}^{n} b_n^2(i, j) \leq \|b\|^2, \qquad (6.4.40)$$

where $n \sigma_n^2(\tau, b)/(n - 1)$ is the *conditional variance* given the lengths of the ties $\tau = (\tau_1; \tau_2) = (\tau_{11}, ..., \tau_{1d_1}; \tau_{21}, ..., \tau_{2d_2})$ under the null hypothesis of independence $\mathcal{H}_0^i$ of the following *averaged scores linear rank statistic*

$$S_n^\tau(b) := \frac{1}{\sqrt{n}} \sum_{i=1}^{n} b_n^\tau(R_{1i}, R_{2i}). \qquad (6.4.41)$$

For the proof of

$$\mathrm{Var}_0\big[\,S_n^\tau(b)\mid\tau\,\big] \;=\; \frac{n}{n-1}\,\sigma_n^2(\tau,b) \tag{6.4.42}$$

we define *randomized ranks* $R_i^* = (R_{1i}^*, R_{2i}^*)$ according to ( $\forall\, i = 1, ..., n$ )

$$R_{\kappa i}^* \;=\; \sum_{j=1}^n 1\big(\,R_{\kappa j} + U_{\kappa j} \le R_{\kappa i} + U_{\kappa i}\,\big), \qquad \kappa = 1, 2, \tag{6.4.43}$$

where $U_{11}, ..., U_{1n}, U_{21}, ..., U_{2n}$ are i.i.d. random variables with uniform distribution on the unit interval $(0,1)$ such that the randomization vector $U = (U_{11}, ..., U_{2n})$ and the observation vector $X = (X_1, ..., X_n)$ are stochastically independent. Then, for given $\kappa = 1, 2$, the randomized ranks $(R_{\kappa 1}^*, ..., R_{\kappa n}^*)$ and the lengths of the ties $(\tau_{\kappa 1}, ..., \tau_{\kappa d_\kappa})$ are stochastically independent, cf. Theorem 29A of Hájek (1969), and because of (6.4.37) and (6.4.38) the averaged scores linear rank statistic $S_n^\tau(b)$ has the representation

$$S_n^\tau(b) \;=\; \frac{1}{\sqrt{n}} \sum_{i=1}^n b_n^\tau(R_{1i}^*, R_{2i}^*). \tag{6.4.44}$$

Using (6.4.44) and (6.4.39) an easy evaluation yields the assertion (6.4.42).


### 6.4.4 Theorem

*Assume $b \in L_2\big((0,1)^2\big)$ with the property (6.4.35).*

*a) Under the general null hypothesis $\mathcal{H}_0^i = \{F \in \mathcal{F}_2 : F = G \times H\}$ the randomized linear rank statistic*

$$S_n^*(b) \;=\; \frac{1}{\sqrt{n}} \sum_{i=1}^n b_n(R_{1i}^*, R_{2i}^*) \tag{6.4.45}$$

*with $b_n(i, j)$ as defined in (6.0.9) has the limiting law*

$$\mathcal{L}\big[\,S_n^*(b)\mid\mathcal{H}_0^i\,\big] \;\overset{\mathcal{L}}{\longrightarrow}\; \mathcal{N}\big(0, \|b\|^2\big). \tag{6.4.46}$$

*b) Under the $(\mathcal{J}_1, \mathcal{J}_2)$ –part $\mathcal{H}_{0n}^i(\mathcal{J}_1, \mathcal{J}_2) = \{F \in \mathcal{F}_2(\mathcal{J}_1, \mathcal{J}_2) : F = G \times H\}$ of the general null hypothesis $\mathcal{H}_0^i$ the averaged scores linear rank statistic $S_n^\tau(b)$ defined in (6.4.41) has the limiting law*

$$\mathcal{L}\big[\,S_n^\tau(b)\mid\mathcal{H}_{0n}^i(\mathcal{J}_1, \mathcal{J}_2)\,\big] \;\overset{\mathcal{L}}{\longrightarrow}\; \mathcal{N}\big(0, \|L_{\mathcal{J}_1,\mathcal{J}_2}b\|^2\big), \tag{6.4.47}$$

*where $L_{\mathcal{J}_1,\mathcal{J}_2} : L_2(0,1) \to L_2(0,1)$ is defined as*

$$L_{\mathcal{J}_1,\mathcal{J}_2} \;=\; L_{\mathcal{J}_1,\emptyset} \circ L_{\emptyset,\mathcal{J}_2}$$

*with* ( $\forall h \in L_2\big((0,1)^2\big)$ , $\forall 0 < s, t < 1$ )

$$(L_{\mathcal{J}_1,\bullet} h)(s,t) \; = \; \big( L_{\mathcal{J}_1} h(\cdot, t) \big)(s), \qquad (L_{\bullet,\mathcal{J}_2} h)(s,t) \; = \; \big( L_{\mathcal{J}_2} h(s, \cdot) \big)(t),$$

*and* $L_{\mathcal{J}_1}$ *resp.* $L_{\mathcal{J}_2}$ *as defined in (3.3.49).*

*Obviously the property (6.4.35) of* $b$ *holds true for* $L_{\mathcal{J}_1,\mathcal{J}_2} b$ *, too.*

*Additionally we have the following convergence in* $\mathcal{H}^i_{0n}(\mathcal{J}_1, \mathcal{J}_2)$ *–probability,*

$$\sigma_n^2(\tau, b) \; := \; \frac{1}{n^2} \sum_{i=1}^{n} \sum_{j=1}^{n} \big( b_n^\tau(i,j) \big)^2 \; \longrightarrow \; \| L_{\mathcal{J}_1,\mathcal{J}_2} b \|^2 , \qquad (6.4.48)$$

*where the averaged scores* $b_n^\tau(i,j)$ *are defined in formula (6.4.37).*

*Proof of a)* : $R_{11}^*, ..., R_{1n}^*$ resp. $R_{21}^*, ..., R_{2n}^*$ may be viewed as the usual ranks of i.i.d. random variables $Y_1^*, ..., Y_n^*$ resp. $Z_1^*, ..., Z_n^*$ with continuous distribution functions. Additionally, under the null hypothesis $\mathcal{H}_0^i$ , we may assume $(Y_1^*, ..., Y_n^*)$ and $(Z_1^*, ..., Z_n^*)$ to be stochastically independent. Therefore the limiting law (6.4.46) is the well-known limiting law of linear rank statistics under the continuous null hypothesis, cf. Theorem 6.1.2. $\square$

Before proving part b) let's give some remarks:

The vector $\tau = (\tau_1; \tau_2) = (\tau_{11}, ..., \tau_{1d_1}; \tau_{21}, ..., \tau_{2d_2})$ of the tie-lengths of the $Y_i$ 's and the $Z_i$ 's defines a pair $(\mathcal{J}_{1n}^{\tau_1}, \mathcal{J}_{2n}^{\tau_2})$ of elements $\mathcal{J}_{1n}^{\tau_1}, \mathcal{J}_{2n}^{\tau_2} \in \mathcal{J}_{0,1}^*$ according to

$$\mathcal{J}_{1n}^{\tau_1} \; := \; \Big\{ \, (T_{10}/n, T_{11}/n] \, , ..., (T_{1(d_1-1)}/n, T_{1d_1}/n] \, \Big\} ,$$

$$\mathcal{J}_{2n}^{\tau_2} \; := \; \Big\{ \, (T_{20}/n, T_{21}/n] \, , ..., (T_{2(d_2-1)}/n, T_{2d_2}/n] \, \Big\} . \qquad (6.4.49)$$

Using the above definition $L_{\mathcal{J}_1,\mathcal{J}_2}$ with $\mathcal{J}_1 = \mathcal{J}_{1n}^{\tau_1}$ and $\mathcal{J}_2 = \mathcal{J}_{2n}^{\tau_2}$ we get from (6.4.37) and (6.4.44) the equalities

$$S_n^\tau(b) \; = \; S_n^*\big( L_{\mathcal{J}_{1n}^{\tau_1},\mathcal{J}_{2n}^{\tau_2}} \, b \big) \qquad (6.4.50)$$

and

$$\| L_{\mathcal{J}_{1n}^{\tau_1},\mathcal{J}_{2n}^{\tau_2}} \, b \|^2 \; = \; \frac{1}{n^2} \sum_{i=1}^{n} \sum_{j=1}^{n} \big( b_n^\tau(i,j) \big)^2 . \qquad (6.4.51)$$

If $\hat{G}_n$ and $\hat{H}_n$ denote the empirical distribution functions of $Y_1, ..., Y_n$ and $Z_1, ..., Z_n$ , respectively, and if $\mathcal{I}(\hat{G}_n)$ and $\mathcal{I}(\hat{H}_n)$ denote the corresponding families of jump-intervals, i.e.

$$\mathcal{I}(\hat{G}_n) \; = \; \Big\{ \, (\hat{G}_n(x_-), \hat{G}_n(x)] : x \in \mathbb{R} \, \Big\} ,$$

$$\mathcal{I}(\hat{H}_n) \; = \; \Big\{ \, (\hat{H}_n(x_-), \hat{H}_n(x)] : x \in \mathbb{R} \, \Big\} ,$$

then we have in addition

$$\mathcal{I}(\hat{G}_n) = \mathcal{J}_{1n}^{\tau_1}, \qquad \mathcal{I}(\hat{H}_n) = \mathcal{J}_{2n}^{\tau_2}. \tag{6.4.52}$$

Especially, the families $\mathcal{I}(\hat{G}_n), \mathcal{I}(\hat{H}_n) \in \mathcal{J}_{0,1}^*$ only depend on $\tau = (\tau_1; \tau_2)$ .

*Proof of part b)* : Assume $F = G \times H \in \mathcal{H}_0^i(\mathcal{J}_1, \mathcal{J}_2)$ , and let $\hat{G}_n$ and $\hat{H}_n$ denote the empirical distribution functions of $Y_1, ..., Y_n$ resp. $Z_1, ..., Z_n$ . Then formula (6.4.52), $\|\hat{G}_n - G\|_\infty \to 0$ and $\|\hat{H}_n - H\|_\infty \to 0$ in probability, and Lemma 7.5.6 imply

$$\|L_{\mathcal{J}_{1n}^{\tau_1}, \mathcal{J}_{2n}^{\tau_2}} b - L_{\mathcal{J}_1, \mathcal{J}_2} b\| \overset{n \to \infty}{\longrightarrow} 0 \quad \text{in } G \times H - \text{probability}. \tag{6.4.53}$$

Because of (6.4.51) and since the right-hand side of (6.4.51) is distribution-free under $\mathcal{H}_{0n}^i(\mathcal{J}_1, \mathcal{J}_2)$ , c.f. Proposition 6.4.3, the proof of (6.4.48) is concluded.

For the proof of (6.4.47) we prove

$$S_n^\tau(b) - S_n^*(L_{\mathcal{J}_1, \mathcal{J}_2} b) \overset{n \to \infty}{\longrightarrow} 0 \quad \text{in } \mathcal{H}_{0n}^i(\mathcal{J}_1, \mathcal{J}_2) - \text{probability}, \tag{6.4.54}$$

and apply part a) with $L_{\mathcal{J}_1, \mathcal{J}_2} b$ instead of $b$ .

For the final proof of (6.4.54) we apply (6.4.50) and the independence of $(R_{11}^*, ..., R_{1n}^*, R_{21}^*, ..., R_{2n}^*)$ and $\tau = (\tau_1; \tau_2)$ under $\mathcal{H}_{0n}^i(\mathcal{J}_1, \mathcal{J}_2)$ in order to get

$$E_0[\,(S_n^\tau(b) - S_n^*(L_{\mathcal{J}_1, \mathcal{J}_2} b))^2 \mid \tau\,] = E_0[\,(S_n^*(L_{\mathcal{J}_{1n}^{\tau_1}, \mathcal{J}_{2n}^{\tau_2}} b - L_{\mathcal{J}_1, \mathcal{J}_2} b))^2 \mid \tau\,]$$

$$\leq 2 \|L_{\mathcal{J}_{1n}^{\tau_1}, \mathcal{J}_{2n}^{\tau_2}} b - L_{\mathcal{J}_1, \mathcal{J}_2} b\|^2. \tag{6.4.55}$$

Therefore (6.4.53) implies (6.4.54). □

For any given level $0 < \alpha < 1$ Theorem 6.4.4 implies that the *randomized linear rank test*

$$\psi_n^*(b) = 1\Big( S_n^*(b) \geq u_\alpha \|b\| \Big) \tag{6.4.56}$$

and the *averaged scores linear rank test*

$$\psi_n^\tau(b) = 1\Big( S_n^\tau(b) \geq u_\alpha \sqrt{\frac{1}{n^2} \sum_{i=1}^n \sum_{j=1}^n (b_n^\tau(i,j))^2} \Big) \tag{6.4.57}$$

are asymptotically of level $\alpha$ for testing the null hypopthesis $\mathcal{H}_{0n}^i(\mathcal{J}_1, \mathcal{J}_2)$ , if the condition $\|L_{\mathcal{J}_1, \mathcal{J}_2} b\| > 0$ holds true.

**Remark:** *In Theorem 6.4.4 and in the definition of the asymptotic tests (6.4.56) and (6.4.57) the score function $b \in L_2((0,1)^2)$ with the property (6.4.35) may be substituted by any $b_n \in L_2((0,1)^2)$ with (6.4.35) such that $\|b_n - b\| \to 0$ as $n \to \infty$ .*

The next theorem will prove $\psi_n^\tau(b)$ to be asymptotically optimal for local asymptotic alternatives of the form (6.4.34), i.e. we do asymptotically best if the score function $b$ is the limit of the (unknown) directions $b_n$ .

### 6.4.5 Theorem

*Assume $\mathcal{J}_1, \mathcal{J}_2 \in \mathcal{J}_{0,1}^*$ and let $(G_n \times H_n,\, n \geq 1)$ be any sequence of hypothesis points in $\mathcal{H}_0^i(\mathcal{J}_1, \mathcal{J}_2)$ , i.e. $G_n \in \mathcal{F}_1(\mathcal{J}_1)$ and $H_n \in \mathcal{F}_1(\mathcal{J}_2)$ . Let $(F_n,\, n \geq 1)$ be the corresponding sequence of local asymptotic alternatives $F_n = G_n \times H_n + n^{-1/2} B_n \circ (G_n, H_n)$ as defined in formula (6.4.34) with some given limiting direction $b \in L_2((0,1)^2)$ which fulfils condition (6.4.35).*

*a) For any score function $h \in L_2((0,1)^2)$ such that the corresponding condition (6.4.35) for $h$ is fulfilled we have the limiting law ( as $n \longrightarrow \infty$ )*

$$\mathcal{L}[\, S_n^\tau(h) \mid F_n\,] \xrightarrow{\;\mathcal{L}\;} \mathcal{N}(\, \langle L_{\mathcal{J}_1,\mathcal{J}_2} h, b \rangle,\ \|L_{\mathcal{J}_1,\mathcal{J}_2} h\|^2\,), \tag{6.4.58}$$

*where $L_{\mathcal{J}_1,\mathcal{J}_2} h$ is defined in Theorem 6.4.4.*

*b) If $\|b\| > 0$ and if the score function $h \in L_2((0,1)^2)$ fulfils the condition $L_{\mathcal{J}_1,\mathcal{J}_2} h = b$ $[\lambda_2 - a.e.]$ , then the averaged scores linear rank test $\psi_n^\tau(h)$ defined by formula (6.4.54) is asymptotically optimal at level $\alpha$ for testing the null hypothesis $\mathcal{H}_0^i(\mathcal{J}_1, \mathcal{J}_2)$ versus $(F_n,\, n \geq 1)$ .*

*Proof:* Because of (6.4.36) and Proposition 6.4.3 we may assume without loss of generality

$$G_n = G \in \mathcal{F}_1(\mathcal{J}_1), \qquad H_n = H \in \mathcal{F}_1(\mathcal{J}_2) \qquad \forall\, n \geq 1. \tag{6.4.59}$$

On the basis of $G$ and $H$ we define randomized random variables $X_i^* = (Y_i^*, Z_i^*)$, $i = 1, ..., n$ , according to

$$Y_i^* = T_1(Y_i, U_{1i}) \quad \text{and} \quad Z_i^* = T_2(Z_i, U_{2i}), \tag{6.4.60}$$

where $T_\kappa : \mathbb{R} \times [0,1] \to [0,1]$ , $\kappa = 1, 2$ , are defined by

$$T_1(y, u) = G(y_-) + u\big(G(y) - G(y_-)\big),$$
$$T_2(y, u) = H(y_-) + u\big(H(y) - H(y_-)\big), \tag{6.4.61}$$

and where $U = (U_{11}, ..., U_{2n})$ is the randomization vector used in the definition (6.4.43) of the randomized ranks $R^* = (R_1^*, ..., R_n^*)$. Then Lemma 3.3.9 implies that the randomized ranks (6.4.43) may be viewed with probability 1 as the usual ranks of the randomized random variables (6.4.60), i.e. for $i = 1, ..., n$ we have

$$R_{1i}^* = \sum_{j=1}^{n} 1(Y_j^* \leq Y_i^*) \qquad \text{and} \qquad R_{2i}^* = \sum_{j=1}^{n} 1(Z_j^* \leq Z_i^*), \qquad (6.4.62)$$

and $X_1^*, ..., X_n^*$ are i.i.d. random variables with values in $[0,1]^2$. Under the null hypopthesis $B_n = 0$ the components $Y_i^*$ and $Z_i^*$ are stochastically independent with the same absolutely continuous distribution function $G^*(y) = H^*(y) = y \quad \forall\, y \in [0,1]$. Under the alternative $F_n = G \times H + n^{-1/2} B_n \circ (G, H)$ the random variable $X_i^* = (Y_i^*, Z_i^*)$ has the absolutely continuous distribution function $F_n^*(s,t) = st + n^{-1/2} B_n(s,t) \quad \forall\, s,t \in [0,1]$, the proof of which is similar to the proof of Lemma 3.3.9.c because of the assumptions $G \in \mathcal{F}_1(\mathcal{J}_1)$, $H \in \mathcal{F}_1(\mathcal{J}_2)$, and $B_n \in \mathcal{B}_n^i(\mathcal{J}_1, \mathcal{J}_2)$.

Especially we see that the sequence $\left( \mathcal{L}[\, (X_1^*, ..., X_n^*) \mid F_n\, ], \; n \geq 1 \right)$ is contiguous to the sequence $\left( \mathcal{L}[\, (X_1^*, ..., X_n^*) \mid G \times H\, ], \; n \geq 1 \right)$, cf. Section 7.1.

a) Using the above results we may apply Theorem 6.1.2 to the linear rank statistic $S_n^*(h_{\mathcal{J}_1, \mathcal{J}_2})$ with $h_{\mathcal{J}_1, \mathcal{J}_2} := L_{\mathcal{J}_1, \mathcal{J}_2} h$ in order to get the limiting law

$$\mathcal{L}[\, S_n^*(h_{\mathcal{J}_1, \mathcal{J}_2}) \mid F_n\, ] \;\xrightarrow{\;\mathcal{L}\;}\; \mathcal{N}\left( \langle h_{\mathcal{J}_1, \mathcal{J}_2}, b \rangle, \; \|h_{\mathcal{J}_1, \mathcal{J}_2}\|^2 \right). \qquad (6.4.63)$$

Additionally the above contiguity and formula (6.4.53) imply

$$S_n^{\tau}(h) - S_n^*(h_{\mathcal{J}_1, \mathcal{J}_2}) \;\xrightarrow{\; n \to \infty\;}\; 0 \qquad \text{in} \quad F_n - \text{probability}. \qquad (6.4.64)$$

Combining (6.4.64) and (6.4.63) we've proved the assertion (6.4.58).

b) The proof of Proposition 6.4.1 and Corollary 6.4.2 imply

$$\frac{dF_n}{d(G \times H)} \circ (G^{-1}, H^{-1}) = 1 + n^{-1/2} b_n \qquad [\lambda_2 - a.e.], \qquad (6.4.65)$$

and with probability 1 we have the equality

$$X_i = (Y_i, Z_i) = \left( G^{-1}(Y_i^*), H^{-1}(Z_i^*) \right). \qquad (6.4.66)$$

Therefore we get with probability 1

$$\frac{dF_n}{d(G \times H)}(X_i) = 1 + n^{-1/2} b_n(Y_i^*, Z_i^*), \qquad i = 1, ., .., n, \qquad (6.4.67)$$

which implies that the (optimal) log-likelihood-ratio statistic $L_n$ for testing
the simple hypothesis $G \times H$ versus the simple alternative $F_n$ has the form

$$L_n = \sum_{i=1}^{n} \log\left(1 + n^{-1/2} b_n(Y_i^*, Z_i^*)\right)$$

$$= \frac{1}{\sqrt{n}} \sum_{i=1}^{n} b_n(Y_i^*, Z_i^*) - \frac{1}{2} \|b\|^2 + \Delta_n, \qquad (6.4.68)$$

where $\Delta_n \to 0$ in $(G \times H)$ –probability, and because of contiguity,

$$\Delta_n \overset{n\to\infty}{\Longrightarrow} 0 \quad \text{in} \quad F_n - \text{probability} . \qquad (6.4.69)$$

Additionally we get from Behnen (1972), cf. Theorem 6.1.2,

$$E_0\left(\frac{1}{\sqrt{n}} \sum_{i=1}^{n} b_n(Y_i^*, Z_i^*) - S_n^*(b)\right)^2 \overset{n\to\infty}{\Longrightarrow} 0, \qquad (6.4.70)$$

and therefore from contiguity and (6.4.68)

$$L_n + \frac{1}{2} \|b\|^2 - S_n^*(b) \overset{n\to\infty}{\Longrightarrow} 0 \quad \text{in} \quad F_n - \text{probability} . \qquad (6.4.71)$$

Thus the assumption $h_{\mathcal{J}_1, \mathcal{J}_2} = b \ [\lambda_2 - a.e.]$, formula (6.4.64), and part a)
conclude the proof. $\square$

Notice, because of $B_n \in \mathcal{B}_n^i(\mathcal{J}_1, \mathcal{J}_2)$ and $\|b_n - b\| \to 0$ the limiting direction
$b$ has the property

$$L_{\mathcal{J}_1, \mathcal{J}_2} b = b \qquad [\lambda_2 - a.e.]. \qquad (6.4.72)$$

In addition we've proved that the randomized random variables $X_1^*, ..., X_n^*$ are
i.i.d. with $\lambda_2$ –density $f_n^* = 1 + n^{-1/2} b_n$ on $[0, 1]^2$ . Therefore the results of
Section 6.2 and Section 3.3 imply that the conditional expectation

$$\hat{B}_{n1}(s,t) = E\left[ \hat{B}_{n1}^*(s,t) \mid R \right], \qquad s,t \in [0, 1], \qquad (6.4.73)$$

of the *randomized rank process*

$$\hat{B}_{n1}^*(s,t) := \frac{1}{\sqrt{n}} \sum_{i=1}^{n} \left( 1\left(R_{1i}^* \le [ns]\right) - \frac{[ns]}{n} \right)\left( 1\left(R_{2i}^* \le [nt]\right) - \frac{[nt]}{n} \right) \quad (6.4.74)$$

may serve as a basic estimator of $B_n$ in the omnibus case. In a first step
we'll smooth $\hat{B}_{N1}$ by square-wise bilinear interpolation, i.e. we define the

$C([0,1]^2)$ –valued rank process $\hat{B}_{n2}$ according to

$$\hat{B}_{n2}(s,t) = (1 - ns + [ns])\,(1 - nt + [nt])\,\hat{B}_{n1}\big(\tfrac{[ns]}{n}, \tfrac{[nt]}{n}\big)$$

$$+ (ns - [ns])\,(1 - nt + [nt])\,\hat{B}_{n1}\big(\tfrac{[ns]+1}{n}, \tfrac{[nt]}{n}\big)$$

$$+ (1 - ns + [ns])\,(nt - [nt])\,\hat{B}_{n1}\big(\tfrac{[ns]}{n}, \tfrac{[nt]+1}{n}\big)$$

$$+ (ns - [ns])\,(nt - [nt])\,\hat{B}_{n1}\big(\tfrac{[ns]+1}{n}, \tfrac{[nt]+1}{n}\big). \qquad (6.4.75)$$

Since the conditional expectation is linear, we also get the properties (6.2.6) and (6.2.7) with the $\lambda_2$ –derivative $\hat{b}_{n2}$ of $\hat{B}_{n2}$ according to, cf. (6.2.8) and Lemma 3.3.9.d,

$$\hat{b}_{n2}(s,t) =$$

$$\frac{1}{\sqrt{n}} \sum_{i=1}^{n} \Big(n\, P[\, R_{1i}^* = [ns]+1 \mid R\,] - 1\Big)\Big(n\, P[\, R_{2i}^* = [nt]+1 \mid R\,] - 1\Big)$$

$$= \frac{1}{\sqrt{n}} \sum_{i=1}^{n} \Big(\frac{n}{R_{1i} - R_{1i}^0}\, 1\big(\frac{R_{1i}^0}{n} \le s < \frac{R_{1i}}{n}\big) - 1\Big)$$

$$\Big(\frac{n}{R_{2i} - R_{2i}^0}\, 1\big(\frac{R_{2i}^0}{n} \le t < \frac{R_{2i}}{n}\big) - 1\Big), \qquad (6.4.76)$$

where $R_{1i}^0$ and $R_{2i}^0$ are defined by

$$R_{\kappa i}^0 = \sum_{j=1}^{n} 1(R_{\kappa j} < R_{\kappa i}), \qquad 1 \le i \le n, \quad \kappa = 1,2. \qquad (6.4.77)$$

As in the continuous case of Section 6.2 we'll finally substitute the unknown optimal score function $b$ of the averaged scores linear rank statistic

$$S_n^\tau(b) = \frac{1}{\sqrt{n}} \sum_{i=1}^{n} \frac{n^2}{(R_{1i} - R_{1i}^0)(R_{2i} - R_{2i}^0)} \int_{R_{1i}^0/n}^{R_{1i}/n} \int_{R_{2i}^0/n}^{R_{2i}/n} b\, d\lambda_2$$

$$= \langle b, \hat{b}_{n2} \rangle \qquad (6.4.78)$$

by kernel estimators of the type $\mathcal{K}_a\hat{b}_{n2}$ as defined in formula (6.2.9). Because of (6.4.78) the resulting (nonlinear) *omnibus rank statistic* $S_n^\tau(\mathcal{K}_a\hat{b}_{n2})$ may be written in the form (6.2.11) with $\hat{b}_{n2}$ given in (6.4.76). If no ties are present, obviously (6.4.76) and (6.2.8) are identical.

In *case of the one-sided testing problem* $\mathcal{H}_0^i$ versus the alternative of positive quadrant dependence $\mathcal{A}_d^0$ we start with the adaptation

$$\hat{B}_{n1}^0 = \hat{B}_{n1}\, 1(\hat{B}_{n1} > 0) \tag{6.4.79}$$

of $\hat{B}_{n1}$ corresponding to the side-condition $B_n \geq 0$ .

Then we define the square-wise bilinear interpolation $\hat{B}_{n2}^0 \geq 0$ of $\hat{B}_{n1}^0$ according to

$$\begin{aligned}
\hat{B}_{n2}^0(s,t) = {}& \left(1 - ns + [ns]\right)\left(1 - nt + [nt]\right) \hat{B}_{n1}^0\left(\frac{[ns]}{n}, \frac{[nt]}{n}\right) \\
& + \left(ns - [ns]\right)\left(1 - nt + [nt]\right) \hat{B}_{n1}^0\left(\frac{[ns]+1}{n}, \frac{[nt]}{n}\right) \\
& + \left(1 - ns + [ns]\right)\left(nt - [nt]\right) \hat{B}_{n1}^0\left(\frac{[ns]}{n}, \frac{[nt]+1}{n}\right) \\
& + \left(ns - [ns]\right)\left(nt - [nt]\right) \hat{B}_{n1}^0\left(\frac{[ns]+1}{n}, \frac{[nt]+1}{n}\right). \tag{6.4.80}
\end{aligned}$$

Again $\hat{B}_{n2}^0$ is a $C\left([0,1]^2\right)$ –valued rank process which also has the properties (6.2.16) and (6.2.17) with $\lambda_2$ –derivative $\hat{b}_{n2}^0$ as in formula (6.2.18). In this case the unknown optimal score function $b$ of the averaged scores linear rank statistic (6.4.78) will be substituted by the corresponding kernel estimator $\mathcal{K}_a \hat{b}_{n2}^0$ as defined in formula (6.2.19). The resulting (nonlinear) *one-sided rank statistic* $S_n^\tau(\mathcal{K}_a \hat{b}_{n2}^0)$ may be written in the form (6.2.20), and for smooth kernels we may use the approximation corresponding to formula (6.2.21).

**Evaluation of the Tests**

It's clear that in case of discontinuous underlying distribution functions the ranks $R_i = (R_{1i}, R_{21})$ , $1 \leq i \leq n$ , are not distribution free under the general null hypothesis $\mathcal{H}_0^i : F = G \times H$ , c.f. Proposition 6.4.3. However, the independence of the vectors $(Y_1, ..., Y_n)$ and $(Z_1, ..., Z_n)$ under $\mathcal{H}_0^i$ implies the independence of the vectors $\tau_1$ and $\tau_2$ and also the independence of the vectors $(R_{11}, ..., R_{1n})$ and $(R_{21}, ..., R_{2n})$ . Therefore Lemma 3.3.10 can be applied separately to $Y_1, ..., Y_n$ and $Z_1, ..., Z_n$ , and thus formula (3.3.99) implies the representation ( $\forall\, 1 \leq i \leq n$ )

$$\begin{aligned}
R_{1i} &= \sum_{j=1}^{d_1} T_{1j}\, 1(T_{1(j-1)} < R_{1i}^* \leq T_{1j}), \\
R_{2i} &= \sum_{j=1}^{d_2} T_{2j}\, 1(T_{2(j-1)} < R_{2i}^* \leq T_{2j}), 
\end{aligned} \tag{6.4.81}$$

where $R_{1i}^*$ is the rank of the randomized random variable $Y_i^*$ in $Y_1^*, ..., Y_n^*$, where $R_{2i}^*$ is the rank of the randomized random variable $Z_i^*$ in $Z_1^*, ..., Z_n^*$, and where $Y_1^*, ..., Y_n^*, Z_1^*, ..., Z_n^*$ are i.i.d. random variables with uniform distribution on the unit interval $(0, 1)$, c.f. (6.4.62).

Since all underlying rank processes are invariant under any permutation of $(R_{11}, R_{21}), ..., (R_{1n}, R_{2n})$, cf. definition (6.4.74), we may rearrange the pairs $(R_{1i}, R_{2i})$, $1 \leq i \leq n$, according to increasing values of the first components. Let's denote the resulting pairs by $(\tilde{R}_{11}, \tilde{R}_{21}), ..., (\tilde{R}_{1n}, \tilde{R}_{2n})$. Then formula (6.4.81) and the stochastic independence under $\mathcal{H}_0^i$ of the vectors $(R_{11}^*, ..., R_{1n}^*)$, $(R_{21}^*, ..., R_{2n}^*)$, $\tau_1$, $\tau_2$ imply the equalities

$$\tilde{R}_{1i} = \sum_{j=1}^{d_1} T_{1j}\, 1(T_{1(j-1)} < i \leq T_{1j}), \qquad i = 1, ..., n, \tag{6.4.82}$$

$$\mathcal{L}_{\mathcal{H}_0^i}[(\tilde{R}_{21}, ..., \tilde{R}_{2n}) \mid \tau] = \mathcal{L}\Big[\Big(\sum_{k=1}^{d_2} T_{2k}\, 1(T_{2(k-1)} < Q_i \leq T_{2k}),\ 1 \leq i \leq n\Big)\Big],$$

where the $T_{\kappa j} = \tau_{\kappa 1} + \cdots + \tau_{\kappa j}$ are fixed and where $(Q_1, ..., Q_n)$ is uniformly distributed on the permutations of $(1, ..., n)$.

As a consequence we'll use the conditional tests [ given $\tau = (\tau_1; \tau_2)$ ] based on the nonlinear rank statistic $S_n(a, K)$ (in the omnibus case) and on the nonlinear rank statistic $S_n^0(a, K)$ (in the one-sided case), i.e.

$$\psi_{n\alpha} = 1\Big(S_n(a, K) > q_\alpha(\tau)\Big),$$

$$\psi_{n\alpha}^0 = 1\Big(S_n^0(a, K) > q_\alpha^0(\tau)\Big). \tag{6.4.83}$$

Because of (6.4.82) there is no principal problem to evaluate the exact conditional critical values $q_\alpha(\tau)$ and $q_\alpha^0(\tau)$ or the conditional $p$-values of the respective test statistics $S_n(a, K)$ and $S_n^0(a, K)$. But the evaluation may become very time-consuming for larger sample sizes.

Starting with an observed $\tau = (\tau_1; \tau_2)$ and using a generator of random permutations we utilize the representation (6.4.82) in order to simulate the conditional $p$-values of $S_n^0(a, K)$ for the Parzen-2 kernel (3.1.49) and the bandwidth $a = 0.40$. An illustration of the practical evaluation of the processes $\hat{B}_{n2}$ and $\hat{B}_{n2}^0$ is given in Numerical Example 2.7.1.

### The Asymptotic Distribution

Since the respective properties (6.2.24) and (6.2.25) hold true in the general case, too, we may use the representations (6.2.27) and (6.2.29) together with

the following limiting theorem in order to derive the asymptotics of the above
rank statistics. The details are omitted.

In order to formulate the limiting law of $\hat{B}_{n2}$ let's define ( $\forall\ \mathcal{J}_1, \mathcal{J}_2 \in \mathcal{J}_{0,1}^*$ )

$$T_{\mathcal{J}_1,\mathcal{J}_2} = T_{\mathcal{J}_1,\emptyset} \circ T_{\emptyset,\mathcal{J}_2} , \tag{6.4.84}$$

where $T_{\mathcal{J}_1,\emptyset}, T_{\emptyset,\mathcal{J}_2} : C\big([0,1]^2\big) \to C\big([0,1]^2\big)$ are defined by

$$(T_{\mathcal{J}_1,\emptyset}f)(s,t) = \begin{cases} f(s,t), & \text{if } s \in [0,1] \setminus \cup\{J : J \in \mathcal{J}_1\}, \\ \frac{y-s}{y-x}\,f(x,t) + \frac{s-x}{y-x}\,f(y,t), & \text{if } s \in (x,y] \in \mathcal{J}_1, \end{cases} \tag{6.4.85}$$

and

$$(T_{\emptyset,\mathcal{J}_2}f)(s,t) = \begin{cases} f(s,t) & \text{if } t \in [0,1] \setminus \cup\{J : J \in \mathcal{J}_2\}, \\ \frac{v-t}{v-u}\,f(s,u) + \frac{t-u}{v-u}\,f(s,v), & \text{if } t \in (u,v] \in \mathcal{J}_2. \end{cases} \tag{6.4.86}$$

Obviously, for each $f \in C\big([0,1]^2\big)$ and each $\mathcal{J} \in \mathcal{J}_{0,1}^*$ the functions $T_{\mathcal{J},\emptyset}f$
and $T_{\emptyset,\mathcal{J}}f$ defined in (6.4.85) and (6.4.86) are elements of $C\big([0,1]^2\big)$ and the
corresponding mappings $T_{\mathcal{J},\emptyset}$ and $T_{\emptyset,\mathcal{J}}$ are linear on $C\big([0,1]^2\big)$. Apparently
we get the inequalities

$$\|T_{\mathcal{J},\emptyset}f\|_\infty \leq \|f\|_\infty, \quad \|T_{\emptyset,\mathcal{J}}f\|_\infty \leq \|f\|_\infty \quad \forall\, f \in C\big([0,1]^2\big). \tag{6.4.87}$$

Together with the linearity this implies the continuity of $T_{\mathcal{J},\emptyset}$ and $T_{\emptyset,\mathcal{J}}$.
Therefore $T_{\mathcal{J}_1,\mathcal{J}_2}W_0$ is a well-defined $C\big([0,1]^2\big)$ –valued process if $W_0$ is the
Gaussian process defined in Theorem 6.2.1.


### 6.4.6 Theorem

*Under the assumption of Theorem 6.4.5 and using the above notations we have
the following limiting law ( $n \to \infty$ ) in $C\big([0,1]^2\big)$ ,*

$$\mathcal{L}\big[\, \hat{B}_{n2} \mid G_n \times H_n + \frac{1}{\sqrt{n}}\, B_n \circ (G_n, H_n) \,\big] \xrightarrow{\mathcal{L}} \mathcal{L}\big[\, T_{\mathcal{J}_1,\mathcal{J}_2}(W_0 + B) \,\big], \tag{6.4.88}$$

*where $B(s,t) = \int_0^s \int_0^t b\, d\lambda_2, \quad 0 \leq s,t \leq 1$ .*

*Under the special null hypothesis $F = G \times H \in \mathcal{H}_0^i(\mathcal{J}_1, \mathcal{J}_2)$ we have in
$\big( C([0,1]^2), \mathcal{B}([0,1]^2) \big)$ the conditional limiting law*

$$\mathcal{L}_{G \times H}[\hat{B}_{n2} \mid \tau] \xrightarrow{\mathcal{L}} \mathcal{L}[T_{\mathcal{J}_2,\mathcal{J}_2}W_0] \quad [a.s.]. \tag{6.4.89}$$

*Proof:*    In analogy to (6.2.5) and (6.2.4) let $\hat{B}_{n2}^{*}$ be the square-wise bilinear interpolation of $\hat{B}_{n1}^{*}$. For $0 \le i \le d_1$ and $0 \le j \le d_2$ we have the equalities $\hat{B}_{n2}^{*}(T_{1i}/n, T_{2j}/n) = \hat{B}_{n2}(T_{1i}/n, T_{2j}/n)$. Since $\hat{B}_{n2}$ can be obtained from $\hat{B}_{n1}^{*}$ by bilinear interpolation over the squares $[T_{1(i-1)}/n, T_{1i}/n] \times [T_{2(j-1)}/n, T_{2j}/n]$, we get the representation

$$\hat{B}_{n2} \; = \; T_{\mathcal{J}_{1n}^{\tau_1}, \mathcal{J}_{2n}^{\tau_2}} \, \hat{B}_{n2}^{*}, \tag{6.4.90}$$

where $\mathcal{J}_{1n}^{\tau_1}$ and $\mathcal{J}_{2n}^{\tau_2}$ are defined in formula (6.4.49).

Additionally we'll need the following simple consequence of Lemma 7.5.4:

If $(G, H)$ and $(\tilde{G}, \tilde{H})$ are pairs of arbitrary distribution functions on $\mathbb{R}$ with $\|G - \tilde{G}\|_\infty < \varepsilon/2$ and $\|H - \tilde{H}\|_\infty < \varepsilon/2$ for arbitrary $\varepsilon \in (0, 1)$, then we have the inequality

$$\|T_{\mathcal{I}(G), \mathcal{I}(H)} f - T_{\mathcal{I}(\tilde{G}), \mathcal{I}(\tilde{H})} f\|_\infty \; \le \; 46\, \omega(f, \varepsilon) \quad \forall\, f \in C([0,1]^2). \tag{6.4.91}$$

Indeed, putting

$$h(s,t) \; = \; \big(T_{\mathcal{I}(H)} f(s, \cdot)\big)(t), \quad \text{and} \quad \tilde{h}(s,t) \; = \; \big(T_{\mathcal{I}(\tilde{H})} f(s, \cdot)\big)(t),$$

for all $0 \le s, t \le 1$, with $T_{\mathcal{I}(H)}$ and $T_{\mathcal{I}(\tilde{H})}$ defined in (7.5.9) we have

$$T_{\mathcal{I}(G), \bullet} h \; = \; T_{\mathcal{I}(G), \mathcal{I}(H)} f, \quad \text{and} \quad T_{\mathcal{I}(G), \bullet} \tilde{h} \; = \; T_{\mathcal{I}(G), \mathcal{I}(\tilde{H})} f \,.$$

Therefore (6.4.87) and (7.5.12) imply

$$\|T_{\mathcal{I}(G), \mathcal{I}(H)} f - T_{\mathcal{I}(G), \mathcal{I}(\tilde{H})} f\|_\infty$$

$$\le \|h - \tilde{h}\|_\infty \; \le \; \sup_{0 \le s \le 1} \; 23\, \omega\big(f(s, \cdot), \varepsilon\big) \; \le \; 23\, \omega(f, \varepsilon).$$

A similar consideration for $(\mathcal{I}(G), \mathcal{I}(H))$ and $(\mathcal{I}(\tilde{G}), \mathcal{I}(H))$ and an application of the triangle inequality conclude the proof of (6.4.91).

Because of (6.4.52), (6.4.90), $\mathcal{I}(G_n) = \mathcal{J}_1$, $\mathcal{I}(H_n) = \mathcal{J}_2$, and (6.4.91), we get the inequality

$$P\{\, \|\hat{B}_{n2} - T_{\mathcal{J}_1, \mathcal{J}_2} \hat{B}_{n2}^{*}\|_\infty \ge \eta \,\} \; \le \; P\{\, \omega(\hat{B}_{n2}^{*}, \varepsilon) \ge \eta/46 \,\}$$
$$+ P\{\, \|\hat{G}_n - G_n\|_\infty \ge \varepsilon/2 \,\} + P\{\, \|\hat{H}_n - H_n\|_\infty \ge \varepsilon/2 \,\} \tag{6.4.92}$$

for any $\eta > 0$ and any $\varepsilon \in (0, 1)$. The continuity of $f \to \omega(f, \varepsilon)$, the continuity of $W_0 + B$, $\|\hat{G}_n - G_n\|_\infty \to 0$ and $\|\hat{H}_n - H_n\|_\infty \to 0$ in probability, and Theorem 6.2.1 imply

$$\|\hat{B}_{n2} - T_{\mathcal{J}_1, \mathcal{J}_2} \hat{B}_{n2}^{*}\|_\infty \; \longrightarrow \; 0 \quad \text{in probability.}$$

Hence assertion (6.4.88) is implied by Theorem 6.2.1 and the continuity of $T_{\mathcal{J}_1,\mathcal{J}_2}$.

Under $F = G \times H \in \mathcal{H}_0^i(\mathcal{J}_1, \mathcal{J}_2)$ we have

$$\|\hat{G}_n - G\|_\infty \to 0 \quad [a.s.] \qquad \text{and} \qquad \|\hat{H}_n - H\|_\infty \to 0 \quad [a.s].$$

For any fixed sequence ( $(\hat{G}_n, \hat{H}_n)$, $n \geq 1$ ) with $\|\hat{G}_n - G\|_\infty \to 0$ and $\|\hat{H}_n - H\|_\infty \to 0$ let $\mathcal{J}_{1n} = \mathcal{I}(\hat{G}_n)$ , $\mathcal{J}_{2n} = \mathcal{I}(\hat{H}_n)$ , and $(\hat{\tau}_{1n}; \hat{\tau}_{2n})$ be the corresponding (non-random) sequences of jump-interval families and tie-lengths vectors. Then, because of

$$\mathcal{L}_{\mathcal{H}_0^i}[\hat{B}_{n2} \mid \tau = \hat{\tau}_n] = \mathcal{L}_0[T_{\mathcal{J}_{1n},\mathcal{J}_{2n}}\hat{B}_{n2}^*], \qquad (6.4.93)$$

$\mathcal{I}(G) = \mathcal{J}_1$ , $\mathcal{I}(H) = \mathcal{J}_2$ , formula (6.4.91), and Theorem 6.2.1, we get

$$\mathcal{L}_{G \times H}[\hat{B}_{n2} \mid \tau = \hat{\tau}_n] = \mathcal{L}_0[T_{\mathcal{J}_{1n},\mathcal{J}_{2n}}\hat{B}_{n2}^*] \xrightarrow{\mathcal{L}} \mathcal{L}[T_{\mathcal{J}_1,\mathcal{J}_2}W_0].$$

Since $W_0$ is a centered Gaussian process, the definitions (6.4.84) to (6.4.86) imply that $T_{\mathcal{J}_1,\mathcal{J}_2}W_0$ is a centered Gaussian process, too. There is no difficulty to evaluate the explicit covariance structure from (6.2.31). $\square$

# Chapter 7

# Appendix

The Appendix will contain some results and proofs which don't fit into the main body of the text and where it's difficult to find a suitable reference of the result used in the present setting.

Additionally Section 7.10 of the Appendix will contain tables of simulated critical values for all the rank tests discussed in Chapter 2.

## 7.1  Proof of Theorem 3.0.1

Let's begin with the definition of *contiguity* of the sequence $(Q_N, N \geq 1)$ with respect to the sequence $(P_N, N \geq 1)$, where $Q_N$ and $P_N$ are probability measures on the measurable space $(\Omega_N, \mathcal{A}_N), N \geq 1$.

### 7.1.1 Definition

*The sequence $(Q_N, N \geq 1)$ is contiguous to the sequence $(P_N, N \geq 1)$, iff for any sequence $(A_N \in \mathcal{A}_N, N \geq 1)$ we have the implication*

$$\lim_{N \to \infty} P_N(A_N) = 0 \implies \lim_{N \to \infty} Q_N(A_N) = 0. \qquad (7.1.1)$$

Contiguity implies that every sequence of random variables which converges to zero in $P_N$ –probability, converges to zero in $Q_N$ –probability, too. In addition, the following $\varepsilon - \delta$ –reformulation of (7.1.1) is used repeatedly in the exposition of this book.

### 7.1.2 Lemma

*The sequence* $(Q_N, N \geq 1)$ *is contiguous to the sequence* $(P_N, N \geq 1)$, *if and only if for each* $\varepsilon > 0$ *there exists some* $\delta = \delta_\varepsilon > 0$ *such that the following implication holds true for all sequences* $(A_N \in \mathcal{A}_N, N \geq 1)$,

$$\limsup_{N \to \infty} P_N(A_N) < \delta \implies \limsup_{N \to \infty} Q_N(A_N) < \varepsilon. \tag{7.1.2}$$

*Proof:* a) We assume contiguity of the sequence $(Q_N, N \geq 1)$ with respect to $(P_N, N \geq 1)$ and prove the assertion (7.1.2) by contradiction: If the assertion is false, then there exists some $\varepsilon_0 > 0$ such that for each $k \geq 1$ there exists a sequence $(A_{N,k} \in \mathcal{A}_N, N \geq 1)$ which fulfils the following two conditions,

$$P_N(A_{N,k}) < \frac{1}{k} \qquad \forall\, N > N_0(k), \tag{7.1.3}$$

$$Q_N(A_{N,k}) > \varepsilon_0 \quad \text{for infinitely many } N \geq 1. \tag{7.1.4}$$

Because of (7.1.4) and (7.1.3) we may choose a sequence $1 \leq N_1 < N_2 < N_3 < \cdots$ such that

$$Q_{N_k}(A_{N_k,k}) > \varepsilon_0 \quad \text{and} \quad P_{N_k}(A_{N_k,k}) < \frac{1}{k} \qquad \forall\, k \geq 1. \tag{7.1.5}$$

Now define the sequence $(A_N^* \in \mathcal{A}_N, N \geq 1)$ according to

$$A_N^* = \begin{cases} A_{N_k,k}, & \text{if } N = N_k \text{ for some } k \geq 1, \\ \emptyset, & \text{otherwise.} \end{cases} \tag{7.1.6}$$

Obviously (7.1.5) and (7.1.6) imply

$$\lim_{N \to \infty} P_N(A_N^*) = 0 \quad \text{and} \quad \limsup_{N \to \infty} Q_N(A_N^*) \geq \varepsilon_0 > 0,$$

which contradicts the contiguity assumption.

b) For proving the reverse direction let's assume $(A_N \in \mathcal{A}_N, N \geq 1)$ and $\lim_{N \to \infty} P_N(A_N) = 0$. According to assumption (7.1.2) we get for each $\varepsilon > 0$ the implication $\limsup_{N \to \infty} Q_N(A_N) < \varepsilon$, and thus $\lim_{N \to \infty} Q_N(A_N) = 0$.
$\square$

*Now we are in the position to prove Theorem 3.0.1:*

Using the terminology of Theorem 3.0.1 let $Q_N$ denote the joint distribution of $(X_1, ..., X_N)$ under the alternative

$$(F_N, G_N) = (H_N + \varrho c_{N1} B_N \circ H_N, \ H_N + \varrho c_{NN} B_N \circ H_N) \in H_{1N}(B_N), \tag{7.1.7}$$

i.e. we have $0 < \varrho \leq 1$ and $H_N \in \mathcal{F}_1^c$ , $B_N \in \mathcal{B}_N^0$ (for all $N \geq 1$ ) such that

$$b_N = B_N' \overset{N \to \infty}{\longrightarrow} b \quad \text{in} \quad L_2(0,1) \tag{7.1.8}$$

for given $b \in L_2^0(0,1)$ . In addition let $P_N$ denote the joint distribution of $(X_1, ..., X_N)$ under the corresponding null hypothesis point $(H_N, H_N) \in \mathcal{F}_1^c \times \mathcal{F}_1^c$ . Clearly, the probability measure $Q_N$ is dominated by the probability measure $P_N$ and the corresponding Radon-Nikodym derivative is given by

$$L_N := \frac{dQ_N}{dP_N}(X_1, ..., X_N) = \prod_{i=1}^{N}\Big(1 + \varrho c_{Ni} b_N(U_i)\Big), \tag{7.1.9}$$

where $U_i := U_{Ni} := H_N(X_i)$, $i = 1, ..., N$ . Under $P_N$ the random variables $U_1, ..., U_N$ are i.i.d. with uniform distribution on the interval $(0,1)$ .

Our first goal is an expansion of $\log L_N$ under $P_N$ in the form

$$\log L_N = \varrho \sum_{i=1}^{N} c_{Ni} b_N(U_i) - \frac{1}{2}\varrho^2 \|b\|^2 + o_{P_N}(1), \tag{7.1.10}$$

where $o_{P_N}(1)$ denotes a remainder term tending to zero in $P_N$ –probability as $N \to \infty$ .

As a preliminary step of the proof of (7.1.10) let's prove

$$\sum_{i=1}^{N} c_{Ni}^2 b_N^2(U_i) \longrightarrow \|b\|^2 \quad \text{in} \quad P_N - \text{probability as} \quad N \to \infty. \tag{7.1.11}$$

On one hand the law of large numbers for i.i.d. random variables implies

$$\sum_{i=1}^{N} c_{Ni}^2 b^2(U_i) = \frac{n}{N}\Big(\frac{1}{m}\sum_{i=1}^{m} b^2(U_i)\Big) + \frac{m}{N}\Big(\frac{1}{n}\sum_{i=m+1}^{m+n} b^2(U_i)\Big) \tag{7.1.12}$$

$$\overset{N \to \infty}{\longrightarrow} (1 - \eta)\|b\|^2 + \eta\|b\|^2 = \|b\|^2 \quad \text{in} \quad P_N - \text{probability},$$

on the other hand we have

$$E_{P_N}|\sum_{i=1}^{N} c_{Ni}^2 \big(b_N^2(U_i) - b^2(U_i)\big)| \leq \sum_{i=1}^{N} c_{Ni}^2 E_{P_N}|b_N^2(U_1) - b^2(U_1)|$$

$$= E_{P_N}\big(|b_N(U_1) - b(U_1)| \cdot |b_N(U_1) + b(U_1)|\big)$$

$$\leq \|b_N - b\| \cdot \big(\|b_N\| + \|b\|\big) \overset{N \to \infty}{\longrightarrow} 0.$$

Combining these results proves (7.1.11).

As a next step we'll prove

$$\lim_{N\to\infty}\sum_{i=1}^{N} P_N\{c_{Ni}^2 b_N^2(U_i) \geq \varepsilon\} = 0 \quad \forall\, \varepsilon > 0, \tag{7.1.13}$$

which especially implies

$$Z_N := \max_{1\leq i\leq N} |c_{Ni}b_N(U_i)| \overset{N\to\infty}{\longrightarrow} 0 \quad \text{in} \quad P_N - \text{probability.} \tag{7.1.14}$$

For the proof of (7.1.13) notice that the convergence to zero in $P_N$ – probability of the random variables $c_{N1}^2 b_N^2(U_1)$ and $c_{NN}^2 b_N^2(U_N)$ implies

$$\lim_{N\to\infty} \max_{1\leq i\leq N} |m_{Ni}| = 0, \tag{7.1.15}$$

where $m_{Ni}$ is any median of $\mathcal{L}[c_{Ni}^2 b_N^2(U_i) \mid P_N]$. Therefore (7.1.11) and Theorem 10.1.1 of Chow and Teicher (1978) prove the assertion (7.1.13).

Now, by Taylor expansion of $\log(1+t)$ for $-1/2 \leq t \leq 1/2$ we easily get

$$\log(1+t) = t - \frac{1}{2}t^2 + t^3 R(t), \quad \text{where} \quad \sup_{|t|\leq 1/2} |R(t)| \leq \frac{8}{3}. \tag{7.1.16}$$

Therefore, on the set $\{\max_{1\leq i\leq N} |c_{Ni}b_N(U_i)| \leq 1/2\} = \{Z_N \leq 1/2\}$ we have the expansion

$$\log L_N = \varrho \sum_{i=1}^{N} c_{Ni}b_N(U_i) - \frac{1}{2}\varrho^2 \sum_{i=1}^{N} c_{Ni}^2 b_N^2(U_i) + R_N, \tag{7.1.17}$$

where

$$|R_N| \leq \frac{8}{3}\varrho^3 Z_N \sum_{i=1}^{N} c_{Ni}^2 b_N^2(U_i) \quad \text{on} \quad \{Z_N \leq 1/2\}. \tag{7.1.18}$$

Since $\lim_{N\to\infty} P_N\{Z_N \leq 1/2\} = 1$ , cf. (7.1.14), the assertion (7.1.10) follows from (7.1.17), (7.1.12), (7.1.18), and (7.1.14).

In addition the equality

$$E_{P_N}\left(\sum_{i=1}^{N} c_{Ni}g(U_i)\right)^2 = \|g\|^2 \quad \forall\, g \in L_2^0(0,1) \tag{7.1.19}$$

implies

$$\sum_{i=1}^{N} c_{Ni}\big(b_N(U_i) - g_N(U_i)\big) \overset{N\to\infty}{\longrightarrow} 0 \quad \text{in} \quad P_N - \text{probability} \tag{7.1.20}$$

for any sequence $(g_N \in L_2(0,1), N \geq 1)$ such that $g_N \to b$ in $L_2(0,1)$ as $N \to \infty$.

Our next goal is the proof of the contiguity of $(Q_N, N \geq 1)$ with respect to $(P_N, N \geq 1)$:

According to Levy's central limit theorem for i.i.d. random variables we get the following limiting law under $P_N$ as $N \to \infty$,

$$\sum_{i=1}^{N} c_{Ni} b(U_i) = \sqrt{\frac{n}{N}} \Big( \frac{1}{\sqrt{m}} \sum_{i=1}^{m} b(U_i) \Big) + \sqrt{\frac{m}{N}} \Big( \frac{1}{\sqrt{n}} \sum_{i=m+1}^{m+n} b(U_i) \Big)$$

$$\xrightarrow{\mathcal{L}} \mathcal{N}(0, \|b\|^2). \tag{7.1.21}$$

The limiting law (7.1.21) and formulae (7.1.10) and (7.1.20) imply the limiting law (as $N \to \infty$)

$$\mathcal{L}[\log L_N \mid P_N] \xrightarrow{\mathcal{L}} \mathcal{N}(-\frac{1}{2}\varrho^2 \|b\|^2, \varrho^2 \|b\|^2). \tag{7.1.22}$$

Therefore Corollary VI.1.2 (Le Cam's first lemma) of Hájek and Šidák (1967) proves the contiguity of $(Q_N, N \geq 1)$ with respect to $(P_N, N \geq 1)$.

In a third step we get from Sections V.1.6 and V.1.5 of Hájek and Šidák (1967) for each $g \in L_2^0(0,1)$ as $N \to \infty$

$$S_N(g) - \sum_{i=1}^{N} c_{Ni} g(U_i) \longrightarrow 0 \text{ in } P_N - \text{probability}. \tag{7.1.23}$$

Since $(Q_N, N \geq 1)$ is contiguous to $(P_N, N \geq 1)$, assertion (7.1.23) holds true in $Q_N$ –probability, too. Using the Cramér–Wold device we easily prove $(\forall b, g \in L_2^0(0,1))$

$$\mathcal{L}\Big[\begin{pmatrix} \sum c_{Ni} g(U_i) \\ \sum c_{Ni} b(U_i) \end{pmatrix} \mid P_N\Big] \xrightarrow{\mathcal{L}} \mathcal{N}\Big( \begin{pmatrix} 0 \\ 0 \end{pmatrix}, \begin{pmatrix} \|g\|^2 & <g,b> \\ <g,b> & \|b\|^2 \end{pmatrix} \Big) \tag{7.1.24}$$

and, because of (7.1.23) and (7.1.10),

$$\mathcal{L}\Big[ \begin{pmatrix} S_N(g) \\ \log L_N \end{pmatrix} \mid P_N \Big] \xrightarrow{\mathcal{L}}$$

$$\mathcal{N}\Big( \begin{pmatrix} 0 \\ -\frac{1}{2}\varrho^2 \|b\|^2 \end{pmatrix}, \begin{pmatrix} \|g\|^2 & \varrho <g,b> \\ \varrho <g,b> & \varrho^2 \|b\|^2 \end{pmatrix} \Big). \tag{7.1.25}$$

Therefore Lemma VI.1.4 (Le Cam's third lemma) of Hájek and Šidák (1967) implies the limiting law (as $N \to \infty$)

$$\mathcal{L}[S_N(g) \mid Q_N] \xrightarrow{\mathcal{L}} \mathcal{N}(\varrho <g,b>, \|g\|^2). \tag{7.1.26}$$

In case of  $\|g\| > 0$  formula (7.1.26) obviously implies

$$\lim_{N \to \infty} P_{(\varrho B_N, H_N)}\{S_N(g) \geq u_\alpha \|g\|\} = 1 - \Phi\big(u_\alpha - \varrho\frac{<g,b>}{\|g\|}\big), \qquad (7.1.27)$$

which is formula (3.0.11) if we redefine  $B_N, b,$  and  $g$  in an obvious manner.
The proof of Theorem 3.0.1 is concluded, if we prove the optimum property
of the sequence  $(\psi_N(b), N \geq 1)$  given in part (a) of the theorem in case of
$\|b\| > 0.$  Let

$$\psi^*_{N\alpha} = 1(\log L_N > k^*_{N\alpha}) + \gamma_{N\alpha}1(\log L_N = k^*_{N\alpha}) \qquad (7.1.28)$$

denote the (optimal) Neyman–Pearson test at given level  $0 < \alpha < 1$  for testing the null hypothesis  $P_N$  versus the alternative  $Q_N$.  Then the asymptotic
optimum property of the rank test  $\psi_N(b)$  is understood as asymptotic equiv-
alence of  $(\psi_N(b), N \geq 1)$  and  $(\psi^*_{N\alpha}, N \geq 1)$  under  $(P_N, N \geq 1)$  as well as
under  $(Q_N, N \geq 1),$  i.e. we have to prove

$$\lim_{N \to \infty} E[\psi_N(b) \mid P_N] = \alpha \qquad (7.1.29)$$

and

$$\lim_{N \to \infty} \big(E[\psi_N(b) \mid Q_N] - E[\psi^*_{N\alpha} \mid Q_N]\big) = 0. \qquad (7.1.30)$$

Assertion (7.1.29) is obvious from  $\mathcal{L}[S_N(b) \mid P_N] \xrightarrow{\mathcal{L}} \mathcal{N}(0, \|b\|^2) ,\ \|b\| > 0 ,$
and  $\psi_N(b) = 1(S_N(b) \geq u_\alpha\|b\|).$  For the proof of (7.1.30) we get from (7.1.22)
and (7.1.28)

$$\lim_{N \to \infty} k^*_{N\alpha} = \varrho u_\alpha - \frac{1}{2}\varrho^2\|b\|^2 \qquad (7.1.31)$$

and

$$\lim_{N \to \infty} P_N\{\log L_N = k^*_{N\alpha}\} = 0. \qquad (7.1.32)$$

Thus, contiguity of  $(Q_N, N \geq 1)$  with respect to  $(P_N, N \geq 1)$  and

$$\log L_N + \frac{1}{2}\varrho^2\|b\|^2 - \varrho S_N(b) \xrightarrow{N \to \infty} 0 \ \text{ in } \ Q_N - \text{probability} \qquad (7.1.33)$$

imply

$$\lim_{N \to \infty} E[\psi^*_{N\alpha} \mid Q_N] = \lim_{N \to \infty} Q_N\{S_N(b) > u_\alpha\}$$
$$= 1 - \Phi(u_\alpha - \varrho\|b\|) = \lim_{N \to \infty} E[\psi_N(b) \mid Q_N]. \qquad (7.1.34)$$

## 7.2 Proof of Proposition 3.2.1

The proof follows the lines of Barlow et al. (1972), p. 314-318.

a) Using the notation of Proposition 3.2.1 let $(v_n, n \geq 1)$ be a sequence in V such that

$$\|b - v_n\| \overset{n \to \infty}{\longrightarrow} \inf_{v \in V} \|b - v\| =: \gamma. \tag{7.2.1}$$

Using the equality

$$\|u + v\|^2 + \|u - v\|^2 = 2(\|u\|^2 + \|v\|^2) \quad \forall\, u, v \in V, \tag{7.2.2}$$

we get $(\forall\, m, n \geq 1)$

$$\frac{1}{4}\|v_m - v_n\|^2 = \frac{1}{2}\|v_m - b\|^2 + \frac{1}{2}\|v_n - b\|^2 - \|\frac{1}{2}(v_m + v_n) - b\|^2. \tag{7.2.3}$$

Since the convexity of V implies $(1/2)(v_m + v_n) \in V$, the definition of $\gamma$ in (7.2.1) yields the inequality

$$\|\frac{1}{2}(v_m + v_n) - b\|^2 \geq \gamma^2. \tag{7.2.4}$$

Combining (7.2.1), (7.2.3), and (7.2.4) proves

$$\|v_m - v_n\| \longrightarrow 0 \quad \text{as} \quad m, n \to \infty, \tag{7.2.5}$$

i.e. the sequence $(v_n, n \geq 1)$ is a Cauchy sequence in $\mathcal{H}$, converging to some $b_0 \in V$, since $V$ is assumed to be a closed subset of $\mathcal{H}$. Especially we get

$$\|b - b_0\| = \gamma. \tag{7.2.6}$$

In order to prove the uniqueness of $b_0 \in V$ with (7.2.6) let's take any $b_1 \in V$ such that $\|b - b_1\| = \gamma$. Then formulae (7.2.3) and (7.2.4) with $v_m$ and $v_n$ substituted by $b_0$ and $b_1$, respectively, imply

$$0 \leq \|b_0 - b_1\|^2 \leq 2\gamma^2 + 2\gamma^2 - 4\gamma^2 = 0, \tag{7.2.7}$$

which proves $b_1 = b_0$. Thus, $\pi_V b := b_0$ fulfils the assertions of part (a).

b) As a first step of the proof we'll prove the following assertion:

Given any element $b_0 \in V$ we have the equivalence

$$\left[\, b_0 = \pi_V b \quad \Longleftrightarrow \quad <b - b_0, b_0 - v> \geq 0 \quad \forall\, v \in V \,\right]. \tag{7.2.8}$$

For the proof of (7.2.8) note that the convexity of $V$ implies

$$v_\alpha := \alpha v + (1 - \alpha)b_0 \in V \quad \forall\, 0 \leq \alpha \leq 1 \tag{7.2.9}$$

for any given $v \in V$. Additionally we'll utilize the equality

$$\|b - v_\alpha\|^2 = \|(b - b_0) + \alpha(b_0 - v)\|^2$$
$$= \|b - b_0\|^2 + 2\alpha < b - b_0, b_0 - v > + \alpha^2 \|b_0 - v\|^2. \tag{7.2.10}$$

Now, for the proof of the first implication, let's assume $b_0 = \pi_V b$. Then part (a) and (7.2.9) imply $\|b - b_0\|^2 = \inf\{\|b - v_\alpha\|^2 : 0 \le \alpha \le 1\}$, i.e. the real function $g(\alpha) := \|b - v_\alpha\|^2$, $0 \le \alpha \le 1$, achieves its minimum at the point $\alpha = 0$. Hence formula (7.2.10) implies that the derivative $2 < b - b_0, b_0 - v >$ of $g$ at the point $\alpha = 0$ is nonnegative.

For the proof of the converse implication let's assume $< b - b_0, b_0 - v > \ge 0$ $\forall v \in V$. Then formula (7.2.10) implies

$$\|b - b_0\| \le \|b - (\alpha v + (1 - \alpha)b_0)\| \quad \forall \, 0 \le \alpha \le 1 \quad \forall v \in V. \tag{7.2.11}$$

Putting $v = \pi_V b \in V$ and $\alpha = 1$ yields $\|b - b_0\| \le \|b - \pi_V b\|$. Therefore part (a) implies $b_0 = \pi_V b$.

Now let's utilize the equivalence (7.2.8) in order to prove part (b):

In the first part assume $b_0 = \pi_V b$. Since $V$ is a cone we may apply (7.2.8) with $v = \beta b_0$ for arbitrary $\beta \ge 0$. This yields

$$0 \le < b - b_0, b_0 - \beta b_0 > = (1 - \beta) < b - b_0, b_0 > \quad \forall \, \beta \ge 0, \tag{7.2.12}$$

which implies $< b - b_0, b_0 > = 0$ and thus formula (3.2.8). In addition, formula (7.2.8) implies the inequality

$$0 = < b - b_0, b_0 > \ge < b - b_0, v > \quad \forall v \in V, \tag{7.2.13}$$

which proves formula (3.2.9).

For the proof of the converse part assume $b_0 \in V$ with $< b_0, b > = < b_0, b_0 >$ and $< b, v > \le < b_0, v >$ $\forall v \in V$. Obviously these assumptions imply

$$< b - b_0, b_0 - v > \ge 0 \quad \forall v \in V. \tag{7.2.14}$$

Therefore (7.2.8) implies $b_0 = \pi_V b$. $\quad \square$

## 7.3    A characterization of monotone functions

A function $g \in L_2(0,1)$ is called $\lambda - a.s.$ nonincreasing if there is a nonincreasing function $g_0 : (0,1) \to \mathbb{R}$ with the property

$$g = g_0 \quad [\lambda - a.s.], \tag{7.3.1}$$

where $\lambda$ denotes the Lebesgue measure on the unit interval.

As in Section 3.2 let $V_0$ denote the set of elements $b \in L_2^0(0,1)$ which have nonpositive integrals, i.e.

$$V_0 = \{b \in L_2^0(0,1) : \ B(t) = \int_0^t b \, d\lambda \le 0 \quad \forall \, 0 \le t \le 1\}. \tag{7.3.2}$$

### 7.3.1 Lemma

For any $g \in L_2(0,1)$ the following two assertions are equivalent:

$$< g, b > \ \le \ 0 \qquad \forall \, b \in V_0, \tag{7.3.3}$$

$$g \quad \text{is} \quad \lambda - a.s. \quad \text{nonincreasing.} \tag{7.3.4}$$

*Proof:*   a) Assume (7.3.4). In order to prove (7.3.3) we may assume without loss of generality that $g$ is a nonincreasing and rightcontinuous function with the property $g(1/2) = 0$. Let $\mu$ denote the Borel–measure on $(0,1)$ defined by

$$\mu((x,y]) = g(x) - g(y) \ge 0 \quad \forall \, 0 < x \le y < 1. \tag{7.3.5}$$

Then Fubini's theorem implies $(\forall \, b \in V_0)$

$$< g, b > = \int_0^1 b(x)g(x) \, \lambda(dx)$$

$$= \int b(x)1(0 < x < \tfrac{1}{2}) \int 1(x < y \le \tfrac{1}{2}) \, \mu(dy) \, \lambda(dx)$$

$$- \int b(x)1(\tfrac{1}{2} \le x < 1) \int 1(\tfrac{1}{2} < y \le x) \, \mu(dy) \, \lambda(dx)$$

$$= \int 1(0 < y \le \tfrac{1}{2})\Big(\int_0^y b \, d\lambda\Big) \mu(dy) - \int 1(\tfrac{1}{2} < y < 1)\Big(\int_y^1 b \, d\lambda\Big) \mu(dy)$$

$$= \int 1(0 < y < 1)B(y) \, \mu(dy) \ \le 0, \tag{7.3.6}$$

since

$$B(y) = \int_0^y b \, d\lambda = - \int_y^1 b \, d\lambda \leq 0 \qquad \forall \, 0 \leq y \leq 1. \qquad (7.3.7)$$

b) In order to prove the reverse implication let's define the product – measurable set

$$M_g := \{(x,y): \ 0 < x < y < 1, \ g(x) < g(y)\}. \qquad (7.3.8)$$

Then, as a first step, we'll prove the assertion

$$\left[ \ < g, b > \ \leq 0 \ \ \forall \, b \in V_0 \ \implies \ (\lambda \times \lambda)(M_g) = 0 \ \right]. \qquad (7.3.9)$$

In a second step we conclude the proof by proving the assertion

$$\left[ (\lambda \times \lambda)(M_g) = 0 \ \implies \ g \ \text{is} \ \lambda - a.s. \ \text{nonincreasing} \right]. \qquad (7.3.10)$$

For the proof of (7.3.9) let's assume $(\lambda \times \lambda)(M_g) > 0$. This implies the following chain of inequalities,

$$0 < \int (g(y) - g(x))1((x,y) \in M_g) \, d(\lambda \times \lambda)(x,y)$$

$$= \int g(y)\lambda\{x: \ (x,y) \in M_g\} \, d\lambda(y) - \int g(x)\lambda\{y: \ (x,y) \in M_g\} \, d\lambda(x)$$

$$= \int g(t)b(t) \, d\lambda(t) \ = \ < g, b >, \qquad (7.3.11)$$

where $b: (0,1) \to \mathbb{R}$ is defined by

$$b(t) = \lambda\{x: \ (x,t) \in M_g\} - \lambda\{y: \ (t,y) \in M_g\}, \qquad 0 < t < 1. \qquad (7.3.12)$$

Obviously, the function $b$ is bounded and measurable. Because of $\int b \, d\lambda = (\lambda \times \lambda)(M_g) - (\lambda \times \lambda)(M_g) = 0$ we get $b \in L_2^0(0,1)$. Taking any $t \in (0,1)$ we get in addition

$$B(t) := \int_0^t b \, d\lambda \qquad (7.3.13)$$

$$= (\lambda \times \lambda)\{(x,y) \in M_g: \ y \leq t\} - (\lambda \times \lambda)\{(x,y) \in M_g: \ x \leq t\}$$

and

$$\{(x,y) \in M_g: \ y \leq t\} = \{(x,y): \ 0 < x < y < 1, \ g(x) < g(y), \ y \leq t\}$$

$$\subset \{(x,y): \ 0 < x < y < 1, \ g(x) < g(y), \ x \leq t\}$$

$$= \{(x,y) \in M_g: \ x \leq t\}. \qquad (7.3.14)$$

The combination of (7.3.11), (7.3.13), (7.3.14), and $b \in L_2^0(0,1)$ yields $b \in V_0$ and $<g, b> > 0$, which concludes the proof of (7.3.9).

In order to prove (7.3.10) let's assume $(\lambda \times \lambda)(M_g) = 0$, and let $L_g \subset (0,1)$ denote the set of Lebesgue-points of $g$. We'll utilize the well-known properties

$$\lambda(L_g) = 1, \quad \lim_{0 < \varepsilon \to 0}\left(\frac{1}{\varepsilon}\int_x^{x+\varepsilon} g \, d\lambda\right) = g(x) \quad \forall \, x \in L_g. \tag{7.3.15}$$

For any $x, y \in L_g$ with $0 < x < y < 1$ and any $\varepsilon > 0$ such that $x + \varepsilon < y$ and $y + \varepsilon < 1$ the assumption $(\lambda \times \lambda)(M_g) = 0$ and formula (7.3.15) imply

$$0 \geq \frac{1}{\varepsilon^2}\int \big(g(t) - g(s)\big) 1(y < t < y + \varepsilon) 1(x < s < x + \varepsilon) \, d(\lambda \times \lambda)(s,t)$$

$$= \frac{1}{\varepsilon}\int_y^{y+\varepsilon} g \, d\lambda - \frac{1}{\varepsilon}\int_x^{x+\varepsilon} g \, d\lambda \xrightarrow{\varepsilon \to 0} g(y) - g(x), \tag{7.3.16}$$

i.e. the function $g$ is nonincreasing on the set $L_g$. Therefore

$$g_0(t) := \sup\{g(s) : \, s \in L_g, \, s \geq t\}, \qquad 0 < t < 1, \tag{7.3.17}$$

defines a nonincreasing function $g_0 : (0,1) \to \mathbb{R}$ with $g_0(t) = g(t) \, \forall \, t \in L_g$. Because of $\lambda(L_g) = 1$ this proves $g$ to be a $\lambda - a.s.$ nonincreasing function.

$\square$

## 7.4   Proof of formulae (3.2.32) and (3.2.33)

Using the $k$ –sample terminology of Section 7.6 we get

$$\|\sqrt{\frac{N}{mn}}\,\hat{B}_{N2} - (F-G)\circ H_N^{-1}\|_\infty \tag{7.4.1}$$

$$\leq \|\hat{F}_{N1} - F_1\circ H_N^{-1}\|_\infty + \|\hat{F}_{N2} - F_2\circ H_N^{-1}\|_\infty.$$

In addition (7.6.14) and (7.6.12) imply

$$\|\hat{F}_{Ni} - F_i\circ H_N^{-1}\|_\infty \leq \frac{1}{n_i} + 3\|\tilde{F}_{Ni} - F_i\circ H_N^{-1}\|_\infty$$

$$+ 2\sum_{\kappa=1}^{2}\frac{n_\kappa}{n_i}\|\tilde{F}_{N\kappa} - F_\kappa\circ H_N^{-1}\|_\infty. \tag{7.4.2}$$

According to the Glivenko–Cantelli theorem the right hand side of (7.4.2) converges to zero in $(F_1, F_2)$ –probability.  Therefore (7.4.1) and (7.4.2) prove assertion (3.2.32).

For the proof of (3.2.33) we get

$$\|F\circ H_N^{-1} - F\circ H^{-1}\|_\infty = \|F\circ H_N^{-1}\circ H_N - F\circ H^{-1}\circ H_N\|_\infty$$

$$= \|F - F\circ H^{-1}\circ H_N\|_\infty = \|F\circ H^{-1}\circ H - F\circ H^{-1}\circ H_N\|_\infty$$

$$= \sup_{x\in\mathbb{R}}\left(F\circ H^{-1}(H(x)\vee H_N(x)) - F\circ H^{-1}(H(x)\wedge H_N(x))\right)$$

$$\leq \frac{1}{\eta}\sup_{x\in\mathbb{R}}\left((\eta F + (1-\eta)G)\circ H^{-1}(H(x)\vee H_N(x))\right.$$

$$\left. - (\eta F + (1-\eta)G)\circ H^{-1}(H(x)\wedge H_N(x))\right)$$

$$= \frac{1}{\eta}\sup_{x\in\mathbb{R}}\left((H(x)\vee H_N(x)) - (H(x)\wedge H_N(x))\right)$$

$$= \frac{1}{\eta}\|H_N - H\|_\infty = \frac{1}{\eta}\|(\frac{m}{N} - \eta)(F-G)\|_\infty$$

$$= \frac{1}{\eta}|\frac{m}{N} - \eta|\cdot\|F-G\|_\infty \leq \frac{1}{\eta}|\frac{m}{N} - \eta|. \tag{7.4.3}$$

Completely similar we get the inequality

$$\|G \circ H_N^{-1} - G \circ H^{-1}\|_\infty \;\le\; \frac{1}{1-\eta}\,|\frac{m}{N} - \eta|. \qquad (7.4.4)$$

Combining (7.4.3) and (7.4.4) proves

$$\|(F-G) \circ H_N^{-1} - (F-G) \circ H^{-1}\|_\infty \;\le\; \frac{|m/N - \eta|}{\eta(1-\eta)} \;\xrightarrow{N\to\infty}\; 0, \qquad (7.4.5)$$

since $m/N \to \eta \in (0,1)$ as $N \to \infty$.

# 7.5  Linear interpolation

According to Proposition 3.3.12 the averaged two-sample rank process $\hat{B}_{N2}$ may be viewed as a linear interpolation of the randomized rank process $\hat{B}_{N2}^*$ between the random arguments $0 = T_0/N < T_1/N < \cdots < T_d/N = 1$, and the process $\hat{B}_{N2}^*$ is tight. Lemma 7.5.2 will prove that the modulus of continuity of $\hat{B}_{N2}$ is bounded by five times the modulus of continuity of $\hat{B}_{N2}^*$. Therefore Theorem 8.2 of Billingsley (1968) implies the tightness of the process $\hat{B}_{N2}$ .

In a first step we'll prove the following auxiliary result.

**7.5.1 Lemma**

*Let $f : [0,1] \to \mathrm{I\!R}$ be a continuous function. Assume $0 < \delta < 1$ and $0 \le a < b \le 1$ such that $\delta \le b - a$ . Define*

$$\omega(f, \delta) := \sup\{|f(s) - f(t)| : \; s,t \in [0,1], \; |s-t| \le \delta\}. \qquad (7.5.1)$$

*Then the following inequality holds true,*

$$\omega(f, \delta) \ge \frac{\delta}{2} \cdot \frac{|f(b) - f(a)|}{b - a}. \qquad (7.5.2)$$

*Proof:*  For $k = 1, 2, ..., k_0 := [2(b - a)/\delta] \ge 2$ define the interval $I_k := [a+(k-1)\delta/2, a+k\delta/2)$ with the conventions $I_{k_0} := [b-\delta/2, b]$ if $a+k_0\delta/2 = b$ and $I_{k_0+1} := [a + k_0\delta/2, b]$ if $a + k_0\delta/2 < b$ . Obviously the $I_k$ 's build a disjoint partition of the closed interval $[a, b]$ .

Let $G : [a, b] \to \mathrm{I\!R}$ denote the line on $[a, b]$ defined by $G(a) = f(a)$ and $G(b) = f(b)$ , i.e.

$$G(x) = f(a) + (x - a)\big(f(b) - f(a)\big)/(b - a), \qquad a \le x \le b.$$

If in each $I_k$ there is a point $t_k$ with the property $f(t_k) = G(t_k)$, then we get $\delta/2 \le t_2 - a \le \delta$ and $f(t_2) - f(a) = G(t_2) - G(a)$, which implies

$$\omega(f, \delta) \ge |f(t_2) - f(a)| = (t_2 - a)\frac{|G(t_2) - G(a)|}{t_2 - a}$$

$$= (t_2 - a)\frac{|G(b) - G(a)|}{b - a} \ge \frac{\delta}{2} \cdot \frac{|f(b) - f(a)|}{b - a}.$$

Thus, in this case, the inequality (7.5.2) holds true.

Now let's assume the existence of some $I_k$ such that $f(t) \neq G(t) \; \forall \, t \in I_k$. This implies the existence of points $s_1$ and $s_2$ such that $a \leq s_1 < s_2 \leq b$, $s_2 - s_1 \geq \delta/2$, $f(s_1) = G(s_1)$, $f(s_2) = G(s_2)$, and either $f(s) \leq G(s) \; \forall \, s \in (s_1, s_2)$ or $f(s) \geq G(s) \; \forall \, s \in (s_1, s_2)$.

Under the additional assumption $f(b) \geq f(a)$ we get in the first case

$$\omega(f, \delta) \geq f(s_2) - f(s_2 - \frac{\delta}{2}) \geq G(s_2) - G(s_2 - \frac{\delta}{2})$$

$$= \frac{\delta}{2} \cdot \frac{f(b) - f(a)}{b - a} = \frac{\delta}{2} \cdot \frac{|f(b) - f(a)|}{b - a},$$

and in the second case

$$\omega(f, \delta) \geq f(s_1 + \frac{\delta}{2}) - f(s_1) \geq G(s_1 + \frac{\delta}{2}) - G(s_1)$$

$$= \frac{\delta}{2} \cdot \frac{f(b) - f(a)}{b - a} = \frac{\delta}{2} \cdot \frac{|f(b) - f(a)|}{b - a}.$$

This proves (7.5.2) in case of $f(b) \geq f(a)$.

Finally, under the additional assumption $f(b) \leq f(a)$ we get in the first case

$$\omega(f, \delta) \geq f(s_1) - f(s_1 + \frac{\delta}{2}) \geq G(s_1) - G(s_1 + \frac{\delta}{2})$$

$$= -\frac{\delta}{2} \cdot \frac{f(b) - f(a)}{b - a} = \frac{\delta}{2} \cdot \frac{|f(b) - f(a)|}{b - a},$$

and in the second case

$$\omega(f, \delta) \geq f(s_2 - \frac{\delta}{2}) - f(s_2) \geq G(s_2 - \frac{\delta}{2}) - G(s_2)$$

$$= -\frac{\delta}{2} \cdot \frac{f(b) - f(a)}{b - a} = \frac{\delta}{2} \cdot \frac{|f(b) - f(a)|}{b - a}.$$

This concludes the proof of (7.5.2). $\square$

### 7.5.2 Lemma

*Let $f : [0, 1] \to \mathbb{R}$ be a continuous function. Assume $0 = T_0 < T_1 < \cdots < T_r = 1$ for some $1 \leq r < \infty$ and define $\bar{f} : [0, 1] \to \mathbb{R}$ according to*

$$\bar{f} \text{ is linear on } [T_{i-1}, T_i] \quad \forall \, i = 1, ..., r,$$

$$\bar{f}(T_i) = f(T_i) \qquad \forall \, i = 0, 1, ..., r. \tag{7.5.3}$$

*If we define $\omega(f, \delta)$ and $\omega(\bar{f}, \delta)$ as in formula (7.5.1), then the following assertion holds true,*

$$\omega(\bar{f}, \delta) \leq 5 \, \omega(f, \delta) \qquad \forall \, 0 < \delta < 1. \tag{7.5.4}$$

*Proof:*  As a first step we'll prove some auxiliary results.

Assume $1 \leq i \leq r$, $T_{i-1} \leq s < T_i$, and $T_i - s \leq \delta$. If the inequality $T_i - T_{i-1} \leq \delta$ holds true, then assumption (7.5.3) and definition (7.5.1) imply

$$|\bar{f}(T_i) - \bar{f}(s)| \leq |\bar{f}(T_i) - \bar{f}(T_{i-1})| = |f(T_i) - f(T_{i-1})| \leq \omega(f, \delta).$$

If the reverse inequality $\delta < T_i - T_{i-1}$ holds true, then assumption (7.5.3) implies

$$\frac{|\bar{f}(T_i) - \bar{f}(s)|}{T_i - s} = \frac{|\bar{f}(T_i) - \bar{f}(T_{i-1})|}{T_i - T_{i-1}} = \frac{|f(T_i) - f(T_{i-1})|}{T_i - T_{i-1}}$$

and an application of Lemma 7.5.1 with $a = T_{i-1}$ and $b = T_i$ yields

$$|\bar{f}(T_i) - \bar{f}(s)| \leq (T_i - s)\frac{2}{\delta}\omega(f, \delta) \leq 2\omega(f, \delta), \qquad \text{since } T_i - s \leq \delta.$$

Combining the results we've proved the assertion

$$|\bar{f}(T_i) - \bar{f}(s)| \leq 2\omega(f, \delta), \qquad \text{if } T_{i-1} \leq s \leq T_i \text{ and } T_i - s \leq \delta. \qquad (7.5.5)$$

A completely similar proof yields the assertion

$$|\bar{f}(s) - \bar{f}(T_{i-1})| \leq 2\omega(f, \delta), \qquad \text{if } T_{i-1} \leq s \leq T_i \text{ and } s - T_{i-1} \leq \delta. \qquad (7.5.6)$$

For the proof of (7.5.4) let's take an arbitrary pair $(s, t)$ such that

$$0 \leq s < t \leq 1 \qquad \text{and} \qquad t \leq s + \delta. \qquad (7.5.7)$$

Then it's sufficient to prove the inequality

$$|\bar{f}(s) - \bar{f}(t)| \leq 5\omega(f, \delta). \qquad (7.5.8)$$

In order to prove (7.5.8) we assume (7.5.7) and consider different cases.

*Case (1):*  Assume $T_{i-1} \leq s < s + \delta \leq T_i$ for some $1 \leq i \leq r$.

Since $\delta \leq T_i - T_{i-1}$, an application of Lemma 7.5.1 with $a = T_{i-1}$ and $b = T_i$ and assumption (7.5.3) imply

$$\frac{2}{\delta}\omega(f, \delta) \geq \frac{|f(T_i) - f(T_{i-1})|}{T_i - T_{i-1}} = \frac{|\bar{f}(T_i) - \bar{f}(T_{i-1})|}{T_i - T_{i-1}}.$$

Since $\bar{f}$ is linear on $[T_{i-1}, T_i]$ we get for any $t \in (s, s + \delta)$ the inequality

$$\frac{2}{\delta}\omega(f, \delta) \geq \frac{|\bar{f}(t) - \bar{f}(s)|}{t - s} \geq \frac{|\bar{f}(t) - \bar{f}(s)|}{\delta},$$

which especially proves (7.5.8) in Case (1).

*Case (2):* For some $1 \leq i \leq r$ assume $T_{i-1} \leq s < T_i < s+\delta$ and $s < t \leq T_i$ .

Since $T_i - s \leq \delta$ and $T_i - t \leq \delta$ we get from (7.5.5)

$$|\bar{f}(s) - \bar{f}(t)| \leq |\bar{f}(s) - \bar{f}(T_i)| + |\bar{f}(T_i) - \bar{f}(t)| \leq 4\omega(f,\delta),$$

which proves (7.5.8) in Case (2).

*Case (3):* For some $1 \leq i \leq r$ assume $T_{i-1} \leq s < T_i < s + \delta$ and $T_i < t \leq T_{i+1}$ .

Since $T_i - s \leq \delta$ and $t - T_i \leq \delta$ we get from (7.5.5) and (7.5.6)

$$|\bar{f}(s) - \bar{f}(t)| \leq |\bar{f}(s) - \bar{f}(T_i)| + |\bar{f}(T_i) - \bar{f}(t)| \leq 4\omega(f,\delta),$$

which proves (7.5.8) in Case (3).

*Case (4):* For some $1 \leq i < j < r$ assume $T_{i-1} \leq s < T_i < s + \delta$ and $T_j < t \leq T_{j+1}$ .

Since $T_i - s \leq \delta$ , $t - T_j \leq \delta$ , and $|T_j - T_i| \leq \delta$ we get from (7.5.5), (7.5.6), and assumption (7.5.3) the following chain of inequalities

$$|\bar{f}(s) - \bar{f}(t)| \leq |\bar{f}(s) - \bar{f}(T_i)| + |\bar{f}(T_i) - \bar{f}(T_j)| + |\bar{f}(T_j) - \bar{f}(t)|$$

$$\leq 2\omega(f,\delta) + |f(T_i) - f(T_j)| + 2\omega(f,\delta)$$

$$\leq 5\omega(f,\delta),$$

which proves (7.5.8) in Case (4).

Since, under the assumption (7.5.7), no other cases are possible, the proof of Lemma 7.5.2 is concluded. $\square$

### 7.5.3 Corollary

*Let $\mathcal{J}$ be any family of pairwise disjoint, half-open subintervals $(x,y]$ of the unit interval $[0,1]$ , cf. Definition 3.3.1, and let $C = C[0,1]$ be the space of all continous real functions of $[0,1]$ .*

*Define the linear mapping $T_{\mathcal{J}} : C \rightarrow C$ by ( $\forall f \in C$ )*

$$(T_{\mathcal{J}}f)(t) = \begin{cases} f(t), & \text{if } t \in [0,1] \setminus \cup \{(x,y] \in \mathcal{J}\}, \\ \frac{y-t}{y-x} f(x) + \frac{t-x}{y-x} f(y), & \text{if } t \in (x,y] \in \mathcal{J}, \end{cases} \qquad (7.5.9)$$

*Then we have the inequalities ( $\forall f \in C$ )*

$$\|T_{\mathcal{J}}f\|_\infty \leq \|f\|_\infty \qquad (7.5.10)$$

and

$$\omega(T_{\mathcal{J}}f,\delta) \ \leq\ 11\,\omega(f,\delta) \qquad \forall\, 0 < \delta < 1, \tag{7.5.11}$$

where $\omega(g,\delta)$ is defined in formula (7.5.1).

*Proof*: Put $\bar{f} = T_{\mathcal{J}}f$ .

Since $\bar{f}$ is a linear interpolation of $f$ , the inequality (7.5.10) is immediate.

For the proof of (7.5.11) let $0 \leq s \leq t \leq 1$ be given with $t - s \leq \delta$ .

If $(s,t]$ is a subset of some element $(x,y] \in \mathcal{J}$ , then Lemma 7.5.2 yields the inequality $|\bar{f}(t) - \bar{f}(s)| \leq 5\,\omega(f,\delta)$ .

If $s,t \in A := [0,1] \setminus \cup\{(x,y] \in \mathcal{J}\}$ we have $\bar{f}(s) = f(s)$ and $\bar{f}(t) = f(t)$ , hence $|\bar{f}(t) - \bar{f}(s)| \leq \omega(f,\delta)$ .

If $s \in A$ , $t \in (x,y] \in \mathcal{J}$ we have

$$|\bar{f}(t) - \bar{f}(s)| \leq |\bar{f}(t) - \bar{f}(x)| + |\bar{f}(x) - \bar{f}(s)| \leq 5\,\omega(f,\delta) + \omega(f,\delta) ,$$

since $\bar{f}(x) = f(x)$ and $\bar{f}(s) = f(s)$ , and because of Lemma 7.5.2.

If $s \in (x_1,y_1] \in \mathcal{J}$ , $t \in (x_2,y_2] \in \mathcal{J}$ we finally have

$$|\bar{f}(s) - \bar{f}(t)| \leq |\bar{f}(s) - \bar{f}(y_1)| + |\bar{f}(y_1) - \bar{f}(x_2| + |\bar{f}(x_2) - \bar{f}(t)|$$

$$\leq 5\,\omega(f,\delta) + \omega(f,\delta) + 5\,\omega(f,\delta) \ . \ \square$$

### 7.5.4 Lemma

*For any distribution function $F$ on $\mathbb{R}$ let $\mathcal{I}(F)$ be the corresponding family of jump-intervals, i.e. $\mathcal{I}(F) = \{ (F(x_-), F(x)] : x \in \mathbb{R} \}$ .*

*Then, for any $\varepsilon \in (0,1)$ and arbitrary distribution functions $F$ and $G$ with $\|G - F\|_\infty < \varepsilon/2$ we have*

$$\|T_{\mathcal{I}(G)}f - T_{\mathcal{I}(F)}f\|_\infty \ \leq\ 23\,\omega(f,\varepsilon) \qquad \forall\, f \in C[0,1]. \tag{7.5.12}$$

*Proof*: Assume $f \in C[0,1]$ and put $f_1 := T_{\mathcal{I}(F)}f$ and $f_2 := T_{\mathcal{I}(G)}f$ .

Let $(s,t] := (F(x_-), F(x)]$ be some jump-interval of $F$ with $t - s \geq \varepsilon$ , if there is any such interval. This implies $\bar{s} := G(x - 0) \in (s - \varepsilon/2, s + \varepsilon/2)$ and $\bar{t} := G(x) \in (t - \varepsilon/2, t + \varepsilon/2)$ . Since $f_1$ and $f_2$ both are linear on the interval $[s+\varepsilon/2, t-\varepsilon/2]$ the absolute difference $|f_1 - f_2|$ over $[s+\varepsilon/2, t-\varepsilon/2]$ becomes maximal at $s + \varepsilon/2$ or at $t - \varepsilon/2$ . Therefore,

$$A := \sup_{s-\varepsilon/2 \leq u \leq t+\varepsilon/2} |f_1(u) - f_2(u)|$$

$$= \max\left( \sup_{s-\varepsilon/2 \leq u \leq s+\varepsilon/2} |f_1(u) - f_2(u)| , \ \sup_{t-\varepsilon/2 \leq u \leq t+\varepsilon/2} |f_1(u) - f_2(u)| \right).$$

Now let's consider the sup over $s - \varepsilon/2 \leq u \leq s + \varepsilon/2$ : Because of $f_1(s) = f(s)$ and $f_2(\bar{s}) = f(\bar{s})$ the triangle inequality and Corollary 7.5.3 imply

$$|f_1(u) - f_2(u)| \leq |f_1(u) - f_1(s)| + |f(s) - f(\bar{s})| + |f_2(\bar{s}) - f_2(u)|$$

$$\leq 11\,\omega(f,\varepsilon) + \omega(f,\varepsilon) + 11\,\omega(f,\varepsilon)\,.$$

Together with a completely similar argument for the sup over $t - \varepsilon/2 \leq u \leq t + \varepsilon/2$ this yields $A \leq 23\,\omega(f,\varepsilon)$ .

Obviously the same bound holds true for all jump-intervals of $F$ with $F(x) - F(x_-) \geq \varepsilon$ .

Now assume $u \in [0,1] \setminus \cup \{ (F(x_-), F(x)] : x \in \mathbb{R},\ F(x) - F(x_-) \geq \varepsilon \}$ . Then there exists some $y$ such that $|u - s| < \varepsilon/2$ for $s := F(y)$ and $|s - t| < \varepsilon/2$ for $t := G(y)$ . Therefore the triangle inequality and Lemma 7.5.2 imply

$$|f_1(u) - f_2(u)| \leq |f_1(u) - f_1(s)| + |f(s) - f(t)| + |f_2(t) - f_2(u)|$$

$$\leq 5\,\omega(f,\varepsilon) + \omega(f,\varepsilon) + 5\,\omega(f,\varepsilon)\,,$$

which concludes the proof. $\square$

### 7.5.5 Lemma

*For any family $\mathcal{J}$ of pairwise disjoint, half-open subintervals $(s,t]$ of $[0,1]$ , cf. Definition 3.3.1, we define the linear mapping $L_{\mathcal{J}} : L_2(0,1) \to L_2(0,1)$ by ($\forall f \in L_2(0,1)$)*

$$(L_{\mathcal{J}}f)(u) \;=\; \begin{cases} f(u), & \text{if } u \in (0,1] \setminus \cup\{(s,t] \in \mathcal{J}\}, \\[2mm] \frac{1}{t-s} \int_s^t f\, d\lambda, & \text{if } u \in (s,t] \in \mathcal{J}. \end{cases} \qquad (7.5.13)$$

*For any distribution function $F$ on $\mathbb{R}$ let $\mathcal{I}(F)$ be the corresponding family of jump-intervals, i.e. $\mathcal{I}(F) = \{ (F(x_-), F(x)] : x \in \mathbb{R} \}$ .*

*Then, for arbitrary distribution functions $F$ and $F_N$ , $N \geq 1$ , such that $\|F_N - F\|_\infty \to 0$ as $N \to \infty$ , we have*

$$L_{\mathcal{I}(F_N)}f \;\overset{N \to \infty}{\longrightarrow}\; L_{\mathcal{I}(F)}f \qquad \text{in } L_2(0,1) \;\; (\forall f \in L_2(0,1))\,. \qquad (7.5.14)$$

*Proof*: Put $\mathcal{J}_N := \mathcal{I}(F_N)$ and $\mathcal{J} := \mathcal{I}(F)$ and assume $f \in L_2(0,1)$ .

Notice, $L_{\mathcal{J}}f$ may be conceived as a version of the conditional expectation $E[f \mid \mathcal{F}]$ of $f$ on the probability space $\big((0,1), \mathbb{B} \cap (0,1), \lambda \big)$ with respect to the $\sigma$-algebra $\mathcal{F}$ generated by the intervals $(s,t] \in \mathcal{J}$ and by the Borel sets in $(0,1) \setminus \cup \{ (s,t] \in \mathcal{J} \}$ .

Since every $f \in L_2(0,1)$ may be approximated in $L_2(0,1)$ by a sequence of polynomials and since $E\big(E[g \mid \mathcal{F}]\big)^2 \leq Eg^2 \quad \forall\, g \in L_2(0,1)$ we may assume that $f$ is some polynomial or even that $f(u) = u^a$, $0 \leq u \leq 1$, for some $a \geq 0$.

Put $\bar{f} = L_{\mathcal{J}} f$, $\bar{f}_N = L_{\mathcal{J}_N} f$ and $g = \bar{f}_N - \bar{f}$. Then $\|\bar{f}\|_\infty \leq 1$, $\|\bar{f}_N\|_\infty \leq 1$ and $\|g\|_\infty \leq 1$.

For $0 < \varepsilon < 1/2$ put $\mathcal{J}(\varepsilon) = \{\,(s,t] \in J : t - s \geq \varepsilon/2\,\}$; the family $\mathcal{J}(\varepsilon)$ may be empty. Since $(0,1) \setminus \cup\{\,(s,t] \in \mathcal{J}(\varepsilon)\}$ only contains intervals. $J \in \mathcal{J}$ with $\lambda(J) < \varepsilon/2$, there exists a partition $\mathcal{P}(\varepsilon)$ of $(0,1]$ into a finite number $r = r(\varepsilon) \geq 1$ of nonvoid intervals $(s,t]$, $s,t \in \overline{F(\mathbb{R})}$, with the following property: If $t - s \geq \varepsilon$ then $s = F(x_-)$ and $t = F(x)$ for some $x \in \mathbb{R}$.

Choose $0 < \eta < \varepsilon/4$ such that $\eta < (t-s) \quad \forall\,(s,t] \in \mathcal{P}(\varepsilon)$ and choose $N \in \mathbb{N}$ such that $\|F_N - F\|_\infty < \eta/r$. Then, for any $(s,t] \in \mathcal{P}(\varepsilon)$ there exist some points $\bar{s}, \bar{t} \in \overline{F_N(\mathbb{R})}$ with $|s - \bar{s}| < \eta/r$ and $|t - \bar{t}| < \eta/r$.

By definition of $L_{\mathcal{J}}$ and $L_{\mathcal{J}_N}$ and by the monotonicity of $f$ we have $f(s) \leq \bar{f}(u) \leq f(t) \quad \forall\, u \in (s,t]$ and $f(\bar{s}) \leq \bar{f}_N(u) \leq f(\bar{t}) \quad \forall\, u \in (\bar{s}, \bar{t}]$.

If $t - s < \varepsilon$ we get $|g(u)| \leq 2\omega(f,\varepsilon) \quad \forall\, u \in (s \vee \bar{s}, t \wedge \bar{t})$ implying

$$\int_s^t g^2 \, d\lambda = \int_s^{s \vee \bar{s}} g^2 \, d\lambda + \int_{s \vee \bar{s}}^{t \wedge \bar{t}} g^2 \, d\lambda + \int_{t \wedge \bar{t}}^t g^2 \, d\lambda$$

$$\leq \frac{2\eta}{r} + (t - s)\big(2\omega(f,\varepsilon)\big)^2. \tag{7.5.15}$$

On the other hand, if $t - s \geq \varepsilon$, $t = F(x)$, $s = F(x_-)$, we may choose $\bar{t} = F_N(x)$ and $\bar{s} = F_N(x_-)$. Then $\bar{f}(u) = (\int_s^t f \, d\lambda)/(t - s) \quad \forall\, u \in (s,t]$ and $\bar{f}_N(u) = (\int_{\bar{s}}^{\bar{t}} f \, d\lambda)/(\bar{t} - \bar{s}) \quad \forall\, u \in (\bar{s}, \bar{t}]$. Since

$$\left| \frac{1}{t-s} \int_s^t f \, d\lambda - \frac{1}{\bar{t}-\bar{s}} \int_{\bar{s}}^{\bar{t}} f \, d\lambda \right|$$

$$\leq \frac{|(\bar{t}-\bar{s}) - (t-s)|}{(t-s)(\bar{t}-\bar{s})} + \frac{1}{t-s}\left| \int_s^t f \, d\lambda - \int_{\bar{s}}^{\bar{t}} f \, d\lambda \right|$$

$$\leq \frac{4\eta/r}{\varepsilon^2} + \frac{2\eta/r}{\varepsilon} \leq \frac{6\eta}{\varepsilon^2}$$

we get in the same way as in (7.5.15) the inequality

$$\int_s^t g^2 \, d\lambda \leq 2\eta/r + (t - s)\big(6\eta/\varepsilon^2\big)^2. \tag{7.5.16}$$

Combining (7.5.15) and (7.5.16) yields

$$\int_0^1 g^2 \, d\lambda = \sum_{(s,t]\in\mathcal{P}(\varepsilon)} \int_s^t g^2 \, d\lambda \leq 2\eta + \big(2\omega(f,\varepsilon)\big)^2 + 2\eta + \big(6\,\eta/\varepsilon^2\big)^2.$$

Consequently,

$$\limsup_{N\to\infty} \int_0^1 (\bar{f}_N - \bar{f})^2 \, d\lambda \leq \Big(2\omega(f,\varepsilon)\Big)^2 \qquad \forall\, \varepsilon \in (0, 1/2),$$

which implies (7.5.14) since $\omega(f,\varepsilon) \to 0$ as $\varepsilon \to 0$. $\square$

For treating bivariate distribution functions in Chapter 6 we need an extension of Lemma 7.5.5 to functions on $[0,1]^2$.

### 7.5.6 Lemma

*Let $\mathcal{J}_2, \mathcal{J}_2$ be two families of pairwise disjoint, half-open subintervals $(s,t]$ of $[0,1]$. We define a linear mapping $L_{\mathcal{J}_1,\mathcal{J}_2} : L_2\big((0,1)^2\big) \to L_2\big((0,1)^2\big)$ in analogy to (7.5.13) by $L_{\mathcal{J}_1,\mathcal{J}_2} = L_{\mathcal{J}_1,\{\emptyset\}} \circ L_{\{\emptyset\},\mathcal{J}_2}$ with ( $\forall\, 0 < u,v < 1$ )*

$$(L_{\mathcal{J}_1,\{\emptyset\}}f)(u,v) = \begin{cases} f(u,v), & \text{if } u \in (0,1] \setminus \cup\{(s,t] \in \mathcal{J}_1\}, \\ \frac{1}{t-s}\int_s^t f(u,v)\,\lambda(du), & \text{if } u \in (s,t] \in \mathcal{J}_1, \end{cases} \tag{7.5.17}$$

*and similarly $L_{\{\emptyset\},\mathcal{J}_2}$.*

*Then, for any distribution functions $G$, $H$ and $G_n$, $H_n$, $n \geq 1$, such that $\|G_n - G\|_\infty \to 0$ and $\|H_n - H\|_\infty \to 0$ as $n \to \infty$, we have*

$$L_{\mathcal{I}(G_n),\mathcal{I}(H_n)}f \xrightarrow{n\to\infty} L_{\mathcal{I}(G),\mathcal{I}(H)}f \qquad \text{in } L_2\big((0,1)^2\big) \tag{7.5.18}$$

*for all $f \in L_2\big((0,1)^2\big)$, where $\mathcal{I}(\cdot)$ is defined as in Lemma 7.5.5.*

*Proof:*
Notice, $L_{\mathcal{J}_1,\mathcal{J}_2}f$ may be conceived as a version of the conditional expectation $E[f \mid \mathcal{F}]$ of $f$ on the probability space $\big((0,1)^2, \mathbb{B}^2 \cap (0,1)^2, \lambda^2\big)$ with respect to the $\sigma$-algebra $\mathcal{F}$ which is generated by the following system of Borel sets in $(0,1)^2$,

$$\Big\{ B_1 \times B_2 : B_\kappa \in \mathcal{J}_\kappa \text{ or } B_\kappa \in \big((0,1)\setminus\cup\{I : I \in \mathcal{J}_\kappa\}\big)\cap\mathbb{B}, \ \kappa = 1,2 \Big\}.$$

Since every $f \in L_2\big((0,1)^2\big)$ may be approximated in $L_2\big((0,1)^2\big)$ by polynomials of the form

$$\sum_{i=1}^k d_i \, u^{a_i} v^{b_i}, \qquad d_i \in \mathbb{R}, \quad a_i, b_i \in \{0,1,2,\ldots\},$$

we may assume $f = f_1 \times f_2$ , i.e. $f(u,v) = f_1(u)\,f_2(v)$ , with $f_1(u) = u^a$ and $f_2(v) = v^b$ for some $a, b \geq 0$ . In this case, however, we have

$$L_{\mathcal{J}_1,\mathcal{J}_2} f = (L_{\mathcal{J}_1} f_1) \times (L_{\mathcal{J}_2} f_2) \, .$$

Applying formula (7.5.14) we get

$$L_{\mathcal{I}(G_n)} f_1 \; \overset{n\to\infty}{\longrightarrow} \; L_{\mathcal{I}(G)} f_1 \qquad \text{in } L_2(0,1),$$

$$L_{\mathcal{I}(H_n)} f_2 \; \overset{n\to\infty}{\longrightarrow} \; L_{\mathcal{I}(H)} f_2 \qquad \text{in } L_2(0,1),$$

and hence

$$\| L_{\mathcal{I}(G_n),\mathcal{I}(H_n)} f - L_{\mathcal{I}(G),\mathcal{I}(H)} f \|$$

$$\leq \| L_{\mathcal{I}(G_n)} f_1 - L_{\mathcal{I}(G)} f_1 \| \, \| f_2 \| + \| f_1 \| \, \| L_{\mathcal{I}(H_n)} f_2 - L_{\mathcal{I}(H)} f_2 \| \; \overset{n\to\infty}{\longrightarrow} \; 0 \, ,$$

which implies (7.5.18). $\square$

## Lemma 7.5.7

*Let $M_+^r$ be the set of positive definite $r \times r$ –matrices with the topology of $\mathbb{R}^{r\times r}$. Then the function $f : \mathbb{R}^r \times M_+^r \to \mathbb{R}$ , defined by*

$$f(x,\Gamma) = \sup\{\, 2\vartheta^T x - \vartheta^T \Gamma \vartheta : \vartheta \geq 0 \,\}, \quad x \in \mathbb{R}^r, \quad \Gamma \in M_+^r , \qquad (7.5.19)$$

*is jointly continuous in x and $\Gamma$ .*

*Proof:*
According to formula (3.2.77) we have $f(x,\Gamma) = \big( g(x,\Gamma) \vee 0 \big)^2$ with $g(x,\Gamma) = \sup\{\, \vartheta^T x : \vartheta \geq 0, \; \vartheta^T \Gamma \vartheta = 1 \,\}$ . Therefore it suffices to prove that the function $g(x,\Gamma)$ is jointly continuous in $x$ and $\Gamma$ .

For the proof consider a sequence $(x_n,\Gamma_n)$, $n \geq 1$, in $\mathbb{R}^r \times M_+^r$ with $(x_n,\Gamma_n) \to (x_0,\Gamma_0) \in \mathbb{R}^r \times M_+^r$. Since $\Gamma_n \to \Gamma_0$, there exists some $k > 0$ such that

$$\bigcup_{n=0}^{\infty} \{\, \vartheta \in \mathbb{R}^r : \vartheta^T \Gamma_n \vartheta = 1 \,\} \subset \{\, \vartheta \in \mathbb{R}^r : |\vartheta| \leq k \,\} \, .$$

This implies the inequality

$$g(x_n,\Gamma_n) = \sup\{\, \vartheta^T(x_n - x_0) + \vartheta^T x_0 : \vartheta \geq 0, \; \vartheta^T \Gamma_n \vartheta = 1 \,\}$$

$$\leq k\,|x_n - x_0| + g(x_0,\Gamma_n) \, .$$

By interchanging $x_n$ and $x_0$ and by combining the two inequalities we get

$$|g(x_n,\Gamma_n) - g(x_0,\Gamma_n)| \leq k\,|x_n - x_0| \longrightarrow 0 \, .$$

Moreover,

$$g(x_0, \Gamma_n) = \sup\Big\{ \Big(1 - \frac{1}{\sqrt{\vartheta^T \Gamma_0 \vartheta}}\Big)\vartheta^T x_0 + \frac{\vartheta^T x_0}{\sqrt{\vartheta^T \Gamma_0 \vartheta}} : \vartheta \geq 0, \ \vartheta^T \Gamma_n \vartheta = 1 \Big\}$$

$$\leq k\,|x_0|\,\sup\Big\{ \Big|1 - \sqrt{\frac{\vartheta^T \Gamma_n \vartheta}{\vartheta^T \Gamma_0 \vartheta}}\Big| : |\vartheta| \leq k \Big\} + g(x_0, \Gamma_0).$$

Since

$$\Big|1 - \sqrt{\frac{\vartheta^T \Gamma_n \vartheta}{\vartheta^T \Gamma_0 \vartheta}}\Big| = \Big| \sqrt{\eta^T \Gamma_0 \eta} - \sqrt{\eta^T \Gamma_n \eta} \Big| \qquad \text{with} \quad \eta = \vartheta/\sqrt{\vartheta^T \Gamma_0 \vartheta},$$

where $\eta$ belongs to the compact set $\{\eta : \eta^T \Gamma_0 \eta = 1\}$, we get from $\Gamma_n \to \Gamma_0$ the inequality $g(x_0, \Gamma_n) \leq g(x_0, \Gamma_0) + o(1)$. The same argument with $\Gamma_n$ and $\Gamma_0$ interchanged yields $g(x_0, \Gamma_n) \to g(x_0, \Gamma_0)$. Combining the results we've proved $g(x_n, \Gamma_n) \to g(x_0, \Gamma_0)$. $\square$

# 7.6   Proof of inequality (4.2.22)

Let $X_{11}, ..., X_{1n_1}, X_{21}, ..., X_{2n_2}, ..., X_{k1}, ..., X_{kn_k}$ be independent real random variables, where the $i$-th sample $X_{i1}, ..., X_{in_i}$ is i.i.d. with continuous distribution function $F_i$, $i = 1, ..., k$. Let $N = n_1 + n_2 + \cdots + n_k$ denote the total sample size and define the mixture

$$H_N = \frac{n_1}{N} F_1 + \frac{n_2}{N} F_2 + \cdots + \frac{n_k}{N} F_k. \qquad (7.6.1)$$

According to Section 1.3 the random variables $H_N(X_{i1}), \cdots, H_N(X_{in_i})$ are i.i.d. with absolutely continuous distribution function $F_i \circ H_N^{-1}$ on $[0, 1]$ and Lebesgue density $f_{Ni}$ according to (4.2.7) and (4.2.8). Especially we get

$$0 \leq f_{Ni} \leq N/n_i \quad [\lambda - a.e.]. \qquad (7.6.2)$$

In the sequel let $\tilde{F}_{Ni}$ denote the empirical distribution function of the transformed random variables $H_N(X_{ij})$, $j = 1, ..., n_i$, of the $i$-th sample,

$$\tilde{F}_{Ni}(t) = \frac{1}{n_i} \sum_{j=1}^{n_i} 1(H_N(X_{ij}) \leq t), \qquad 0 \leq t \leq 1. \qquad (7.6.3)$$

Then the mixture $\tilde{H}_N$ of $\tilde{F}_{N1}, ..., \tilde{F}_{Nk}$ corresponding to (7.6.1) is the empirical distribution function of the pooled transformed sample,

$$\tilde{H}_N(t) = \sum_{i=1}^{k} \frac{n_i}{N} \tilde{F}_{Ni}(t) = \frac{1}{N} \sum_{i=1}^{k} \sum_{j=1}^{n_i} 1(H_N(X_{ij}) \leq t), \ \ 0 \leq t \leq 1. \qquad (7.6.4)$$

Finally, let $\breve{F}_{Ni}$ denote the empirical distribution function of the normed ranks $R_{i1}/N, ..., R_{in_i}/N$ of the $i$-th sample,

$$\breve{F}_{Ni}(t) = \frac{1}{n_i} \sum_{j=1}^{n_i} 1(\frac{R_{ij}}{N} \leq t), \qquad 0 \leq t \leq 1. \qquad (7.6.5)$$

Obviously the definition (4.2.19) of the rank $R_{ij}$ and formula (7.6.4) imply the equality

$$\tilde{H}_N(H_N(X_{ij})) = R_{ij}/N \quad [a.s.] \qquad (7.6.6)$$

Using (7.6.6) and the elementary inequality

$$|1(x \leq t) - 1(y \leq t)| \leq 1(t - |y - x| < y \leq t + |y - x|) \qquad (7.6.7)$$

we get the following chain of inequalities ( $\forall\, 1 \le i \le k \quad \forall\, 0 \le t \le 1$ ),

$$|\breve{F}_{Ni}(t) - \tilde{F}_{Ni}(t)| \le \frac{1}{n_i} \sum_{j=1}^{n_i} |1(\frac{R_{ij}}{N} \le t) - 1(H_N(X_{ij}) \le t)|$$

$$\le \frac{1}{n_i} \sum_{j=1}^{n_i} 1(t - |H_N(X_{ij}) - \frac{R_{ij}}{N}| < H_N(X_{ij}) \le t + |H_N(X_{ij}) - \frac{R_{ij}}{N}|)$$

$$\le \frac{1}{n_i} \sum_{j=1}^{n_i} 1(t - \|I - \tilde{H}_N\|_\infty < H_N(X_{ij}) \le t + \|I - \tilde{H}_N\|_\infty)$$

$$= \tilde{F}_{Ni}(t + \|I - \tilde{H}_N\|_\infty) - \tilde{F}_{Ni}(t - \|I - \tilde{H}_N\|_\infty), \tag{7.6.8}$$

where $I : [0,1] \to [0,1]$ denotes the identity $I(t) \equiv t$ .

Therefore a comparison of $\tilde{F}_{Ni}(s)$ with $(F_i \circ H_N^{-1})(s) = \int_0^s f_{Ni}\, d\lambda$ and inequality (7.6.2) yield

$$|\breve{F}_{Ni}(t) - \tilde{F}_{Ni}(t)| \le 2\|\tilde{F}_{Ni} - F_i \circ H_N^{-1}\|_\infty + \int_{t-\|I-\tilde{H}_N\|_\infty}^{t+\|I-\tilde{H}_N\|_\infty} f_{Ni}\, d\lambda$$

$$\le 2\|\tilde{F}_{Ni} - F_i \circ H_N^{-1}\|_\infty + 2\frac{N}{n_i}\|I - \tilde{H}_N\|_\infty. \tag{7.6.9}$$

Since (7.6.1) implies

$$I = \sum_{\kappa=1}^{k} \frac{n_\kappa}{N} F_\kappa \circ H_N^{-1} \tag{7.6.10}$$

we get from (7.6.4) the additional inequality

$$\|I - \tilde{H}_N\|_\infty \le \sum_{\kappa=1}^{k} \frac{n_\kappa}{N} \|\tilde{F}_{N\kappa} - F_\kappa \circ H_N^{-1}\|_\infty. \tag{7.6.11}$$

Using the triangle inequality, (7.6.9), and (7.6.11) we arrive at the final inequality

$$\|\breve{F}_{Ni} - F_i \circ H_N^{-1}\|_\infty \le 3\|\tilde{F}_{Ni} - F_i \circ H_N^{-1}\|_\infty$$

$$+ 2\sum_{\kappa=1}^{k} \frac{n_\kappa}{n_i} \|\tilde{F}_{N\kappa} - F_\kappa \circ H_N^{-1}\|_\infty, \tag{7.6.12}$$

which proves (4.2.22). $\square$

Notice, if $n_i/N \to \eta_i \in (0,1)$ as $N \to \infty$ for all $i = 1, ..., k$, then (7.6.12) and the Kolmogorov–Smirnov theorem imply

$$\sup_{B \in \mathcal{B}^0_{kN}} \sup_{H \in \mathcal{F}^c_1} P_{(B,H)}\{\max_{1 \leq i \leq k} \|\sqrt{N}(\breve{F}_{Ni} - I) - B_i\|_\infty > x_N\} \longrightarrow 0,$$

$$(7.6.13)$$

$$\text{as} \quad N \to \infty \quad \text{and} \quad x_N \to \infty.$$

If $\hat{F}_{Ni}$ denotes the piecewise linearized version (4.2.23) of $\breve{F}_{Ni}$ then we have in addition

$$\|\hat{F}_{Ni} - \breve{F}_{Ni}\|_\infty \leq \frac{1}{n_i}, \qquad i = 1, ..., k. \tag{7.6.14}$$

## 7.7 · Proof of Theorem 4.2.2

For each $\kappa = 1, ..., k$ let's define the $k$ –sample coefficients $c_{\kappa i}$ according to

$$c_{\kappa i} = \left(\delta_{\kappa i} - \frac{n_\kappa}{N}\right)\frac{1}{\sqrt{n_\kappa}}, \qquad i = 1, ..., k, \tag{7.7.1}$$

where $\delta_{\kappa i}$ denotes Kronecker's delta. Then we have the properties ( $\forall\, 1 \leq \kappa \leq 1$)

$$\sum_{i=1}^{k}\sum_{j=1}^{n_i} c_{\kappa i} = 0, \tag{7.7.2}$$

$$\sum_{i=1}^{k}\sum_{j=1}^{n_i} c_{\kappa i}^2 = 1 - \frac{n_\kappa}{N} \overset{N\to\infty}{\longrightarrow} 1 - \eta_\kappa > 0, \tag{7.7.3}$$

$$0 < \max_{1\leq i\leq k} c_{\kappa i}^2 \leq \frac{1}{n_\kappa} \overset{N\to\infty}{\longrightarrow} 0. \tag{7.7.4}$$

Using the definition (4.2.24) of $d_N(r,t)$ we may rewrite each rank process $\hat{B}_{N\kappa}$ in the following form,

$$\hat{B}_{N\kappa}(t) = \sqrt{N}(\hat{F}_{N\kappa}(t) - t) = \sqrt{N}\left(\hat{F}_{N\kappa}(t) - \frac{1}{N}\sum_{i=1}^{N} d_N(i,t)\right)$$

$$= \sqrt{N}\left(\frac{1}{n_\kappa}\sum_{j=1}^{n_\kappa} d_N(R_{\kappa j}, t) - \frac{1}{N}\sum_{i=1}^{k}\sum_{j=1}^{n_i} d_N(R_{ij}, t)\right)$$

$$= \frac{\sqrt{N}}{n_\kappa}\sum_{i=1}^{k}\left(\delta_{\kappa i} - \frac{n_\kappa}{N}\right)\sum_{j=1}^{n_i} d_N(R_{ij}, t)$$

$$= \sqrt{\frac{N}{n_\kappa}}\sum_{i=1}^{k}\sum_{j=1}^{n_i} c_{\kappa i}\, d_N(R_{ij}, t). \tag{7.7.5}$$

Now let's utilize the results of Section V.3.5 of Hájek and Šidák (1967). Obviously their definition (1) of $a_N(i,t)$ is related to our definition (4.2.24) of $d_N(r,t)$ via

$$d_N(r,t) = 1 - a_N(r,t) \qquad \forall\, 0 \leq t \leq 1 \quad \forall\, r = 1, ..., N. \tag{7.7.6}$$

Therefore (7.7.5) and (7.7.2) imply the equality

$$\hat{B}_{N\kappa}(t) = -\sqrt{\frac{N}{n_\kappa}}\sum_{i=1}^{k}\sum_{j=1}^{n_i} c_{\kappa i}\, a_N(R_{ij}, t), \qquad 0 \leq t \leq 1. \tag{7.7.7}$$

Hence, for each $\kappa = 1, ..., k$ and $N \to \infty$, the properties (7.7.2) to (7.7.4) and the Theorem V.3.5 of Hájek and Šidák (1967) yield the following limiting law under the null hypothesis of randomness $\mathcal{H}_0^r$,

$$\sqrt{\frac{n_\kappa}{N}}\; \hat{B}_{N\kappa} \;\xrightarrow{\mathcal{L}}\; W_0 \quad \text{in} \;\; C[0,1]. \tag{7.7.8}$$

This implies that the sequence $((\hat{B}_{N1}, ..., \hat{B}_{Nk}),\; N \geq 1)$ is tight in $(C[0,1])^k$ under $\mathcal{H}_0^r$ and, because of contiguity, under any sequence of $(B_N, H_N)$ – alternatives of the type (4.2.21), too.  Therefore the proof of Theorem 4.2.2 is concluded, if we verify the convergence in law of all finite dimensional distributions of the proces $(\hat{B}_{N1}, ..., \hat{B}_{Nk})$ to the corresponding finite dimensional distributions of the Gaussian process $(W_1 + B_1, ..., W_k + B_k)$ defined in Theorem 4.2.2.

Defining the processes

$$\tilde{B}_{N\kappa}(t) = \frac{1}{\sqrt{\eta_\kappa}} \sum_{i=1}^{k} \sum_{j=1}^{n_i} c_{\kappa i}\; 1(R_{ij} \leq Nt), \qquad 0 \leq t \leq 1, \tag{7.7.9}$$

we get from (7.7.5) and (4.2.24)

$$|\hat{B}_{N\kappa}(t) - \tilde{B}_{N\kappa}(t)|$$

$$\leq |\hat{B}_{N\kappa}(t) - \sqrt{\frac{\eta_{N\kappa}}{\eta_\kappa}}\,\hat{B}_{N\kappa}(t)| + \frac{1}{\sqrt{\eta_\kappa}} \sum_{i=1}^{k} \sum_{j=1}^{n_i} |\,c_{\kappa i}|\; 1(R_{ij} = [Nt] + 1)$$

$$\leq |1 - \sqrt{\frac{\eta_{N\kappa}}{\eta_\kappa}}\,|\cdot \|\hat{B}_{N\kappa}\|_\infty + \frac{1}{\sqrt{\eta_\kappa}} \max_{1 \leq i \leq k} |c_{\kappa i}|. \tag{7.7.10}$$

Therefore $\eta_{N\kappa} \to \eta_\kappa$ , (7.7.8), and (7.7.4) imply that we may consider the finite dimensional distributions of the process $(\tilde{B}_{N1}, ..., \tilde{B}_{Nk})$ instead of the corresponding finite dimensional distributions of the original process $(\hat{B}_{N1}, ..., \hat{B}_{Nk})$.

In the next step we'll compare $\tilde{B}_{N\kappa}(t)$ with

$$W_{N\kappa}(t) := \frac{1}{\sqrt{\eta_\kappa}} \sum_{i=1}^{k} \sum_{j=1}^{n_i} c_{\kappa i}\; 1(H_N(X_{ij}) \leq t). \tag{7.7.11}$$

Since the $R_{ij}$ 's are $[a.s.]$ the ranks of the $H_N(X_{ij})$ 's and since under the null hypothesis of randomness $\mathcal{H}_0^r$ the components of

$$(U_{11}, ..., U_{kn_k}) := (H_N(X_{11}), ..., H_N(X_{kn_k})) \tag{7.7.12}$$

are i.i.d. random variables with uniform distribution on $[0,1]$, we get under $\mathcal{H}_0^r$ from (7.7.2) and (7.7.3)

$$\eta_\kappa E_0\big(\tilde{B}_{N\kappa}(t) - W_{N\kappa}(t)\big)^2 = E_0\Big(\sum_{i=1}^{k}\sum_{j=1}^{n_i} c_{\kappa i}\big(1(\frac{R_{ij}}{N} \le t) - 1(U_{ij} \le t)\big)\Big)^2$$

$$= (1 - \frac{n_\kappa}{N})E_0\Big(1(\frac{R_{11}}{N} \le t) - 1(U_{11} \le t)\Big)^2$$

$$- (1 - \frac{n_\kappa}{N})E_0\Big(1(\frac{R_{11}}{N} \le t) - 1(U_{11} \le t)\Big)\Big(1(\frac{R_{12}}{N} \le t) - 1(U_{12} \le t)\Big)$$

$$\le 2\, E_0\Big(1(\frac{R_{11}}{N} \le t) - 1(U_{11} \le t)\Big)^2$$

$$\le 4\, E_0\Big(\varphi(\frac{R_{11}}{N}) - \varphi_N(\frac{R_{11}}{N})\Big)^2 + 4\, E_0\Big(\varphi_N(\frac{R_{11}}{N}) - \varphi(U_{11})\Big)^2, \quad (7.7.13)$$

where $\varphi : [0,1] \to \mathbb{R}$ and $\varphi_N : [0,1] \to \mathbb{R}$ are defined according to

$$\varphi(u) = 1(u \le t), \qquad 0 \le u \le 1, \qquad (7.7.14)$$

and

$$\varphi_N(u) = \sum_{i=1}^{N} E_0[\varphi(U_{11}) \mid R_{11} = i]\, 1(\frac{i-1}{N} < u \le \frac{i}{N}). \qquad (7.7.15)$$

Then on one hand Theorem V.1.4.a und Theorem V.1.4.b of Hájek and Šidák (1967) imply

$$E_0\big(\varphi_N(\frac{R_{11}}{N}) - \varphi(U_{11})\big)^2 \xrightarrow{N\to\infty} 0 \qquad (7.7.16)$$

and

$$\int_0^1 \big(\varphi_N(u) - \varphi(u)\big)^2\, du \xrightarrow{N\to\infty} 0, \qquad (7.7.17)$$

on the other hand we get

$$E_0\big(\varphi(\frac{R_{11}}{N}) - \varphi_N(\frac{R_{11}}{N})\big)^2 = \frac{1}{N}\sum_{i=1}^{N}\big(\varphi(\frac{i}{N}) - \varphi_N(\frac{i}{N})\big)^2$$

$$= \sum_{i=1}^{N}\int_{(i-1)/N}^{i/N} \big(\varphi(\frac{i}{N}) - \varphi_N(\frac{i}{N})\big)^2\, du$$

$$\le \int_0^1 \big(\varphi(u) - \varphi_N(u)\big)^2\, du + \frac{1}{N}, \qquad (7.7.18)$$

since $\varphi$ and $\varphi_N$ are constant on each interval $(\frac{i-1}{N}, \frac{i}{N}]$ with $i \neq [Nt]+1$ and since $\left(\varphi(\frac{i}{N}) - \varphi_N(\frac{i}{N})\right)^2 \leq 1$. Combining (7.7.13) to (7.7.18) and using contiguity we've proved for each $\kappa = 1, ..., k$ and any $0 \leq t \leq 1$ that $\tilde{B}_{N\kappa}(t) - W_{N\kappa}(t)$ converges to zero in $\mathcal{H}_0^r$ –probability and in $(B_N, H_N)$ –probability, too. Therefore it suffices to consider the finite dimensional distributions of the process $W_N = (W_{N1}, ..., W_{Nk})$.

Utilizing the Cramér–Wold device this means that we have to prove the following limiting law under $(B_N, H_N)$ –alternatives of the type (4.2.21),

$$Z_N := \sum_{\kappa=1}^{k} \sum_{\varrho=1}^{r} a_{\kappa\varrho} \, W_{N\kappa}(t_{\kappa\varrho})$$

$$\xrightarrow{\mathcal{L}} \sum_{\kappa=1}^{k} \sum_{\varrho=1}^{r} a_{\kappa\varrho} \left(W_\kappa(t_{\kappa\varrho}) + B_\kappa(t_{\kappa\varrho})\right) =: Z \tag{7.7.19}$$

with arbitrary $1 \leq r < \infty$, $0 \leq t_{\kappa\varrho} \leq 1$, and $a_{\kappa\varrho} \in \mathbb{R}$. Since $W = (W_1, ..., W_k)$ is assumed to be a k-dimensional centered Gaussian process with covariance structure (4.2.38) the real random variable $Z$ has a normal distribution with expectation $\mu$ according to

$$\mu = \sum_{\kappa=1}^{k} \sum_{\varrho=1}^{r} a_{\kappa\varrho} \, B_\kappa(t_{\kappa\varrho}) \tag{7.7.20}$$

and variance $\sigma^2$ according to

$$\sigma^2 = \sum_{\kappa=1}^{k} \sum_{\varrho=1}^{r} \sum_{\tau=1}^{k} \sum_{\nu=1}^{r} a_{\kappa\varrho} \, a_{\tau\nu} \left(t_{\kappa\varrho} \wedge t_{\tau\nu} - t_{\kappa\varrho} t_{\tau\nu}\right) \left(\frac{\delta_{\kappa\tau}}{\sqrt{\eta_\kappa \eta_\tau}} - 1\right). \tag{7.7.21}$$

Now let's apply Lindeberg's central limit theorem in order to prove (7.7.19) by proving

$$\mathcal{L}[\, Z_N \mid (B_N, H_N)\,] \xrightarrow{\mathcal{L}} \mathcal{N}(\mu, \sigma^2) \tag{7.7.22}$$

with $\mu$ and $\sigma^2$ given in (7.7.20) and (7.7.21).

For each $N$ the random variables

$$Z_{ij} := \sum_{\kappa=1}^{k} \sum_{\varrho=1}^{r} \frac{a_{\kappa\varrho}}{\sqrt{\eta_\kappa}} \, c_{\kappa i} \, 1(U_{ij} \leq t_{\kappa\varrho}), \quad 1 \leq j \leq n_i, \, 1 \leq i \leq k, \tag{7.7.23}$$

are stochastically independent, and according to (7.7.11) and (7.7.12) we have the following representation of $Z_N$,

$$Z_N = \sum_{i=1}^{k} \sum_{j=1}^{n_i} Z_{ij}. \tag{7.7.24}$$

For each $i = 1, ..., k$ the random variables $U_{i1}, ..., U_{in_i}$ are i.i.d. with absolutely continuous distribution function $I + B_{Ni}/\sqrt{N}$, and the underlying sequence $[\,(B_N, H_N) \in \mathcal{B}^0_{kN} \times \mathcal{F}^c_1,\ N \geq k\,]$ has the property (4.2.21), which implies

$$\max_{1 \leq i \leq k} \|B_{Ni} - B_i\|_\infty \ \leq\ \max_{1 \leq i \leq k} \|b_{Ni} - b_i\| \ =: c_N \ \overset{N \to \infty}{\longrightarrow}\ 0. \qquad (7.7.25)$$

In addition, the assumtion $B_N \in \mathcal{B}^0_{kN}$ implies the property

$$\frac{n_1}{N} B_{N1} + \frac{n_2}{N} B_{N2} + \cdots + \frac{n_k}{N} B_{Nk} = 0. \qquad (7.7.26)$$

Now let's compute $\sigma^2_N = \mathrm{Var}\,[\,Z_N \mid (B_N, H_N)\,]$ and $\mu_N = E_N Z_N = E[\,Z_N \mid (B_N, H_N)\,]$. In a first step we get from (7.7.24)

$$\sigma^2_N = \sum_{i=1}^{k} \sum_{j=1}^{n_i} \mathrm{Var}\,[\,Z_{ij} \mid (B_N, H_N)\,]. \qquad (7.7.27)$$

In a second step we get from (7.7.23)

$$\mathrm{Var}\,[\,Z_{ij} \mid (B_N, H_N)\,]$$
$$= \sum_{\kappa=1}^{k} \sum_{\varrho=1}^{r} \sum_{\tau=1}^{k} \sum_{\nu=1}^{r} \frac{a_{\kappa\varrho}}{\sqrt{\eta_\kappa}} \frac{a_{\tau\nu}}{\sqrt{\eta_\tau}}\, c_{\kappa i}\, c_{\tau i}\, d_{ij}(t_{\kappa\varrho}, t_{\tau\nu}), \qquad (7.7.28)$$

where

$$d_{ij}(s,t) = \mathrm{Cov}\,[\,1(U_{ij} \leq s), 1(U_{ij} \leq t) \mid (B_N, H_N)\,]$$
$$= (I + B_{Ni}/\sqrt{N})(s \wedge t) - (I + B_{Ni}/\sqrt{N})(s)(I + B_{Ni}/\sqrt{N})(t)$$
$$= s \wedge t - st + \frac{1}{\sqrt{N}}\, R_{Ni}(s,t) \qquad (7.7.29)$$

and, according to (7.7.25),

$$|R_{Ni}(s,t)| \leq 4\, \|B_{Ni}\|_\infty \ \leq\ 4\, c_N + 4 \max_{1 \leq i \leq k} \|B_i\|_\infty \leq c_0 < \infty. \qquad (7.7.30)$$

This implies

$$\sigma^2_N = \sum_{\kappa=1}^{k} \sum_{\varrho=1}^{r} \sum_{\tau=1}^{k} \sum_{\nu=1}^{r} \frac{a_{\kappa\varrho}}{\sqrt{\eta_\kappa}} \frac{a_{\tau\nu}}{\sqrt{\eta_\tau}} \left[ (t_{\kappa\varrho} \wedge t_{\tau\nu} - t_{\kappa\varrho} t_{\tau\nu}) \sum_{i=1}^{k} n_i\, c_{\kappa i}\, c_{\tau i} \right.$$
$$\left. + \frac{1}{\sqrt{N}} \sum_{i=1}^{k} n_i\, c_{\kappa i}\, c_{\tau i}\, R_{Ni}(t_{\kappa\varrho}, t_{\tau\nu}) \right]$$

$$= \sum_{\kappa=1}^{k}\sum_{\varrho=1}^{r}\sum_{\tau=1}^{k}\sum_{\nu=1}^{r} a_{\kappa\varrho}\, a_{\kappa\nu}\big(t_{\kappa\varrho}\wedge t_{\tau\nu} - t_{\kappa\varrho}t_{\tau\nu}\big)\Big(\frac{\delta_{\kappa\tau}}{\sqrt{\eta_{\kappa}\eta_{\tau}}} - \sqrt{\frac{\eta_{N\kappa}\eta_{N\tau}}{\eta_{\kappa}\eta_{\tau}}}\,\Big)$$

$$+ \frac{1}{\sqrt{N}}\, R_{N},$$

where the sequence $(R_N, N \geq k)$ is bounded.  Especially we've proved

$$\sigma_N^2 \overset{N\to\infty}{\longrightarrow} \sigma^2. \tag{7.7.31}$$

As to the evaluation of $\mu_N$ we get from (7.7.2), (7.7.1), and (7.7.26)

$$\mu_N = \sum_{\kappa=1}^{k}\sum_{\varrho=1}^{r}\frac{a_{\kappa\varrho}}{\sqrt{\eta_\kappa}} \sum_{i=1}^{k}\sum_{j=1}^{n_i} c_{\kappa i}\Big(t_{\kappa\varrho} + \frac{1}{\sqrt{N}}B_{Ni}(t_{\kappa\varrho})\Big)$$

$$= \frac{1}{\sqrt{N}}\sum_{\kappa=1}^{k}\sum_{\varrho=1}^{r}\frac{a_{\kappa\varrho}}{\sqrt{\eta_\kappa}}\sum_{i=1}^{k}\frac{n_i}{\sqrt{n_\kappa}}\Big(\delta_{\kappa i} - \frac{n_\kappa}{N}\Big)B_{Ni}(t_{\kappa\varrho})$$

$$= \sum_{\kappa=1}^{k}\sum_{\varrho=1}^{r} a_{\kappa\varrho}\sqrt{\frac{\eta_{N\kappa}}{\eta_\kappa}}\, B_{N\kappa}(t_{\kappa\varrho}). \tag{7.7.32}$$

Therefore $\eta_{N\kappa} \longrightarrow \eta_\kappa$ and (7.7.25) prove

$$\mu_N \overset{N\to\infty}{\longrightarrow} \sum_{\kappa=1}^{k}\sum_{\varrho=1}^{r} a_{\kappa\varrho}\, B_\kappa(t_{\kappa\varrho}) = \mu. \tag{7.7.33}$$

If we assume $\sigma^2 = 0$, then (7.7.31) and (7.7.33) obviously imply (7.7.22).  If we assume $\sigma^2 > 0$, then the inequality $(\forall\, i = 1,...,k \quad \forall\, j = 1,...,n_i)$

$$\sqrt{N}\,|Z_{ij}| \leq \sum_{\kappa=1}^{k}\sum_{\varrho=1}^{r}\frac{|a_{\kappa\varrho}|}{\sqrt{\eta_\kappa}}\sqrt{\frac{N}{n_\kappa}} \overset{N\to\infty}{\longrightarrow} \sum_{\kappa=1}^{k}\sum_{\varrho=1}^{r}\frac{|a_{\kappa\varrho}|}{\eta_\kappa} < \infty \tag{7.7.34}$$

and (7.7.31) obviously prove Lindeberg's condition $(\forall\, \varepsilon > 0)$

$$\frac{1}{\sigma_N^2}\sum_{i=1}^{k}\sum_{j=1}^{n_i} E_N\big[\,(Z_{ij} - E_N Z_{ij})^2\, 1(|Z_{ij} - E_N Z_{ij}| \geq \varepsilon\sigma_N)\,\big] \overset{N\to\infty}{\longrightarrow} 0, \tag{7.7.35}$$

where $E_N$ denotes the expectation under $(B_N, H_N)$.  In this case (7.7.31), (7.7.33), and Lindeberg's central limit theorem imply (7.7.22).  Therefore the proof of Theorem 4.2.2 is complete.

## 7.8 Proof of Theorem 5.2.1

Define $T : C[0,1] \to C[0,1]$ according to

$$(Tf)(t) = f(1-t) \qquad \forall \, 0 \le t \le 1 \quad \forall \, f \in C[0,1]. \qquad (7.8.1)$$

Apparently $T$ is linear and continuous on $C[0,1]$ endowed with the sup-norm topology, and $T \circ T$ is the identity on $C[0,1]$. Additionally we have $W^* = TW$. Therefore it suffices to prove

$$\mathcal{L}[\, T\hat{B}_{n2} \mid F_n \,] \xrightarrow{\mathcal{L}} \mathcal{L}[\, W + \varrho T\tilde{B} \,]. \qquad (7.8.2)$$

For the proof of (7.8.2) we get from (5.2.4) and (5.2.3) $\quad \forall \, 0 \le t \le 1$

$$(T\hat{B}_{n2})(t) = -\frac{1}{\sqrt{n}} \sum_{i=1}^{n} \Big( \, 1(n+1-R_i^+ \le [nt])$$
$$+ (nt - [nt])1(n+1-R_i^+ = [nt]+1) \Big) \, \text{sign}(X_i). \qquad (7.8.3)$$

The random variables $R^+$, $\text{sign}(X_1), \dots, \text{sign}(X_n)$ are stochastically independent if the null hypothesis of symmetry $\mathcal{H}_0^s$ holds true. Therefore (5.1.5) and (7.8.3) imply

$$\mathcal{L}[\, T\hat{B}_{n2} \mid \mathcal{H}_0^2 \,] = \mathcal{L}[\, W_n \,], \qquad (7.8.4)$$

where $W_n$ is the piecewise linear partial sum process

$$W_n(t) = \frac{1}{\sqrt{n}} \Big( \sum_{i=1}^{[nt]} Z_i + (nt - [nt]) \, Z_{[nt]+1} \Big), \qquad 0 \le t \le 1, \qquad (7.8.5)$$

with i.i.d. random variables $Z_1, \dots, Z_n$ such that $1/2 = P\{Z_i = -1\} = P\{Z_i = 1\}$. Thus Theorem 10.1 of Billingsley (1967) (Donsker's theorem) implies

$$\mathcal{L}[\, T\hat{B}_{n2} \mid \mathcal{H}_0^s \,] \xrightarrow{\mathcal{L}} \mathcal{L}[\, W \,], \qquad (7.8.6)$$

which especially proves (7.8.2) in case of $\varrho = 0$.

For the proof of (7.8.2) in case of $0 < \varrho \le 1$ we notice that (7.8.6) implies the tightness of the sequence $(\mathcal{L}[T\hat{B}_{n2} \mid \mathcal{H}_0^s], \, n \ge 1)$. Additionally the sequence $(\mathcal{L}[(X_1, \dots, X_n) \mid F_n], \, n \ge 1)$ is contiguous to the sequence $(\mathcal{L}[X_1, \dots, X_n) \mid H_n], \, n \ge 1)$, cf. Section 7.1, and $H_n \in \mathcal{H}_0^s \quad \forall \, n \ge 1$. Therefore Theorem 8.2 of Billingsley (1967) implies the tightness of the sequence $\mathcal{L}[T\hat{B}_{n2} \mid F_n], \, n \ge 1)$, too. Hence the proof of (7.8.2) is concluded, if we prove the finite dimensional distributions of $T\hat{B}_{n2}$ under $F_n$ to converge in distribution to the corresponding finite dimensional distributions of $W + \varrho T\tilde{B}$.

Utilizing $(T\hat{B}_{n2})(0) = 0$ and the Cramér-Wold device it suffices to prove

$$\mathcal{L}[\sum_{\kappa=1}^{k} y_\kappa \, (T\hat{B}_{n2})(t_\kappa) \mid F_n ] \xrightarrow{\mathcal{L}} \mathcal{L}[\sum_{\kappa=1}^{k} y_\kappa \, (W + \varrho T\tilde{B})(t_\kappa) ], \qquad (7.8.7)$$

where $k \geq 1$, $y_1, ..., y_k \in \mathbb{R}$, and $0 < t_1 < \cdots < t_k \leq 1$.

Since the process $W$ is a centered Gaussian process with covariance function $EW(s)W(t) = s \wedge t \quad \forall \, s,t \in [0,1]$, the right hand side of formula (7.8.7) is a normal distribution with mean

$$\mu := \varrho \sum_{\kappa=1}^{k} (T\tilde{B})(t_\kappa) = \varrho \sum_{\kappa=1}^{k} 2B(1 - t_\kappa/2) \qquad (7.8.8)$$

and variance

$$\sigma^2 := \sum_{\kappa=1}^{k} \sum_{\tau=1}^{k} y_\kappa \, y_\tau \, (t_\kappa \wedge t_\tau). \qquad (7.8.9)$$

Considering the left hand side of (7.8.7) we get from (7.8.3) the representation

$$Z_n := \sum_{\kappa=1}^{k} y_\kappa \, (T\hat{B}_{n2})(t_\kappa) = \frac{1}{\sqrt{n}} \sum_{i=1}^{n} a_n(R_i^+) \, \mathrm{sign}(X_i), \qquad (7.8.10)$$

where the scores $a_n(i)$ have the form

$$a_n(i) = \qquad\qquad\qquad\qquad\qquad\qquad\qquad\qquad\qquad\qquad\qquad\quad (7.8.11)$$

$$-\sum_{\kappa=1}^{k} y_\kappa \, \Big( 1(n + 1 - i \leq [nt_\kappa]) + (nt_\kappa - [nt_\kappa]) \, 1(n + 1 - i = [nt_\kappa] + 1) \Big).$$

Obviously (7.8.11) implies

$$\int_0^1 \Big( a_n(1 + [nx]) - h(x) \Big)^2 \, dx \xrightarrow{n \to \infty} 0, \qquad (7.8.12)$$

where

$$h(x) := -\sum_{\kappa=1}^{k} y_\kappa \, 1(x \geq 1 - t_\kappa), \qquad 0 \leq x \leq 1. \qquad (7.8.13)$$

Therefore Theorem 2.1 of Behnen (1972) implies the limiting law

$$\mathcal{L}[ Z_n \mid F_n ] \xrightarrow{\mathcal{L}} \mathcal{N}\Big( \varrho \int_0^1 h(x) \, b(\tfrac{1}{2} + \tfrac{1}{2}x) \, dx, \; \|h\|^2 \Big). \qquad (7.8.14)$$

Because of

$$\|h\|^2 \;=\; \sum_{\kappa=1}^{k}\sum_{\tau=1}^{k} y_\kappa\, y_\tau\, (t_\kappa \wedge t_\tau) \;=\; \sigma^2 \tag{7.8.15}$$

and

$$\varrho \int_0^1 h(x)\, b(\tfrac{1}{2}+\tfrac{1}{2}x)\, dx = 2\varrho \int_{1/2}^1 h(2x-1)\, b(x)\, dx$$

$$= -2\varrho \sum_{\kappa=1}^{k} y_\kappa \int_{1/2}^1 1(\, 2x-1 \geq 1-t_\kappa\,)\, b(x)\, dx$$

$$= -2\varrho \sum_{\kappa=1}^{k} y_\kappa \left( B(1) - B(1 - t_\kappa/2) \right)$$

$$= \;\; 2\varrho \sum_{\kappa=1}^{k} y_\kappa\, B(1 - t_\kappa/2) \;=\; \mu \tag{7.8.16}$$

formula (7.8.14) concludes the proof of (7.8.7). Thus the proof of Theorem 5.2.1
is complete.

# 7.9  Proof of inequality (6.2.33)

For each  $i = 1, ..., n$  and  $0 \le t \le 1$  let's define

$$a_n(i,t) := 1\big(i \le [nt]\big) + \big(nt - [nt]\big)\, 1\big(i = [nt] + 1\big) - t, \qquad (7.9.1)$$

where  $[x]$  denotes the integer part of  $x$ . Using formula (6.2.5) and definition (7.9.1) we get

$$\hat{B}_{n2}(s,t) = n^{-1/2} \sum_{i=1}^{n} a_n(R_{1i}, s)\, a_n(R_{2i}, t) \qquad (7.9.2)$$

and therefore  ( $\forall\, 0 \le s \le u \le 1,\ 0 \le t \le v \le 1$ )

$$\hat{B}_{n2}(u,v) - \hat{B}_{n2}(u,t) - \hat{B}_{n2}(s,v) + \hat{B}_{n2}(s,t)$$

$$= n^{-1/2} \sum_{i=1}^{n} \Big(a_n(R_{1i}, u) - a_n(R_{1i}, s)\Big) \Big(a_n(R_{2i}, v) - a_n(R_{2i}, t)\Big)$$

$$= n^{-1/2} \sum_{i=1}^{n} d_n(R_{1i}, u, s)\, d_n(R_{2i}, v, t), \qquad (7.9.3)$$

where

$$d_n(i, u, s) := a_n(i, u) - a_n(i, s). \qquad (7.9.4)$$

Since  $(R_{11}, ..., R_{1n})$  and  $(R_{21}, ..., R_{2n})$  are stochastically independent under the null hypothesis  $\mathcal{H}_0^i$  the  $\Delta$ -difference (7.9.3) has the same distribution under  $\mathcal{H}_0^i$  as

$$\Delta_n(s,t,u,v) := n^{-1/2} \sum_{i=1}^{n} d_n(i, u, s)\, d_n(R_{2i}, v, t). \qquad (7.9.5)$$

Therefore the proof of inequality (6.2.33) is concluded if we prove the following inequality (  $\forall\, 0 \le s \le u \le 1,\ 0 \le t \le v \le 1,\ n \ge 4$  )

$$E_0 \Delta_n^4(s,t,u,v) \le k_0 \Big((u - s)(v - t)\Big)^{3/2} \qquad (7.9.6)$$

where  $k_0$  is a finite constant which doesn't depend on  $s, t, u, v$ , and  $n$ .
For the proof of (7.9.6) we'll use the abbreviations

$$d_{1i} := d_n(i, u, s), \qquad\qquad d_{2i} := d_n(i, v, t), \qquad (7.9.7)$$

and for $\kappa = 1, 2$,

$$S_{\kappa 1} := \sum_1^n d_{\kappa i}\, d_{\kappa j}\, d_{\kappa q}\, d_{\kappa r} \qquad (\text{ pairwise disjoint indices }),$$

$$S_{\kappa 2} := \sum_1^n d_{\kappa i}^2\, d_{\kappa j}\, d_{\kappa q} \qquad (\text{ pairwise disjoint indices }),$$

$$S_{\kappa 3} := \sum_1^n d_{\kappa i}^2\, d_{\kappa j}^2\, 1(i \neq j),$$

$$S_{\kappa 4} = \sum_1^n d_{\kappa i}^3\, d_{\kappa j}\, 1(i \neq j),$$

$$S_{\kappa 5} := \sum_{i=1}^n d_{\kappa i}^4\, . \tag{7.9.8}$$

Since $(R_{21}, ..., R_{2n})$ is uniformly distributed on the permutations of $(1, ..., n)$ we obviously get

$$n(n-1)(n-2)(n-3)\, E_0\Big(\, d_{2R_{21}}\, d_{2R_{22}}\, d_{2R_{23}}\, d_{2R_{24}}\, \Big) \;=\; S_{21},$$

$$n(n-1)(n-2)\, E_0\Big(\, d_{2R_{21}}^2\, d_{2R_{22}}\, d_{2R_{23}}\, \Big) \;=\; S_{22},$$

$$n(n-1)\, E_0\Big(\, d_{2R_{21}}^2\, d_{2R_{22}}^2\, \Big) \;=\; S_{23},$$

$$n(n-1)\, E_0\Big(\, d_{2R_{21}}^3\, d_{2R_{22}}\, \Big) \;=\; S_{24},$$

$$n\, E_0\Big(\, d_{2R_{21}}^4\, \Big) \;=\; S_{25}, \tag{7.9.9}$$

and therefore

$$n^3 E_0 \Delta_n^4(s, t, u, v)$$

$$= \frac{1}{(n-1)(n-2)(n-3)}\, S_{11}\, S_{21} + \frac{6}{(n-1)(n-2)}\, S_{12}\, S_{22}$$

$$+ \frac{3}{n-1}\, S_{13}\, S_{23} + \frac{4}{n-1}\, S_{14}\, S_{24} + S_{15}\, S_{25}. \tag{7.9.10}$$

In the next step we'll use (7.9.1) and (7.9.4) in order to get the following properties $(\,\forall\, 0 \leq s \leq u \leq 1)$,

$$\sum_{i=1}^n a_n(i, s) = 0, \qquad\qquad \sum_{i=1}^n d_n(i, u, s) = 0, \tag{7.9.11}$$

$$d_n(i, u, s)$$

$$= 1([ns] < i \le [nu]) - (ns - [ns])\, 1(i = [ns] + 1)$$

$$+ (nu - [nu])\, 1(i = [nu] + 1) - (u - s),$$

$$= \begin{cases} (nu - ns)\, 1(i = [ns] + 1) - (u - s), & \text{if } [ns] = [nu], \\[2mm] (1 - ns + [ns])\, 1(i = [ns] + 1) + 1([ns] + 1 < i \le [nu]) \\[1mm] \quad + (nu - [nu])\, 1(i = [nu] + 1) - (u - s), & \text{if } [ns] < [nu]. \end{cases} \qquad (7.9.12)$$

Thus, in the case $[ns] = [nu]$, i.e. $0 \le nu - ns \le 1$, we get

$$\sum_{i=1}^{n} d_n^2(i, u, s) \le (nu - ns)^2 + n\,(u - s)^2$$

$$\le (nu - ns) + n\,(u - s) \; = \; 2n\,(u - s)$$

and similarly

$$\sum_{i=1}^{n} d_n^4(i, u, s) \le (nu - ns)^4 + n\,(u - s)^4$$

$$\le 2\,(nu - ns)^4 \; \le \; 2\,n^{3/2}\,(u - s)^{3/2},$$

in the case $[ns] + 1 = [nu]$, i.e. $0 \le u - s < 2/n$, we have

$$\sum_{i=1}^{n} d_n^2(i, u, s) \le \Big( (1 - ns + [ns]) - (u - s) \Big)^2 + \Big( (nu - [nu]) - (u - s) \Big)^2$$

$$+ n\,(u - s)^2$$

$$\le (nu - ns)^2 + (nu - ns)^2 + (nu - ns)^2$$

$$\le 6n\,(u - s)$$

and similarly

$$\sum_{i=1}^{n} d_n^4(i, u, s) \; \le \; 3\,(nu - ns)^4 \; \le \; 3\,(2)^{5/2}\,n^{3/2}\,(u - s)^{3/2},$$

finally in the case $[ns] + 2 \le [nu]$, i.e. $u - s > 1/n$, we get

$$\sum_{i=1}^{n} d_n^2(i, u, s) \le ([nu] + 1 - [us]) + n\,(u - s)^2$$

$$\le (nu + 1 - ns + 1) + n\,(u - s)$$

$$\le 2n\,(u - s) + 2 \; \le \; 4n\,(u - s)$$

and similarly

$$\sum_{i=1}^{n} d_n^4(i, u, s) \leq (nu + 1 - ns + 1) + n\,(u - s)^4$$

$$\leq 4n\,(u - s) \;\leq\; 4n\,(u - s)\,\sqrt{n\,(u - s)}$$

$$\leq 4n^{3/2}\,(u - s)^{3/2}.$$

Combining the three cases we've proved ( $\forall\, 0 \leq s \leq u \leq 1$ )

$$\sum_{i=1}^{n} d_n^2(i, u, s) \;\leq\; 6n\,(u - s) \tag{7.9.13}$$

and

$$\sum_{i=1}^{n} d_n^4(i, u, s) \;\leq\; 24\,n^{3/2}\,(u - s)^{3/2}. \tag{7.9.14}$$

As a consequence we get from (7.9.7) and (7.9.8) the inequalities

$$|S_{15}| \leq 4\,n^{3/2}\,(u - s)^{3/2}, \tag{7.9.15}$$

$$|S_{14}| = \left|\, \sum_{i=1}^{n} d_{\kappa i}^3 \sum_{j=1}^{n} d_{\kappa j} - S_{15} \,\right|$$

$$= |S_{15}| \;\leq\; 4n^{3/2}\,(u - s)^{3/2}, \tag{7.9.16}$$

$$|S_{13}| \leq \left(\, \sum_{i=1}^{n} d_{1i}^2 \,\right)^2 \;\leq\; 36\,n^2\,(u - s)^2, \tag{7.9.17}$$

$$|S_{12}| = \left|\, 2\,S_{15} - \left(\, \sum_{i=1}^{n} d_{1i}^2 \,\right)^2 \,\right|$$

$$\leq 2\,|S_{15}| + 36\,n^2\,(u - s)^2 \;\leq\; 44\,n^2\,(u - s)^{3/2}, \tag{7.9.18}$$

$$|S_{11}| = |\, 6\,S_{12} + 3\,S_{13} + 4\,S_{14} + S_{15} \,|$$

$$\leq 264\,n^2\,(u - s)^{3/2} + 108\,n^2\,(u - s)^2$$

$$+ 16\,n^{3/2}\,(u - s)^{3/2} + 4\,n^{3/2}\,(u - s)^{3/2}$$

$$\leq 392\,n^2\,(u - s)^{3/2}. \tag{7.9.19}$$

Substituting $(u - s)$ by $(v - t)$ we get completely similar inequalities for $|S_{2\tau}|$ , $\tau = 1, ..., 5$ . Therefore formula (7.9.10) implies the inequality (7.9.6), i.e.

$$E_0 \Delta_n^4(s,t,u,v)$$

$$\leq (392)^2 \, \frac{n}{(n-1)(n-2)(n-3)} \, (u-s)^{3/2} \, (v-t)^{3/2}$$

$$+ 6 \, (44)^2 \, \frac{n}{(n-1)(n-2)} \, (u-s)^{3/2} \, (v-t)^{3/2}$$

$$+ 3 \, (36)^2 \, \frac{n}{n-1} \, (u-s)^2 \, (v-t)^2$$

$$+ 4 \, (4)^2 \, \frac{1}{n-1} \, (u-s)^{3/2} \, (v-t)^{3/2}$$

$$+ 16 \, (u-s)^{3/2} \, (v-t)^{3/2}$$

$$\leq k_0 \, \Big( (u-s)(v-t) \Big)^{3/2} \tag{7.9.20}$$

for some suitable constant $k_0 < \infty$. $\quad\square$

## 7.10 Tables

For any rank statistic $S$ defined in Chapter 2 this section will provide a table of simulated critical values $k_\alpha$ according to $P_0\{S > k_\alpha\} = \alpha$, where $P_0$ denotes the distribution under the respective **continuous null hypothesis.**

Given the sample size we've used the generator RANDOM of Turbo-Pascal in order to produce 5000 Monte Carlo values of the rank statistic $S$. The tabulated critical values $k_\alpha$ are the upper $\alpha$-quantiles of the observed Monte Carlo $S$-sample.

In the case of tied observations the continuous model is invalid. If the fraction of tied observations is small, the given (continuous) critical values may serve as approximations of the exact conditional critical values given $\tau$. If the fraction of tied observations is substantial, one should use the provided programs for the simulation of the exact conditional $p$-values.

## Table 2.1.A
## Omnibus Rank Test for 2 Samples

*Simulated critical values* $k_\alpha$ *according to* $P_0\{ S > k_\alpha \} = \alpha$ *for the omnibus two-sample rank statistic* $S$ *defined in formula (6) of Section 2.1, under the continuous null hypothesis* $\mathcal{H}_0^r : F = G$ *( no ties ). The underlying bandwidth is* $a = 0.40$ *and the Monte Carlo sample size is* 5000.

| $m$ | $n$ | 0.01 | 0.02 | 0.03 | 0.04 | 0.05 | 0.06 | 0.07 | 0.08 | 0.09 | 0.10 |
|---|---|---|---|---|---|---|---|---|---|---|---|
| 10 | 10 | 8.50 | 7.53 | 7.05 | 6.68 | 6.34 | 6.11 | 5.84 | 5.68 | 5.47 | 5.34 |
| 10 | 20 | 8.55 | 7.60 | 7.07 | 6.64 | 6.38 | 6.13 | 5.88 | 5.70 | 5.53 | 5.39 |
| 10 | 30 | 8.66 | 7.59 | 7.04 | 6.67 | 6.39 | 6.12 | 5.87 | 5.68 | 5.51 | 5.36 |
| 10 | 40 | 9.09 | 8.07 | 7.47 | 6.93 | 6.56 | 6.26 | 5.97 | 5.74 | 5.54 | 5.38 |
| 10 | 50 | 9.34 | 8.08 | 7.46 | 6.93 | 6.55 | 6.23 | 5.97 | 5.78 | 5.58 | 5.41 |
| 20 | 10 | 9.11 | 7.98 | 7.38 | 6.95 | 6.58 | 6.22 | 5.98 | 5.76 | 5.58 | 5.40 |
| 20 | 20 | 8.84 | 7.81 | 7.31 | 6.86 | 6.43 | 6.14 | 5.92 | 5.70 | 5.52 | 5.37 |
| 20 | 30 | 9.01 | 8.01 | 7.36 | 6.84 | 6.51 | 6.18 | 5.91 | 5.68 | 5.50 | 5.32 |
| 20 | 40 | 8.72 | 7.69 | 7.08 | 6.66 | 6.38 | 6.11 | 5.89 | 5.66 | 5.50 | 5.31 |
| 20 | 50 | 9.17 | 7.94 | 7.23 | 6.73 | 6.41 | 6.14 | 5.90 | 5.69 | 5.53 | 5.36 |
| 30 | 10 | 8.77 | 7.82 | 7.27 | 6.86 | 6.58 | 6.31 | 6.10 | 5.92 | 5.70 | 5.47 |
| 30 | 20 | 8.89 | 7.74 | 7.30 | 6.85 | 6.53 | 6.27 | 6.02 | 5.85 | 5.64 | 5.47 |
| 30 | 30 | 9.18 | 8.26 | 7.63 | 7.11 | 6.60 | 6.26 | 6.04 | 5.81 | 5.57 | 5.29 |
| 30 | 40 | 9.32 | 8.25 | 7.70 | 7.20 | 6.74 | 6.48 | 6.20 | 5.88 | 5.67 | 5.49 |
| 30 | 50 | 9.24 | 8.01 | 7.54 | 7.11 | 6.69 | 6.37 | 6.05 | 5.79 | 5.61 | 5.42 |
| 40 | 10 | 8.91 | 7.79 | 7.27 | 6.75 | 6.32 | 6.02 | 5.80 | 5.58 | 5.41 | 5.23 |
| 40 | 20 | 9.08 | 8.10 | 7.38 | 6.94 | 6.61 | 6.31 | 6.10 | 5.86 | 5.66 | 5.49 |
| 40 | 30 | 8.90 | 7.99 | 7.43 | 7.04 | 6.60 | 6.28 | 5.98 | 5.77 | 5.60 | 5.40 |
| 40 | 40 | 9.41 | 8.21 | 7.65 | 7.00 | 6.58 | 6.22 | 5.95 | 5.74 | 5.53 | 5.34 |
| 40 | 50 | 9.21 | 8.19 | 7.65 | 7.10 | 6.70 | 6.45 | 6.15 | 5.90 | 5.66 | 5.47 |
| 50 | 10 | 8.80 | 7.78 | 7.18 | 6.69 | 6.32 | 6.06 | 5.83 | 5.59 | 5.37 | 5.22 |
| 50 | 20 | 9.08 | 7.96 | 7.34 | 6.89 | 6.58 | 6.24 | 5.99 | 5.79 | 5.61 | 5.46 |
| 50 | 30 | 9.19 | 8.14 | 7.47 | 7.04 | 6.67 | 6.36 | 6.11 | 5.91 | 5.68 | 5.51 |
| 50 | 40 | 9.13 | 7.85 | 7.33 | 6.98 | 6.59 | 6.32 | 6.02 | 5.73 | 5.56 | 5.37 |
| 50 | 50 | 9.15 | 8.14 | 7.48 | 6.97 | 6.59 | 6.21 | 6.00 | 5.81 | 5.62 | 5.42 |

### Table 2.1.B
### One-sided Rank Test for 2 Samples

*Simulated critical values $k_\alpha^0$ according to $P_0\{S^0 > k_\alpha^0\} = \alpha$ for the one-sided two-sample rank statistic $S^0$ defined in formula (13) of Section 2.1 under the continuous null hypothesis $\mathcal{H}_0^r : F = G$ ( no ties ). The underlying bandwidth is $a = 0.40$ and the Monte Carlo sample size is $5000$.*

| $m$ | $n$ | 0.01 | 0.02 | 0.03 | 0.04 | 0.05 | 0.06 | 0.07 | 0.08 | 0.09 | 0.10 |
|---|---|---|---|---|---|---|---|---|---|---|---|
| 10 | 10 | 7.03 | 6.26 | 5.56 | 5.13 | 4.86 | 4.58 | 4.35 | 4.19 | 4.02 | 3.86 |
| 10 | 20 | 7.02 | 6.12 | 5.49 | 5.10 | 4.82 | 4.51 | 4.25 | 4.00 | 3.84 | 3.69 |
| 10 | 30 | 7.24 | 6.41 | 5.77 | 5.32 | 5.00 | 4.66 | 4.31 | 4.06 | 3.90 | 3.73 |
| 10 | 40 | 7.85 | 6.43 | 5.59 | 5.07 | 4.69 | 4.47 | 4.22 | 3.98 | 3.79 | 3.63 |
| 10 | 50 | 7.37 | 6.16 | 5.53 | 5.05 | 4.72 | 4.38 | 4.05 | 3.86 | 3.65 | 3.50 |
| 20 | 10 | 7.40 | 6.22 | 5.66 | 5.17 | 4.78 | 4.49 | 4.25 | 4.07 | 3.91 | 3.73 |
| 20 | 20 | 7.60 | 6.29 | 5.70 | 5.23 | 4.93 | 4.64 | 4.42 | 4.16 | 3.97 | 3.81 |
| 20 | 30 | 7.55 | 6.47 | 5.80 | 5.35 | 4.87 | 4.58 | 4.32 | 4.09 | 3.88 | 3.71 |
| 20 | 40 | 7.38 | 6.34 | 5.61 | 5.17 | 4.89 | 4.59 | 4.35 | 4.14 | 3.98 | 3.78 |
| 20 | 50 | 7.26 | 6.20 | 5.65 | 5.23 | 4.94 | 4.52 | 4.23 | 4.04 | 3.85 | 3.70 |
| 30 | 10 | 7.39 | 6.35 | 5.65 | 5.16 | 4.85 | 4.55 | 4.32 | 4.16 | 3.98 | 3.82 |
| 30 | 20 | 7.28 | 6.28 | 5.77 | 5.22 | 4.82 | 4.50 | 4.30 | 4.07 | 3.86 | 3.68 |
| 30 | 30 | 7.82 | 6.60 | 5.75 | 5.11 | 4.79 | 4.48 | 4.25 | 3.99 | 3.80 | 3.65 |
| 30 | 40 | 7.48 | 6.54 | 5.75 | 5.27 | 4.86 | 4.54 | 4.26 | 4.05 | 3.81 | 3.63 |
| 30 | 50 | 7.59 | 6.43 | 5.74 | 5.32 | 4.97 | 4.57 | 4.27 | 4.06 | 3.81 | 3.62 |
| 40 | 10 | 7.48 | 6.05 | 5.48 | 5.02 | 4.67 | 4.36 | 4.10 | 3.91 | 3.67 | 3.52 |
| 40 | 20 | 7.63 | 6.54 | 5.85 | 5.43 | 5.03 | 4.76 | 4.47 | 4.25 | 4.03 | 3.82 |
| 40 | 30 | 7.44 | 6.37 | 5.65 | 5.10 | 4.81 | 4.54 | 4.25 | 3.99 | 3.84 | 3.70 |
| 40 | 40 | 7.43 | 6.33 | 5.70 | 5.20 | 4.75 | 4.40 | 4.18 | 3.92 | 3.73 | 3.57 |
| 40 | 50 | 7.70 | 6.55 | 5.78 | 5.29 | 4.85 | 4.51 | 4.21 | 3.96 | 3.78 | 3.56 |
| 50 | 10 | 7.18 | 6.20 | 5.50 | 5.03 | 4.69 | 4.39 | 4.14 | 3.88 | 3.71 | 3.53 |
| 50 | 20 | 7.51 | 6.56 | 5.77 | 5.33 | 4.99 | 4.58 | 4.35 | 4.10 | 3.87 | 3.71 |
| 50 | 30 | 7.63 | 6.59 | 5.95 | 5.48 | 5.05 | 4.71 | 4.38 | 4.20 | 4.01 | 3.83 |
| 50 | 40 | 7.43 | 6.42 | 5.77 | 5.28 | 4.91 | 4.64 | 4.39 | 4.15 | 3.94 | 3.75 |
| 50 | 50 | 7.50 | 6.33 | 5.75 | 5.31 | 4.85 | 4.60 | 4.30 | 4.07 | 3.87 | 3.73 |

## Table 2.1.C
## Projection Rank Test for 2 Samples

*Simulated critical values $k_\alpha^\pi$ according to $P_0\{S^\pi > k_\alpha^\pi\} = \alpha$ for the two-sample projection rank statistic $S^\pi$ defined in formula (23) of Section 2.1 under the continuous null hypothesis $\mathcal{H}_0^\tau : F = G$ ( no ties ). The Monte Carlo sample size is 5000.*

| $m$ | $n$ | 0.01 | 0.02 | 0.03 | 0.04 | 0.05 | 0.06 | 0.07 | 0.08 | 0.09 | 0.10 |
|----|----|------|------|------|------|------|------|------|------|------|------|
| 10 | 10 | 7.24 | 6.01 | 5.20 | 4.76 | 4.43 | 4.13 | 3.89 | 3.69 | 3.47 | 3.30 |
| 10 | 20 | 6.93 | 6.03 | 5.42 | 4.85 | 4.41 | 4.06 | 3.79 | 3.56 | 3.35 | 3.15 |
| 10 | 30 | 7.31 | 6.05 | 5.49 | 4.89 | 4.43 | 4.18 | 3.93 | 3.74 | 3.45 | 3.26 |
| 10 | 40 | 7.51 | 6.24 | 5.34 | 4.79 | 4.39 | 4.10 | 3.77 | 3.55 | 3.29 | 3.10 |
| 10 | 50 | 7.26 | 5.89 | 5.13 | 4.59 | 4.18 | 3.79 | 3.50 | 3.25 | 3.10 | 2.90 |
| 20 | 10 | 7.46 | 6.04 | 5.31 | 4.77 | 4.34 | 4.10 | 3.82 | 3.57 | 3.39 | 3.23 |
| 20 | 20 | 7.28 | 6.00 | 5.42 | 4.98 | 4.58 | 4.31 | 3.96 | 3.68 | 3.50 | 3.34 |
| 20 | 30 | 7.35 | 6.22 | 5.42 | 4.84 | 4.48 | 4.17 | 3.84 | 3.58 | 3.41 | 3.22 |
| 20 | 40 | 7.17 | 5.91 | 5.24 | 4.83 | 4.57 | 4.18 | 3.86 | 3.61 | 3.36 | 3.19 |
| 20 | 50 | 7.08 | 6.01 | 5.31 | 4.75 | 4.41 | 4.04 | 3.76 | 3.51 | 3.32 | 3.10 |
| 30 | 10 | 7.46 | 6.09 | 5.43 | 5.00 | 4.59 | 4.25 | 3.97 | 3.68 | 3.48 | 3.25 |
| 30 | 20 | 7.44 | 6.13 | 5.49 | 4.85 | 4.42 | 4.06 | 3.74 | 3.52 | 3.34 | 3.12 |
| 30 | 30 | 7.70 | 6.28 | 5.35 | 4.79 | 4.34 | 3.99 | 3.75 | 3.48 | 3.25 | 3.03 |
| 30 | 40 | 7.75 | 6.34 | 5.40 | 4.78 | 4.33 | 4.05 | 3.70 | 3.49 | 3.26 | 3.10 |
| 30 | 50 | 6.99 | 6.01 | 5.38 | 4.92 | 4.55 | 4.17 | 3.87 | 3.60 | 3.45 | 3.19 |
| 40 | 10 | 6.82 | 5.78 | 5.27 | 4.77 | 4.37 | 4.03 | 3.68 | 3.40 | 3.22 | 2.99 |
| 40 | 20 | 7.66 | 6.40 | 5.58 | 5.02 | 4.66 | 4.24 | 3.98 | 3.71 | 3.44 | 3.23 |
| 40 | 30 | 7.60 | 6.01 | 5.25 | 4.79 | 4.36 | 4.01 | 3.76 | 3.51 | 3.30 | 3.09 |
| 40 | 40 | 7.40 | 6.08 | 5.21 | 4.65 | 4.18 | 3.91 | 3.62 | 3.40 | 3.20 | 3.02 |
| 40 | 50 | 7.41 | 6.33 | 5.39 | 4.73 | 4.29 | 3.94 | 3.70 | 3.50 | 3.29 | 3.12 |
| 50 | 10 | 6.83 | 5.88 | 5.05 | 4.50 | 4.19 | 3.90 | 3.66 | 3.44 | 3.24 | 3.09 |
| 50 | 20 | 7.39 | 6.26 | 5.59 | 4.95 | 4.59 | 4.20 | 3.90 | 3.63 | 3.43 | 3.22 |
| 50 | 30 | 7.62 | 6.35 | 5.71 | 5.08 | 4.68 | 4.19 | 3.96 | 3.70 | 3.55 | 3.32 |
| 50 | 40 | 7.46 | 6.03 | 5.52 | 4.88 | 4.52 | 4.24 | 3.98 | 3.67 | 3.50 | 3.27 |
| 50 | 50 | 7.24 | 6.13 | 5.36 | 4.90 | 4.45 | 4.07 | 3.80 | 3.56 | 3.41 | 3.24 |

**Table 2.2.A**
**Rank Test for Dispersion about the Median**

*Simulated critical values $k_\alpha^\mu$ according to $P_0\{S^\mu > k_\alpha^\mu\} = \alpha$ for the dispersion rank statistic $S^\mu$ with $\mu = 1/2$ as defined in formula (4) of Section 2.2, under the continuous null hypothesis $\mathcal{H}_0^r : F = G$ ( no ties ). The underlying bandwidth is $a = 0.40$ and the Monte Carlo sample size is $5000$.*

| $m$ | $n$ | 0.01 | 0.02 | 0.03 | 0.04 | 0.05 | 0.06 | 0.07 | 0.08 | 0.09 | 0.10 |
|---|---|---|---|---|---|---|---|---|---|---|---|
| 10 | 10 | 6.58 | 5.68 | 5.18 | 4.81 | 4.50 | 4.31 | 4.07 | 3.92 | 3.75 | 3.61 |
| 10 | 20 | 7.19 | 6.29 | 5.56 | 5.08 | 4.79 | 4.51 | 4.25 | 4.05 | 3.89 | 3.68 |
| 10 | 30 | 7.36 | 6.18 | 5.69 | 5.32 | 4.97 | 4.67 | 4.47 | 4.28 | 4.11 | 3.91 |
| 10 | 40 | 7.78 | 6.79 | 6.10 | 5.52 | 5.11 | 4.79 | 4.45 | 4.20 | 3.95 | 3.78 |
| 10 | 50 | 7.99 | 6.70 | 6.01 | 5.56 | 5.02 | 4.76 | 4.55 | 4.25 | 3.97 | 3.79 |
| 20 | 10 | 6.31 | 5.35 | 4.92 | 4.48 | 4.20 | 3.95 | 3.76 | 3.60 | 3.47 | 3.32 |
| 20 | 20 | 6.68 | 5.77 | 5.24 | 4.84 | 4.63 | 4.38 | 4.15 | 3.92 | 3.75 | 3.58 |
| 20 | 30 | 7.37 | 6.19 | 5.52 | 5.06 | 4.75 | 4.44 | 4.17 | 3.94 | 3.74 | 3.58 |
| 20 | 40 | 6.78 | 6.03 | 5.54 | 5.08 | 4.71 | 4.37 | 4.17 | 3.94 | 3.78 | 3.59 |
| 20 | 50 | 7.45 | 6.18 | 5.55 | 5.05 | 4.69 | 4.43 | 4.19 | 3.98 | 3.81 | 3.67 |
| 30 | 10 | 6.26 | 5.52 | 5.04 | 4.61 | 4.25 | 3.97 | 3.76 | 3.59 | 3.41 | 3.29 |
| 30 | 20 | 6.84 | 5.90 | 5.32 | 4.91 | 4.58 | 4.31 | 4.09 | 3.89 | 3.72 | 3.59 |
| 30 | 30 | 7.34 | 6.03 | 5.39 | 4.92 | 4.56 | 4.26 | 4.03 | 3.81 | 3.65 | 3.49 |
| 30 | 40 | 7.53 | 6.32 | 5.61 | 5.17 | 4.65 | 4.37 | 4.16 | 3.91 | 3.75 | 3.61 |
| 30 | 50 | 7.50 | 6.38 | 5.59 | 5.06 | 4.63 | 4.37 | 4.13 | 3.95 | 3.74 | 3.58 |
| 40 | 10 | 6.08 | 5.22 | 4.79 | 4.46 | 4.16 | 3.94 | 3.71 | 3.54 | 3.39 | 3.27 |
| 40 | 20 | 6.57 | 5.91 | 5.09 | 4.70 | 4.31 | 4.06 | 3.90 | 3.74 | 3.56 | 3.40 |
| 40 | 30 | 6.99 | 5.87 | 5.23 | 4.89 | 4.54 | 4.29 | 4.04 | 3.80 | 3.61 | 3.44 |
| 40 | 40 | 7.24 | 6.10 | 5.42 | 4.87 | 4.45 | 4.20 | 3.91 | 3.73 | 3.54 | 3.36 |
| 40 | 50 | 7.26 | 6.08 | 5.48 | 4.95 | 4.50 | 4.18 | 3.96 | 3.76 | 3.57 | 3.41 |
| 50 | 10 | 5.86 | 5.11 | 4.59 | 4.29 | 4.07 | 3.84 | 3.62 | 3.44 | 3.30 | 3.18 |
| 50 | 20 | 6.66 | 5.84 | 5.27 | 4.73 | 4.39 | 4.13 | 3.87 | 3.64 | 3.51 | 3.37 |
| 50 | 30 | 6.70 | 5.85 | 5.28 | 4.81 | 4.48 | 4.16 | 3.96 | 3.79 | 3.59 | 3.43 |
| 50 | 40 | 7.09 | 5.89 | 5.34 | 4.85 | 4.53 | 4.31 | 4.04 | 3.83 | 3.61 | 3.48 |
| 50 | 50 | 7.20 | 6.11 | 5.47 | 4.92 | 4.57 | 4.27 | 4.03 | 3.76 | 3.57 | 3.43 |

**Table 2.2.B**
**Projection Rank Test for Dispersion about the Median**

*Simulated critical values $k_\alpha^\pi$ according to $P_0\{ S^\pi > k_\alpha^\pi \} = \alpha$ for the projection rank statistic $S^\pi$ for dispersion about the median as defined in Section 2.2, under the continuous null hypothesis $\mathcal{H}_0^r : F = G$ ( no ties ). The Monte Carlo sample size is $5000$.*

| $m$ | $n$ | 0.01 | 0.02 | 0.03 | 0.04 | 0.05 | 0.06 | 0.07 | 0.08 | 0.09 | 0.10 |
|---|---|---|---|---|---|---|---|---|---|---|---|
| 10 | 10 | 5.99 | 5.23 | 4.68 | 4.40 | 4.10 | 3.87 | 3.63 | 3.44 | 3.26 | 3.12 |
| 10 | 20 | 7.63 | 6.33 | 5.76 | 5.37 | 4.99 | 4.70 | 4.31 | 4.08 | 3.87 | 3.66 |
| 10 | 30 | 7.93 | 6.91 | 6.21 | 5.86 | 5.42 | 5.01 | 4.76 | 4.53 | 4.27 | 4.05 |
| 10 | 40 | 9.15 | 7.90 | 7.05 | 6.31 | 5.81 | 5.42 | 5.07 | 4.72 | 4.41 | 4.22 |
| 10 | 50 | 10.28 | 8.23 | 7.47 | 6.81 | 6.16 | 5.67 | 5.39 | 5.07 | 4.76 | 4.36 |
| 20 | 10 | 6.03 | 4.92 | 4.37 | 3.99 | 3.70 | 3.42 | 3.23 | 3.06 | 2.88 | 2.72 |
| 20 | 20 | 6.71 | 5.59 | 5.07 | 4.67 | 4.36 | 4.08 | 3.85 | 3.66 | 3.43 | 3.20 |
| 20 | 30 | 7.56 | 6.48 | 5.71 | 5.16 | 4.85 | 4.56 | 4.29 | 4.05 | 3.83 | 3.64 |
| 20 | 40 | 7.67 | 6.57 | 5.88 | 5.32 | 4.91 | 4.57 | 4.31 | 4.05 | 3.80 | 3.53 |
| 20 | 50 | 8.36 | 6.81 | 6.01 | 5.52 | 5.05 | 4.69 | 4.43 | 4.20 | 3.95 | 3.72 |
| 30 | 10 | 6.29 | 5.32 | 4.46 | 3.94 | 3.62 | 3.33 | 3.12 | 2.89 | 2.70 | 2.57 |
| 30 | 20 | 6.89 | 5.64 | 4.97 | 4.42 | 4.11 | 3.83 | 3.60 | 3.39 | 3.27 | 3.15 |
| 30 | 30 | 7.37 | 6.02 | 5.28 | 4.86 | 4.48 | 4.20 | 3.90 | 3.58 | 3.35 | 3.21 |
| 30 | 40 | 7.83 | 6.49 | 5.71 | 5.23 | 4.86 | 4.49 | 4.22 | 3.92 | 3.68 | 3.48 |
| 30 | 50 | 7.56 | 6.37 | 5.77 | 5.28 | 4.85 | 4.49 | 4.24 | 3.95 | 3.66 | 3.48 |
| 40 | 10 | 5.91 | 4.88 | 4.29 | 3.83 | 3.49 | 3.22 | 3.02 | 2.83 | 2.65 | 2.49 |
| 40 | 20 | 6.54 | 5.44 | 4.85 | 4.41 | 4.05 | 3.77 | 3.56 | 3.37 | 3.18 | 3.04 |
| 40 | 30 | 6.62 | 5.48 | 4.93 | 4.56 | 4.22 | 3.98 | 3.77 | 3.51 | 3.32 | 3.17 |
| 40 | 40 | 7.06 | 5.92 | 5.19 | 4.74 | 4.33 | 4.03 | 3.72 | 3.49 | 3.29 | 3.15 |
| 40 | 50 | 7.90 | 6.47 | 5.65 | 5.14 | 4.74 | 4.41 | 4.14 | 3.87 | 3.67 | 3.40 |
| 50 | 10 | 5.59 | 4.64 | 4.12 | 3.69 | 3.43 | 3.14 | 2.88 | 2.71 | 2.57 | 2.45 |
| 50 | 20 | 6.45 | 5.18 | 4.57 | 4.21 | 3.92 | 3.68 | 3.44 | 3.27 | 3.08 | 2.94 |
| 50 | 30 | 6.83 | 5.60 | 5.08 | 4.63 | 4.23 | 3.96 | 3.69 | 3.52 | 3.34 | 3.18 |
| 50 | 40 | 7.39 | 6.06 | 5.33 | 4.80 | 4.41 | 4.15 | 3.92 | 3.72 | 3.54 | 3.33 |
| 50 | 50 | 7.18 | 5.95 | 5.42 | 4.88 | 4.53 | 4.16 | 3.96 | 3.76 | 3.55 | 3.32 |

## Table 2.3.A
## Omnibus Rank Test for 3 Samples

*Simulated critical values $k_\alpha$ under the continuous null hypothesis $\mathcal{H}_0^r : F_1 = F_2 = F_3$ ( no ties ) according to $P_0\{ S > k_\alpha \} = \alpha$ for the three-sample omnibus rank statistic $S$ defined in formula (7) of Section 2.3. The underlying bandwidth is $a = 0.40$ and the Monte Carlo sample size is $5000$.*

| $n_1$ | $n_2$ | $n_3$ | 0.01 | 0.02 | 0.03 | 0.04 | 0.05 | 0.06 | 0.07 | 0.08 | 0.09 | 0.10 |
|---|---|---|---|---|---|---|---|---|---|---|---|---|
| 10 | 10 | 10 | 13.59 | 12.50 | 11.67 | 11.13 | 10.59 | 10.26 | 10.00 | 9.72 | 9.56 | 9.31 |
| 10 | 10 | 20 | 13.84 | 12.51 | 11.83 | 11.31 | 10.81 | 10.44 | 10.16 | 9.87 | 9.62 | 9.41 |
| 10 | 10 | 30 | 13.61 | 12.29 | 11.48 | 11.01 | 10.52 | 10.17 | 9.84 | 9.62 | 9.34 | 9.12 |
| 10 | 20 | 10 | 13.60 | 12.48 | 11.48 | 10.95 | 10.53 | 10.18 | 9.85 | 9.62 | 9.35 | 9.18 |
| 10 | 20 | 20 | 14.28 | 12.52 | 11.69 | 11.04 | 10.62 | 10.24 | 9.93 | 9.67 | 9.43 | 9.23 |
| 10 | 20 | 30 | 13.73 | 12.56 | 11.73 | 11.12 | 10.68 | 10.31 | 9.98 | 9.72 | 9.46 | 9.25 |
| 10 | 30 | 10 | 13.70 | 12.46 | 11.74 | 11.11 | 10.59 | 10.27 | 9.99 | 9.74 | 9.48 | 9.23 |
| 10 | 30 | 20 | 13.93 | 12.30 | 11.60 | 11.15 | 10.80 | 10.34 | 10.01 | 9.78 | 9.50 | 9.24 |
| 10 | 30 | 30 | 13.71 | 12.50 | 11.47 | 10.94 | 10.59 | 10.30 | 10.00 | 9.75 | 9.51 | 9.28 |
| 20 | 10 | 10 | 13.31 | 12.19 | 11.38 | 10.91 | 10.56 | 10.24 | 9.97 | 9.72 | 9.46 | 9.26 |
| 20 | 10 | 20 | 13.46 | 12.17 | 11.59 | 10.94 | 10.52 | 10.22 | 9.94 | 9.73 | 9.52 | 9.34 |
| 20 | 10 | 30 | 13.73 | 12.21 | 11.47 | 11.03 | 10.64 | 10.32 | 10.04 | 9.79 | 9.54 | 9.39 |
| 20 | 20 | 10 | 13.87 | 12.44 | 11.76 | 11.11 | 10.72 | 10.38 | 10.07 | 9.74 | 9.55 | 9.31 |
| 20 | 20 | 20 | 13.86 | 12.58 | 11.70 | 11.15 | 10.64 | 10.33 | 10.08 | 9.84 | 9.60 | 9.39 |
| 20 | 20 | 30 | 13.77 | 12.38 | 11.52 | 10.98 | 10.58 | 10.25 | 9.96 | 9.71 | 9.47 | 9.28 |
| 20 | 30 | 10 | 13.85 | 12.52 | 11.76 | 11.02 | 10.60 | 10.28 | 10.00 | 9.77 | 9.50 | 9.27 |
| 20 | 30 | 20 | 13.91 | 12.71 | 11.89 | 11.27 | 10.88 | 10.43 | 10.10 | 9.84 | 9.60 | 9.37 |
| 20 | 30 | 30 | 14.22 | 12.78 | 11.85 | 11.35 | 10.80 | 10.49 | 10.20 | 9.93 | 9.71 | 9.44 |
| 30 | 10 | 10 | 13.86 | 12.66 | 11.99 | 11.39 | 10.88 | 10.44 | 10.10 | 9.83 | 9.59 | 9.36 |
| 30 | 10 | 20 | 13.47 | 12.35 | 11.63 | 11.11 | 10.66 | 10.23 | 9.91 | 9.64 | 9.41 | 9.24 |
| 30 | 10 | 30 | 13.88 | 12.72 | 11.92 | 11.20 | 10.68 | 10.33 | 9.98 | 9.71 | 9.51 | 9.32 |
| 30 | 20 | 10 | 13.33 | 12.41 | 11.68 | 11.11 | 10.75 | 10.34 | 9.95 | 9.69 | 9.46 | 9.23 |
| 30 | 20 | 20 | 14.16 | 12.87 | 11.80 | 11.35 | 10.85 | 10.47 | 10.16 | 9.81 | 9.57 | 9.34 |
| 30 | 20 | 30 | 13.72 | 12.55 | 11.84 | 11.14 | 10.69 | 10.33 | 10.03 | 9.79 | 9.51 | 9.29 |
| 30 | 30 | 10 | 14.26 | 12.73 | 11.75 | 11.18 | 10.70 | 10.33 | 10.07 | 9.77 | 9.48 | 9.24 |
| 30 | 30 | 20 | 13.79 | 12.36 | 11.64 | 10.98 | 10.57 | 10.21 | 9.85 | 9.54 | 9.29 | 9.08 |
| 30 | 30 | 30 | 13.98 | 12.54 | 11.65 | 11.22 | 10.83 | 10.45 | 10.17 | 9.89 | 9.59 | 9.29 |

## Table 2.3.B
## Trend Rank Test for 3 Samples

*Simulated critical values $k_\alpha^0$ under the continuous null hypothesis $\mathcal{H}_0^r : F_1 = F_2 = F_3$ ( no ties ) according to $P_0\{S^0 > k_\alpha^0\} = \alpha$ for the three-sample trend rank statistic $S^0$ defined in formula (15) of Section 2.3. The underlying bandwidth is $a = 0.40$ and the Monte Carlo sample size is $5000$.*

| $n_1$ | $n_2$ | $n_3$ | 0.01 | 0.02 | 0.03 | 0.04 | 0.05 | 0.06 | 0.07 | 0.08 | 0.09 | 0.10 |
|---|---|---|---|---|---|---|---|---|---|---|---|---|
| 10 | 10 | 10 | 9.93 | 8.74 | 7.91 | 7.36 | 6.99 | 6.63 | 6.27 | 5.95 | 5.75 | 5.54 |
| 10 | 10 | 20 | 10.05 | 8.54 | 7.81 | 7.30 | 6.86 | 6.48 | 6.15 | 5.90 | 5.68 | 5.44 |
| 10 | 10 | 30 | 10.11 | 8.83 | 7.90 | 7.29 | 6.83 | 6.43 | 6.16 | 5.94 | 5.70 | 5.48 |
| 10 | 20 | 10 | 9.65 | 8.53 | 7.82 | 7.38 | 6.96 | 6.65 | 6.34 | 6.01 | 5.77 | 5.57 |
| 10 | 20 | 20 | 9.65 | 8.54 | 7.78 | 7.18 | 6.76 | 6.45 | 6.11 | 5.82 | 5.60 | 5.39 |
| 10 | 20 | 30 | 10.05 | 8.69 | 7.96 | 7.37 | 6.87 | 6.51 | 6.21 | 5.95 | 5.69 | 5.47 |
| 10 | 30 | 10 | 10.49 | 8.94 | 8.03 | 7.46 | 6.85 | 6.54 | 6.21 | 5.98 | 5.79 | 5.58 |
| 10 | 30 | 20 | 10.21 | 8.87 | 8.04 | 7.51 | 6.94 | 6.64 | 6.33 | 6.11 | 5.85 | 5.63 |
| 10 | 30 | 30 | 10.01 | 8.84 | 7.87 | 7.25 | 6.87 | 6.51 | 6.25 | 6.03 | 5.82 | 5.57 |
| 20 | 10 | 10 | 9.99 | 8.74 | 7.84 | 7.33 | 6.97 | 6.69 | 6.37 | 6.11 | 5.86 | 5.64 |
| 20 | 10 | 20 | 9.85 | 8.74 | 8.05 | 7.49 | 7.08 | 6.68 | 6.27 | 6.03 | 5.84 | 5.62 |
| 20 | 10 | 30 | 10.22 | 9.12 | 8.36 | 7.54 | 7.06 | 6.65 | 6.27 | 6.00 | 5.72 | 5.50 |
| 20 | 20 | 10 | 10.30 | 8.82 | 8.11 | 7.61 | 7.20 | 6.87 | 6.54 | 6.22 | 5.97 | 5.76 |
| 20 | 20 | 20 | 9.93 | 8.52 | 7.67 | 7.14 | 6.68 | 6.39 | 6.11 | 5.91 | 5.67 | 5.45 |
| 20 | 20 | 30 | 10.22 | 8.93 | 8.00 | 7.47 | 7.03 | 6.64 | 6.19 | 5.94 | 5.72 | 5.45 |
| 20 | 30 | 10 | 10.12 | 8.82 | 8.01 | 7.48 | 7.03 | 6.69 | 6.49 | 6.22 | 5.98 | 5.80 |
| 20 | 30 | 20 | 9.81 | 8.71 | 7.89 | 7.40 | 7.01 | 6.56 | 6.18 | 5.92 | 5.72 | 5.51 |
| 20 | 30 | 30 | 10.20 | 8.62 | 7.90 | 7.38 | 7.07 | 6.67 | 6.29 | 6.00 | 5.77 | 5.56 |
| 30 | 10 | 10 | 10.38 | 9.03 | 8.20 | 7.54 | 7.13 | 6.70 | 6.32 | 6.07 | 5.82 | 5.55 |
| 30 | 10 | 20 | 10.26 | 8.75 | 8.00 | 7.52 | 7.04 | 6.68 | 6.30 | 6.06 | 5.83 | 5.59 |
| 30 | 10 | 30 | 10.94 | 9.52 | 8.66 | 7.99 | 7.43 | 6.95 | 6.55 | 6.27 | 5.96 | 5.69 |
| 30 | 20 | 10 | 9.94 | 8.97 | 8.18 | 7.65 | 7.21 | 6.79 | 6.47 | 6.18 | 5.88 | 5.64 |
| 30 | 20 | 20 | 10.31 | 8.82 | 8.16 | 7.60 | 7.18 | 6.72 | 6.40 | 6.11 | 5.83 | 5.60 |
| 30 | 20 | 30 | 10.36 | 8.94 | 8.06 | 7.51 | 7.05 | 6.73 | 6.38 | 6.04 | 5.78 | 5.54 |
| 30 | 30 | 10 | 10.93 | 9.25 | 8.33 | 7.62 | 7.12 | 6.74 | 6.38 | 6.09 | 5.88 | 5.69 |
| 30 | 30 | 20 | 9.86 | 8.91 | 7.78 | 7.21 | 6.80 | 6.51 | 6.21 | 5.92 | 5.67 | 5.45 |
| 30 | 30 | 30 | 10.03 | 8.79 | 8.07 | 7.43 | 7.01 | 6.69 | 6.33 | 6.09 | 5.79 | 5.55 |

## Table 2.3.C
## Projection Rank Test (Trend) for 3 Samples

*Simulated critical values $k_\alpha^\pi$ under the continuous null hypothesis $\mathcal{H}_0^r : F_1 = F_2 = F_3$ ( no ties ) according to $P_0\{ S^\pi > k_\alpha^\pi \} = \alpha$ for the three-sample projection rank statistic $S^\pi$ defined in Section 2.3. The Monte Carlo sample size is 5000.*

| $n_1$ | $n_2$ | $n_3$ | 0.01 | 0.02 | 0.03 | 0.04 | 0.05 | 0.06 | 0.07 | 0.08 | 0.09 | 0.10 |
|---|---|---|---|---|---|---|---|---|---|---|---|---|
| 10 | 10 | 10 | 7.19 | 6.18 | 5.48 | 5.06 | 4.65 | 4.31 | 4.00 | 3.72 | 3.49 | 3.25 |
| 10 | 10 | 20 | 7.05 | 5.95 | 5.21 | 4.79 | 4.40 | 4.10 | 3.84 | 3.56 | 3.31 | 3.13 |
| 10 | 10 | 30 | 7.33 | 6.05 | 5.26 | 4.67 | 4.34 | 4.04 | 3.73 | 3.51 | 3.27 | 3.08 |
| 10 | 20 | 10 | 7.50 | 6.19 | 5.68 | 5.10 | 4.62 | 4.26 | 3.87 | 3.64 | 3.41 | 3.19 |
| 10 | 20 | 20 | 7.17 | 6.06 | 5.20 | 4.88 | 4.39 | 4.07 | 3.84 | 3.53 | 3.36 | 3.12 |
| 10 | 20 | 30 | 7.74 | 6.44 | 5.73 | 5.15 | 4.73 | 4.31 | 4.01 | 3.77 | 3.53 | 3.32 |
| 10 | 30 | 10 | 7.23 | 6.21 | 5.46 | 4.86 | 4.53 | 4.23 | 3.98 | 3.74 | 3.51 | 3.30 |
| 10 | 30 | 20 | 7.37 | 5.96 | 5.22 | 4.69 | 4.29 | 3.98 | 3.77 | 3.48 | 3.25 | 3.08 |
| 10 | 30 | 30 | 7.36 | 6.13 | 5.34 | 4.73 | 4.38 | 4.08 | 3.85 | 3.52 | 3.31 | 3.13 |
| 20 | 10 | 10 | 7.24 | 6.01 | 5.20 | 4.75 | 4.47 | 4.18 | 3.88 | 3.63 | 3.41 | 3.22 |
| 20 | 10 | 20 | 7.26 | 6.16 | 5.35 | 4.77 | 4.44 | 4.05 | 3.81 | 3.53 | 3.32 | 3.16 |
| 20 | 10 | 30 | 7.41 | 6.17 | 5.48 | 4.99 | 4.48 | 4.09 | 3.79 | 3.47 | 3.24 | 3.04 |
| 20 | 20 | 10 | 7.47 | 6.22 | 5.42 | 4.94 | 4.51 | 4.08 | 3.83 | 3.57 | 3.34 | 3.14 |
| 20 | 20 | 20 | 7.71 | 6.32 | 5.34 | 4.92 | 4.47 | 4.11 | 3.85 | 3.59 | 3.33 | 3.15 |
| 20 | 20 | 30 | 7.90 | 6.35 | 5.42 | 4.94 | 4.57 | 4.24 | 3.87 | 3.62 | 3.40 | 3.19 |
| 20 | 30 | 10 | 7.17 | 5.79 | 5.22 | 4.66 | 4.27 | 3.95 | 3.72 | 3.44 | 3.23 | 3.06 |
| 20 | 30 | 20 | 7.26 | 6.15 | 5.43 | 4.91 | 4.51 | 4.13 | 3.85 | 3.52 | 3.30 | 3.07 |
| 20 | 30 | 30 | 7.32 | 6.05 | 5.37 | 4.85 | 4.36 | 4.01 | 3.79 | 3.50 | 3.30 | 3.07 |
| 30 | 10 | 10 | 8.22 | 6.54 | 5.65 | 5.03 | 4.61 | 4.29 | 3.97 | 3.65 | 3.43 | 3.24 |
| 30 | 10 | 20 | 8.09 | 6.01 | 5.24 | 4.66 | 4.21 | 3.97 | 3.74 | 3.49 | 3.23 | 3.04 |
| 30 | 10 | 30 | 7.63 | 6.39 | 5.37 | 4.87 | 4.46 | 4.13 | 3.85 | 3.62 | 3.40 | 3.19 |
| 30 | 20 | 10 | 7.19 | 5.78 | 5.17 | 4.63 | 4.25 | 3.94 | 3.64 | 3.44 | 3.23 | 3.06 |
| 30 | 20 | 20 | 7.36 | 6.10 | 5.27 | 4.71 | 4.30 | 3.98 | 3.69 | 3.46 | 3.23 | 3.02 |
| 30 | 20 | 30 | 7.53 | 6.16 | 5.55 | 4.97 | 4.48 | 4.08 | 3.77 | 3.52 | 3.30 | 3.10 |
| 30 | 30 | 10 | 7.64 | 6.39 | 5.78 | 5.25 | 4.78 | 4.36 | 4.05 | 3.72 | 3.47 | 3.35 |
| 30 | 30 | 20 | 7.27 | 5.95 | 5.24 | 4.71 | 4.34 | 3.99 | 3.72 | 3.44 | 3.23 | 3.09 |
| 30 | 30 | 30 | 7.71 | 6.40 | 5.54 | 5.05 | 4.58 | 4.20 | 3.92 | 3.59 | 3.38 | 3.18 |

### Table 2.4
### Omnibus Rank Test for 2 Samples on the Circle

*Simulated critical values $k_\alpha$ according to $P_0\{S > k_\alpha\} = \alpha$ for the omnibus rank statistic $S$ for two samples on the circle as defined in Section 2.4, under the continuous null hypothesis $\mathcal{H}_0^r : F_1 = F_2$ ( no ties ). The underlying bandwidth is $a = 0.40$ and the Monte Carlo sample size is $5000$.*

| $m$ | $n$ | 0.01 | 0.02 | 0.03 | 0.04 | 0.05 | 0.06 | 0.07 | 0.08 | 0.09 | 0.10 |
|---|---|---|---|---|---|---|---|---|---|---|---|
| 10 | 10 | 7.51 | 6.77 | 6.30 | 5.96 | 5.57 | 5.33 | 5.14 | 4.98 | 4.81 | 4.58 |
| 10 | 20 | 7.33 | 6.69 | 6.25 | 5.84 | 5.56 | 5.33 | 5.13 | 4.96 | 4.84 | 4.66 |
| 10 | 30 | 7.82 | 6.78 | 6.10 | 5.68 | 5.43 | 5.16 | 4.98 | 4.80 | 4.64 | 4.52 |
| 10 | 40 | 7.79 | 6.76 | 6.23 | 5.82 | 5.56 | 5.29 | 5.06 | 4.90 | 4.73 | 4.58 |
| 10 | 50 | 7.65 | 6.83 | 6.21 | 5.79 | 5.53 | 5.19 | 4.92 | 4.76 | 4.59 | 4.42 |
| 20 | 10 | 7.92 | 6.76 | 6.22 | 5.87 | 5.56 | 5.31 | 5.10 | 4.92 | 4.74 | 4.58 |
| 20 | 20 | 8.26 | 7.14 | 6.47 | 5.98 | 5.66 | 5.43 | 5.19 | 4.94 | 4.78 | 4.62 |
| 20 | 30 | 8.27 | 6.91 | 6.41 | 5.96 | 5.56 | 5.27 | 5.02 | 4.85 | 4.69 | 4.50 |
| 20 | 40 | 8.09 | 6.87 | 6.32 | 5.91 | 5.57 | 5.28 | 5.07 | 4.84 | 4.65 | 4.50 |
| 20 | 50 | 7.99 | 6.98 | 6.18 | 5.89 | 5.62 | 5.37 | 5.13 | 4.91 | 4.72 | 4.55 |
| 30 | 10 | 7.36 | 6.62 | 6.19 | 5.86 | 5.48 | 5.26 | 5.05 | 4.87 | 4.71 | 4.55 |
| 30 | 20 | 7.93 | 6.89 | 6.37 | 5.89 | 5.55 | 5.30 | 5.15 | 4.92 | 4.75 | 4.55 |
| 30 | 30 | 8.21 | 6.90 | 6.37 | 5.98 | 5.69 | 5.36 | 5.15 | 4.93 | 4.73 | 4.59 |
| 30 | 40 | 7.90 | 7.02 | 6.48 | 5.93 | 5.59 | 5.33 | 5.06 | 4.84 | 4.65 | 4.47 |
| 30 | 50 | 7.77 | 6.87 | 6.30 | 5.92 | 5.60 | 5.33 | 5.12 | 4.86 | 4.72 | 4.57 |
| 40 | 10 | 7.98 | 6.94 | 6.40 | 5.98 | 5.71 | 5.42 | 5.19 | 4.96 | 4.76 | 4.59 |
| 40 | 20 | 7.82 | 6.98 | 6.40 | 6.03 | 5.64 | 5.34 | 5.12 | 4.87 | 4.73 | 4.57 |
| 40 | 30 | 7.90 | 6.89 | 6.27 | 5.88 | 5.52 | 5.24 | 5.02 | 4.83 | 4.67 | 4.53 |
| 40 | 40 | 7.85 | 6.86 | 6.24 | 5.90 | 5.60 | 5.35 | 5.09 | 4.85 | 4.71 | 4.54 |
| 40 | 50 | 7.54 | 6.64 | 6.18 | 5.83 | 5.49 | 5.20 | 4.95 | 4.79 | 4.60 | 4.48 |
| 50 | 10 | 8.16 | 6.89 | 6.36 | 5.95 | 5.63 | 5.35 | 5.11 | 4.92 | 4.69 | 4.55 |
| 50 | 20 | 7.59 | 6.79 | 6.18 | 5.82 | 5.51 | 5.26 | 5.07 | 4.90 | 4.67 | 4.51 |
| 50 | 30 | 7.95 | 7.07 | 6.37 | 5.97 | 5.66 | 5.36 | 5.12 | 4.92 | 4.71 | 4.53 |
| 50 | 40 | 7.77 | 6.89 | 6.35 | 5.93 | 5.67 | 5.30 | 5.11 | 4.89 | 4.72 | 4.60 |
| 50 | 50 | 7.99 | 6.86 | 6.32 | 5.94 | 5.65 | 5.41 | 5.14 | 4.95 | 4.74 | 4.56 |

## Table 2.5.A
## Omnibus Rank Test for Type II Censored Observations

*Simulated critical values $k_\alpha$ according to $P_0\{ S > k_\alpha \} = \alpha$ for the omnibus two-sample rank statistic $S$ defined in Section 2.5, under the continuous null hypothesis $\mathcal{H}_0^r : F = G$ ( no ties ). The underlying bandwidth is $a = 0.40$ and the Monte Carlo sample size is 5000 .*

The fraction of censored observations is p = 0.10

| $m$ | $n$ | 0.01 | 0.02 | 0.03 | 0.04 | 0.05 | 0.06 | 0.07 | 0.08 | 0.09 | 0.10 |
|---|---|---|---|---|---|---|---|---|---|---|---|
| 10 | 10 | 8.50 | 7.50 | 7.04 | 6.65 | 6.34 | 6.11 | 5.84 | 5.65 | 5.48 | 5.34 |
| 10 | 20 | 8.55 | 7.63 | 7.07 | 6.64 | 6.39 | 6.14 | 5.84 | 5.67 | 5.51 | 5.38 |
| 10 | 30 | 8.64 | 7.58 | 7.02 | 6.68 | 6.37 | 6.07 | 5.85 | 5.67 | 5.49 | 5.35 |
| 10 | 40 | 9.05 | 8.07 | 7.41 | 6.88 | 6.54 | 6.23 | 5.93 | 5.73 | 5.54 | 5.34 |
| 10 | 50 | 9.14 | 8.07 | 7.47 | 6.91 | 6.51 | 6.20 | 5.95 | 5.74 | 5.56 | 5.37 |
| 20 | 10 | 9.11 | 7.95 | 7.36 | 6.93 | 6.58 | 6.22 | 5.99 | 5.75 | 5.56 | 5.41 |
| 20 | 20 | 8.83 | 7.81 | 7.28 | 6.85 | 6.42 | 6.15 | 5.91 | 5.67 | 5.52 | 5.37 |
| 20 | 30 | 9.04 | 7.98 | 7.36 | 6.78 | 6.49 | 6.13 | 5.91 | 5.68 | 5.47 | 5.29 |
| 20 | 40 | 8.69 | 7.67 | 7.06 | 6.67 | 6.37 | 6.10 | 5.88 | 5.67 | 5.50 | 5.29 |
| 20 | 50 | 9.14 | 7.91 | 7.20 | 6.73 | 6.41 | 6.13 | 5.87 | 5.67 | 5.49 | 5.33 |
| 30 | 10 | 8.79 | 7.84 | 7.23 | 6.82 | 6.57 | 6.29 | 6.07 | 5.89 | 5.68 | 5.47 |
| 30 | 20 | 8.88 | 7.75 | 7.30 | 6.87 | 6.52 | 6.28 | 6.05 | 5.81 | 5.63 | 5.44 |
| 30 | 30 | 9.16 | 8.26 | 7.64 | 7.10 | 6.61 | 6.26 | 6.03 | 5.82 | 5.54 | 5.28 |
| 30 | 40 | 9.32 | 8.24 | 7.64 | 7.15 | 6.75 | 6.44 | 6.19 | 5.87 | 5.66 | 5.48 |
| 30 | 50 | 9.26 | 8.04 | 7.53 | 7.08 | 6.66 | 6.33 | 6.00 | 5.73 | 5.59 | 5.44 |
| 40 | 10 | 8.84 | 7.79 | 7.29 | 6.74 | 6.33 | 6.03 | 5.82 | 5.58 | 5.37 | 5.22 |
| 40 | 20 | 9.04 | 8.12 | 7.41 | 6.94 | 6.61 | 6.31 | 6.07 | 5.86 | 5.63 | 5.48 |
| 40 | 30 | 8.92 | 8.01 | 7.44 | 6.95 | 6.58 | 6.28 | 6.01 | 5.75 | 5.56 | 5.38 |
| 40 | 40 | 9.42 | 8.16 | 7.58 | 6.94 | 6.58 | 6.24 | 5.94 | 5.73 | 5.54 | 5.32 |
| 40 | 50 | 9.23 | 8.20 | 7.59 | 7.11 | 6.72 | 6.42 | 6.18 | 5.87 | 5.67 | 5.43 |
| 50 | 10 | 8.72 | 7.78 | 7.17 | 6.65 | 6.32 | 6.09 | 5.83 | 5.58 | 5.36 | 5.19 |
| 50 | 20 | 9.12 | 7.89 | 7.32 | 6.88 | 6.57 | 6.24 | 6.00 | 5.80 | 5.58 | 5.44 |
| 50 | 30 | 9.16 | 8.09 | 7.48 | 7.03 | 6.65 | 6.35 | 6.12 | 5.86 | 5.65 | 5.49 |
| 50 | 40 | 9.06 | 7.82 | 7.33 | 6.96 | 6.54 | 6.29 | 5.97 | 5.73 | 5.55 | 5.37 |
| 50 | 50 | 9.15 | 8.08 | 7.45 | 6.97 | 6.56 | 6.22 | 5.98 | 5.78 | 5.56 | 5.40 |

## Table 2.5.A
## Continuation

*Simulated critical values $k_\alpha$ according to $P_0\{S > k_\alpha\} = \alpha$ for the omnibus two-sample rank statistic $S$ defined in Section 2.5, under the continuous null hypothesis $\mathcal{H}_0^r : F = G$ ( no ties ). The underlying bandwidth is $a = 0.40$ and the Monte Carlo sample size is 5000.*

The fraction of censored observations is p = 0.20

| $m$ | $n$ | 0.01 | 0.02 | 0.03 | 0.04 | 0.05 | 0.06 | 0.07 | 0.08 | 0.09 | 0.10 |
|----|----|------|------|------|------|------|------|------|------|------|------|
| 10 | 10 | 8.40 | 7.49 | 6.90 | 6.51 | 6.24 | 5.97 | 5.73 | 5.49 | 5.34 | 5.21 |
| 10 | 20 | 8.66 | 7.61 | 7.06 | 6.70 | 6.35 | 6.07 | 5.85 | 5.61 | 5.43 | 5.23 |
| 10 | 30 | 8.45 | 7.76 | 7.25 | 6.79 | 6.39 | 6.12 | 5.83 | 5.63 | 5.48 | 5.31 |
| 10 | 40 | 8.76 | 7.74 | 7.05 | 6.63 | 6.30 | 6.04 | 5.82 | 5.61 | 5.43 | 5.27 |
| 10 | 50 | 8.86 | 7.79 | 7.04 | 6.54 | 6.23 | 5.95 | 5.73 | 5.54 | 5.32 | 5.15 |
| 20 | 10 | 8.71 | 7.73 | 7.16 | 6.71 | 6.46 | 6.13 | 5.86 | 5.67 | 5.49 | 5.25 |
| 20 | 20 | 8.85 | 7.80 | 7.06 | 6.68 | 6.26 | 6.07 | 5.80 | 5.60 | 5.39 | 5.20 |
| 20 | 30 | 8.65 | 7.89 | 7.11 | 6.70 | 6.38 | 6.11 | 5.91 | 5.72 | 5.52 | 5.30 |
| 20 | 40 | 8.75 | 7.77 | 7.05 | 6.59 | 6.23 | 5.93 | 5.66 | 5.48 | 5.23 | 5.10 |
| 20 | 50 | 8.96 | 7.72 | 7.23 | 6.78 | 6.43 | 6.10 | 5.80 | 5.53 | 5.33 | 5.17 |
| 30 | 10 | 8.61 | 7.51 | 6.98 | 6.54 | 6.15 | 5.88 | 5.69 | 5.50 | 5.35 | 5.21 |
| 30 | 20 | 9.04 | 7.91 | 7.27 | 6.87 | 6.47 | 6.13 | 5.85 | 5.60 | 5.44 | 5.31 |
| 30 | 30 | 9.12 | 7.96 | 7.24 | 6.75 | 6.43 | 6.11 | 5.87 | 5.64 | 5.44 | 5.24 |
| 30 | 40 | 8.72 | 7.55 | 7.08 | 6.65 | 6.29 | 6.00 | 5.76 | 5.55 | 5.37 | 5.20 |
| 30 | 50 | 9.09 | 7.71 | 7.09 | 6.71 | 6.42 | 6.10 | 5.82 | 5.60 | 5.38 | 5.21 |
| 40 | 10 | 8.77 | 7.69 | 7.03 | 6.45 | 6.07 | 5.82 | 5.56 | 5.39 | 5.23 | 5.07 |
| 40 | 20 | 8.84 | 7.78 | 7.14 | 6.73 | 6.32 | 6.09 | 5.83 | 5.64 | 5.43 | 5.25 |
| 40 | 30 | 8.81 | 8.00 | 7.27 | 6.75 | 6.33 | 5.90 | 5.68 | 5.49 | 5.27 | 5.14 |
| 40 | 40 | 9.09 | 7.86 | 7.16 | 6.78 | 6.45 | 6.18 | 5.86 | 5.60 | 5.40 | 5.19 |
| 40 | 50 | 8.70 | 7.60 | 7.08 | 6.67 | 6.20 | 5.89 | 5.59 | 5.43 | 5.25 | 5.09 |
| 50 | 10 | 9.18 | 7.83 | 7.13 | 6.67 | 6.33 | 5.95 | 5.71 | 5.51 | 5.30 | 5.15 |
| 50 | 20 | 8.66 | 7.80 | 7.29 | 6.78 | 6.38 | 6.08 | 5.81 | 5.59 | 5.41 | 5.22 |
| 50 | 30 | 8.56 | 7.44 | 6.98 | 6.61 | 6.30 | 5.97 | 5.71 | 5.50 | 5.33 | 5.15 |
| 50 | 40 | 9.01 | 7.90 | 7.19 | 6.76 | 6.33 | 6.07 | 5.80 | 5.53 | 5.33 | 5.16 |
| 50 | 50 | 9.34 | 7.93 | 7.21 | 6.73 | 6.38 | 6.04 | 5.73 | 5.54 | 5.38 | 5.16 |

## Table 2.5.A
## Continuation

*Simulated critical values $k_\alpha$ according to $P_0\{ S > k_\alpha \} = \alpha$ for the omnibus two-sample rank statistic $S$ defined in Section 2.5, under the continuous null hypothesis $\mathcal{H}_0^r : F = G$ ( no ties ). The underlying bandwidth is $a = 0.40$ and the Monte Carlo sample size is 5000.*

The fraction of censored observations is p = 0.30

| $m$ | $n$ | 0.01 | 0.02 | 0.03 | 0.04 | 0.05 | 0.06 | 0.07 | 0.08 | 0.09 | 0.10 |
|---|---|---|---|---|---|---|---|---|---|---|---|
| 10 | 10 | 8.43 | 7.33 | 6.68 | 6.31 | 5.89 | 5.60 | 5.41 | 5.19 | 5.02 | 4.85 |
| 10 | 20 | 8.29 | 7.35 | 6.72 | 6.20 | 5.85 | 5.52 | 5.28 | 5.08 | 4.88 | 4.73 |
| 10 | 30 | 8.01 | 7.16 | 6.58 | 6.16 | 5.78 | 5.54 | 5.35 | 5.11 | 4.92 | 4.79 |
| 10 | 40 | 8.39 | 7.34 | 6.76 | 6.22 | 5.85 | 5.55 | 5.32 | 5.15 | 4.98 | 4.81 |
| 10 | 50 | 8.29 | 7.20 | 6.51 | 6.05 | 5.69 | 5.50 | 5.30 | 5.08 | 4.91 | 4.73 |
| 20 | 10 | 8.07 | 7.16 | 6.67 | 6.26 | 5.99 | 5.75 | 5.48 | 5.26 | 5.06 | 4.88 |
| 20 | 20 | 8.40 | 7.45 | 6.89 | 6.44 | 6.09 | 5.75 | 5.53 | 5.26 | 5.10 | 4.86 |
| 20 | 30 | 8.74 | 7.59 | 6.84 | 6.23 | 5.89 | 5.60 | 5.36 | 5.21 | 5.05 | 4.89 |
| 20 | 40 | 8.69 | 7.55 | 6.87 | 6.45 | 6.07 | 5.71 | 5.44 | 5.22 | 4.98 | 4.84 |
| 20 | 50 | 8.40 | 7.23 | 6.72 | 6.34 | 5.85 | 5.60 | 5.39 | 5.19 | 5.02 | 4.84 |
| 30 | 10 | 8.69 | 7.48 | 6.79 | 6.34 | 5.98 | 5.70 | 5.45 | 5.26 | 5.07 | 4.86 |
| 30 | 20 | 7.96 | 7.09 | 6.44 | 6.05 | 5.75 | 5.49 | 5.23 | 5.06 | 4.92 | 4.76 |
| 30 | 30 | 8.78 | 7.47 | 6.78 | 6.29 | 5.98 | 5.74 | 5.49 | 5.26 | 5.06 | 4.90 |
| 30 | 40 | 8.53 | 7.33 | 6.81 | 6.33 | 5.97 | 5.69 | 5.47 | 5.25 | 5.01 | 4.85 |
| 30 | 50 | 8.37 | 7.50 | 6.89 | 6.40 | 6.00 | 5.60 | 5.38 | 5.20 | 5.00 | 4.88 |
| 40 | 10 | 8.39 | 7.25 | 6.58 | 6.07 | 5.73 | 5.46 | 5.24 | 5.06 | 4.88 | 4.73 |
| 40 | 20 | 8.27 | 7.51 | 6.87 | 6.40 | 6.00 | 5.75 | 5.48 | 5.24 | 5.06 | 4.91 |
| 40 | 30 | 8.37 | 7.30 | 6.69 | 6.35 | 6.03 | 5.67 | 5.40 | 5.18 | 4.98 | 4.83 |
| 40 | 40 | 8.55 | 7.46 | 6.88 | 6.43 | 6.01 | 5.70 | 5.38 | 5.16 | 4.94 | 4.75 |
| 40 | 50 | 8.44 | 7.24 | 6.68 | 6.30 | 5.92 | 5.61 | 5.35 | 5.17 | 4.96 | 4.83 |
| 50 | 10 | 8.40 | 7.38 | 6.52 | 6.07 | 5.70 | 5.45 | 5.21 | 5.00 | 4.85 | 4.71 |
| 50 | 20 | 8.79 | 7.51 | 6.78 | 6.32 | 6.02 | 5.72 | 5.42 | 5.23 | 4.99 | 4.80 |
| 50 | 30 | 8.20 | 7.35 | 6.66 | 6.18 | 5.82 | 5.45 | 5.24 | 5.05 | 4.86 | 4.74 |
| 50 | 40 | 8.67 | 7.22 | 6.65 | 6.22 | 5.89 | 5.61 | 5.37 | 5.19 | 4.99 | 4.81 |
| 50 | 50 | 8.50 | 7.58 | 6.71 | 6.22 | 5.91 | 5.57 | 5.37 | 5.09 | 4.89 | 4.72 |

## Table 2.5.B
## One-sided Rank Test for Type II Censored Observations

*Simulated critical values $k_\alpha^0$ according to $P_0\{ S^0 > k_\alpha^0 \} = \alpha$ for the one-sided two-sample rank statistic $S^0$ defined in Section 2.5, under the continuous null hypothesis $\mathcal{H}_0^r : F = G$ ( no ties ). The underlying bandwidth is $a = 0.40$ and the Monte Carlo sample size is 5000 .*

The fraction of censored observations is p $= 0.10$

| $m$ | $n$ | 0.01 | 0.02 | 0.03 | 0.04 | 0.05 | 0.06 | 0.07 | 0.08 | 0.09 | 0.10 |
|---|---|---|---|---|---|---|---|---|---|---|---|
| 10 | 10 | 6.99 | 6.19 | 5.57 | 5.10 | 4.78 | 4.47 | 4.24 | 3.99 | 3.82 | 3.69 |
| 10 | 20 | 7.46 | 6.31 | 5.73 | 5.34 | 4.87 | 4.54 | 4.33 | 4.10 | 3.93 | 3.74 |
| 10 | 30 | 6.89 | 5.98 | 5.44 | 5.05 | 4.65 | 4.44 | 4.17 | 3.94 | 3.72 | 3.57 |
| 10 | 40 | 7.24 | 6.17 | 5.55 | 5.00 | 4.70 | 4.26 | 3.99 | 3.82 | 3.64 | 3.50 |
| 10 | 50 | 7.55 | 6.31 | 5.64 | 5.13 | 4.83 | 4.54 | 4.18 | 3.96 | 3.75 | 3.58 |
| 20 | 10 | 7.01 | 6.03 | 5.53 | 5.16 | 4.80 | 4.56 | 4.25 | 4.01 | 3.81 | 3.64 |
| 20 | 20 | 7.94 | 6.67 | 6.04 | 5.51 | 5.11 | 4.84 | 4.53 | 4.30 | 4.04 | 3.80 |
| 20 | 30 | 6.98 | 6.18 | 5.67 | 5.20 | 4.89 | 4.57 | 4.28 | 4.06 | 3.88 | 3.68 |
| 20 | 40 | 7.26 | 6.18 | 5.53 | 5.14 | 4.81 | 4.53 | 4.25 | 4.01 | 3.82 | 3.65 |
| 20 | 50 | 7.90 | 6.62 | 5.79 | 5.31 | 4.91 | 4.51 | 4.26 | 4.05 | 3.89 | 3.70 |
| 30 | 10 | 7.34 | 6.14 | 5.60 | 5.17 | 4.84 | 4.59 | 4.33 | 4.13 | 3.92 | 3.78 |
| 30 | 20 | 7.65 | 6.46 | 5.98 | 5.47 | 5.05 | 4.68 | 4.37 | 4.17 | 4.00 | 3.84 |
| 30 | 30 | 7.50 | 6.38 | 5.80 | 5.30 | 4.97 | 4.63 | 4.37 | 4.10 | 3.85 | 3.72 |
| 30 | 40 | 7.40 | 6.28 | 5.51 | 5.09 | 4.70 | 4.35 | 4.06 | 3.86 | 3.67 | 3.52 |
| 30 | 50 | 8.00 | 6.55 | 5.80 | 5.27 | 4.94 | 4.57 | 4.28 | 4.08 | 3.86 | 3.68 |
| 40 | 10 | 7.11 | 6.18 | 5.50 | 5.00 | 4.75 | 4.43 | 4.17 | 4.01 | 3.82 | 3.66 |
| 40 | 20 | 7.10 | 6.12 | 5.51 | 5.04 | 4.71 | 4.45 | 4.21 | 3.98 | 3.77 | 3.61 |
| 40 | 30 | 7.49 | 6.31 | 5.63 | 5.21 | 4.88 | 4.64 | 4.38 | 4.14 | 3.93 | 3.74 |
| 40 | 40 | 7.37 | 6.39 | 5.65 | 5.01 | 4.59 | 4.31 | 4.05 | 3.83 | 3.68 | 3.55 |
| 40 | 50 | 7.47 | 6.49 | 5.82 | 5.30 | 4.98 | 4.57 | 4.28 | 3.99 | 3.78 | 3.62 |
| 50 | 10 | 7.60 | 6.14 | 5.67 | 5.32 | 4.92 | 4.58 | 4.33 | 4.19 | 3.98 | 3.76 |
| 50 | 20 | 7.55 | 6.32 | 5.58 | 5.20 | 4.88 | 4.53 | 4.33 | 4.12 | 3.95 | 3.70 |
| 50 | 30 | 7.75 | 6.32 | 5.65 | 5.14 | 4.75 | 4.49 | 4.32 | 4.08 | 3.94 | 3.71 |
| 50 | 40 | 7.37 | 6.23 | 5.70 | 5.14 | 4.81 | 4.44 | 4.22 | 3.99 | 3.78 | 3.61 |
| 50 | 50 | 7.70 | 6.58 | 5.76 | 5.26 | 4.88 | 4.54 | 4.29 | 4.05 | 3.89 | 3.74 |

**Table 2.5.B**
**Continuation**

*Simulated critical values $k_\alpha^0$ according to $P_0\{S^0 > k_\alpha^0\} = \alpha$ for the one-sided two-sample rank statistic $S^0$ defined in Section 2.5, under the continuous null hypothesis $\mathcal{H}_0^r : F = G$ ( no ties ). The underlying bandwidth is $a = 0.40$ and the Monte Carlo sample size is $5000$.*

The fraction of censored observations is p $= 0.20$

| m | n | 0.01 | 0.02 | 0.03 | 0.04 | 0.05 | 0.06 | 0.07 | 0.08 | 0.09 | 0.10 |
|---|---|------|------|------|------|------|------|------|------|------|------|
| 10 | 10 | 7.03 | 6.02 | 5.43 | 5.02 | 4.57 | 4.36 | 4.14 | 3.94 | 3.80 | 3.63 |
| 10 | 20 | 7.05 | 6.06 | 5.50 | 5.04 | 4.68 | 4.40 | 4.20 | 4.00 | 3.83 | 3.65 |
| 10 | 30 | 7.13 | 6.16 | 5.45 | 4.97 | 4.58 | 4.32 | 4.05 | 3.90 | 3.68 | 3.50 |
| 10 | 40 | 6.70 | 5.88 | 5.37 | 4.90 | 4.57 | 4.23 | 3.98 | 3.72 | 3.51 | 3.39 |
| 10 | 50 | 7.35 | 5.95 | 5.36 | 4.94 | 4.54 | 4.27 | 3.97 | 3.70 | 3.52 | 3.34 |
| 20 | 10 | 7.28 | 6.30 | 5.78 | 5.31 | 4.98 | 4.66 | 4.38 | 4.17 | 3.98 | 3.75 |
| 20 | 20 | 7.66 | 6.37 | 5.69 | 5.16 | 4.82 | 4.49 | 4.19 | 3.95 | 3.73 | 3.53 |
| 20 | 30 | 7.22 | 6.05 | 5.35 | 4.96 | 4.69 | 4.38 | 4.08 | 3.91 | 3.73 | 3.57 |
| 20 | 40 | 7.21 | 6.43 | 5.64 | 5.17 | 4.72 | 4.47 | 4.20 | 4.00 | 3.77 | 3.59 |
| 20 | 50 | 7.51 | 6.58 | 5.52 | 4.99 | 4.55 | 4.29 | 4.04 | 3.84 | 3.65 | 3.48 |
| 30 | 10 | 7.25 | 6.15 | 5.46 | 5.05 | 4.70 | 4.44 | 4.16 | 3.99 | 3.81 | 3.66 |
| 30 | 20 | 7.40 | 6.32 | 5.63 | 5.22 | 4.84 | 4.51 | 4.21 | 4.04 | 3.82 | 3.56 |
| 30 | 30 | 7.79 | 6.33 | 5.67 | 5.21 | 4.78 | 4.48 | 4.27 | 4.02 | 3.85 | 3.64 |
| 30 | 40 | 7.26 | 6.02 | 5.49 | 5.12 | 4.70 | 4.39 | 4.13 | 3.95 | 3.77 | 3.58 |
| 30 | 50 | 7.24 | 6.24 | 5.68 | 5.26 | 4.84 | 4.53 | 4.23 | 3.98 | 3.77 | 3.61 |
| 40 | 10 | 7.77 | 6.43 | 5.70 | 5.24 | 4.87 | 4.57 | 4.34 | 4.12 | 3.93 | 3.73 |
| 40 | 20 | 7.87 | 6.66 | 5.90 | 5.37 | 5.01 | 4.69 | 4.35 | 4.10 | 3.90 | 3.68 |
| 40 | 30 | 7.55 | 6.33 | 5.64 | 5.22 | 4.81 | 4.48 | 4.21 | 3.97 | 3.76 | 3.57 |
| 40 | 40 | 7.12 | 6.18 | 5.51 | 5.03 | 4.65 | 4.32 | 4.00 | 3.77 | 3.56 | 3.41 |
| 40 | 50 | 7.21 | 6.00 | 5.43 | 4.88 | 4.50 | 4.15 | 3.97 | 3.79 | 3.58 | 3.43 |
| 50 | 10 | 7.56 | 6.41 | 5.75 | 5.20 | 4.76 | 4.42 | 4.17 | 3.96 | 3.78 | 3.63 |
| 50 | 20 | 7.74 | 6.49 | 5.76 | 5.42 | 5.01 | 4.70 | 4.45 | 4.16 | 3.90 | 3.70 |
| 50 | 30 | 7.07 | 6.03 | 5.36 | 4.95 | 4.61 | 4.33 | 4.10 | 3.87 | 3.69 | 3.51 |
| 50 | 40 | 7.29 | 6.31 | 5.45 | 5.01 | 4.65 | 4.35 | 4.09 | 3.83 | 3.65 | 3.49 |
| 50 | 50 | 7.35 | 6.32 | 5.59 | 5.08 | 4.72 | 4.44 | 4.17 | 3.88 | 3.67 | 3.49 |

## Table 2.5.B
## Continuation

*Simulated critical values $k_\alpha^0$ according to $P_0\{\,S^0 > k_\alpha^0\,\} = \alpha$ for the one-sided two-sample rank statistic $S^0$ defined in Section 2.5, under the continuous null hypothesis $\mathcal{H}_0^r : F = G$ ( no ties ). The underlying bandwidth is $a = 0.40$ and the Monte Carlo sample size is $5000$.*

The fraction of censored observations is p $= 0.30$

| $m$ | $n$ | 0.01 | 0.02 | 0.03 | 0.04 | 0.05 | 0.06 | 0.07 | 0.08 | 0.09 | 0.10 |
|---|---|---|---|---|---|---|---|---|---|---|---|
| 10 | 10 | 6.58 | 5.93 | 5.34 | 4.98 | 4.52 | 4.26 | 4.02 | 3.76 | 3.63 | 3.52 |
| 10 | 20 | 6.49 | 5.45 | 5.02 | 4.65 | 4.35 | 4.06 | 3.85 | 3.68 | 3.44 | 3.28 |
| 10 | 30 | 6.45 | 5.44 | 5.02 | 4.69 | 4.37 | 4.05 | 3.81 | 3.58 | 3.39 | 3.23 |
| 10 | 40 | 6.52 | 5.39 | 4.89 | 4.48 | 4.12 | 3.83 | 3.57 | 3.39 | 3.21 | 3.04 |
| 10 | 50 | 6.42 | 5.38 | 4.93 | 4.50 | 4.19 | 3.93 | 3.75 | 3.53 | 3.33 | 3.16 |
| 20 | 10 | 7.34 | 6.12 | 5.59 | 5.12 | 4.71 | 4.36 | 4.05 | 3.86 | 3.59 | 3.39 |
| 20 | 20 | 6.84 | 5.76 | 5.16 | 4.71 | 4.36 | 4.06 | 3.85 | 3.65 | 3.47 | 3.26 |
| 20 | 30 | 7.25 | 6.16 | 5.47 | 4.93 | 4.54 | 4.22 | 3.96 | 3.71 | 3.52 | 3.38 |
| 20 | 40 | 6.56 | 5.54 | 4.95 | 4.60 | 4.27 | 3.98 | 3.75 | 3.54 | 3.34 | 3.16 |
| 20 | 50 | 6.68 | 5.55 | 5.03 | 4.64 | 4.28 | 4.02 | 3.82 | 3.61 | 3.44 | 3.26 |
| 30 | 10 | 7.40 | 6.18 | 5.48 | 5.01 | 4.66 | 4.39 | 4.14 | 3.84 | 3.66 | 3.46 |
| 30 | 20 | 6.97 | 5.86 | 5.24 | 4.75 | 4.36 | 4.10 | 3.87 | 3.68 | 3.51 | 3.27 |
| 30 | 30 | 7.38 | 5.93 | 5.19 | 4.82 | 4.51 | 4.16 | 3.93 | 3.69 | 3.50 | 3.37 |
| 30 | 40 | 6.90 | 5.96 | 5.30 | 4.82 | 4.40 | 4.07 | 3.87 | 3.64 | 3.47 | 3.32 |
| 30 | 50 | 6.73 | 5.81 | 5.19 | 4.70 | 4.44 | 4.08 | 3.84 | 3.62 | 3.44 | 3.26 |
| 40 | 10 | 7.18 | 6.28 | 5.57 | 5.10 | 4.68 | 4.24 | 3.99 | 3.77 | 3.53 | 3.35 |
| 40 | 20 | 7.26 | 5.93 | 5.33 | 4.83 | 4.42 | 4.10 | 3.79 | 3.60 | 3.41 | 3.24 |
| 40 | 30 | 7.33 | 6.12 | 5.46 | 4.98 | 4.65 | 4.31 | 4.11 | 3.90 | 3.66 | 3.48 |
| 40 | 40 | 7.39 | 6.31 | 5.53 | 4.99 | 4.67 | 4.29 | 4.05 | 3.80 | 3.62 | 3.43 |
| 40 | 50 | 7.21 | 6.12 | 5.31 | 4.85 | 4.47 | 4.14 | 3.95 | 3.76 | 3.57 | 3.39 |
| 50 | 10 | 7.15 | 6.06 | 5.39 | 4.89 | 4.61 | 4.27 | 4.03 | 3.85 | 3.66 | 3.47 |
| 50 | 20 | 7.25 | 6.23 | 5.54 | 5.05 | 4.52 | 4.15 | 3.90 | 3.61 | 3.42 | 3.25 |
| 50 | 30 | 6.73 | 5.74 | 5.07 | 4.74 | 4.48 | 4.14 | 3.82 | 3.62 | 3.47 | 3.28 |
| 50 | 40 | 7.10 | 5.91 | 5.34 | 4.97 | 4.64 | 4.25 | 4.01 | 3.78 | 3.52 | 3.34 |
| 50 | 50 | 7.00 | 5.78 | 5.23 | 4.69 | 4.42 | 4.09 | 3.81 | 3.55 | 3.35 | 3.19 |

## Table 2.5.C
## Projection Rank Test for Type II Censored Observations

*Simulated critical values* $k_\alpha^\pi$ *according to* $P_0\{ S^\pi > k_\alpha^\pi \} = \alpha$ *for the two-sample projection rank statistic* $S^\pi$ *defined in Section 2.5, under the continuous null hypothesis* $\mathcal{H}_0^r : F = G$ *( no ties ). The Monte Carlo sample size is* 5000 *.*

The fraction of censored observations is p $= 0.10$

| $m$ | $n$ | 0.01 | 0.02 | 0.03 | 0.04 | 0.05 | 0.06 | 0.07 | 0.08 | 0.09 | 0.10 |
|----|----|------|------|------|------|------|------|------|------|------|------|
| 10 | 10 | 7.42 | 6.00 | 5.24 | 4.76 | 4.43 | 4.12 | 3.89 | 3.68 | 3.52 | 3.30 |
| 10 | 20 | 7.06 | 5.87 | 5.16 | 4.68 | 4.36 | 4.08 | 3.84 | 3.62 | 3.39 | 3.21 |
| 10 | 30 | 7.61 | 6.38 | 5.52 | 4.78 | 4.36 | 4.03 | 3.78 | 3.57 | 3.31 | 3.11 |
| 10 | 40 | 7.35 | 6.18 | 5.37 | 4.78 | 4.46 | 4.10 | 3.81 | 3.59 | 3.39 | 3.15 |
| 10 | 50 | 7.08 | 5.90 | 5.25 | 4.76 | 4.24 | 3.88 | 3.68 | 3.38 | 3.19 | 3.02 |
| 20 | 10 | 7.48 | 6.33 | 5.56 | 5.01 | 4.62 | 4.24 | 3.94 | 3.68 | 3.45 | 3.27 |
| 20 | 20 | 6.93 | 5.81 | 5.13 | 4.54 | 4.19 | 3.87 | 3.63 | 3.40 | 3.17 | 3.04 |
| 20 | 30 | 7.37 | 6.02 | 5.16 | 4.71 | 4.37 | 4.00 | 3.70 | 3.45 | 3.23 | 3.06 |
| 20 | 40 | 7.31 | 5.88 | 5.13 | 4.62 | 4.30 | 3.97 | 3.65 | 3.44 | 3.21 | 3.04 |
| 20 | 50 | 7.53 | 6.06 | 5.38 | 4.82 | 4.37 | 4.01 | 3.78 | 3.54 | 3.26 | 3.10 |
| 30 | 10 | 7.23 | 6.28 | 5.38 | 4.82 | 4.49 | 4.23 | 3.99 | 3.77 | 3.52 | 3.32 |
| 30 | 20 | 7.59 | 6.23 | 5.62 | 4.97 | 4.59 | 4.19 | 3.82 | 3.58 | 3.36 | 3.16 |
| 30 | 30 | 7.52 | 6.09 | 5.36 | 4.84 | 4.44 | 4.09 | 3.85 | 3.63 | 3.43 | 3.27 |
| 30 | 40 | 7.52 | 6.38 | 5.72 | 4.95 | 4.46 | 4.03 | 3.70 | 3.41 | 3.17 | 2.99 |
| 30 | 50 | 7.25 | 6.26 | 5.54 | 4.90 | 4.45 | 4.15 | 3.87 | 3.65 | 3.42 | 3.23 |
| 40 | 10 | 7.75 | 6.12 | 5.44 | 4.91 | 4.57 | 4.14 | 3.87 | 3.61 | 3.40 | 3.22 |
| 40 | 20 | 7.36 | 6.26 | 5.41 | 4.93 | 4.51 | 4.16 | 3.87 | 3.56 | 3.35 | 3.15 |
| 40 | 30 | 7.23 | 6.04 | 5.35 | 4.71 | 4.34 | 4.00 | 3.75 | 3.51 | 3.33 | 3.13 |
| 40 | 40 | 7.26 | 5.90 | 5.19 | 4.66 | 4.30 | 3.99 | 3.70 | 3.44 | 3.18 | 3.00 |
| 40 | 50 | 7.56 | 6.27 | 5.55 | 4.95 | 4.51 | 4.10 | 3.82 | 3.56 | 3.36 | 3.15 |
| 50 | 10 | 7.83 | 6.36 | 5.62 | 5.07 | 4.66 | 4.23 | 3.91 | 3.65 | 3.42 | 3.24 |
| 50 | 20 | 8.13 | 6.48 | 5.53 | 4.96 | 4.47 | 4.10 | 3.78 | 3.52 | 3.30 | 3.15 |
| 50 | 30 | 7.30 | 5.98 | 5.25 | 4.65 | 4.28 | 3.99 | 3.68 | 3.39 | 3.21 | 3.09 |
| 50 | 40 | 7.30 | 5.78 | 5.06 | 4.63 | 4.17 | 3.90 | 3.62 | 3.43 | 3.21 | 3.05 |
| 50 | 50 | 7.81 | 6.38 | 5.43 | 4.87 | 4.40 | 4.02 | 3.71 | 3.44 | 3.23 | 3.05 |

**Table 2.5.C**
**Continuation**

*Simulated critical values $k_\alpha^\pi$ according to $P_0\{S^\pi > k_\alpha^\pi\} = \alpha$ for the two-sample projection rank statistic $S^\pi$ defined in Section 2.5, under the continuous null hypothesis $\mathcal{H}_0^\tau : F = G$ ( no ties ). The Monte Carlo sample size is 5000.*

The fraction of censored observations is p $= 0.20$

| $m$ | $n$ | 0.01 | 0.02 | 0.03 | 0.04 | 0.05 | 0.06 | 0.07 | 0.08 | 0.09 | 0.10 |
|---|---|---|---|---|---|---|---|---|---|---|---|
| 10 | 10 | 6.92 | 5.89 | 5.29 | 4.83 | 4.43 | 4.14 | 3.81 | 3.52 | 3.28 | 3.14 |
| 10 | 20 | 7.34 | 6.04 | 5.22 | 4.69 | 4.35 | 4.03 | 3.76 | 3.51 | 3.31 | 3.14 |
| 10 | 30 | 7.85 | 6.41 | 5.57 | 4.95 | 4.50 | 4.19 | 3.89 | 3.64 | 3.47 | 3.30 |
| 10 | 40 | 7.12 | 5.94 | 5.23 | 4.72 | 4.25 | 3.95 | 3.68 | 3.42 | 3.23 | 3.03 |
| 10 | 50 | 7.19 | 5.87 | 5.10 | 4.65 | 4.19 | 3.93 | 3.61 | 3.34 | 3.15 | 2.98 |
| 20 | 10 | 7.26 | 6.16 | 5.46 | 4.88 | 4.40 | 3.96 | 3.74 | 3.51 | 3.28 | 3.13 |
| 20 | 20 | 7.51 | 6.14 | 5.25 | 4.77 | 4.40 | 4.04 | 3.78 | 3.48 | 3.27 | 3.10 |
| 20 | 30 | 7.97 | 6.44 | 5.63 | 4.97 | 4.47 | 4.06 | 3.78 | 3.52 | 3.22 | 3.03 |
| 20 | 40 | 6.99 | 5.57 | 4.87 | 4.49 | 4.13 | 3.84 | 3.54 | 3.33 | 3.14 | 2.96 |
| 20 | 50 | 7.26 | 5.96 | 5.30 | 4.78 | 4.34 | 4.05 | 3.81 | 3.54 | 3.27 | 3.11 |
| 30 | 10 | 7.93 | 6.48 | 5.60 | 5.03 | 4.45 | 4.12 | 3.88 | 3.62 | 3.45 | 3.23 |
| 30 | 20 | 7.47 | 6.39 | 5.60 | 4.98 | 4.56 | 4.16 | 3.83 | 3.58 | 3.29 | 3.09 |
| 30 | 30 | 7.53 | 6.10 | 5.29 | 4.71 | 4.28 | 3.83 | 3.57 | 3.32 | 3.06 | 2.90 |
| 30 | 40 | 7.57 | 6.25 | 5.38 | 4.74 | 4.36 | 4.05 | 3.71 | 3.45 | 3.21 | 3.02 |
| 30 | 50 | 7.46 | 6.12 | 5.42 | 4.96 | 4.49 | 4.19 | 3.81 | 3.54 | 3.32 | 3.15 |
| 40 | 10 | 7.55 | 6.01 | 5.23 | 4.82 | 4.40 | 4.06 | 3.82 | 3.58 | 3.37 | 3.16 |
| 40 | 20 | 7.66 | 6.32 | 5.44 | 4.77 | 4.44 | 4.11 | 3.80 | 3.52 | 3.32 | 3.15 |
| 40 | 30 | 7.11 | 5.74 | 5.10 | 4.61 | 4.28 | 3.95 | 3.70 | 3.53 | 3.31 | 3.15 |
| 40 | 40 | 7.49 | 5.90 | 5.29 | 4.70 | 4.37 | 4.02 | 3.79 | 3.52 | 3.32 | 3.13 |
| 40 | 50 | 7.56 | 6.22 | 5.32 | 4.76 | 4.33 | 3.95 | 3.68 | 3.42 | 3.21 | 3.01 |
| 50 | 10 | 8.18 | 6.49 | 5.65 | 4.97 | 4.60 | 4.25 | 3.99 | 3.75 | 3.52 | 3.31 |
| 50 | 20 | 7.28 | 6.15 | 5.46 | 4.88 | 4.39 | 4.02 | 3.77 | 3.49 | 3.26 | 3.10 |
| 50 | 30 | 7.44 | 6.06 | 5.32 | 4.66 | 4.25 | 3.96 | 3.63 | 3.37 | 3.17 | 3.00 |
| 50 | 40 | 7.46 | 6.04 | 5.28 | 4.86 | 4.45 | 4.09 | 3.75 | 3.53 | 3.28 | 3.08 |
| 50 | 50 | 7.51 | 6.28 | 5.34 | 4.88 | 4.32 | 3.99 | 3.79 | 3.49 | 3.22 | 3.06 |

## Table 2.5.C
## Continuation

*Simulated critical values $k_\alpha^\pi$ according to $P_0\{S^\pi > k_\alpha^\pi\} = \alpha$ for the two-sample projection rank statistic $S^\pi$ defined in Section 2.5, under the continuous null hypothesis $\mathcal{H}_0^r : F = G$ ( no ties ). The Monte Carlo sample size is 5000 .*

The fraction of censored observations is p $= 0.30$

| m | n | 0.01 | 0.02 | 0.03 | 0.04 | 0.05 | 0.06 | 0.07 | 0.08 | 0.09 | 0.10 |
|---|---|------|------|------|------|------|------|------|------|------|------|
| 10 | 10 | 6.69 | 5.52 | 4.93 | 4.53 | 4.18 | 3.96 | 3.67 | 3.43 | 3.21 | 3.05 |
| 10 | 20 | 7.19 | 5.93 | 5.11 | 4.60 | 4.18 | 3.80 | 3.54 | 3.31 | 3.14 | 2.91 |
| 10 | 30 | 6.72 | 5.75 | 4.99 | 4.55 | 4.15 | 3.80 | 3.52 | 3.29 | 3.04 | 2.86 |
| 10 | 40 | 7.38 | 5.81 | 5.09 | 4.51 | 4.13 | 3.81 | 3.54 | 3.32 | 3.15 | 2.96 |
| 10 | 50 | 7.28 | 5.76 | 5.05 | 4.53 | 4.12 | 3.84 | 3.57 | 3.32 | 3.14 | 2.94 |
| 20 | 10 | 7.41 | 6.20 | 5.49 | 4.76 | 4.35 | 4.01 | 3.77 | 3.51 | 3.30 | 3.09 |
| 20 | 20 | 6.99 | 5.82 | 5.10 | 4.53 | 4.10 | 3.76 | 3.45 | 3.27 | 3.09 | 2.89 |
| 20 | 30 | 7.40 | 6.20 | 5.43 | 4.93 | 4.42 | 4.03 | 3.74 | 3.50 | 3.26 | 3.11 |
| 20 | 40 | 7.12 | 5.84 | 5.23 | 4.63 | 4.30 | 3.95 | 3.72 | 3.41 | 3.18 | 2.95 |
| 20 | 50 | 6.74 | 5.79 | 5.16 | 4.59 | 4.25 | 3.89 | 3.62 | 3.38 | 3.18 | 3.04 |
| 30 | 10 | 7.72 | 6.11 | 5.37 | 4.82 | 4.34 | 3.90 | 3.63 | 3.44 | 3.25 | 3.12 |
| 30 | 20 | 7.69 | 6.24 | 5.20 | 4.67 | 4.27 | 3.92 | 3.64 | 3.41 | 3.24 | 3.08 |
| 30 | 30 | 7.74 | 6.23 | 5.31 | 4.70 | 4.30 | 3.91 | 3.62 | 3.40 | 3.18 | 2.99 |
| 30 | 40 | 7.01 | 5.74 | 4.98 | 4.47 | 3.98 | 3.76 | 3.49 | 3.28 | 3.12 | 2.95 |
| 30 | 50 | 7.21 | 5.97 | 5.32 | 4.74 | 4.37 | 4.10 | 3.78 | 3.53 | 3.29 | 3.05 |
| 40 | 10 | 7.28 | 6.05 | 5.47 | 4.90 | 4.48 | 4.11 | 3.79 | 3.55 | 3.34 | 3.15 |
| 40 | 20 | 7.23 | 5.89 | 5.16 | 4.64 | 4.26 | 3.93 | 3.66 | 3.44 | 3.25 | 3.02 |
| 40 | 30 | 7.34 | 5.99 | 5.27 | 4.79 | 4.23 | 3.82 | 3.58 | 3.39 | 3.19 | 2.96 |
| 40 | 40 | 7.29 | 6.00 | 5.26 | 4.69 | 4.20 | 3.89 | 3.66 | 3.42 | 3.23 | 3.03 |
| 40 | 50 | 7.46 | 6.07 | 5.16 | 4.57 | 4.14 | 3.80 | 3.56 | 3.33 | 3.04 | 2.85 |
| 50 | 10 | 7.86 | 6.17 | 5.47 | 4.92 | 4.44 | 4.10 | 3.79 | 3.62 | 3.29 | 3.10 |
| 50 | 20 | 7.55 | 6.22 | 5.35 | 4.89 | 4.44 | 4.12 | 3.69 | 3.47 | 3.23 | 3.00 |
| 50 | 30 | 7.10 | 6.14 | 5.33 | 4.74 | 4.25 | 3.89 | 3.66 | 3.38 | 3.12 | 2.88 |
| 50 | 40 | 6.94 | 5.64 | 5.02 | 4.55 | 4.03 | 3.77 | 3.45 | 3.19 | 3.02 | 2.84 |
| 50 | 50 | 7.44 | 5.98 | 5.20 | 4.64 | 4.32 | 3.98 | 3.65 | 3.38 | 3.14 | 2.96 |

## Table 2.6.A
## Omnibus Rank Test for Testing Symmetry

*Simulated critical values $k_\alpha$ under the continuous null hypothesis $\mathcal{H}_0^s : F = F_-$ (no ties in the absolute values) according to $P_0\{ S > k_\alpha \} = \alpha$ for the omnibus symmetry rank statistic $S$ defined in formula (8) of Section 2.6. The bandwidth is $a = 0.40$ and the Monte Carlo sample size is 5000.*

| $n$ | 0.01 | 0.02 | 0.03 | 0.04 | 0.05 | 0.06 | 0.07 | 0.08 | 0.09 | 0.10 |
|---|---|---|---|---|---|---|---|---|---|---|
| 10 | 8.25 | 8.25 | 7.33 | 6.63 | 6.58 | 6.54 | 6.54 | 6.50 | 6.50 | 6.33 |
| 12 | 9.12 | 7.99 | 7.73 | 7.22 | 7.11 | 6.99 | 6.97 | 6.34 | 6.21 | 6.12 |
| 14 | 8.94 | 8.43 | 7.91 | 7.53 | 7.27 | 7.01 | 6.82 | 6.70 | 6.56 | 6.47 |
| 16 | 9.94 | 8.71 | 8.14 | 7.75 | 7.52 | 7.27 | 7.07 | 6.86 | 6.71 | 6.53 |
| 18 | 9.60 | 8.87 | 8.34 | 7.88 | 7.57 | 7.27 | 7.07 | 6.85 | 6.66 | 6.54 |
| 20 | 10.30 | 9.31 | 8.59 | 8.08 | 7.74 | 7.42 | 7.16 | 6.97 | 6.76 | 6.61 |
| 22 | 10.25 | 9.22 | 8.76 | 8.30 | 7.91 | 7.58 | 7.37 | 7.18 | 6.99 | 6.77 |
| 24 | 10.33 | 9.35 | 8.71 | 8.27 | 7.87 | 7.53 | 7.31 | 7.03 | 6.88 | 6.72 |
| 26 | 10.29 | 9.39 | 8.82 | 8.28 | 7.94 | 7.68 | 7.46 | 7.19 | 6.98 | 6.74 |
| 28 | 10.41 | 9.37 | 8.74 | 8.16 | 7.88 | 7.61 | 7.37 | 7.15 | 6.98 | 6.78 |
| 30 | 10.71 | 9.74 | 8.97 | 8.36 | 8.01 | 7.70 | 7.54 | 7.32 | 7.14 | 6.94 |
| 32 | 10.74 | 9.52 | 8.90 | 8.46 | 8.08 | 7.79 | 7.44 | 7.25 | 6.99 | 6.79 |
| 34 | 10.93 | 9.92 | 9.22 | 8.66 | 8.34 | 7.98 | 7.63 | 7.39 | 7.18 | 6.95 |
| 36 | 10.63 | 9.77 | 8.96 | 8.43 | 8.06 | 7.72 | 7.43 | 7.22 | 6.96 | 6.73 |
| 38 | 10.75 | 9.51 | 8.84 | 8.37 | 7.97 | 7.69 | 7.40 | 7.22 | 7.01 | 6.78 |
| 40 | 10.72 | 9.74 | 9.02 | 8.46 | 8.06 | 7.76 | 7.48 | 7.27 | 7.04 | 6.84 |
| 42 | 10.80 | 9.76 | 9.20 | 8.64 | 8.26 | 7.98 | 7.74 | 7.50 | 7.26 | 7.05 |
| 44 | 10.94 | 9.91 | 9.19 | 8.61 | 8.10 | 7.80 | 7.50 | 7.26 | 7.05 | 6.86 |
| 46 | 10.89 | 9.88 | 9.14 | 8.63 | 8.13 | 7.84 | 7.60 | 7.33 | 7.09 | 6.88 |
| 48 | 11.32 | 9.66 | 9.04 | 8.54 | 8.17 | 7.83 | 7.54 | 7.30 | 7.00 | 6.79 |
| 50 | 10.91 | 9.63 | 9.05 | 8.43 | 8.01 | 7.70 | 7.48 | 7.15 | 6.99 | 6.76 |
| 52 | 11.35 | 9.99 | 9.28 | 8.73 | 8.37 | 8.01 | 7.72 | 7.40 | 7.16 | 6.95 |
| 54 | 11.09 | 9.84 | 9.14 | 8.50 | 8.10 | 7.82 | 7.55 | 7.31 | 7.09 | 6.92 |
| 56 | 10.84 | 9.61 | 8.91 | 8.48 | 8.11 | 7.80 | 7.50 | 7.24 | 7.04 | 6.87 |
| 58 | 11.31 | 10.01 | 9.30 | 8.70 | 8.26 | 7.97 | 7.72 | 7.39 | 7.10 | 6.90 |
| 60 | 11.64 | 9.94 | 9.19 | 8.67 | 8.27 | 7.96 | 7.52 | 7.32 | 7.06 | 6.84 |

**Table 2.6.B**
**One-sided Rank Test for Testing Symmetry**

*Simulated critical values $k_\alpha^0$ under the continuous null hypothesis $\mathcal{H}_0^s : F = F_-$ (no ties in the absolute values) according to $P_0\{ S^0 > k_\alpha^0 \} = \alpha$ for the one-sided symmetry rank statistic $S^0$ defined in formula (15) of Section 2.6. The bandwidth is $a = 0.40$ and the Monte Carlo sample size is $5000$.*

| $n$ | 0.01 | 0.02 | 0.03 | 0.04 | 0.05 | 0.06 | 0.07 | 0.08 | 0.09 | 0.10 |
|---|---|---|---|---|---|---|---|---|---|---|
| 10 | 7.38 | 6.58 | 6.33 | 5.83 | 5.63 | 5.58 | 5.33 | 4.92 | 4.75 | 4.71 |
| 12 | 7.78 | 6.99 | 6.25 | 5.97 | 5.65 | 5.35 | 5.21 | 5.10 | 4.96 | 4.74 |
| 14 | 7.91 | 7.08 | 6.74 | 6.46 | 6.04 | 5.75 | 5.47 | 5.22 | 5.01 | 4.85 |
| 16 | 8.19 | 7.32 | 6.84 | 6.43 | 6.07 | 5.81 | 5.61 | 5.41 | 5.24 | 5.01 |
| 18 | 8.63 | 7.49 | 6.81 | 6.48 | 6.13 | 5.73 | 5.44 | 5.16 | 4.97 | 4.74 |
| 20 | 8.82 | 7.48 | 6.85 | 6.41 | 6.06 | 5.80 | 5.57 | 5.36 | 5.20 | 4.99 |
| 22 | 8.74 | 7.81 | 7.19 | 6.57 | 6.12 | 5.84 | 5.56 | 5.27 | 5.12 | 4.94 |
| 24 | 9.02 | 7.73 | 7.01 | 6.58 | 6.22 | 5.94 | 5.69 | 5.39 | 5.19 | 4.98 |
| 26 | 8.65 | 7.79 | 7.12 | 6.56 | 6.28 | 5.95 | 5.71 | 5.43 | 5.25 | 5.08 |
| 28 | 9.08 | 7.74 | 7.21 | 6.76 | 6.22 | 5.98 | 5.64 | 5.33 | 5.13 | 4.98 |
| 30 | 8.79 | 7.69 | 7.27 | 6.90 | 6.49 | 6.14 | 5.80 | 5.54 | 5.22 | 5.05 |
| 32 | 8.86 | 7.83 | 7.19 | 6.69 | 6.34 | 5.91 | 5.60 | 5.37 | 5.11 | 4.91 |
| 34 | 8.90 | 7.97 | 7.38 | 6.88 | 6.50 | 6.14 | 5.81 | 5.53 | 5.25 | 5.07 |
| 36 | 9.09 | 8.00 | 7.15 | 6.56 | 6.19 | 5.83 | 5.53 | 5.34 | 5.06 | 4.84 |
| 38 | 8.88 | 7.91 | 7.24 | 6.65 | 6.20 | 5.83 | 5.60 | 5.31 | 5.13 | 4.93 |
| 40 | 9.43 | 8.02 | 7.26 | 6.78 | 6.43 | 6.06 | 5.77 | 5.55 | 5.24 | 5.05 |
| 42 | 9.32 | 8.14 | 7.48 | 6.92 | 6.42 | 6.03 | 5.72 | 5.40 | 5.15 | 4.91 |
| 44 | 9.07 | 7.92 | 7.15 | 6.69 | 6.32 | 5.86 | 5.55 | 5.31 | 5.13 | 4.94 |
| 46 | 9.34 | 7.83 | 7.06 | 6.59 | 6.26 | 6.00 | 5.74 | 5.47 | 5.20 | 4.97 |
| 48 | 9.33 | 7.92 | 7.05 | 6.50 | 6.14 | 5.80 | 5.46 | 5.24 | 4.99 | 4.79 |
| 50 | 9.29 | 8.05 | 7.46 | 6.83 | 6.32 | 5.96 | 5.71 | 5.46 | 5.19 | 5.01 |
| 52 | 9.57 | 8.36 | 7.63 | 6.98 | 6.54 | 6.15 | 5.80 | 5.53 | 5.27 | 5.07 |
| 54 | 8.91 | 7.55 | 7.03 | 6.48 | 6.06 | 5.64 | 5.38 | 5.16 | 4.93 | 4.76 |
| 56 | 9.37 | 7.99 | 7.23 | 6.64 | 6.33 | 5.98 | 5.71 | 5.47 | 5.20 | 5.02 |
| 58 | 9.76 | 8.34 | 7.48 | 6.78 | 6.43 | 6.18 | 5.89 | 5.55 | 5.30 | 5.13 |
| 60 | 9.30 | 8.18 | 7.36 | 6.72 | 6.37 | 6.05 | 5.71 | 5.46 | 5.21 | 5.05 |

## Table 2.6.C
## Projection Rank Test for Testing Symmetry

*Simulated critical values $k_\alpha^\pi$ under the continuous null hypothesis $\mathcal{H}_0^s : F = F_-$ (no ties in the absolute values) according to $P_0\{ S^\pi > k_\alpha^\pi \} = \alpha$ for the symmetry projection rank statistic $S^\pi$ defined in formula (25) of Section 2.6. The Monte Carlo sample size is 5000.*

| $n$ | 0.01 | 0.02 | 0.03 | 0.04 | 0.05 | 0.06 | 0.07 | 0.08 | 0.09 | 0.10 |
|---|---|---|---|---|---|---|---|---|---|---|
| 10 | 5.47 | 4.84 | 4.34 | 3.90 | 3.41 | 3.22 | 2.89 | 2.71 | 2.53 | 2.36 |
| 12 | 6.05 | 4.70 | 4.01 | 3.55 | 3.23 | 2.95 | 2.78 | 2.51 | 2.34 | 2.19 |
| 14 | 5.82 | 4.77 | 4.21 | 3.81 | 3.39 | 3.08 | 2.88 | 2.68 | 2.48 | 2.30 |
| 16 | 5.81 | 4.91 | 4.18 | 3.80 | 3.46 | 3.19 | 2.86 | 2.64 | 2.43 | 2.27 |
| 18 | 6.10 | 4.94 | 4.24 | 3.69 | 3.35 | 3.09 | 2.85 | 2.62 | 2.44 | 2.24 |
| 20 | 6.04 | 4.91 | 4.33 | 3.82 | 3.46 | 3.11 | 2.89 | 2.66 | 2.40 | 2.22 |
| 22 | 5.85 | 4.87 | 4.40 | 3.86 | 3.45 | 3.11 | 2.76 | 2.50 | 2.33 | 2.18 |
| 24 | 6.20 | 4.88 | 4.18 | 3.73 | 3.32 | 3.04 | 2.87 | 2.68 | 2.47 | 2.28 |
| 26 | 5.97 | 4.99 | 4.38 | 3.89 | 3.42 | 3.11 | 2.84 | 2.58 | 2.42 | 2.24 |
| 28 | 6.44 | 4.98 | 4.23 | 3.77 | 3.42 | 3.12 | 2.88 | 2.60 | 2.38 | 2.21 |
| 30 | 6.06 | 4.95 | 4.38 | 3.79 | 3.50 | 3.07 | 2.75 | 2.54 | 2.35 | 2.18 |
| 32 | 6.20 | 4.87 | 4.35 | 3.78 | 3.38 | 3.08 | 2.77 | 2.54 | 2.32 | 2.14 |
| 34 | 6.22 | 4.92 | 4.25 | 3.84 | 3.50 | 3.11 | 2.87 | 2.63 | 2.45 | 2.24 |
| 36 | 6.02 | 4.93 | 4.15 | 3.73 | 3.28 | 3.01 | 2.73 | 2.50 | 2.32 | 2.17 |
| 38 | 6.12 | 4.79 | 4.10 | 3.64 | 3.28 | 2.97 | 2.72 | 2.50 | 2.33 | 2.17 |
| 40 | 6.35 | 5.14 | 4.46 | 3.91 | 3.48 | 3.19 | 2.93 | 2.64 | 2.43 | 2.26 |
| 42 | 5.88 | 4.76 | 4.18 | 3.76 | 3.38 | 3.06 | 2.82 | 2.59 | 2.37 | 2.18 |
| 44 | 6.48 | 5.12 | 4.36 | 3.78 | 3.41 | 3.09 | 2.80 | 2.57 | 2.40 | 2.20 |
| 46 | 5.97 | 4.75 | 4.27 | 3.72 | 3.39 | 2.92 | 2.68 | 2.50 | 2.30 | 2.12 |
| 48 | 5.91 | 4.73 | 4.06 | 3.69 | 3.24 | 2.94 | 2.78 | 2.58 | 2.39 | 2.19 |
| 50 | 6.10 | 5.08 | 4.31 | 3.78 | 3.48 | 3.18 | 2.90 | 2.67 | 2.47 | 2.26 |
| 52 | 6.77 | 5.23 | 4.46 | 3.93 | 3.56 | 3.18 | 2.92 | 2.75 | 2.54 | 2.35 |
| 54 | 6.10 | 4.99 | 4.24 | 3.65 | 3.19 | 2.89 | 2.63 | 2.41 | 2.25 | 2.09 |
| 56 | 5.98 | 4.88 | 4.24 | 3.76 | 3.34 | 3.04 | 2.75 | 2.56 | 2.35 | 2.20 |
| 58 | 6.13 | 5.17 | 4.35 | 3.93 | 3.52 | 3.19 | 2.94 | 2.69 | 2.54 | 2.35 |
| 60 | 5.93 | 4.82 | 4.32 | 3.76 | 3.43 | 3.13 | 2.82 | 2.65 | 2.42 | 2.26 |

**Table 2.7.A**
**Omnibus Rank Test for Testing Independence**

*Simulated critical values $k_\alpha$ under the continuous null hypothesis $\mathcal{H}_0^i : F = G \times H$ (no ties) according to $P_0\{S > k_\alpha\} = \alpha$ for the omnibus independence rank statistic $S$ defined in formula (9) of Section 2.7. The underlying bandwidth is $a = 0.40$ and the Monte Carlo sample size is 5000.*

| $n$ | 0.01 | 0.02 | 0.03 | 0.04 | 0.05 | 0.06 | 0.07 | 0.08 | 0.09 | 0.10 |
|---|---|---|---|---|---|---|---|---|---|---|
| 10 | 12.98 | 12.39 | 12.12 | 11.85 | 11.64 | 11.50 | 11.36 | 11.18 | 11.08 | 10.98 |
| 12 | 13.40 | 12.82 | 12.42 | 12.09 | 11.83 | 11.62 | 11.45 | 11.33 | 11.22 | 11.10 |
| 14 | 13.93 | 13.24 | 12.83 | 12.48 | 12.21 | 11.99 | 11.80 | 11.61 | 11.46 | 11.30 |
| 16 | 14.22 | 13.40 | 12.88 | 12.47 | 12.20 | 11.94 | 11.72 | 11.54 | 11.39 | 11.25 |
| 18 | 14.62 | 13.56 | 13.03 | 12.60 | 12.23 | 11.96 | 11.76 | 11.58 | 11.43 | 11.24 |
| 20 | 14.36 | 13.62 | 13.15 | 12.81 | 12.51 | 12.24 | 12.00 | 11.80 | 11.62 | 11.44 |
| 22 | 15.07 | 13.94 | 13.40 | 13.05 | 12.69 | 12.31 | 12.03 | 11.73 | 11.57 | 11.35 |
| 24 | 14.81 | 13.74 | 13.21 | 12.79 | 12.43 | 12.16 | 11.93 | 11.76 | 11.59 | 11.38 |
| 26 | 14.86 | 13.96 | 13.48 | 12.98 | 12.55 | 12.24 | 12.00 | 11.74 | 11.54 | 11.36 |
| 28 | 15.07 | 13.88 | 13.20 | 12.81 | 12.46 | 12.20 | 11.89 | 11.72 | 11.56 | 11.38 |
| 30 | 15.51 | 14.22 | 13.52 | 13.11 | 12.76 | 12.48 | 12.24 | 11.97 | 11.72 | 11.57 |
| 32 | 15.23 | 14.21 | 13.59 | 13.07 | 12.77 | 12.56 | 12.30 | 12.02 | 11.78 | 11.58 |
| 34 | 14.99 | 13.92 | 13.38 | 12.93 | 12.60 | 12.31 | 12.07 | 11.89 | 11.70 | 11.49 |
| 36 | 15.32 | 14.24 | 13.61 | 12.98 | 12.62 | 12.37 | 12.08 | 11.85 | 11.64 | 11.48 |
| 38 | 15.43 | 14.40 | 13.61 | 13.13 | 12.83 | 12.48 | 12.21 | 11.95 | 11.76 | 11.60 |
| 40 | 15.18 | 14.28 | 13.46 | 12.97 | 12.65 | 12.40 | 12.17 | 11.94 | 11.72 | 11.56 |
| 42 | 15.32 | 14.36 | 13.61 | 13.19 | 12.81 | 12.48 | 12.21 | 11.92 | 11.73 | 11.54 |
| 44 | 15.42 | 14.33 | 13.59 | 13.09 | 12.75 | 12.48 | 12.25 | 11.98 | 11.79 | 11.61 |
| 46 | 15.26 | 14.30 | 13.56 | 13.12 | 12.75 | 12.48 | 12.24 | 12.02 | 11.80 | 11.54 |
| 48 | 15.58 | 14.37 | 13.63 | 13.19 | 12.81 | 12.50 | 12.18 | 11.93 | 11.65 | 11.43 |
| 50 | 15.73 | 14.44 | 13.75 | 13.19 | 12.81 | 12.48 | 12.27 | 12.03 | 11.76 | 11.57 |
| 52 | 15.15 | 14.25 | 13.58 | 13.16 | 12.73 | 12.37 | 12.13 | 11.85 | 11.64 | 11.45 |
| 54 | 15.28 | 14.32 | 13.66 | 13.22 | 12.91 | 12.61 | 12.30 | 11.97 | 11.74 | 11.54 |
| 56 | 15.85 | 14.35 | 13.71 | 13.18 | 12.65 | 12.39 | 12.12 | 11.87 | 11.68 | 11.50 |
| 58 | 15.65 | 14.55 | 13.72 | 13.26 | 12.91 | 12.52 | 12.28 | 12.02 | 11.81 | 11.66 |
| 60 | 15.46 | 14.27 | 13.48 | 12.98 | 12.67 | 12.33 | 12.06 | 11.79 | 11.61 | 11.41 |

## Table 2.7.B
### One-sided Rank Test for Testing Independence

*Simulated critical values $k_\alpha^0$ under the continuous null hypothesis $\mathcal{H}_0^i : F = G \times H$ (no ties) according to $P_0\{ S^0 > k_\alpha^0 \} = \alpha$ for the one-sided independence rank statistic $S^0$ defined in formula (16) of Section 2.7. The underlying bandwidth is $a = 0.40$ and the Monte Carlo sample size is $5000$.*

| $n$ | 0.01 | 0.02 | 0.03 | 0.04 | 0.05 | 0.06 | 0.07 | 0.08 | 0.09 | 0.10 |
|---|---|---|---|---|---|---|---|---|---|---|
| 10 | 10.97 | 10.25 | 9.80 | 9.42 | 9.05 | 8.77 | 8.50 | 8.30 | 8.11 | 7.96 |
| 12 | 11.34 | 10.52 | 9.88 | 9.36 | 8.96 | 8.67 | 8.41 | 8.18 | 7.93 | 7.78 |
| 14 | 11.48 | 10.65 | 10.04 | 9.57 | 9.15 | 8.81 | 8.56 | 8.28 | 8.04 | 7.83 |
| 16 | 12.08 | 10.67 | 10.04 | 9.57 | 9.19 | 8.83 | 8.55 | 8.32 | 8.09 | 7.95 |
| 18 | 11.29 | 10.38 | 9.75 | 9.30 | 9.01 | 8.71 | 8.45 | 8.24 | 8.02 | 7.85 |
| 20 | 11.70 | 10.72 | 10.09 | 9.68 | 9.24 | 8.88 | 8.51 | 8.24 | 8.02 | 7.85 |
| 22 | 12.38 | 10.93 | 10.23 | 9.54 | 9.17 | 8.84 | 8.60 | 8.33 | 8.13 | 7.93 |
| 24 | 12.31 | 10.75 | 9.96 | 9.53 | 9.05 | 8.77 | 8.45 | 8.23 | 8.01 | 7.82 |
| 26 | 11.82 | 10.53 | 10.04 | 9.48 | 9.06 | 8.65 | 8.35 | 8.11 | 7.89 | 7.67 |
| 28 | 12.10 | 10.88 | 10.08 | 9.57 | 9.23 | 8.88 | 8.56 | 8.35 | 8.13 | 7.87 |
| 30 | 11.79 | 10.59 | 9.88 | 9.51 | 9.20 | 8.82 | 8.56 | 8.29 | 8.08 | 7.84 |
| 32 | 12.37 | 10.89 | 10.04 | 9.65 | 9.11 | 8.81 | 8.48 | 8.23 | 8.03 | 7.78 |
| 34 | 12.00 | 10.95 | 10.19 | 9.65 | 9.30 | 8.94 | 8.57 | 8.26 | 8.04 | 7.81 |
| 36 | 12.13 | 11.01 | 10.03 | 9.53 | 9.15 | 8.79 | 8.52 | 8.25 | 7.98 | 7.79 |
| 38 | 12.35 | 11.15 | 10.56 | 10.01 | 9.37 | 8.96 | 8.65 | 8.34 | 8.09 | 7.84 |
| 40 | 12.35 | 11.03 | 10.34 | 9.78 | 9.33 | 8.95 | 8.66 | 8.32 | 8.07 | 7.80 |
| 42 | 12.57 | 11.12 | 10.26 | 9.77 | 9.40 | 9.10 | 8.68 | 8.36 | 8.11 | 7.88 |
| 44 | 12.18 | 11.04 | 10.37 | 9.70 | 9.24 | 8.84 | 8.53 | 8.32 | 8.02 | 7.85 |
| 46 | 12.36 | 11.20 | 10.24 | 9.68 | 9.20 | 8.88 | 8.62 | 8.41 | 8.09 | 7.87 |
| 48 | 12.05 | 10.89 | 10.32 | 9.74 | 9.33 | 9.03 | 8.76 | 8.49 | 8.21 | 7.94 |
| 50 | 12.38 | 11.16 | 10.53 | 9.81 | 9.30 | 8.95 | 8.56 | 8.33 | 8.06 | 7.82 |
| 52 | 12.10 | 10.90 | 10.22 | 9.73 | 9.24 | 8.80 | 8.50 | 8.26 | 8.01 | 7.82 |
| 54 | 12.43 | 11.13 | 10.36 | 9.87 | 9.27 | 8.87 | 8.60 | 8.31 | 8.08 | 7.85 |
| 56 | 12.17 | 10.90 | 10.14 | 9.68 | 9.23 | 8.94 | 8.62 | 8.31 | 8.04 | 7.83 |
| 58 | 12.09 | 11.06 | 10.29 | 9.72 | 9.23 | 8.83 | 8.50 | 8.17 | 7.94 | 7.72 |
| 60 | 11.83 | 10.88 | 10.19 | 9.68 | 9.24 | 8.86 | 8.56 | 8.35 | 8.11 | 7.96 |

## Table 2.7.C
## Projection Rank Test for Testing Independence

*Simulated critical values $k_\alpha^\pi$ under the continuous null hypothesis $\mathcal{H}_0^i : F = G \times H$ (no ties) according to $P_0\{S^\pi > k_\alpha^\pi\} = \alpha$ for the independence projection rank statistic $S^\pi$ defined in formula (30) of Section 2.7. The Monte Carlo sample size is $5000$.*

| $n$ | 0.01 | 0.02 | 0.03 | 0.04 | 0.05 | 0.06 | 0.07 | 0.08 | 0.09 | 0.10 |
|---|---|---|---|---|---|---|---|---|---|---|
| 10 | 8.57 | 7.45 | 6.99 | 6.48 | 6.13 | 5.85 | 5.58 | 5.31 | 4.95 | 4.69 |
| 12 | 8.54 | 7.47 | 6.77 | 6.21 | 5.81 | 5.47 | 5.17 | 4.90 | 4.67 | 4.42 |
| 14 | 8.81 | 7.51 | 6.92 | 6.27 | 5.81 | 5.46 | 5.16 | 4.91 | 4.68 | 4.47 |
| 16 | 8.89 | 7.62 | 6.83 | 6.37 | 5.90 | 5.60 | 5.26 | 5.03 | 4.78 | 4.58 |
| 18 | 8.50 | 7.24 | 6.57 | 6.09 | 5.79 | 5.38 | 5.08 | 4.88 | 4.68 | 4.49 |
| 20 | 8.83 | 7.52 | 6.90 | 6.41 | 6.00 | 5.61 | 5.29 | 5.02 | 4.75 | 4.48 |
| 22 | 9.72 | 8.12 | 7.11 | 6.30 | 5.76 | 5.38 | 5.17 | 4.90 | 4.63 | 4.43 |
| 24 | 9.13 | 7.94 | 7.11 | 6.42 | 5.90 | 5.53 | 5.17 | 4.95 | 4.70 | 4.48 |
| 26 | 8.76 | 7.39 | 6.60 | 6.11 | 5.70 | 5.27 | 5.00 | 4.75 | 4.57 | 4.36 |
| 28 | 9.02 | 7.89 | 7.00 | 6.35 | 5.90 | 5.59 | 5.26 | 5.03 | 4.74 | 4.56 |
| 30 | 8.83 | 7.52 | 6.64 | 6.07 | 5.69 | 5.36 | 5.15 | 4.95 | 4.73 | 4.49 |
| 32 | 9.19 | 7.60 | 6.61 | 6.09 | 5.67 | 5.35 | 5.11 | 4.79 | 4.61 | 4.43 |
| 34 | 9.51 | 7.77 | 6.76 | 6.15 | 5.77 | 5.41 | 5.09 | 4.85 | 4.61 | 4.38 |
| 36 | 9.14 | 7.68 | 6.98 | 6.31 | 5.74 | 5.38 | 5.00 | 4.70 | 4.51 | 4.30 |
| 38 | 9.17 | 7.69 | 7.02 | 6.44 | 6.06 | 5.60 | 5.24 | 4.96 | 4.67 | 4.43 |
| 40 | 9.09 | 7.70 | 6.90 | 6.38 | 5.90 | 5.51 | 5.19 | 4.91 | 4.64 | 4.46 |
| 42 | 9.19 | 8.07 | 7.20 | 6.56 | 6.03 | 5.67 | 5.33 | 5.09 | 4.84 | 4.57 |
| 44 | 8.98 | 7.82 | 7.00 | 6.40 | 5.98 | 5.59 | 5.27 | 4.93 | 4.67 | 4.48 |
| 46 | 9.33 | 7.90 | 7.05 | 6.37 | 5.91 | 5.68 | 5.31 | 5.02 | 4.79 | 4.59 |
| 48 | 8.45 | 7.48 | 6.86 | 6.33 | 5.99 | 5.73 | 5.38 | 5.10 | 4.91 | 4.67 |
| 50 | 9.19 | 7.81 | 6.97 | 6.39 | 5.91 | 5.51 | 5.24 | 4.99 | 4.77 | 4.58 |
| 52 | 9.37 | 7.84 | 6.96 | 6.38 | 5.93 | 5.55 | 5.21 | 4.91 | 4.67 | 4.38 |
| 54 | 9.44 | 7.97 | 7.11 | 6.42 | 5.87 | 5.45 | 5.15 | 4.89 | 4.60 | 4.40 |
| 56 | 8.83 | 7.45 | 6.75 | 6.31 | 5.81 | 5.45 | 5.14 | 4.92 | 4.59 | 4.39 |
| 58 | 9.41 | 7.64 | 6.83 | 6.45 | 5.88 | 5.48 | 5.23 | 5.00 | 4.67 | 4.46 |
| 60 | 9.09 | 7.58 | 6.78 | 6.28 | 5.93 | 5.59 | 5.31 | 5.05 | 4.80 | 4.54 |

# Bibliography


[1] Ash, R. A. and Gardener, M. F. (1975). *Topics in Stochastic Processes.* Academic Press, New York.

[2] Barlow, R. E., Bartholomew, D. J., Bremner, J. M., and Brunk, H. D. (1972). *Statistical Inference under Order Restrictions.* Wiley, New York.

[3] Behnen, K. (1972). A characterization of certain rank-order tests with bounds for the asymptotic relative efficiency. Ann. Math. Statist. **43**, 1839–1851.

[4] Behnen, K. (1975). The Randles–Hogg test and an alternative proposal. Comm. Statist. **4**, 203–238.

[5] Behnen, K. (1976). Asymptotic comparison of rank tests for the regression problem when ties are present. Ann. Statist. **4**, 157–174.

[6] Behnen, K. and Hušková, M. (1984). A simple algorithm for the adaptation of scores and power behaviour of the corresponding rank test. Comm. Statist. – Theor. Meth. **13**, 305–325.

[7] Behnen, K. and Neuhaus, G. (1983). Galton's test as a linear rank test with estimated scores and its local asymptotic efficiency. Ann. Statist. **11**, 588–599.

[8] Bickel, P. J. and Wichura, M. J. (1971). Convergence criteria for multiparameter stochastic processes and some applications. Ann. Math. Statist. **42**, 1656–1671.

[9] Billingsley, P. (1968). *Convergence of Probability Measures.* Wiley, New York.

[10] Chow, Y. S. and Teicher, H. (1978). *Probability Theory.* Springer. New York.

[11] Conover, W. J. (1980). *Practical Nonparametric Statistics*, 2nd ed. Wiley, New York.

[12] David, F. N. (1953). A note on the evaluation of the multivariate normal integral. Biometrika **40**, 458–459.

[13] Dunford, N. and Schwartz, J. T. (1958). *Linear Operators. Part I: General Theory.* Interscience, New York.

[14] Durbin, J. and Knott, M. (1972). Components of Cramér–von Mises statistics I. J. Royal Statist. Soc. B **34**, 290–307.

[15] Feller, W. (1968). *An Introduction to Probability Theory and Its Applications*, Vol. 1, 3rd ed. Wiley, New York.

[16] Groeneboom, P. and Pyke, R. (1983). Asymptotic normality of statistics based on the convex minorants of empirical distribution functions. Ann. Probab. **11**, 328–345.

[17] Hájek, J. (1969). *A Course in Nonparametric Statistics.* Holden-Day, San Francisco.

[18] Hájek, J. and Šidák, Z. (1967). *Theory of Rank Tests.* Academic Press, New York.

[19] Hollander, M. and Wolfe, D. A. (1973). *Nonparametric Statistical Methods.* Wiley, New York.

[20] Johnson, N. L. and Kotz, S. (1972). *Distributions in Statistics: Continuous Multivariate Distributions.* Wiley, New York.

[21] Koopmans, L. H. (1974). *The Spectral Analysis of Time Series.* Academic Press, New York.

[22] Kudô, A. (1963). A multivariate analogue of the one-sided test. Biometrika **50**, 403–418.

[23] Lehmann, E. L. (1975). *Nonparametrics: Statistical Methods Based on Ranks.* Holden-Day, San Francisco.

[24] Mehrotra, K. G. and Johnson, R. A. (1976). Asymptotic sufficiency and asymptotically most powerful tests for the two sample censored situation. Ann. Statist. **4**, 589–596.

[25] Neuhaus, G. (1987). Local asymptotics for linear rank statistics with estimated score functions. Ann. Statist. **15**, 491–512.

[26] Nüesch, P. E. (1966). On the problem of testing location in multivariate populations for restricted alternatives. Ann. Math. Statist. **37**, 113–119.

[27] Nüesch, P. E. (1970). Estimation of monotone parameters and the Kuhn–Tucker conditions. Ann. Math. Statist. **41**, 1800.

[28] Randles, R. H. and Hogg, R. V. (1973). Adaptive distribution-free tests. Comm. Satist. **2**, 337–356.

[29] Rosenblatt, M. (1971). Curve estimates. Ann. Math. Statist. **42**, 1815–1842.

[30] Smirnow, W. I. (1966). *Lehrgang der höheren Mathematik, Teil 4*. VEB Deutscher Verlag der Wissenschaften, Berlin.

# Author Index

# Program Disks

A complete set of programs for the evaluation of the test statistics of Chapter 2 and for the simulation of the corresponding conditional $p$ –values is available for the IBM PC (XT or AT) and compatibles on 5.25", 360 KB disks. The disks contain executable programs as well as the corresponding source code for the Turbo Pascal 4.0 compiler.

Two versions are available:

For a PC **with** numerical coprocessor order version 3–519–09282–4 ,

for a PC **without** numerical coprocessor order version 3–519–09281–6 .

| program | description |
| --- | --- |
| TEST21MC.PAS/EXE | evaluation of the test statistics of Section 2.1 and simulation of the corresp. cond. $p$ –values |
| TEST22MC.PAS/EXE | evaluation of the test statistics of Section 2.2 and simulation of the corresp. cond. $p$ –values |
| TEST23MC.PAS/EXE | evaluation of the test statistics of Section 2.3 and simulation of the corresp. cond. $p$ –values |
| TEST24MC.PAS/EXE | evaluation of the test statistics of Section 2.4 and simulation of the corresp. cond. $p$ –values |
| TEST25MC.PAS/EXE | evaluation of the test statistics of Section 2.5 and simulation of the corresp. cond. $p$ –values |
| TEST26MC.PAS/EXE | evaluation of the test statistics of Section 2.6 and simulation of the corresp. cond. $p$ –values |
| TEST27MC.PAS/EXE | evaluation of the test statistics of Section 2.7 and simulation of the corresp. cond. $p$ –values |
| COMMENTS.TXT/EXE | general comments on the use of the programs |

# Teubner Studienbücher zur Statistik

Afflerbach: **Statistik-Praktikum mit dem PC**
198 Seiten. DM 24,80

Behnen/Neuhaus: **Grundkurs Stochastik**
376 Seiten. DM 38,–

v. Collani: **Optimale Wareneingangskontrolle**
150 Seiten. DM 29,80

Dinges/Rost: **Prinzipien der Stochastik**
294 Seiten. DM 36,–

Floret: **Maß- und Integrationstheorie**
360 Seiten. DM 38,–

Kohlas: **Stochastische Methoden des Operations Research**
192 Seiten. DM 26,80

Lehn/Wegmann: **Einführung in die Statistik**
220 Seiten. DM 24,80

Lehn/Wegmann/Rettig: **Aufgabensammlung zur Einführung in die Statistik**
246 Seiten. DM 26,80

Rauhut/Schmitz/Zachow: **Spieltheorie**
400 Seiten. DM 38,–

Topsøe: **Informationstheorie**
88 Seiten. DM 18,80

Uhlmann: **Statistische Qualitätskontrolle**
2. Aufl. 229 Seiten. DM 39,–

Witting: **Mathematische Statistik**
3. Aufl. 223 Seiten. DM 28,80

Wolfsdorf: **Versicherungsmathematik**
Teil 1: Personenversicherung
491 Seiten. DM 42,–
Teil 2: Theoretische Grundlagen, Risikotheorie, Sachversicherung
407 Seiten. DM 38,–

Preisänderungen vorbehalten

 B. G. Teubner Stuttgart

# Mathematische Methoden in der Technik

**Band 1:** **Törnig/Gipser/Kaspar, Numerische Lösung von partiellen Differentialgleichungen der Technik**
183 Seiten. DM 36,–

**Band 2:** **Dutter, Geostatistik**
159 Seiten. DM 34,–

**Band 3:** **Spellucci/Törnig, Eigenwertberechnung in den Ingenieurwissenschaften**
196 Seiten. DM 36,–

**Band 4:** **Buchberger/Kutzler/Feilmeier/Kratz/Kulisch/Rump, Rechnerorientierte Verfahren**
281 Seiten. DM 48,–

**Band 5:** **Babovsky/Beth/Neunzert/Schulz-Reese, Mathematische Methoden in der Systemtheorie: Fourieranalysis**
173 Seiten. DM 36,–

**Band 8:** **Weiß, Stochastische Modelle für Anwender**
192 Seiten. DM 36,–

**Band 9:** **Antes, Anwendungen der Methode der Randelemente in der Elastodynamik und der Fluiddynamik**
196 Seiten. DM 36,–

**Band 10:** Vogt, **Methoden der Statistischen Qualitätskontrolle**
295 Seiten. DM 48,–


In Vorbereitung

**Band 6:** **Krüger/Scheiba, Mathematische Methoden in der Systemtheorie: Stochastische Prozesse**


Preisänderungen vorbehalten

 **B. G. Teubner Stuttgart**